AF633803

Space Technology Library

Volume 46

Editor-in-Chief

James R. Wertz, Microcosm, Inc., El Segundo, USA

The *Space Technology Library* is a series of high-level research books treating a variety of important issues related to space missions. A wide range of space-related topics is covered starting from mission analysis and design, through a description of spacecraft structure to spacecraft attitude determination and control. A number of excellent volumes in the *Space Technology Library* were provided through the US Air Force Academy's Space Technology Series. The quality of the book series is guaranteed through the efforts of its managing editor and well-respected editorial board. Books in the *Space Technology Library* are sponsored by ESA, NASA and the United States Department of Defense.

Frederick A. Leve • Mason A. Peck •
Brian J. Hamilton • William Bialke

Spacecraft Momentum Control Systems

Second Edition

Springer

Frederick A. Leve
Arlington, VA, USA

Brian J. Hamilton
Glendale, AZ, USA

Mason A. Peck
Ithaca, NY, USA

William Bialke
Lochridge Farm
Trumansburg, NY, USA

ISSN 0924-4263 ISSN 2542-8896 (electronic)
Space Technology Library
ISBN 978-3-031-96164-9 ISBN 978-3-031-96165-6 (eBook)
https://doi.org/10.1007/978-3-031-96165-6

© The Editor(s) (if applicable) and The Author(s), under exclusive license to Springer Nature Switzerland AG 2015, 2025

This work is subject to copyright. All rights are solely and exclusively licensed by the Publisher, whether the whole or part of the material is concerned, specifically the rights of translation, reprinting, reuse of illustrations, recitation, broadcasting, reproduction on microfilms or in any other physical way, and transmission or information storage and retrieval, electronic adaptation, computer software, or by similar or dissimilar methodology now known or hereafter developed.
The use of general descriptive names, registered names, trademarks, service marks, etc. in this publication does not imply, even in the absence of a specific statement, that such names are exempt from the relevant protective laws and regulations and therefore free for general use.
The publisher, the authors and the editors are safe to assume that the advice and information in this book are believed to be true and accurate at the date of publication. Neither the publisher nor the authors or the editors give a warranty, expressed or implied, with respect to the material contained herein or for any errors or omissions that may have been made. The publisher remains neutral with regard to jurisdictional claims in published maps and institutional affiliations.

This Springer imprint is published by the registered company Springer Nature Switzerland AG
The registered company address is: Gewerbestrasse 11, 6330 Cham, Switzerland

If disposing of this product, please recycle the paper.

Preface

Space has seen a revolution in the 10 years since Spacecraft Momentum Control Systems first appeared. At that time, what we called NewSpace had taken hold, bringing with it rapid innovation at small scale; commercial space imaging was coming into its own; and horizon goals included asteroid prospecting. Furthermore, we hoped to see a return to the moon via what came to be called NASA's Artemis program. In the years since we've seen asteroid mining placed on the back burner, but many exciting trends have come to the fore. A few years ago the United States stood up the Space Force, whose areas of responsibility include all of CisLunar space. That remarkable shift in emphasis has inspired technologies that include cell phone-like communications in orbit, In-Space Servicing, Assembly, and Manufacturing (ISAM), and a new focus on trusted autonomy with implications for space cybersecurity. So, in 2025 another revolution is underway: Big Space. The successes of the SpaceX Falcon Heavy and the Starship/Super Heavy promise to open up space through orders-of-magnitude lower cost than the most recent generation of government-driven launch vehicles. The result is expected to be CisLunar scale infrastructure, spacecraft built in orbit, and resources shared across cislunar space, leading to a true in-space economy.

These changes comprise some of the reasons for this new edition. For example, we take on the problem of in-space assembly and repair more explicitly as a problem where control-moment gyroscopes provide uniquely valuable capabilities. The continued growth of commercial approaches to space motivate new chapters, for example, one on bearings in space and another on flight experiences with momentum-control systems, with the intent to inform adoption of key technologies in this new generation of space vehicles. This edition now includes significantly expanded material on attitude control, equations of motion, and foundational mathematics to deepen the context for the core momentum-control technologies treated here. Finally, in response to the worldwide increase in demand for newly graduated aerospace engineers, this edition includes example problems that ought to be suitable as exercises for upper-level undergraduate or graduate study of this material. Bill Bialke has joined the team of authors as an expert in reaction-wheel technology and related technical matters. His contributions broaden the scope of

Spacecraft Momentum Control Systems, which the authors expect addresses the interests of several communities: commercial, military, intelligence, and academic. Along with the new appendices and exercises, we hope that this edition's new diagrams and refined nomenclature offer the depth and usability that this important topic deserves.

Arlington, VA, USA Frederick A. Leve
Ithaca, NY, USA Mason A. Peck
Glendale, AZ, USA Brian J. Hamilton
Trumansburg, NY, USA William E. Bialke

Declarations

Competing Interests The authors have no competing interests to declare that are relevant to the content of this manuscript.

Contents

About the Authors

Frederick A. Leve Program Officer at the Air Force Research Laboratory, Air Force Office of Scientific Research, received his PhD in Aerospace Engineering from the University of Florida in 2010. While at the University of Florida, he received the IAF Silver Hermann Oberth Medal and the AIAA Abe Zarem Award for Astronautics. Currently, he runs a basic research program in dynamical systems and control theory and leads the Autonomy Working Group. He has published many papers in the area of attitude dynamics and control, specifically with respect to momentum control systems and also performs research in under-actuated control, fault tolerant control, control allocation, analytical mechanics, and system identification. He is a recipient of the 2014 AFRL Early Career Award.

Mason A. Peck Professor of Astronautics at Cornell University, received his PhD in Aerospace Engineering from the University of California, Los Angeles, in 2001. He has worked as an aerospace engineer since 1994 and has been on the faculty at Cornell since 2004. From late 2011 through early 2014, he was NASA's Chief Technologist. In that role, he served as the agency's chief strategist for technology investment and prioritization and advocate for innovation in aeronautics and space technology. His research lab focuses on fundamental research in space technology that can be advanced through flight experiments. Examples include Violet, a nanosatellite for demonstrating CMG steering laws, and KickSat, the world's first crowdfunded spacecraft. He holds 19 patents in the United States and the EU and has over 100 academic publications. He received the NASA Distinguished Public Service Medal in 2014.

Brian J. Hamilton Former Engineering Fellow at Honeywell Aerospace, received a BSEE with Honors from the University of Illinois, 1976. He has 42 years of experience at Honeywell (formerly, Sperry) and has participated in the development of CMG technology since its infancy. In later years, his research focus has been on CMG array control and steering, and the general application of momentum systems to agile spacecraft attitude control. Other areas of specialty include nonlinear modeling, controls design, system optimization, and active magnetic suspension. He holds 12 patents.

William Bialke Former Chief Engineer of Ithaco Space Systems, received a BSME with Honors from North Dakota State University in 1983. He has devoted his entire career to the development and manufacture of reaction wheels used on over 250 spacecraft, starting with Honeywell, and later as the architect of the Ithaco reaction wheel product line. He has written a number of papers on reaction wheel design tradeoffs, micro-vibration disturbance sources, and bearing failure modes. He was lead investigator of a root-cause analysis of on-orbit reaction wheel failures spanning 16 years, which led to a discovery of electrical discharge as the primary contributor to anomalies and failures in reaction wheel bearings. He holds 6 patents.

Acronyms

CMG	Control Moment Gyroscope
DGCMG	Double-Gimbal Control Moment Gyroscope
IGA	Inner Gimbal Assembly
IV	Induced Vibration
LOS	Line-of-Sight
MBS	Mass Balancing System
MCS	Momentum Control System
rms	Root Mean Square
RPM	Revolutions per Minute
RWA	Reaction Wheel Assembly
SGCMG	Single-Gimbal Control Moment Gyroscope
VSCMG	Variable-Speed Control Moment Gyroscope

Symbols

F_{N}	Inertial reference frame
F_{G_i}	Ith CMG gimbal reference frame
F_{R_i}	Ith CMG rotor reference frame
A	Arbitrary point in spacecraft reference frame
p_i	CMG nominal center of mass reference point at the intersection of the gimbal and rotor axes assumed fixed in the spacecraft body reference frame
C_B	Spacecraft body center of mass point
C_{G_i}	Ith CMG gimbal center of mass point
C_{R_i}	Ith CMG rotor center of mass point
$\boldsymbol{\rho}$	Differential mass vector position from center of mass of a component
$\mathbf{r}_{B/A}$	Vector from point A to spacecraft body center of mass, C_B
$\mathbf{r}_{p_i/A}$	Vector from point A to point p_i
$\mathbf{r}_{G_i/p_i}$	Vector from point p_i to CMG gimbal center of mass point
$\mathbf{r}_{R_i/p_i}$	Vector from point p_i to CMG rotor center of mass point
$\hat{\mathbf{b}}_i$	Spacecraft body coordinate axis 'i'
$\hat{\mathbf{s}}_i$	Ith CMG spin coordinate axis
$\hat{\mathbf{o}}_i$	Ith CMG output torque coordinate axis
$\hat{\mathbf{g}}_i$	Ith CMG gimbal coordinate axis
m_{B}	Spacecraft bus mass
$m_{\text{G},i}$	Ith CMG gimbal mass
$m_{\text{R},i}$	Ith CMG rotor mass
m_{G,p_i}	Ith CMG gimbal particle imbalance mass
m_{R,p_i}	Ith CMG rotor particle imbalance mass
$\mathbf{J}_s$	Spacecraft rigid body inertia dyadic
$\mathbf{J}_t$	Rotor transverse inertia
$\mathbf{J}$	Spacecraft inertia dyadic for the spacecraft's mass center, including the spacecraft as a rigid body and all momentum-control devices
$\mathbf{J}_B^{\text{B}}$	Inertia dyadic of spacecraft about its own center of mass
$\mathbf{J}_A^{\text{B}}$	Inertia dyadic of spacecraft about point A

$\mathbf{J}_{G_i}^{\mathrm{G}_i}$	Inertia dyadic of Ith CMG gimbal about its own center of mass
$\mathbf{J}_{A}^{\mathrm{G}_i}$	Inertia dyadic of Ith CMG gimbal about point A
$\mathbf{J}_{R_i}^{\mathrm{R}_i}$	Inertia dyadic of Ith CMG rotor about its own center of mass
$\mathbf{J}_{A}^{\mathrm{R}_i}$	Inertia dyadic of Ith CMG rotor inertia about point A
$I_{r,i}$	Scalar inertia component of Ith CMG rotor about its spin axis
$J_{gr,i}$	Scalar inertia component of Ith CMG gimbal-wheel assembly about its gimbal axis
$\mathbf{J}_{g,i}$	IGA inertia dyadic, i.e., everything that contributes to the IGA rigid-body inertia in all axes
$J_{g,eff}$	Effective scalar IGA inertia. This inertia includes everything rigid that the gimbal must accelerate along with the output-axis stiffness (K_{OA}) effect: $J_{g,eff} = J_g + \frac{h^2}{K_{OA}}$
$\mathbf{J}_{g,eff}$	Effective IGA inertia dyadic. This inertia includes everything that contributes to the IGA rigid-body inertia in all axes, along with the output-axis stiffness effect along the gimbal axis direction
$\mathbf{h}_{A}^{\mathrm{B}}$	Angular momentum of spacecraft bus about point A
$\mathbf{h}_{B}^{\mathrm{B}}$	Angular momentum of spacecraft bus about its own center of mass
$\mathbf{h}_{p_i}^{\mathrm{G}_i}$	Angular momentum of Ith CMG gimbal about p_i
$\mathbf{h}_{G_i}^{\mathrm{G}_i}$	Angular momentum of Ith CMG gimbal about its own center of mass
$\mathbf{h}_{p_i}^{\mathrm{R}_i}$	Angular momentum of Ith CMG rotor about p_i
$\mathbf{h}_{R_i}^{\mathrm{R}_i}$	Angular momentum of Ith CMG rotor about its own center of mass
$\mathbf{h}_r$	Angular momentum vector of a single CMG rotor
$\mathbf{h}$	Angular momentum vector for a momentum system such that $\mathbf{h} = \sum_{i=1}^{n} \mathbf{h}_{r_i}$
h	Angular momentum body coordinate matrix representation for an array of CMG or RWA
h_s	Superspin
$\tau_{gf,i}$	Internal friction torque of Ith CMG gimbal
$\tau_{rf,i}$	Internal friction torque of Ith CMG rotor
$\tau_{r,i}$	Ith scalar rotor torque for a single-gimbal CMG (for a perfectly aligned and rigid rotor)
$\tau_{g,i}$	Ith scalar gimbal torque for a single-gimbal CMG (for a perfectly aligned and rigid IGA and gimbal)
τ_d	Scalar drag torque
$\boldsymbol{\tau}_o$	Vector output torque for a single-gimbal CMG (for a perfectly aligned and rigid IGA and gimbal). This torque acts on the spacecraft and is therefore is equal in magnitude to but opposite in direction to the torque that acts on the CMG
$\boldsymbol{\omega}^{\mathrm{B/N}}$	Angular velocity vector of spacecraft body with respect to the inertial reference frame
$\boldsymbol{\omega}^{\mathrm{G}_i/\mathrm{B}}$	Angular velocity vector of Ith CMG gimbal with respect to the spacecraft body reference frame

$\boldsymbol{\omega}^{\mathrm{G}_i/\mathrm{R}_i}$	Angular velocity vector of Ith CMG rotor with respect to the Ith CMG gimbal reference frame
ω	Scalar spacecraft angular rate or body coordinates matrix representation of spacecraft angular velocity (context sensitive)
$\mathbf{v}_A$	Translational inertial velocity of point A
Ω_i	Ith scalar CMG rotor rate
$\Omega, \dot{\Omega}$	Matrices of CMG rotor spin rates and accelerations
$\Delta, \dot{\Delta}, \ddot{\Delta}$	Matrices of CMG gimbal angles, rates, and accelerations
$\delta_i, \dot{\delta}_i, \ddot{\delta}_i$	Ith CMG gimbal angle, rate, and acceleration
β	CMG array skew angle
γ	CMG array clocking angle
K_{OA}	Output axis stiffness
K_a	Performance ratio (maximum rate over maximum acceleration)
K_c	Performance ratio (maximum acceleration over maximum jerk)
K_T	Motor torque constant (torque per unit current)
K_{MD}	Motor K_M density (ft-lb/sqrt(W) per lb)
i	Index for a single CMG within a multiple-CMG array
n	Number of CMGs in a multiple-CMG array
m	Singularity measure
α	Singularity parameter and scalar angular acceleration (context sensitive)
j	Scalar angular jerk (context sensitive)
${}^{\mathrm{B}}Q^{\mathrm{A}}$	Direction-cosine matrix that relates the representation of a vector v in A coordinates (${}^{\mathrm{B}}v$) to its representation in A coordinates (${}^{\mathrm{A}}v$)
V	Lyapunov function

Preferred Units

m	Mass	Kg
J	Polar Inertia	Nms^2(=Kg m^2)
h	Angular Momentum	Nms (=Kg m^2/s)
τ	Torque	Nm
ω	Angular Velocity	s^{-1}
K_t	Motor Torque Constant	Nm/A (= V/rad/s)

Chapter 1
Introduction

This introductory chapter surveys the evolution, application, and significance of momentum-control systems in spacecraft. It highlights the transition of these game-changing technologies from exclusive government use into the commercial sector. The chapter outlines the operational principles, advantages, and limitations of control-moment gyroscopes and reaction wheels compared to alternative attitude-control actuators like thrusters and magnetic torquers. It also traces the historical development of momentum-control from early spin-stabilized spacecraft to contemporary three-axis stabilized platforms, and it discusses emerging applications of CMGs in small spacecraft, satellite servicing, and asteroid missions. Finally, it frames the ongoing research challenges, especially in managing singularities within CMG arrays, setting the stage for deeper technical discussions in the subsequent chapters.

1.1 Spacecraft Design, Commercial Space, and Angular Momentum

The WorldView I Spacecraft shown in Fig. 1.1, successfully reached orbit on September 18, 2007.

Ball Aerospace, the prime contractor, and DigitalGlobe, the owner and operator of the satellite, celebrated this event as a technical success and commemorated the occasion with a press release that called it "a major contribution towards the advancement of the commercial remote-sensing industry by providing higher collection capabilities, more frequent revisit time, and greater imaging flexibility" (Ball Aerospace [1]). This satellite is one of several that now provide commercial earth imagery for customers that include Google Earth. It is an agile satellite, meaning that the satellite achieves comparatively high angular rates and accelerations. As a commercial agile spacecraft, it is the first of its kind.

© The Author(s), under exclusive license to Springer Nature Switzerland AG 2025

F. A. Leve et al., *Spacecraft Momentum Control Systems*, Space Technology Library 46, https://doi.org/10.1007/978-3-031-96165-6_1

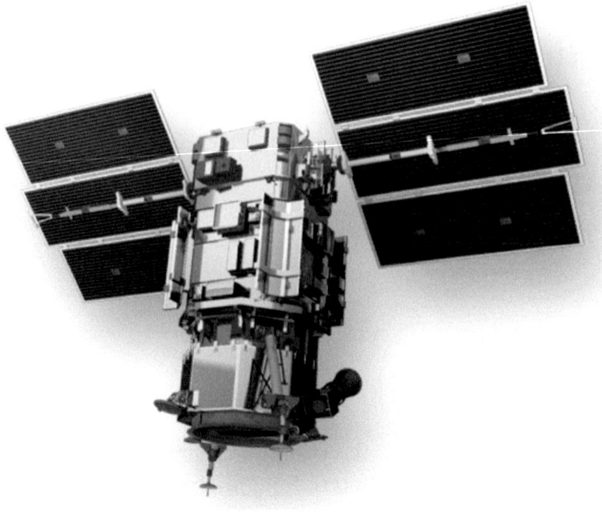

Fig. 1.1 WorldView I Satellite (Image courtesy of DigitalGlobe Inc.)

A technology known as the control-moment gyroscope (CMG) makes Worldview uniquely agile. Until WorldView's launch, the U.S. and Russian governments were the only owner/operators of spacecraft that use this technology. Examples include NASA's Skylab, the Soviet/Russian space station MIR, and the International Space Station. These momentum-control devices enable spacecraft to absorb large external torque disturbances and to slew payloads quickly from one attitude to another. Power is a precious resource on even the largest spacecraft. So, the fact that CMGs apply these high torques with tens to hundreds of times greater power efficiency than other momentum actuators makes them an appealing choice in the design of contemporary earth-observation spacecraft.

Other commercial agile satellites, such as Orbview (See Fig. 1.2), Digital Globe's GeoEye (Fig. 1.3), QuickBird (Fig. 1.4) and Iokonos (Fig. 1.5), depend not on CMGs but high-torque reaction wheel assemblies (RWAs) as attitude-control actuators. By 2007, RWAs had been used for decades, in a variety of applications, despite that CMGs were known to offer orders-of-magnitude higher torque for the power. With the launch of WorldView I, momentum-control technologies that had been used only for government-sponsored spacecraft programs entered the commercial realm. Honeywell International provides the CMGs for the WorldView spacecraft. Ball was Honeywell's first commercial customer for its M95 CMGs, the smallest flight-qualified class of CMG available at that time.

Now, over a decade later, CMG technologies are not only mainstream in commercial Earth observation, they are also appearing in new applications. CMGs have found their way into small spacecraft with scientific objectives, such as Cornell University's Violet nanosatellite shown in Fig. 1.6.

Fig. 1.2 OrbView 4 Satellite (Image courtesy of Orbital)

Fig. 1.3 QuickBird Satellite (Image courtesy of DigitalGlobe Inc.)

Violet is a 50 kg spacecraft with an ultraviolet spectrometer that is designed to make observations of the Earth's upper atmosphere to help astronomers calibrate observations of exoplanets. It has eight small CMGs, built by Goodrich Corporation. Honeybee Robotics has created golf-ball size CMGs suitable for 10–40 kg nanosatellites. The University of Florida built and launched similarly small CMGs (see Sect. 4.8). And Honeywell's Miniature Momentum Control System (MMCS)

Fig. 1.4 Ikonos Spacecraft (Image courtesy of GeoEye Inc.)

Fig. 1.5 GeoEye Spacecraft (Image courtesy of DigitalGlobe Inc.)

Fig. 1.6 Violet Nanosatellite (Image courtesy of Cornell University)

provides a plug-and-play solution for 100–1000 kg spacecraft that require high-precision pointing performance (see Sect. 2.1.4).

1.2 Momentum Control Devices and Attitude Control Systems

Spacecraft attitude control involves sensors and actuators, along with a feedback-control architecture that embraces the subtleties of spaceflight dynamics. Sensors such as star trackers and rate gyroscopes measure spacecraft motion, which on-board computers compare to desired kinematics. To correct this error, i.e. the difference between the measured and desired state, actuators such as momentum-control devices and thrusters apply torques to the spacecraft. The nature of these actuators—their torque and momentum capabilities, as well as their precision and speed of response—determines their usefulness for the range of missions that spacecraft are intended to achieve.

This book focuses on CMGs and RWAs, which are known as momentum-control devices. A spacecraft typically includes several such devices in an array. This array of actuators, along with electronics, high-level software, structural components, and possibly vibration isolators, comprises a momentum-control system (MCS). In turn, an MCS is part of the attitude-control system for a spacecraft. Figure 1.7 is a diagram that shows the relationships among these many nested elements.

Momentum devices produce torque by changing their stored angular momentum, realized in a spinning disc. Because momentum is a vector quantity, the product of

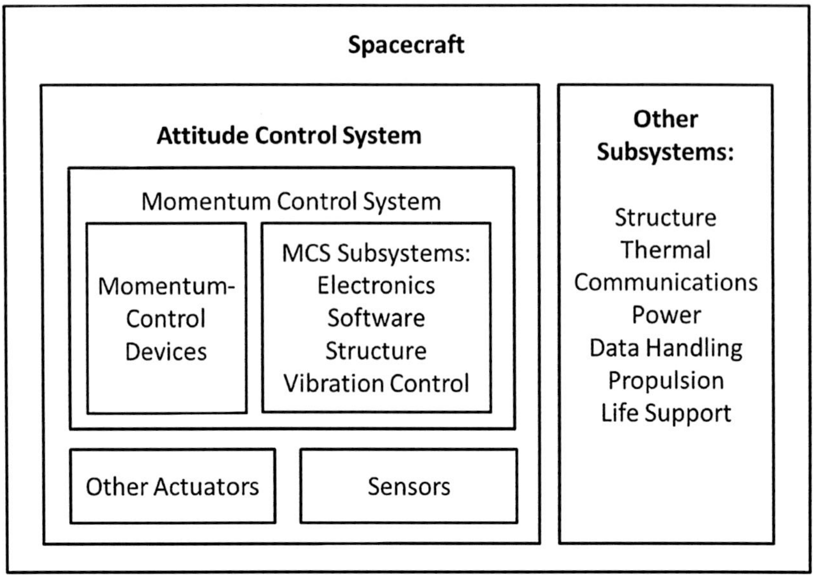

Fig. 1.7 Attitude-control centric view of spacecraft subsystems

angular velocity and inertia, there are two ways to effect this change. The simplest way is exploited in an RWA. The vector direction of the angular momentum of an RWA bolted to the spacecraft is constant–fixed in a spacecraft-fixed reference frame. The length or magnitude of the momentum vector changes as the wheel spins faster or slower. In the case of a CMG, the wheel speed (and thus the length of the momentum vector) is relatively constant, but a gimbal tilts the spinning rotor to change the direction of the momentum vector. The trades between these two technologies are many, and this book discusses them at some length. In each case, the device can store momentum up to a design maximum. That maximum value exists because a rotor of a given size is designed to spin at some maximum speed, typically determined by the tensile-strength limit of the rotor's material and its geometry. Going beyond that speed would incur excessive mechanical stress and fatigue, resulting in failure. Diagrams of a CMG and RWA are shown in Figs. 1.8 and 1.9.

The spinning rotors in an MCS serve two functions: momentum storage and torque application. From elementary physics, the MCS accumulates momentum as it imparts torque to the spacecraft (or reacts torque from the spacecraft, depending on one's perspective). This torque is internal, in the sense that the angular momentum of the overall spacecraft is a constant, regardless of what the MCS is doing. The MCS and the spacecraft body exchange angular momentum, but none is created. In this respect, an MCS is fundamentally different from a reaction-control system comprised of thrusters or jets. Momentum devices offer clear benefits in spacecraft design: years of operation without expending resources, the highest precision of any

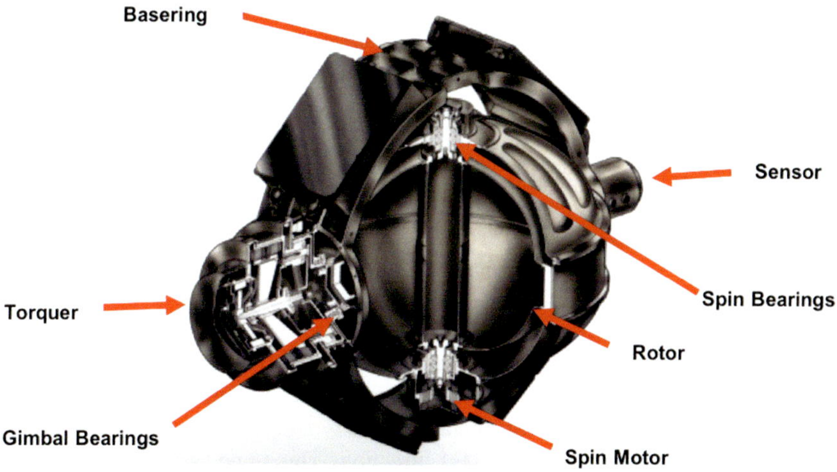

Fig. 1.8 Key components of a single-gimbal CMG (Image courtesy of Honeywell, Inc.)

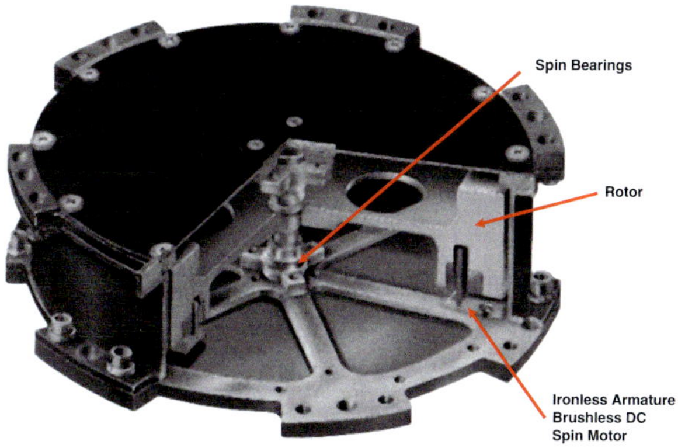

Fig. 1.9 Key components of a reaction wheel (Bialke [2])

actuator, and freedom to place these actuators anywhere in the bus structure. But these benefits come with limitations. In a thruster-based system, the thrusters can apply torque in a given direction until the expendables have been exhausted, which may be hours to months. However, a momentum-control system can apply torque in a single direction for only a limited time because the hardware can only accumulate a limited amount of momentum. Once the rotors have reached their maximum spin speeds or the gimbal angles have swung fully toward a given direction, the MCS cannot apply further torque to the spacecraft. Managing the momentum state of the MCS to prevent saturation or related singularities is a unique aspect of implementing momentum-control devices. Chapters 8 and 9 address this aspect in detail.

Propellant is a finite, expendable resource, while the electrical current required to change the momentum in a RWA or a CMG is typically renewable, through solar power. Therefore, thrusters are typically not used when internal torques would suffice. Another downside to thrusters is that they apply either coarse, impulsive torques or precise but long-duration ones. Chemical thrusters offer the former performance, and electric thrusters the latter. Attitude-control requirements often demand better pointing than chemical thrusters alone can provide, and they demand faster responses (higher bandwidth) than electric thrusters are capable of. Only an MCS is suited to this combination of precision and bandwidth. Table 1.1 summarizes these options.

An MCS also differs from a set of magnetic torque-coils that might be used in Earth orbit. Such devices impart a torque to the spacecraft without using expendable resources, torquing against the geomagnetic field, but the electromagnetic torque available cannot provide instantaneous three-axis attitude control because it can never lie along the geomagnetic field direction. The same is true of a gravity-gradient boom, which exploits the gradient of the Earth's gravity to cause a spacecraft to point toward nadir like a float on a fishing line. An MCS typically works alongside these external-torque actuators in order to provide a complete attitude-control solution: internal torques for high-agility and/or three-axis momentum storage, and occasional momentum unloading (or dumps) via thrusters or torque coils. This fundamental difference among attitude-control actuators drives spacecraft operations and has a direct impact on both the sizing of the MCS, and how it is used in practice. Chapter 3 addresses these issues in depth.

1.3 A Brief History of Spin

Momentum control has been a central element of successful space-systems architectures since the beginning of the space age. Some of the earliest spacecraft, such as Hughes' Syncom geostationary communications satellite, inherited a spin rate from the rockets that took them to orbit. However, earlier efforts along these lines had not been completely successful. The first successfully launched U.S. spacecraft were the Explorer series, each of which famously, and unexpectedly, fell over into a so-called flat spin after it separated from the rocket spinning around its longitudinal, skinny axis. The skinny axis of a rigid body has the least moment of inertia. It turns out that spinning in this direction is a maximum-energy state, from which the spacecraft's dynamics can only degrade thanks to dissipative effects, such as structural damping and internal fluid motions. In the case of Explorer I and subsequent Explorer spacecraft, long whip antennas extended from the body and dissipated this kinetic energy as they flexed, causing the spacecraft to fall over like an exhausted toy top after several hours in orbit.

In the early days of the Space Age, this skinny axis was known as the "roll" axis, using an aircraft convention that originated in nautical terminology. The coordinate axis often assigned to this axis was "Z," and even today a typical large spacecraft

Table 1.1 Comparison of attitude-control actuators

Actuators	Typical attitude-control applications	Agility ratio rate/acc (s) (smaller ratio is more agile)	Precision (°)	ACS control bandwidth (Hz)	Torque/power (Nm/W)
Spin stabilization	Low-precision omnidirectional RF payloads and magnetospheric science	N/A	0.1–1 in 2 axes	N/A	N/A
Gravity-gradient boom	Coarse earth observation and technology demonstrations	N/A	5–20 in 2 axes	N/A	N/A
Magnetic torquers	Momentum dumps, nanosatellites in LEO, gravity-gradient augmentation	Low agility	1–10 in 2 axes	0.01–0.1	0.001–0.01
Momentum wheels	Astronomy, communications	N/A	0.1–1 in 3 axes	0.01–0.1	0.001–0.01
Reaction wheels	Astronomy, communications, Earth observation	5–500	0.1–1 in 3 axes	0.01–1	0.001–0.1
Control-moment gyros	Earth imaging and radar, satellite servicing, asteroid grappling	0.5–5 (Most agile)	Better than 0.01 in 3 axes	0.1–1	1–10
Electric thrusters	GEO stationkeeping	Low agility	0.01–0.1 in 3 axes	0.001–0.01	0.0001–0.001
Chemical thrusters	Missile defense, Interplanetary transfer, spin-speed adjustment and reorientation, momentum dumps	High agility	1–10 in 3 axes	0.1–1	Very high

associates its Z axis with the longitudinal, or skinny, axis of the rocket that takes it to orbit. Syncom and similar spacecraft learned from that mistake. They were built as squat cylinders, making the Z or roll axis the axis of maximum inertia, and therefore a minimum energy state. Syncom was therefore stable in Z spin, avoiding Explorer IV's fate.

This so-called passive spin stabilization ensured some degree of passive attitude control for spacecraft of the early 1960s. Soon thereafter, RCA and Hughes began to exploit the dual-spin spacecraft principle: part of the spacecraft spins (known as

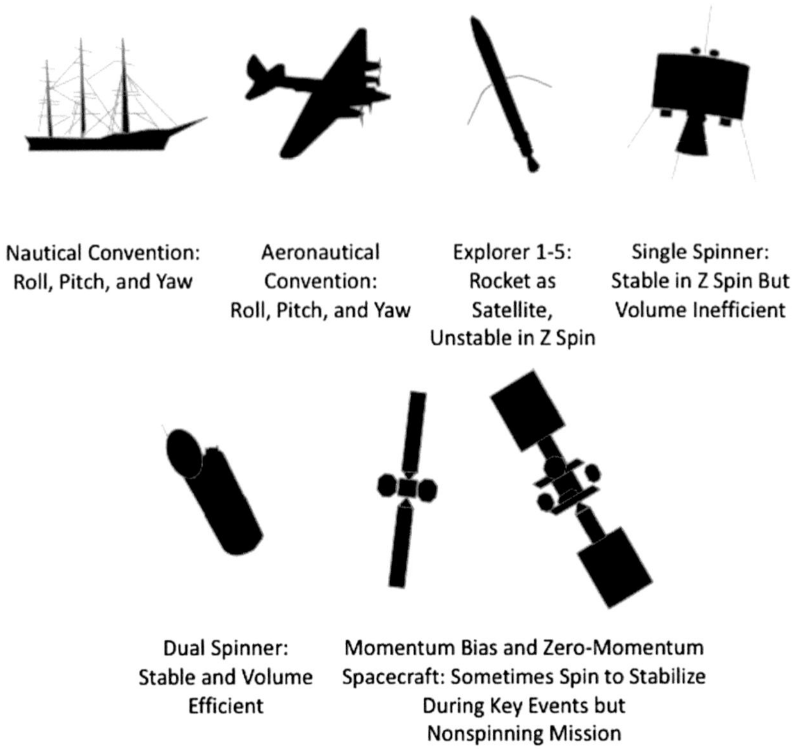

Fig. 1.10 Evolution of spinning spacecraft dynamics

the rotor), and part of it remains inertially fixed (known as the stator or platform). With the correct distribution of energy dissipation and mass between the rotor and the stator, the spinning portion ensures attitude stabilization of a conveniently launched, slender body that would otherwise be unstable. This innovation led to larger satellites, with more solar-panel area for more powerful payloads.

The evolution of spinning spacecraft dynamics is shown in Fig. 1.10.

The rotor of dual-spin spacecraft from the 1970s and 1980s soon gave way to momentum wheels. Those wheels serve precisely the same purpose as the rotor in a dual-spin spacecraft but, with higher speed and lower inertia, take up considerably less space and lead to higher-power, three-axis stabilized spacecraft buses. Some momentum wheels included a tip/tilt platform, such as on the Hughes/Boeing 601 class commercial satellites, allowing the fixed momentum to vary a little and control attitude with active feedback. The so-called momentum bias in these spacecraft ensured a stable attitude even in the presence of certain failures. The 1990s saw a general trend toward zero-momentum commercial satellites, including Lockheed Martin's A2100, the Boeing 702 class spacecraft, and others from Space Systems Loral, Ball Aerospace, and Orbital Sciences. These spacecraft use reaction wheels, with nominally zero momentum, for feedback control of attitude dynamics. Zero-momentum architectures are widely regarded as the most versatile, and lowest in power among the many ways in which angular momentum can be used in spaceflight, but with the risk that any failure that disables attitude control also leads to uncertain attitude motions.

While this trend from passive to active MCS control through RWAs continued in the commercial world, CMGs were prevalent among DoD spacecraft. Many journal publications from the late 1960s through the 1990s attest to innovations in CMG-based attitude control at Lockheed Martin, Boeing, and elsewhere. Throughout the decades, a key research question has remained: how does one distribute torque actuation among an array of CMGs, consisting of four or more, in order to avoid what are known as kinematic singularities among the actuators? It is known that no general-purpose solution allows unlimited freedom to traverse momentum space without torque limitation. Consequently, this problem has been solved for many special cases and in a broadly general, theoretical sense. However, a general-purpose, practicable solution that requires no future knowledge but that also exploits all possible momentum capability from a CMG array remains an important research objective.

1.4 The Proliferation of Momentum Control

From the 1960s through about 2000, large U.S. prime contractors such as Lockheed Martin, Boeing, and Northrop Grumman, dominated the aerospace industry's use of CMGs in space. Now, U.S. companies new to CMGs are pursuing satellite projects that take advantage of these capable actuators: as of this writing ATK (the former Swales in Beltsville, MD) and General Dynamics (the former Spectrum Astro in Gilbert, AZ), among others, have contemplated CMG enabled satellites.

RWAs have been used in optical payloads for space in the past for optical payloads that are sometimes known as reactionless coelostat telescopes (Loewenthal [3]). CMGs have more recently proven to provide lower-power reactionless robotic motions. A likely application of CMG robotics is in satellite servicing and asteroid manipulation because both applications benefit from a CMG's ability to absorb high external torques. An asteroid-mining or servicing spacecraft such as NASA's Asteroid Redirect Mission shown in Fig. 1.11 or Vivisat's Mission Extension Vehicle for Satellite Servicing shown in Fig. 1.12 would be able to keep its central body pointed without its central body experiencing torques from its robotic arms.

Government-funded human-space applications of CMGs are well established. They include not only Skylab, Mir, and ISS, but also the Astronaut Maneuvering Unit from the 1970s, which included six small CMGs. Someday we may see commercial human-space uses, such as hotels stabilized with CMGs.

And CMGs are moving beyond space applications. These devices turn out to be useful in ships at sea, including roll and pitch stabilization of small craft, as

Fig. 1.11 NASA's Asteroid Redirect Mission (ARM) (Image courtesy of NASA)

Fig. 1.12 Vivisat's mission extension vehicle for satellite servicing (Image courtesy of Vivisat)

in the case of the SeaKeeper gyro. They are also being considered for automotive applications. Lit Motors, a San Francisco startup, intends to produce small two-wheel vehicles balanced by a feedback-control system centered on CMGs (see Chap. 2).

In the U.S., commercial space is growing exponentially, with companies like SpaceX providing significantly lower cost access to space, Planet Labs launching dozens of CubeSats for global, persistent Earth imaging, and companies such as Deep Space Industries and Planetary Resources Inc. setting out to mine asteroids. An unintended consequence of the U.S. International Traffic in Arms Regulations (ITAR) is that larger U.S. companies have been forced to yield market share in some cutting-edge satellite technologies to overseas providers, although the US is generally seen as leading the world in CMG technology. Long the province only of U.S. companies, sophisticated momentum control is now being implemented throughout the world. For example, Surrey Satellite Technology Ltd. in the U.K. collaborated with Turkey to launch Bilsat, an early small-sat demonstrator of CMGs. EADS (in its earlier incarnation as Astrium) first built a CMG in the 1990s and continues to market it, and their subsidiary Thales/Alenia has been working on CMG-enabled spacecraft since the mid-2000s.

1.5 About This Book

This context motivates a thorough look at the technology of momentum control systems. The authors intend this book to serve as a technical resource for engineers, and that is why it covers a breadth of topics related to momentum control. It also provides depth in certain areas, such as the dynamics of momentum-control devices, requirements for momentum-control systems, and a thorough treatment of CMG singularities that are entirely absent from the general texts available on spacecraft attitude dynamics, control, and astrodynamics. While of likely interest to that small, deeply idiosyncratic community of momentum-control experts, some of this unique material also ought to appeal to those with a more academic perspective. The aging of the U.S. aerospace engineering community means that new spacecraft engineers have fewer opportunities for mentorship and may have to take on more responsibility for challenging satellite design work than might have been expected a decade or two ago. Some of the material in this book represents sage advice collected from the experts, those who have been working in the field for many years. The technical content of this book begins in Chap. 2, with an overview of the applications of angular momentum: various spacecraft-mission architectures and operations concepts, including agile Earth-observation, robotic servicing, human space, and some of the terrestrial applications. Chapter 3 develops requirements for spacecraft momentum-control systems. This discussion covers requirements flow-down from spacecraft-level issues such as day-in-the-life kinematics statistics, and other systems-engineering issues that ultimately trade risk in the context of size, weight, and power.

Chapter 4 begins a series of chapters on the core mathematics and physics of momentum devices. It offers a detailed analysis of the dynamics of spacecraft with RWAs and CMGs. Some attention is given to non-rigid effects, such as flexibility in rotors and gimbals, mass imbalances, friction, hysteresis, and materials issues.

Chapter 5 offers a mathematical treatment of the singularities associated with CMG array architectures. Chapter 6 describes the many options for these architectures and offers recommendations for the most promising approaches for their use in spacecraft attitude control. Chapter 7 describes system identification methods for identifying spacecraft and actuator parameters. Chapter 8 provides a brief expository on some practical attitude control methods. Chapter 9 explains how to steer the CMGs around or within singularities for this wide variety of arrays. Appendices offer further detail on the integral equations of motion for unbalanced rotors and the impact of singularities on attitude controllability, and other mathematical preliminaries. These chapters bring together decades of academic literature and practical experience from around the world on these subtle matters. Because these chapters constitute the most mathematically demanding material in the book, it is no surprise that these topics are of current academic research interest.

Chapter 10 moves deeper within the momentum devices. It summarizes the inner-loop control of the rotors and gimbals, addressing issues such as sensors, control of the rotor spin rate and gimbal rate, and experimental parameter identification. Chapter 11 summarizes key concepts for motors in space. Chapter 12 provides some information on bearing types and selection. Chapter 13 describes ways in which to represent the behavior of momentum devices in practical systems through computer modeling and test-bed validation. Finally, Chap. 14 wraps the book up with some details on a recent flight experiment of a 4-CMG array onboard the Orbital Cygnus spacecraft.

The mathematical content here requires some background in vector mechanics and linear algebra. The notation the authors have adopted is explained and is meant to articulate the issues rigorously. It is consistent with many U.S. journals and academic practice. We expect that most of the chapters will be accessible to a general audience with an engineering background.

Chapter 2
Applications

Virtually all spacecraft incorporate some form of momentum control. Even the simplest spin-stabilized spacecraft—nothing more than a rigid body stabilized by virtue of its own spin—might be considered to benefit from a form of momentum control. Dual-spin spacecraft incorporate a spinning "rotor" and a non-spinning "platform," which achieves attitude stabilization through a similar but more versatile approach. Contemporary spacecraft use a wide range of momentum-control technologies—momentum wheels, reaction wheels, and several types of control-moment gyroscopes—with an ever-increasing number of applications. The applications discussed here focus on contemporary spacecraft, from small autonomous satellites to large human-space vehicles and earth-observation systems. However, the hardware and basic principles have found uses outside the field of aerospace engineering, largely in commercial applications. Examples include stabilization of trains, automobiles, and boats. As spinoffs from the high-tech, high-reliability world of aerospace make their way into other fields, one can expect to find reaction wheels and CMGs in even more surprising places.

2.1 Spacecraft Applications

2.1.1 GEO Communications Spacecraft

Perhaps the most commercially successful space application for momentum systems is the geosynchronous communications satellite. Arthur C. Clarke's famous 1945 Wireless World article popularized the idea of a communications spacecraft that hovers above the Earth by orbiting at Earth's spin rate. Thanks to that early start, GEO spacecraft were among the first to be profitable, beginning with spin-stabilized spacecraft in the early 1960s such as Hughes' Syncom 2 (1963) and dual-spin satellites from RCA Astronautics and Hughes (among others). Momentum wheels

© The Author(s), under exclusive license to Springer Nature Switzerland AG 2025

F. A. Leve et al., *Spacecraft Momentum Control Systems*, Space Technology Library 46, https://doi.org/10.1007/978-3-031-96165-6_2

and reaction wheels became the preferred actuators for attitude stabilization of these ever larger spacecraft beginning in the 1980s, and that basic architecture has continued into the present.

The Hughes/Boeing 601 class spacecraft serve as an instructive example of a momentum-wheel architecture. The earliest examples include the Optus series of spacecraft (1987), but variants on this design continue to launch and operate successfully to this day. These spacecraft point an antenna farm—a payload comprised of many transponders and reflectors—toward the Earth. Pointing accuracy of a fraction of a degree is the goal for GEO spacecraft in general, and the 601 class is typical in this regard.

The solar array on the 601 spacecraft provides up to 4.8kW of power. This level of power performance is unattainable from the smaller solar arrays of a cylindrical dual-spinner. Three-axis attitude control is required for these panels to point at the sun while the payload points at the Earth. Zammit [4] and Smay [5] offer considerable detail on this attitude-control architecture, which is summarized here. To achieve this performance, the spacecraft incorporates a momentum-wheel platform. This platform includes a momentum wheel and a tip/tilt table (or gimbals) capable of redirecting the wheel's angular momentum vector with small-angle motions. The wheel is aligned nominally with the spacecraft's pitch axis, which runs through the two solar wings and is directed along the Earth's north/south axis. This orientation keeps a momentum bias in a direction that passively stabilizes the spacecraft in an event of a failure of the attitude-control sub-system. The wheel is kept at a roughly constant speed (responding only to small errors in pitch angle). The jackscrews on the momentum-wheel platform (shown in Fig. 2.1) allow wheel to absorb disturbance torques in the roll and yaw directions.

The platform must precess this stored-momentum vector at one full rotation per orbit, since its direction is inertially constant while the spacecraft itself rotates around the Earth to keep its payload pointed at a fixed location on the surface. The

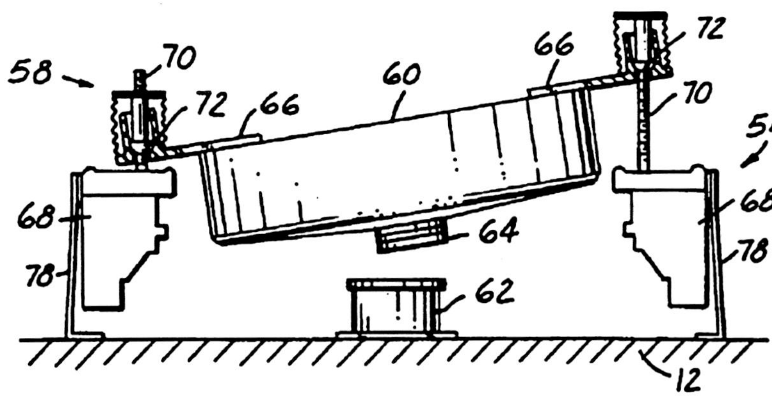

Fig. 2.1 Momentum wheel platform (Yuan and Wittmann [6])

combined momentum bias and three-axis attitude control represents a successful superposition of 1970s-era passive stabilization with more modern three-axis control techniques.

Other GEO Spacecraft are known as "zero-momentum" satellites because their reaction wheels do not store momentum that is intended for passive stabilization. For this reason, a momentum wheel and a reaction wheel are essentially the same hardware; they merely have different applications in the spacecraft attitude-control architecture. An example of a zero-momentum satellite is Boeing's 702 class spacecraft shown in Fig. 2.2, which was introduced in the late 1990s. Modern 702 spacecraft have solar arrays with wingspans larger than that of a Boeing 737 aircraft, and with power up to 18 kW (Stribling [7]). This larger size demands more headroom in the reaction-wheel array torque and momentum envelopes—i.e., the attitude control subsystem cannot successfully react to the torque disturbances and store the resulting momentum with the very limited capability of momentum-wheel platforms. Instead, contemporary zero-momentum spacecraft use comparatively large reaction wheels, each offering up to 100 Nms of momentum-and as much as 2 Nm of torque (Bialke and Stromswold [8]).

The zero-momentum approach also offers greater versatility in the operations concept, for example in the case where a spacecraft is not in a perfectly equatorial orbit. Such a satellite might drift north/south during its lifetime to eliminate the cost of propellant for north/south stationkeeping (Lim and Salvatore [9]). In doing so,

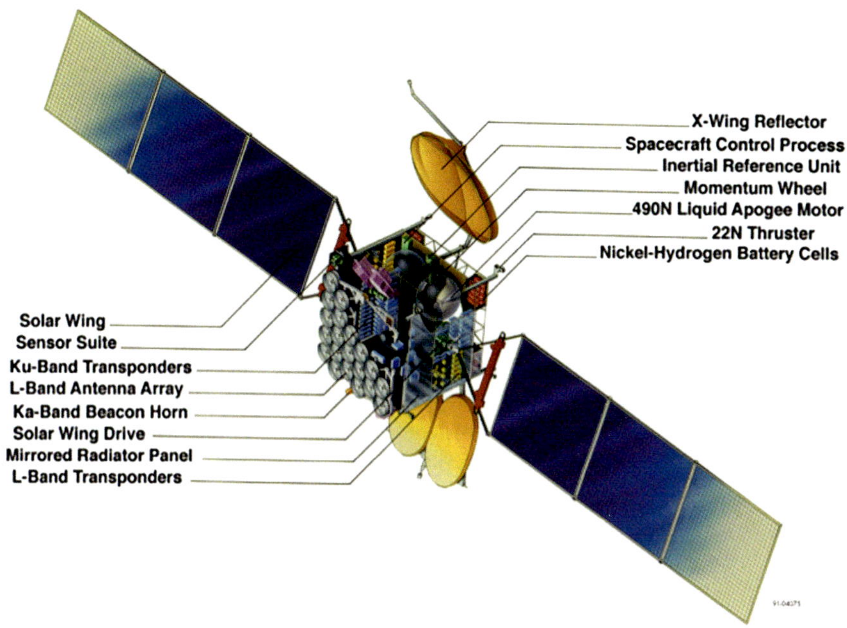

Fig. 2.2 Boeing 702 Class Spacecraft (Image courtesy of Boeing)

the spacecraft also cannot simply align its pitch axis normal to the orbit plane at all times, as is the case for the 601. Instead, three-axis attitude control provides torque to follow a more subtle attitude trajectory through the reaction wheels and stores whatever momentum accumulates due to disturbance torques (from the environment, the payload, and other effects). The attitude-control subsystem dumps that momentum occasionally through reaction-control jet activity, whether chemical thrusters or electric thrusters.

Zero-momentum spacecraft use an array of reaction wheels, often four of them. These four wheels are oriented so that the failure of any single wheel results in an array whose wheel spin axes still robustly span three dimensions. There are many approaches to the design of these arrays, some of which are discussed in Chap. 6.

2.1.2 Agile Spacecraft Missions and ConOps

The concept of operation, or ConOps, for a spacecraft lays out the method by which operators use the spacecraft to achieve its mission. The ConOps of a commercial GEO spacecraft is comparatively straightforward, sometimes requiring only one operator to watch over several such satellites at a time. Others, such as spacecraft that perform Earth imaging in LEO, are more demanding. Their subtler ConOps involve frequent acquisition and loss of signal as the spacecraft flies in and out of view of ground stations. The momentum control system is at the core of how these spacecraft function and how well they perform.

Earth imaging spacecraft such as Ball Aerospace's WorldView-3 use an array of several CMGs as a momentum system (Poli et al. [10]). These spacecraft typically operate in LEO, where they are as close as possible to the Earth's surface to optimize the size of objects on the ground that their optics can resolve. However, they are also high enough that the atmospheric drag can be overcome by an affordable amount of propellant, for a useful lifetime. This process of optimizing the design to trade off among altitude, payload performance, lifetime, and propellant is beyond the scope of this book, and it does not very directly involve the momentum system. The reason is that the specific altitude within typical LEO orbits only slightly influences the speed with which the spacecraft must reorient, or slew, to capture images.

The attitude-control subsystem has two key objectives: (1) perform large-angle, high-speed slews to direct the payload toward a point of interest on the ground and (2) maintain the attitude with sufficient accuracy and precision to allow that payload to capture imagery as it passes over the point of interest. Objective (1) demands high torque and momentum storage. Torque accelerates or decelerates the spacecraft, and momentum storage establishes the peak angular velocity that the spacecraft can reach. In addition, the rate at which this torque can change is important: it influences angular jerk, the derivative of acceleration, which turns out to drive the design of the gimbal of a CMG. The total duration of these slews depends on rate, acceleration, and jerk, as well as the flexible response of the spacecraft, fluid slosh, and other behaviors. However, it is the MCS capability that primarily determines the speed with which an Earth-imaging spacecraft can move from target to target. This agility directly translates into profitability: the faster the slews, the more data can be collected, and the more product the spacecraft can deliver to its customers.

These two objectives work against each other, to a large extent. Fine precision, low-disturbance attitude control is hard enough. It demands very low gimbal rates to produce gentle torques that do not exceed jitter requirements. And yet this goal must be achieved with actuators that are also capable of imparting hundreds to thousands of Nm of torque. So, the dynamic range of the CMG gimbals, the dynamic and static balance of the rotors, and the structural stiffness of the MCS, are key parameters that flow down from the ConOps. Over the years, they have led to CMG technology that is necessarily both precise and powerful.

2.1.3 Space Stations

Despite that the vast majority of CMGs have been used in Earth observation spacecraft, the most well-known applications for CMGs are in human space: Skylab, Mir, and the International Space Station.

The International Space Station (ISS) serves as an instructive example. Its MCS incorporates four double-gimbal CMGs in the Zarya module. The two gimbals per CMG enable the ISS's control system to redirect the momentum from each CMG toward any point on a sphere, with the result that the momentum envelope of the ISS's CMG array is a sphere four times the radius of any one of its four CMGs' angular momentum. Figure 2.3 shows this arrangement. The ConOps of the ISS includes a zero-propellant maneuver, in which the combination of gravity-gradient torque, atmospheric drag, and solar radiation pressure are largely balanced by the attitude of the body and the orientation of its vast solar array (Bedrossian et al. [11]).

Despite this innovative approach to minimizing momentum accumulation, the CMGs still need to absorb disturbances. They do so with extremely large rotors (4760 Nms) and powerful gimbal motors, producing 258 Nm output torque (L-3 [12]). The large momentum-storage capability is understandable: ISS is the largest spacecraft ever built, weighing over 400,000 kg, and one would expect large actuators for a large vehicle. But the gimbal-motor torque is disproportionately high compared to single-gimbal CMGs. It results from the double-gimbal configuration. The reason is that in such a CMG, the motion of one gimbal tilts the rotor's momentum and causes a gyroscopic torque that the second gimbal must react. In contrast, a single-gimbal CMG imparts this gyroscopic torque directly to the CMG's base ring, and from there into a stiff spacecraft bus structure, without the need for electrical power to constrain the motion. This significant savings in size, weight, and power motivate the nearly universal choice of single-gimbal CMGs in space applications.

Heritage is a powerful justification for any spacecraft design choice. ISS uses double-gimbal CMGs in part because Skylab did. In the 1960s, when Skylab was being designed, there had been only limited experience with CMGs, and no NASA

Fig. 2.3 The four ISS CMGs shown in 2000 at Kennedy Space Center's space station processing facility (Image courtesy of NASA)

project had ever used them. At that time, the problems with kinematic singularities associated in a single-gimbal CMG array were believed to be insurmountable—that one could not fly a spacecraft stably with such actuators. This belief persisted for long enough to determine Skylab's double-gimbal CMG architecture and establish the heritage for ISS. A straightforward sizing exercise suggests that an array of contemporary single-gimbal CMGs could also serve the needs of ISS and would benefit from the decades of technology development that have made CMGs a reliable solution for high-torque spacecraft.

The ISS CMGs are an outstanding engineering accomplishment. They are successfully controlling ISS and have been doing so for years. Nevertheless, they also experienced failures in the early 2000s related to electronics and to spin bearings. It is not surprising that as the first (and only) of their kind, these CMGs would represent some technical risk. The Skylab CMGs also experienced failures, the first occurring only months after launch: again, a bearing-related failure, which is the most common mechanical failure mode for a momentum actuator. This apparent trend—two space stations, two sets of CMG failures—should not be taken as evidence of CMG reliability in general. In fact, contrary to what these examples may suggest, no spacecraft has ever had its lifetime shortened by a failure of a single-gimbal CMG.

2.1.4 Small Spacecraft

Small spacecraft per se do not represent an application area. However, they are often considered unique in that they hold significant promise for highly responsive design, build, and test. The DoD's Operationally Responsive Space (ORS) office focuses on small spacecraft for this reason, among others (Foust [13]). One of its goals is to reduce the time to develop and launch a spacecraft by orders of magnitude: months or days instead of the current practice of over a decade.

Small scale also offers increased pointing agility, but not simply for geometric reasons. In fact, scaling every feature of a large, agile satellite's geometry down to a smallsat scale does not change its agility at all. The reason is that the spacecraft's rigid-body inertia drops with the fifth power of length (all things being equal), and so does the inertia in its momentum devices. So, the ratio of momentum to actuator inertia—roughly vehicle angular rate—is preserved for simple length scaling. The same is true for angular acceleration and jerk.

But there is an advantage. A smaller rotor can spin faster because the tensile stress in the rotor drops for smaller radii. Similarly, the structural natural frequencies of the momentum-device increase for smaller scale, all things being equal. Therefore, for a given rotor- and gimbal-speed control-loop bandwidth, the control/structure interaction is less. Alternatively, these control loops can be increased in bandwidth and remain lower than the structural frequencies. So, with higher rotor speeds and higher-bandwidth inner loops in the momentum devices, small spacecraft can achieve greater agility than their larger analogues. Momentum devices for such systems must be designed to accommodate the higher base-rate effects, as described in Chap. 3. They also may expend more electrical power to maintain rotor speed than a similar device with a lower-speed rotor.

Nevertheless, such designs are possible, as evidenced by the Violet spacecraft. This satellite is one of the entrants in the University Nanosatellite Program of the Air Force Research Lab. Violet is capable of at least $10°/s$, $10°/s^2$, and $50°/s^3$, thanks to its comparatively low inertia (about $2\,kg\text{-}m^2$ in any axis) and its Goodrich/Ithaco small CMGs. It may be capable of as high as $40°/s$, $40°/s^2$, and $60°/s^3$ (Gersh and Peck [14]). Figure 2.4 shows a drawing of the interior (left) and a photo of two of the CMGs in the partly integrated spacecraft (right).

By providing an in-orbit testbed for steering algorithms, the Violet project is designed to enable guest investigators to prove out next-generation CMG steering laws. In this respect, Violet's mission was designed to complement that of AFRL's Advanced PnP Technologies (APT) satellite, the spacecraft that had formerly been known as TACSAT-5 (Department of the Air Force [15]). Among other objectives, APT was meant as an experimental platform for Honeywell's plug-and-play momentum-control system, their Mini-MCS. Violet's CMG experiments are distinct but complementary to those articulated in the 2009 TACSAT-5 Broad Agency Announcement.

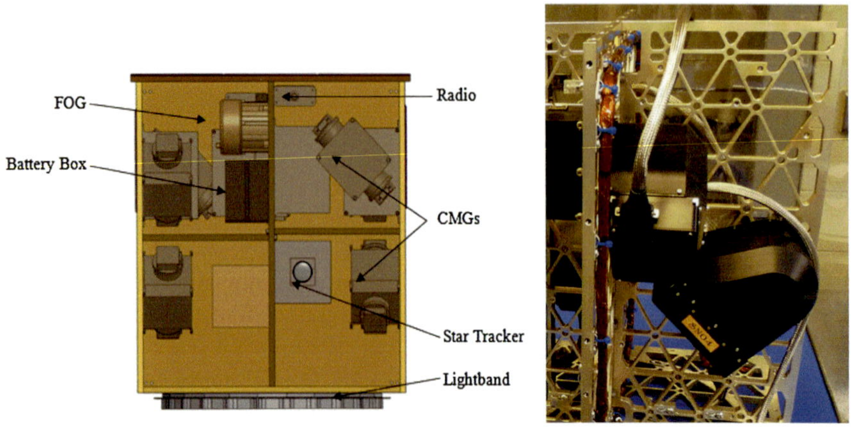

Fig. 2.4 Violet interior (left) and photo of two CMGs mounted on bus structural panels

- APT's objective in this area was to experimentally validate the MMCS, focusing on its suitability for plug-and-play integration in a responsive, agile spacecraft bus. In doing so it evaluated a specific CMG array architecture appropriate for responsive space. In contrast, Violet is designed to investigate the general problem of high-performance CMG steering, without explicit relevance to responsiveness.
- Violet's experimental objectives include CMG array architectures of four, five, and six CMGs that are different from the APT array. The result is that the two spacecraft together represent a broad range of array architectures.
- APT's CMGs and its bus were more than an order of magnitude larger than Violet's. They occupy different operational spaces for tactical spacecraft. So, taken together, the two spacecrafts' CMG experiments represented a broad assessment of CMG steering performance across a range of spacecraft scale.
- Violet is designed to experiment with very high agility (10–40°/s), while APT was meant to evaluate its CMG performance for a maximum base rate of 3°/s. These kinematics may be relevant for different missions, but the larger dynamic range of the two spacecraft taken together would have made the CMG experiments relevant for a wider range of applications than would either spacecraft on its own.
- APT's MMCS was designed so that the CMGs are mechanically aligned relative to each other with high precision, and this alignment is unaffected by its soft isolation mount. In contrast, Violet's CMGs are individually isolated and mounted. The two spacecraft will therefore be able to evaluate CMG steering in the presence of these two architectures to determine experimentally the relative benefits.

Despite APT's cancelation, the Honeywell mini-MCS remains a promising prospect for small-spacecraft applications because it fills a capability gap between

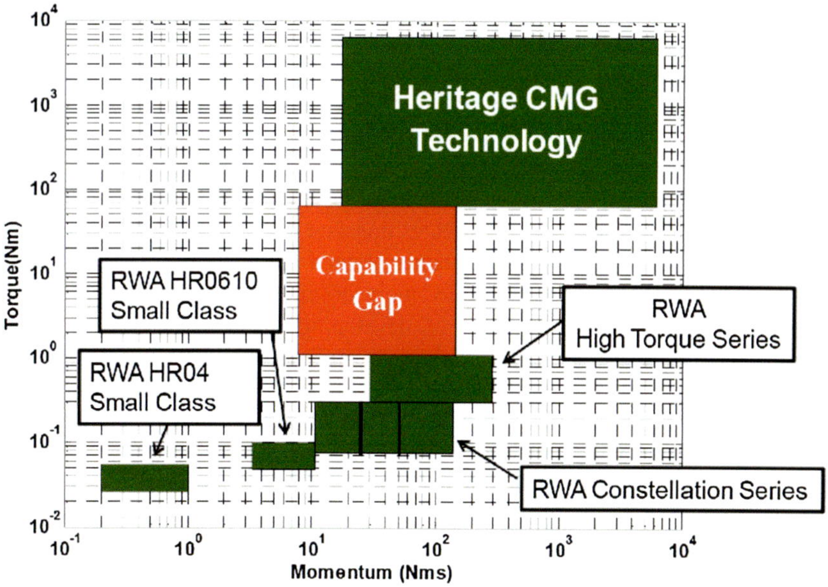

Fig. 2.5 Torque vs. momentum for some common CMGs and reaction wheels (McMickell et al. [16])

Fig. 2.6 Isometric cut-away view of miniature momentum control system (McMickell et al. [16])

the very small scale and the very large. This gap is indicated in Fig. 2.5. Figure 2.6 is a drawing of Honeywell's mini-MCS.

TacSat-2's mass properties and agility requirements sized the MMCS: 370 kg, with a maximum inertia of 190 kg-m^2. A slew-rate requirement of 3.0°/s and angular acceleration of 1.5°/s^2 was meant to accommodate planned ORS missions (McMickell et al. [16]). The integrated momentum devices provide a ready solution for momentum control in a small spacecraft, filling that gap and enabling a small spacecraft to achieve the responsiveness that ORS seeks.

So-called momentum control assemblies comprise a set of at least three devices (CMGs and/or RWAs). An MCS benefits from combining actuators, electronics and software into an integrated package. In this package, the devices are pre-aligned and vibration-isolated, with integrated control and steering algorithms, health-monitoring, and fault tolerant software (Hamilton [17] and Hamilton and Underhill [18]). These combined attributes translate to a system that is lower mass, has better vibration rejection, and is better aligned and held to greater tolerances than separately integrated and controlled momentum devices.

Honeywell was at the forefront of developing MCS technology. With the advancements in hybrid bearings, ironless armature spin motors, and modern singularity avoidance algorithms (Chap. 9), Honeywell was able dramatically reduce the mass, volume, and power draw of a momentum control system while keeping performance constant and in many cases even increasing (Davis [19]).

Examples of two current MCS designs developed by Honeywell are the MCS8 (formally called mini-MCS) and MCS200. These systems consist of a four-roof CMG array with electronics. They are shown in Figs. 2.7 and 2.8 (McMickell et al. [16]).

Fig. 2.7 MCS8 (Courtesy of Honeywell Defense and Space)

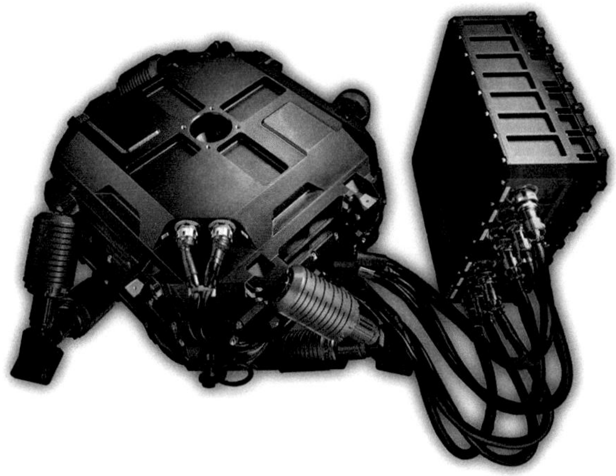

Fig. 2.8 MCS200 (Courtesy of Honeywell Defense and Space)

2.1.5 *Satellite Servicing*

Technologies for satellite servicing have seen accelerated growth in recent years. Thanks in part to Congressional support for this area within NASA's technology budget, the Satellite Servicing Capabilities Office at Goddard Spaceflight Center has demonstrated a number of successes in fluid transfer and mechanical grappling in orbit for spacecraft that include commercial satellites. This capability can extend the life of existing satellites and enable failed satellites to complete their missions. When satellite servicing becomes common, requirements on satellite reliability may be relaxed, making spacecraft more affordable.

The work has its origins in Shuttle-era repair missions beginning with Solar Max (1984). Several other commercial spacecraft retrievals and repairs followed, culminating in the series of Hubble Space Telescope repair missions in the 1990s. Currently at least two commercial companies are working in this area: ITT Excelis and ViviSat. The latter is a collaboration between ATK and Space Systems/Loral. Furthermore, DARPA has been sponsoring work on in-orbit assembly of spacecraft from components reclaimed from other spacecraft, known as the Phoenix Program.

When the servicing spacecraft makes contact with its client (or target) such as in Fig. 2.9, the angular momentum of the two becomes that of a single, joined, rigid body and manifests as angular velocity of that body. It is possible, in fact nearly certain, that the mass center of the two spacecraft would be offset from the point of contact and from their relative velocity. Consequently, the angular momentum of the pair can be nonzero after contact, even if neither has angular momentum about their respective mass centers before contact.

Presumably only one spacecraft actively controls its attitude at the moment of contact. Its attitude control responds to this change in angular velocity by

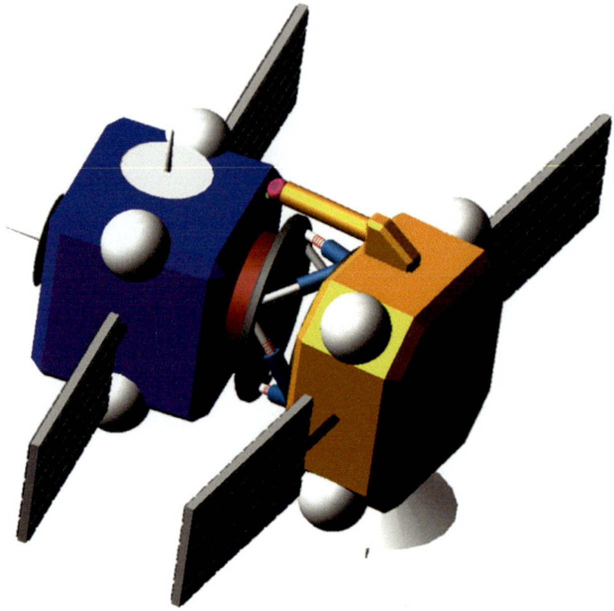

Fig. 2.9 Servicing satellite docking with tumbling defunct satellite

commanding the momentum system to apply torque. The suddenness of this change requires high torque and suggests that an array of single-gimbal CMGs may be the best choice for the momentum system of a satellite-servicing vehicle. To date, no such momentum system has been implemented. Perhaps as these capabilities evolve, in the form of a system like NASA's proposed RESTORE spacecraft, the need for CMGs in this application will become evident (Long et al. [20]).

Even more forward-looking concepts for on-orbit assembly at NASA have been considered, for example the assembly of large space structures where the launch of the entire system is impossible. An example of such a mission is the assembly of large space structures from a multitude of heterogeneous, smaller satellites, as suggested in Fig. 2.10. Such missions require precise attitude control in addition to a high torque and angular momentum capacity, both possibly achieved through an array of CMGs.

2.1.6 Asteroid Grappling

The problem of how to manipulate a resident space object is a common thread between satellite servicing and asteroid grappling. An important difference is that the mass properties and hard points for grappling are likely well known in the case of satellite servicing. However, an asteroid has poorly understood mass properties

Fig. 2.10 On-orbit assembly through heterogeneous satellite platforms

and may be nothing but a weakly bound collection of rubble. A satellite servicing mission may be able to feed-forward the contact disturbance. However, the unknown mass properties of the asteroid make this prospect less likely in asteroid rendezvous scenarios.

NASA's proposed Asteroid Redirect Mission is intended as a low-cost demonstration of critical technologies for Mars exploration, using existing hardware wherever possible. It also may open the door to asteroid mining and related applications of in-situ resource utilization (Strange et al. [21]). The uncertainties inherent in asteroid manipulation and the importance of mission assurance in the NASA context once again point to the need for a CMG-based attitude control system for asteroid applications of the future.

2.1.7 Space Robotics

Robotics in space benefit from managing the angular momentum of a multibody system, such as a robotic arm, so that the spacecraft on which the arm is mounted feels minimal reaction. In the absence of a ground against which to react, such robotic systems pose a number of technical challenges. There are several ways to achieve reactionlessness, but a novel approach is to embed momentum devices within the multibody system, using those internal moments to steer the various links. An example is shown in Fig. 2.11.

This approach has been proposed for coelostat telescopes, which are optical systems that may include a bent path. These jointed telescopes benefit from reactionless motors, such as a reaction wheel whose momentum absorbs the motion of the telescope body. The result is a mechanically quieter system that reduces vibration of spacecraft components such as solar arrays, leading to lower optical jitter.

CMGs have also been proposed as a means to manage the momentum within a multibody system, such as a robotic arm with a camera mounted as its end-effector. A scissored pair of CMGs replaces a single reaction wheel in this concept,

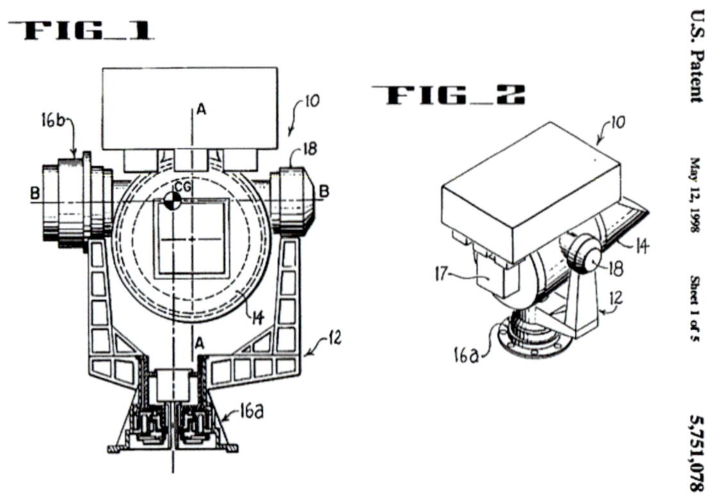

Fig. 2.11 Reactionless gimbal (Loewenthal [3])

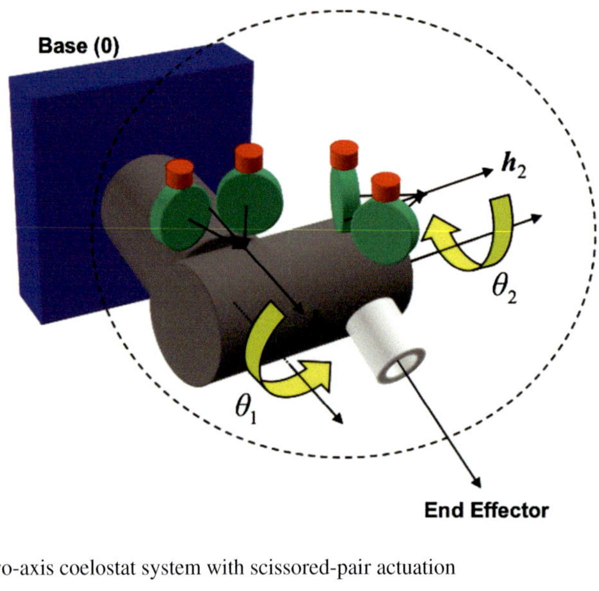

Fig. 2.12 Two-axis coelostat system with scissored-pair actuation

enabling far more agile motions for a given electrical power (Carpenter, 2009). Figure 2.12 illustrates this concept. Figure 2.13 shows a prototype of a three-link system (Carpenter and Peck [22]).

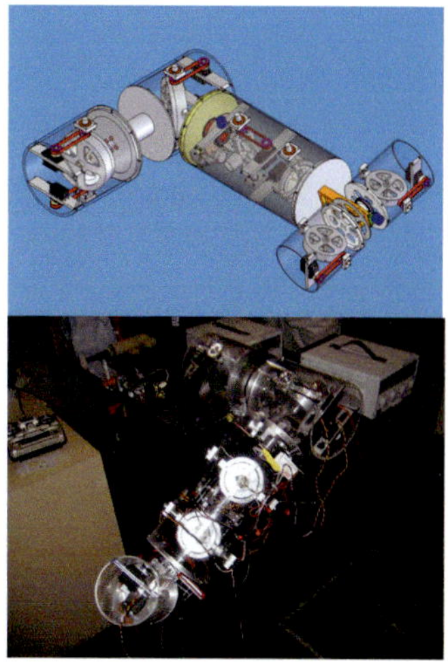

Fig. 2.13 CAD model of three-link robotic system with CMG scissored pairs (top) and prototype (bottom)

2.2 Terrestrial Applications

Early technological innovation in momentum devices for space benefited from terrestrial advances in tribology and materials science. In fact, in the 1800s precursors of modern CMGs were used for nautical applications. The Sperry Gyroscope Company (founded in 1910) is the ancestor of today's Honeywell Defense and Space Electronics Systems, where most contemporary CMGs are built. In recent years, space technology has begun to inform terrestrial applications in automotive and nautical domains, bringing momentum control full circle.

2.2.1 Brennan's Monorail

An early terrestrial application of CMGs is Louis Brennan's monorail concept (Brown [23]). Figure 2.14 shows two rotors, which resemble a scissored pair, capable of stabilizing a two-wheel cart. Gearing between the two rotors directs the gyroscopic torque along the rail direction, preventing the train from tipping over (Brennan [24]).

Brennan's monorail was featured at the Japan-British Exhibition at the White City, London, in 1910. The car carried 50 passengers on a circular track at up to 20 mph. Brennan built two such vehicles, but the concept ultimately failed to develop commercially.

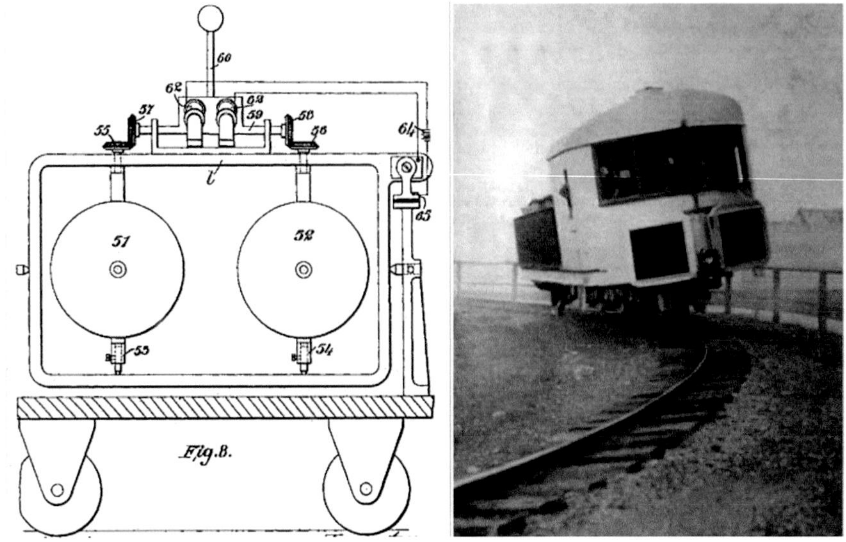

Fig. 2.14 Brennan's gyroscopic monorail. Patent (left), and prototype (right)

Fig. 2.15 Wolseley's two-wheeled wonder car

2.2.2 *Wolseley's Two-Wheeled Wonder Car*

As far back as 1914, CMGs were used to stabilize two wheeled vehicles. One such example was the Wolseley's Two-wheeled Wonder Car shown in Fig. 2.15 built in 1914. The stabilizing gyro concept for the car which consisted of 40-inch gyroscope spun at 3000RPM, connected to weighted pendulums was finished in 1913 and the car was made public in 1914.

Fig. 2.16 (left) Internal scissored pair and (right) Lit Motors C-1 vehicle (Image courtesy of Lit Motors)

2.2.3 *Lit Motors*

Lit Motors' C-1 vehicle is a modern-day application of Brennan's principles in many respects. However, it contains a number of innovations in the feedback control of these momentum devices that were completely unknown 100 years ago. Its two-wheel automobile combines the speed and efficiency of a motorcycle with the accessibility of a sedan, thanks to roll stabilization provided by a scissored pair of single-gimbal CMGs. Figure 2.16 shows the C-1 in motion and its internal scissored pair.

2.2.4 *Nautical Roll and Pitch Stabilization*

The use of spinning rotors to stabilize roll (sometimes pitch) of a sea vessel goes back to the nineteenth century. These devices were passive, transmitting base rate to torque that was taken out through friction in the gimbal. In 1931 the luxury vessel Conte Di Savoia was fitted with three large gyros meant to provide a comfortable environment for its passengers. Figure 2.17 shows two of these devices.

Seakeeper Inc. currently produces a line of actively controlled CMGs for a range of seagoing vessels. They are designed to achieve 70–90% roll reduction on boats up to 20 tons, vessels 30–50'. It is a substantial advancement over its nineteenth and twentieth century predecessors, with digital feedback control of the gimbal and evacuated rotor housing. These CMGs are the largest in existence, far out-performing even the ISS CMGs in torque and momentum with up to 35,000 Nm of output torque and 78,000 Nms of angular momentum. Figure 2.18 shows five installed in a yacht.

Fig. 2.17 Two of the Conti Di Savoia roll stabilizers

Fig. 2.18 Five Seakeeper gyros (Image courtesy of Seakeeper, Inc.)

2.3 Chapter Summary

Virtually all spacecraft incorporate some form of momentum control. Even the simplest spin-stabilized spacecraft—nothing more than a rigid body stabilized by virtue of its own spin—might be considered to benefit from a form of momentum control. Dual-spin spacecraft incorporate a spinning "rotor" and a non-spinning "platform," which achieves attitude stabilization through a similar but more versatile approach. Contemporary spacecraft use a wide range of momentum-control technologies—momentum wheels, reaction wheels, and several types of control-moment gyroscopes—with an ever-increasing number of applications. The applications discussed here focus on contemporary spacecraft, from small autonomous satellites to large human-space vehicles and earth-observation systems. However, the hardware and basic principles have found uses outside the field of aerospace engineering, largely in commercial applications. Examples include stabilization of trains, automobiles, and boats. As spinoffs from the high-tech, high-reliability world of aerospace make their way into other fields, one can expect to find reaction wheels and CMGs in even more surprising places.

2.4 Exercises

1. A LEO satellite needs to perform high-agility, precise attitude maneuvers in three axes: 3°/s rate, 3°/s^2 acceleration, and 1e-4° attitude precision. Cost is not a concern. Which actuator technology is likely to be chosen to accomplish this objective?

   (a) Reaction wheels
   (b) Control-moment gyroscopes
   (c) Thrusters
   (d) Magnetic torque rods

Chapter 3
Requirements Development for Momentum Control Systems

This chapter addresses the subject of how to flow spacecraft agility requirements down to an array of momentum devices and provides guidelines for choosing the appropriate technology. We shall see that the design trades in selecting and sizing momentum devices span orders of magnitude in performance, complexity and cost. Therefore, to arrive at an efficient and cost-effective design, it is important not to over-specify the momentum-system requirements.

3.1 Quantifying Agility Requirements

We begin the process by quantifying an “agility” metric. A typical day in the life of a spacecraft attitude control system generally allocated to two modes of operation. The first consists of tracking a target or commanded attitude profile. This mode drives requirements for stability and precision but generally does not influence the sizing of the momentum system because vehicle accelerations and rates are low in this mode. The second mode of operation, executing slew maneuvers between targets or attitudes, is where the need for agility resides. A more agile vehicle completes these maneuvers in a shorter period of time, and thus requires some combination of more vehicle rate and more vehicle acceleration. These kinematics drive the sizing of the momentum system.

Euler tells us that a rigid body can rotate about a single eigenaxis to slew from any attitude to any other, and thus for quantifying agility requirements, we shall employ eigenaxis slews (Fig. 3.1). The figure shows the line of sight (LOS) associated with the spacecraft’s payload as it slews from its initial orientation to its final orientation. This simplifies the problem to a single degree of freedom.

For a first cut at the requirements, we need to know only the angle of rotation about the eigenaxis and how long the spacecraft has to perform the maneuver. The requirements can be represented in a plane whose abscissa is time and ordinate

© The Author(s), under exclusive license to Springer Nature Switzerland AG 2025
F. A. Leve et al., *Spacecraft Momentum Control Systems*, Space Technology Library 46, https://doi.org/10.1007/978-3-031-96165-6_3

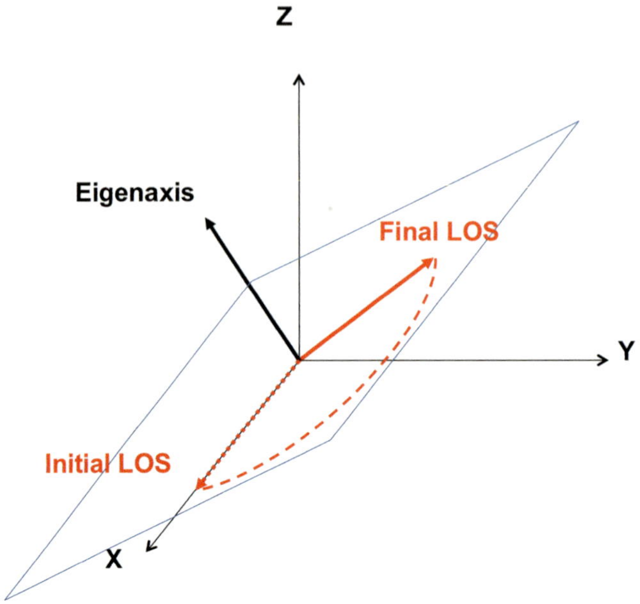

Fig. 3.1 Eigenaxis slew

is slew angle. For a typical day in the life, a spacecraft's slew maneuvers are represented by a collection of points in this plane. To size an MCS, we must compare these points to the capability of the MCS in that plane.

3.1.1 Slew Angle vs. Time

To analyze the problem, we must make some assumption regarding the nature of the slew maneuver profile. One example might be bang-bang acceleration. In this scenario, the vehicle accelerates as quickly as possible to some maximum velocity, then decelerates as quickly as possible at the end of the maneuver to return to the rest state. Clearly, this approach results in the fastest possible slew within the limits on the available acceleration and velocity.

There are reasons why we might choose to use a less aggressive slew profile. For example, an aggressive slew might "ring up" flexible modes in the spacecraft structure, lengthening settling time and thereby compromising the benefit of the rapid slew. In these cases, it is common to remove various frequencies from the slew by pre-filtering the commanded motion or choosing a smooth angle-vs-time profile. An example of such a slew is the versine discussed in Chap. 4, a 1-cos(t) function. A versine slew exhibits longer slew times given the same limits on acceleration (α) and velocity (ω) when compared to the bang-bang, but the frequency content may

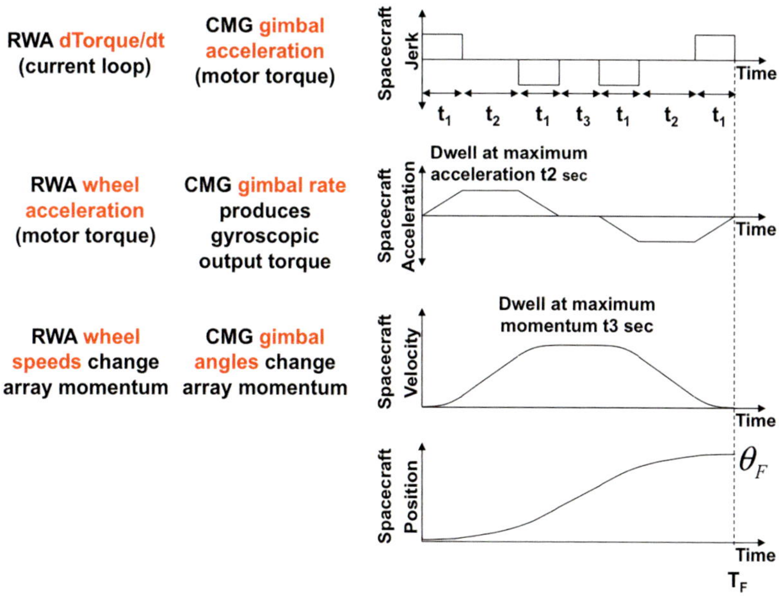

Fig. 3.2 The bang-bang jerk slew maneuver

exclude the natural frequencies of the spacecraft structure. Another less aggressive approach is to limit vehicle jerk (j), the first derivative of acceleration. We shall see that in the case of CMGs, the hardware limits vehicle jerk because the rate of the CMG gimbal (and thus the vehicle acceleration) cannot be changed instantaneously. For these reasons, the analysis here focuses on a bang-bang jerk motion. This work can be modified to accommodate the chosen slew profile for a given mission, or appropriate margin can be added to these basic results.

The plots of Fig. 3.2 illustrate the bang-bang jerk slew profile, in which a finite spacecraft jerk is assumed. When this jerk is applied for t_1 seconds, the spacecraft reaches its maximum acceleration capability.

The spacecraft dwells at this peak acceleration if necessary for t_2 seconds, so that the velocity reaches the maximum available. The spacecraft then dwells at maximum velocity for t_3 seconds, if necessary, then begins a mirror-image deceleration. At the end of a slew T_F seconds in length, the spacecraft arrives at the displacement θ_F with zero velocity, acceleration, and jerk.

The text to the left of the plot identifies the physical behavior of the CMG or RWA that corresponds to jerk, acceleration or velocity of the vehicle. For example, consider scalar angular acceleration, α which we know from one-dimensional Newtonian mechanics ($\tau = J\alpha$) results from torque τ. In the case of a reaction wheel, this torque comes from the spin motor (which causes wheel acceleration in the opposite direction). However, for a CMG, the output torque is gyroscopic, following the vector relationship $\boldsymbol{\tau} = \boldsymbol{\omega}^{\mathrm{G/B}} \times \mathbf{h}_r$. The vector $\boldsymbol{\omega}^{\mathrm{G/B}}$ in this equation is gimbal angular rate, as discussed in Chap. 4. This shift by one time derivative from the CMG kinematics to the spacecraft kinematics is a characteristic of CMGs which must be understood in the sizing of the actuators. The maneuver plotted in Fig. 3.2 is characterized by the following relationships, given the fixed jerk, j:

$$T_F = 4t_1 + 2t_2 + t_3$$
$$\theta_F = j(t_1)(t_1 + t_2)(2t_1 + t_2 + t_3) \tag{3.1}$$

A plot of total slew angle θ_F vs T_F consists of three distinct segments. A very short slew might consist only of four periods of jerk-limited motion, t_1 seconds each, with $t_2 = t_3 = 0$. In this case, t_1 is insufficient for the acceleration to reach its limit, and the equations reduce to

$$T_F = 4t_1$$
$$\theta_F = j(t_1)(t_1)(2t_1) = j\frac{T_F^3}{32} \tag{3.2}$$

Such a short slew is called "jerk-limited" because the limits on acceleration and velocity are not reached, and therefore only the jerk limit is relevant. Figure 3.3 plots θ_F vs T_F for a random choice of acceleration and jerk limits, and with the

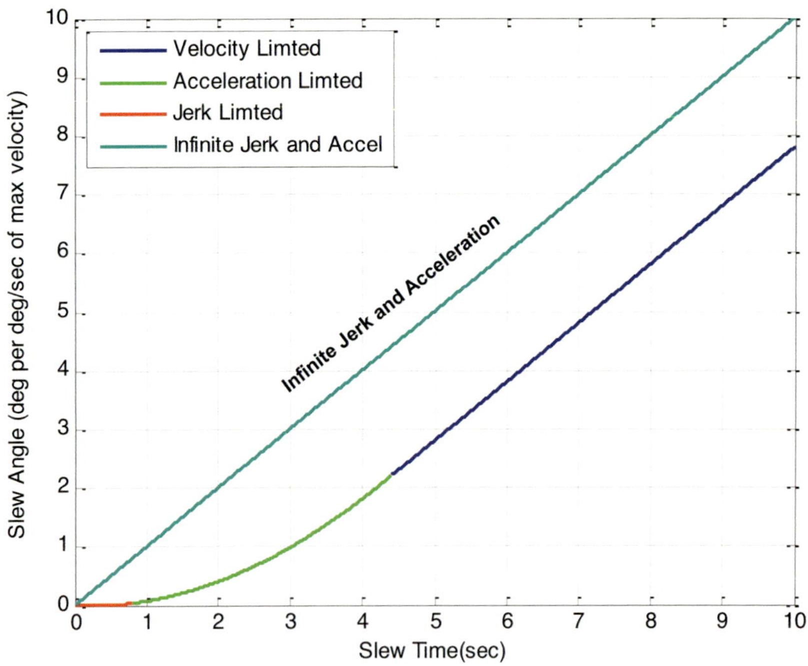

Fig. 3.3 Angle vs. time

velocity limit normalized. The angle during the jerk-limited segment is cubic in T_F and is shown in red. For longer slews, t_1 reaches its maximum value (because the spacecraft reaches its acceleration limit), and t_2 begins to increase from 0. In this region, the slew is "acceleration limited" because the time is constrained only by the maximum acceleration; the maximum velocity is never reached. In this region, shown as green on the plot, the relationship is quadratic in T_F. Finally, for still longer slews, the spacecraft reaches maximum velocity and dwells there for t_3 seconds. For motion that falls within the "velocity-limited" segment (blue in the plot), larger displacement requires additional time, which increases linearly in inverse proportion to the velocity limit.

The unachievable, ideal case is indicated with the text "infinite jerk and acceleration." The horizontal distance between the dark blue segment and this ideal line is the additional time required due to limitations on acceleration and jerk. In momentum system design, there is often a trade to be made between torque and momentum, i.e., spacecraft acceleration and velocity. In view of the preceding discussion, that trade is influenced by time available for the slew. Very short slews may place more value on acceleration, and therefore actuator torque, while longer slews almost always benefit from higher vehicle velocity, and therefore momentum (Hamilton and Underhill [18] and Hamilton [17]).

3.1.2 *Performance Ratios: Vehicle*

We now introduce the concept of performance ratios as a convenient way to score the agility of a vehicle. Using these metrics allows us to discuss certain key time-based quantities independently of the actual angular deflection.

The previous section established that the time t_1 is bounded by the ratio of maximum acceleration to maximum jerk. Of similar importance is the ratio of maximum velocity to maximum acceleration. Both ratios have dimensions of time, which proves useful. With this in mind, we shall define two parameters, Kc and Ka:

$$Kc = \text{max acceleration / max jerk}$$

$$Ka = \text{max velocity / max acceleration}$$

Kc is the time required for the spacecraft to reach maximum acceleration given maximum jerk, and Ka is the time it takes to reach maximum velocity given maximum acceleration. By manipulating Eqs. (3.1), we can identify the times at which the transitions between the segments occur as a function of the ratios, as shown in Fig. 3.4.

The jerk and acceleration limits are not shown in Fig. 3.3 but are now indicated in the lower left of Fig. 3.4 for the specific case of $Kc = 0.2$, $Ka = 2$. In the formulas, ω_{SC} is the vehicle rate, ω. At this point in the requirements-analysis process, limits

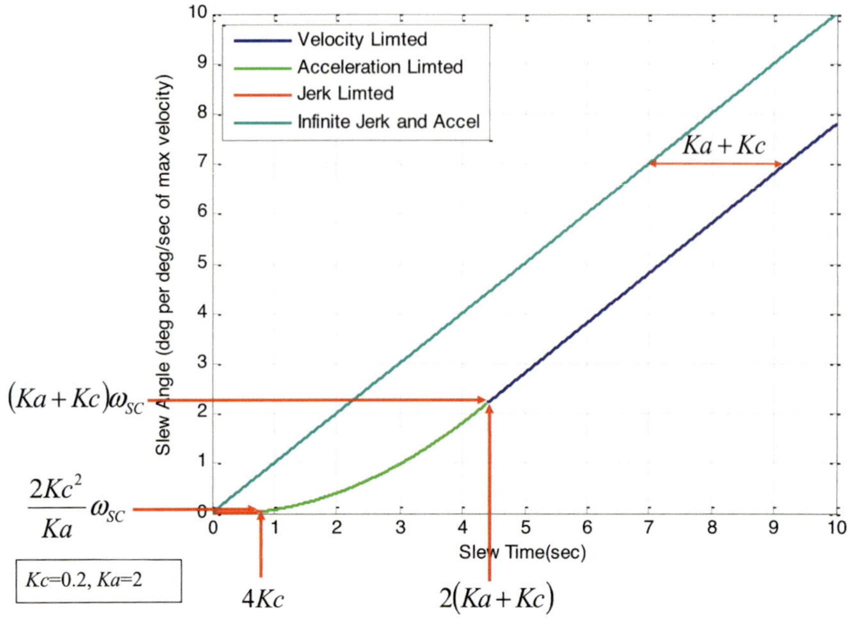

Fig. 3.4 Segment transition points

on velocity, acceleration and jerk are not yet known. And yet, because we know the ratios among them, we can say exactly how much longer a large angle slew (blue) takes in practice compared to the ideal slew (cyan) that does not account for hardware limitations: $Ka + Kc$ or 2.2 s in this specific example. Furthermore, any slew lasting less than 4.4 s cannot reach the maximum design velocity and thus is acceleration limited. Similarly, any slew under 0.8 s is jerk-limited. Given information about how rapidly the spacecraft is required to complete its slews, we now have powerful insight into how to select quantities that directly drive the hardware design.

3.1.3 Performance Ratios: Momentum System

We now recall that for a vehicle slewing about a single axis, $H = J\omega$ and $\tau = J\alpha$, where in this context, J is the vehicle inertia about the eigenaxis. So if we define a quantity Ka to be the ratio of vehicle ω/α, then it is also the ratio H/τ. This ratio is important because momentum and torque are the key performance parameters of the array of momentum devices. Similarly, as in the annotations of Fig. 3.2, if Kc is the ratio of vehicle α/j, it is also the ratio of the CMG gimbal $\dot{\delta}/\ddot{\delta}$. With these

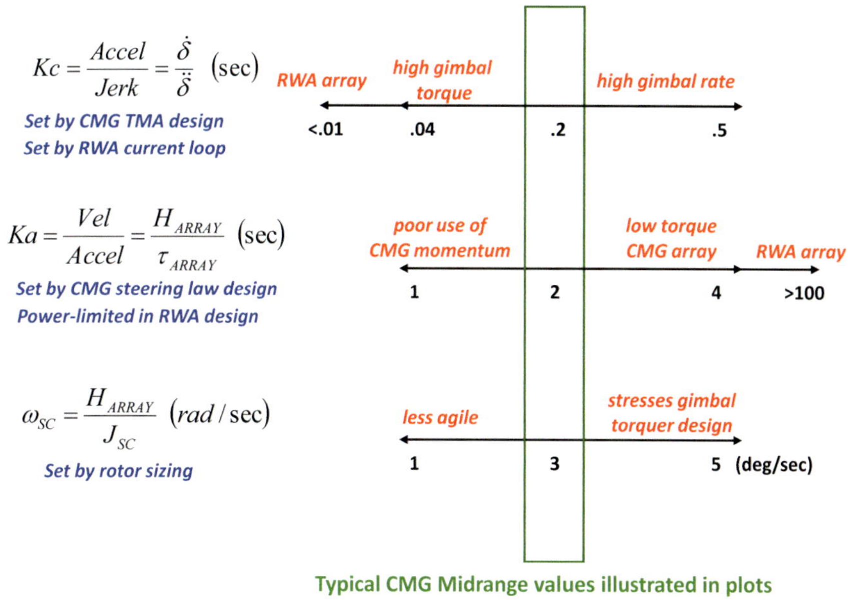

Fig. 3.5 Typical performance ratios

general principles in mind, let us examine typical values of these ratios for realistic momentum devices, as illustrated in Fig. 3.5.

The ratio Kc represents the time the vehicle takes to reach maximum acceleration (torque). For an RWA, where achieving the necessary torque requires only applying current to the motor, this happens very quickly (Kc is very small). However, for a CMG, achieving the necessary vehicle torque requires accelerating the gimbal to its maximum rate. For typical CMGs, reaching this peak gimbal rate takes around 200 ms. This time is driven by the gimbal torque-motor design.

The ratio Ka represents the time the vehicle takes to reach maximum velocity (given maximum acceleration). For an RWA, where torque is limited by power considerations, this number is typically large (>100 s), i.e. an RWA can take minutes to accelerate to its maximum speed. However, a CMG can change its momentum direction over the angular range of interest in seconds. The low values of Ka possible using CMGs make agile vehicles practical.

Some caution must be exercised in choosing Ka and Kc to be very far from these typical values. We know that performance is driven by the sum of these two, and Ka is generally larger. One might therefore jump to the conclusion that reducing Ka as much as possible is the right choice, and this might be achieved by increasing gimbal rate (more torque). The fallacy here is made evident by recognizing that increasing gimbal rate also increases the value of Kc, thus diminishing the return. The H and τ available at the array level are also influenced by how the array is controlled (see Chap. 9), and this is another way to influence Ka.

3.1.4 Vehicle Velocity

The shape of the curve in Fig. 3.4 and the times associated with the different segments are completely determined by the ratios Ka and Kc. The final quantity referred to in Fig. 3.5 is the maximum vehicle velocity. This quantity scales the ordinate (slew angle) axis of Fig. 3.4, completing the picture. We now have a way to directly relate these three parameters to the slew angle vs. time requirements.

It is useful to note that these relationships hold for spacecraft of similar agility regardless of their size. Recalling that $H = J\omega$, the chosen vehicle velocity combined with the inertia of the vehicle directly influences the net momentum that the array of CMGs or RWAs must produce. This requirement in turn influences the size of the momentum devices, and thus the size, weight, and power of the momentum system.

3.1.5 Summary

This section established the general approach for translating vehicle agility into momentum system requirements. Once we know the momentum and Ka required, we can begin the process of momentum device selection. Other factors will also enter into these trades, as explained below.

3.2 Momentum Device Technology

Momentum devices fall into two broad categories. The first, the RWA is a single degree-of-freedom device consisting of a rotor and a motor. In this device the rotor speed is manipulated, producing a momentum vector of varying length, whose direction is fixed in the body frame. The other category, the CMG, is any of various devices employing motorized gimbal mechanisms to redirect the rotor spin axis. In this device the rotor is typically spun at a constant speed, resulting in a momentum vector that has fixed length, but varying direction. Either type of device can be used to produce attitude control torques by exchanging momentum with the spacecraft. But we shall see that there are practical advantages to using certain device types for certain applications. This section discusses each of the devices. Subsequent sections examine each of the key trades and offer guidelines for selection.

3.2.1 Torque, Momentum and De-mystifying Precession

Before we can discuss these devices, we must digress briefly to review some fundamental physics. Of interest is what happens to a body when it experiences a torque. Chapter 4 provides the foundational mechanics. This section seeks to complement those mathematics with physical insight.

We begin with a body at rest, for example, an RWA rotor with zero spin speed. Its angular momentum, **h**, is zero. By definition torque is the rate of change of momentum in an inertial frame, $\boldsymbol{\tau} = \frac{d\mathbf{h}}{dt}$, and both are vector quantities. If we apply a constant torque in the y direction for time dt, as shown on the left side of Fig. 3.6, the resulting angular momentum is $d\mathbf{h}$ in the y direction. If we continue to apply the torque for another dt seconds, and then another, the net angular momentum vector continues to grow in the y direction. Since the angular momentum of a body is related to its inertia and rate by $\mathbf{h} = \mathbf{J}\boldsymbol{\omega}$, body rate increases by $d\boldsymbol{\omega}$ for each dt time increment. This increase is an acceleration, $\boldsymbol{\alpha}$, following the familiar equation:

$$\boldsymbol{\tau} = \frac{d\mathbf{h}}{dt} = \mathbf{J}\frac{d\boldsymbol{\omega}}{dt} = \mathbf{J}\boldsymbol{\alpha} \tag{3.3}$$

Now, what if the body is not initially at rest? We next consider the case in which the body is initially spinning about the specific axis (call it x), such as in the gimbal mount of a CMG. Its initial angular momentum, $\mathbf{h}_0$, is large (relative to the magnitude of $d\mathbf{h}$) in the x direction. If we apply a constant torque as before about the y axis for time dt, the resulting angular momentum is the sum of the initial $\mathbf{h}_0$ and the $d\mathbf{h}$ in the y direction. If we continue to apply the torque for another increment

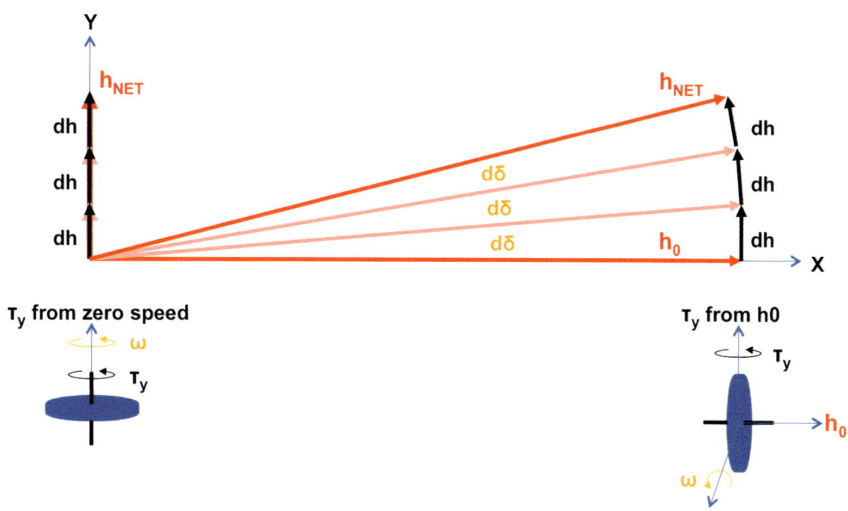

Fig. 3.6 Increments of momentum

dt, and then another, the net angular momentum vector continues to change and begin to arc over toward y, as illustrated in the right side of Fig. 3.6.

The figure shows how the net momentum vector varies for this case. As we can see, the net momentum started in the x direction and is rotating about an axis out of the plane of the page. With the passage of each dt, the net vector rotates through an additional angle $d\boldsymbol{\delta}$ in accordance with $d\mathbf{h} = d\boldsymbol{\delta} \times \mathbf{h}$. This result says that the torque results in a body rate (not an acceleration as before), and that rate is about an axis out of the plane of the page. Specifically,

$$\boldsymbol{\tau} = \frac{d\mathbf{h}}{dt} = \frac{d\boldsymbol{\delta}}{dt} \times \mathbf{h} = \boldsymbol{\omega} \times \mathbf{h} \tag{3.4}$$

Equation (3.4) is the classical precession result from undergraduate studies of gyroscopic dynamics. For many, the spatial (cross product) and temporal (one time derivative) differences between $\boldsymbol{\tau} = \mathbf{J}\boldsymbol{\alpha}$ and $\boldsymbol{\tau} = \boldsymbol{\omega} \times \mathbf{h}$ are inexplicable and simply taken on faith. But we can now see that both behaviors are the direct result of the same $\frac{d\mathbf{h}}{dt}$ when a torque is applied. This is a recurring theme: when studying momentum devices, it is much more instructive to visualize momentum vectors than torque.

3.2.2 The Reaction Wheel

This momentum device operates in a single degree of freedom, with the spin axis of the rotor fixed to the vehicle structure by spin bearings as illustrated in Fig. 3.7. The

Fig. 3.7 Reaction wheel

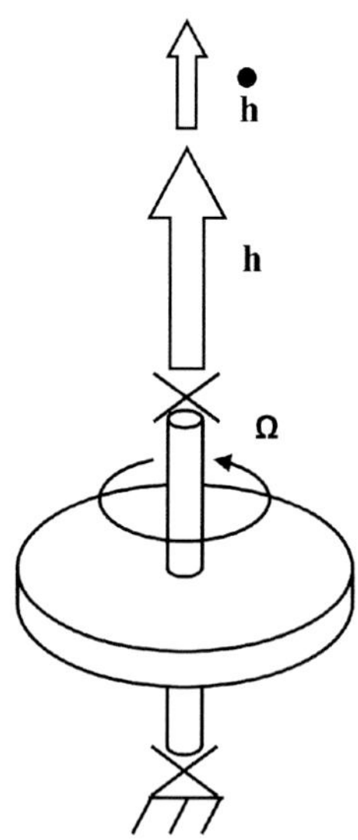

output torque is produced directly by the same motor that controls rotor speed. That is, the motor acts on the rotor, and reacts on the spacecraft–hence the name "reaction wheel." The action on the rotor causes it to accelerate as $\boldsymbol{\tau} = \mathbf{J}\boldsymbol{\alpha}$, as discussed in the previous section.

In the production of torque, this machine uses mechanical shaft power equal to torque times the spin speed of the rotor: $P = \tau \cdot \omega$. As such, it uses the greatest energy possible in changing its angular momentum. Typical large RWAs have maximum rotor speeds in the range from 2000 to 6000 rpm, with small RWAs going even higher. At 2000 rpm, 1 Nm of torque (0.74 ft-lb) requires 209 W of power, without even considering the heat power dissipated in the motor, electronics and bearings. To reduce the power, one could modify the rotor design by reducing its speed and increasing its inertia (size, weight), but these trades have disadvantages of their own. Therefore reaction wheels capable of greater than 1 Nm of torque are rarely seen.

3.2.3 *The Double-Gimbal Control Moment Gyroscope (DGCMG)*

Unlike the RWA, which fixes the spin axis to the vehicle using bearings, this device allows the rotor's spin axis to rotate in the spacecraft reference frame using a double-gimbal mount, as illustrated in Fig. 3.8. Also unlike the RWA, the rotor's spin speed is constant during normal operation.

In this configuration, the rotor acts as a free gyroscope. It has an angular momentum vector **h**, directed along its spin axis, with magnitude $J_r\Omega_r$. When

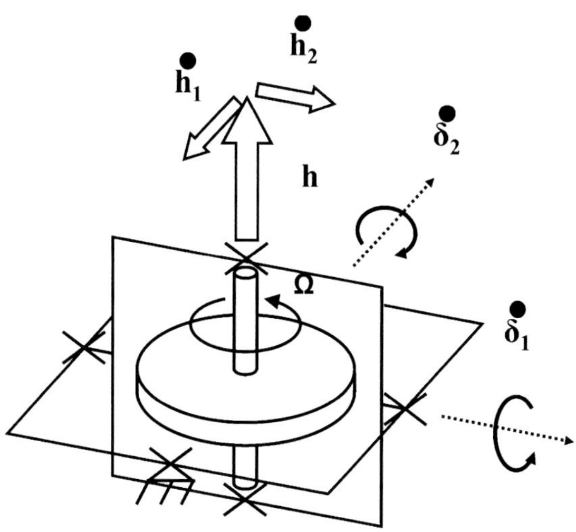

Fig. 3.8 Double-gimbal control moment gyroscope

torque acts on the rotor, it precesses in accordance with $\boldsymbol{\tau} = \boldsymbol{\omega} \times \mathbf{h}$ as derived in Sect. 3.2.1.

As in the RWA, the output torque of this device (the torque applied to the spacecraft) is directly the result of motor reactions. The outer gimbal motor reacts directly, while the inner gimbal motor reacts through the outer gimbal bearings. As such, the motors that position these gimbals must be sized to produce the entire output torque for which the device is designed in addition to their own control torques.

However, unlike the RWA, the rotation speed of these motors is very low compared to the spin speed of the rotor. We recall that the RWA motor is severely limited in torque capability due to the mechanical power ($P = \tau \cdot \omega$). The CMG does not have this limitation, and can therefore produce greater output torques.

It can be shown (see Chap. 11) that, holding the power density and K_M density constant, both the power and weight of a typical motor increase as the 2/3 power of torque. As the motor size increases, so do the gimbals and their associated structure. Introducing gearing can help. However, eventually it becomes a challenge to design a double-gimbal CMG with reasonable and prudent structural mode frequencies and we again reach a limitation on torque production. This limit generally falls in the rough vicinity of 50–100 Nm. Perhaps the largest DGCMGs ever built are those produced for the International Space Station, which are rated at 300 Nm.

3.2.4 The Single-Gimbal Control Moment Gyroscope (SGCMG)

The single-gimbal CMG (SGCMG), as illustrated in Fig. 3.9, can be thought of as the DGCMG of Fig. 3.8 with the outer gimbal (about δ_2, above) removed. As a

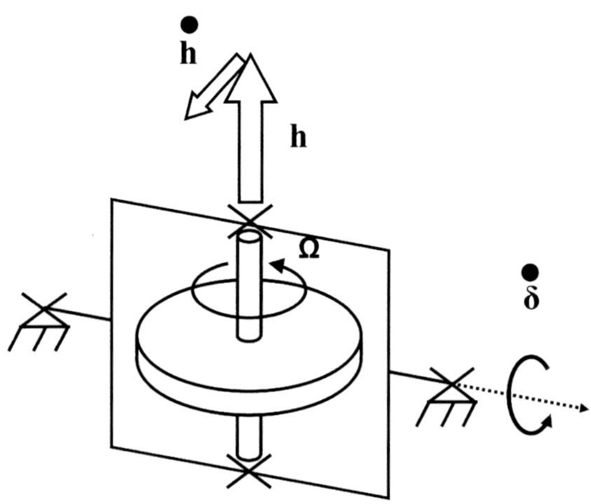

Fig. 3.9 Single-gimbal control moment gyroscope

result, the rotor spin axis is constrained to move in the plane normal to δ. If we torque on the rotor with this single inner gimbal, we would expect the rotor to precess about the axis of the outer gimbal (δ_2 in a DGCMG). Since that gimbal has been removed, a constraint torque arises that eliminates motion in that axis, acting on the spacecraft through the inner gimbal bearings. This torque performs no mechanical work but is sufficient to cause the rotor instead to precess about the gimbal axis at the gimbal rate.

This curious byproduct of gyroscopic physics results in a machine that possesses significant "torque amplification." This term describes a device with a relatively small gimbal motor that can produce very large output torques. Indeed, if such a device were attached to a fixed test frame and had very stiff gimbals (imposing the constraint), one could not even tell by examining the dynamics of the gimbal axis that the rotor was spinning. The motor torque required is simply that necessary to accelerate the gimbal inertia to produce the desired gimbal rate. The output torque felt by the spacecraft is then the constraint torque $\boldsymbol{\tau} = -\dot{\boldsymbol{\delta}} \times \mathbf{h}$, where $\dot{\boldsymbol{\delta}}$ is the gimbal rate and $\mathbf{h}$ is the rotor momentum vector.

This torque-amplification effect makes the single-gimbal CMG the clear power-efficient choice for the most agile vehicles, requiring the largest torque output. Large SGCMGs are in service in spacecraft applications with output torques well in excess of 1000 Nm. Very large CMGs for terrestrial applications exist with output torque in excess of 100,000 Nm that only require a few kW to operate.

3.2.5 *Vehicle Rate and the SGCMG Gimbal Torquer*

The preceding section stated that the dynamics of a SGCMG gimbal is independent of rotor momentum for the case of a very stiff gimbal fixed to a test frame. We now consider the implications of having the device mounted to a spacecraft, which can be in motion. We call this spacecraft motion the "base rate" of the CMG.

Specifically, we consider the case in which the vehicle is rotating with an inertial velocity ω_s about the output axis of the SGCMG (dashed black in Fig. 3.10). Those same gimbal bearings that provided the constraint in the earlier discussion must now force the rotor to rotate with velocity ω_s about the output axis along with the vehicle. This configuration is illustrated in Fig. 3.10.

Because this is a gyroscope, if the bearings were to torque the rotor about the (dashed) output axis, the rotor would precess about the gimbal axis. To prevent this gimbal rotation, enough gimbal axis torque must be applied to make the rotor instead precess (along with the vehicle) about the output axis, thus unloading the bearings. If the gimbal torque motor is not strong enough to apply this torque, control of the gimbal is lost.

Only a torque about the gimbal axis can cause the gyroscope to move with the vehicle, and this torque must come from the gimbal torque motor. For large, agile vehicles, the base rate effect can quickly become the dominant sizing driver, often requiring the torque motor to utilize a gearbox and/or draw a lot of power.

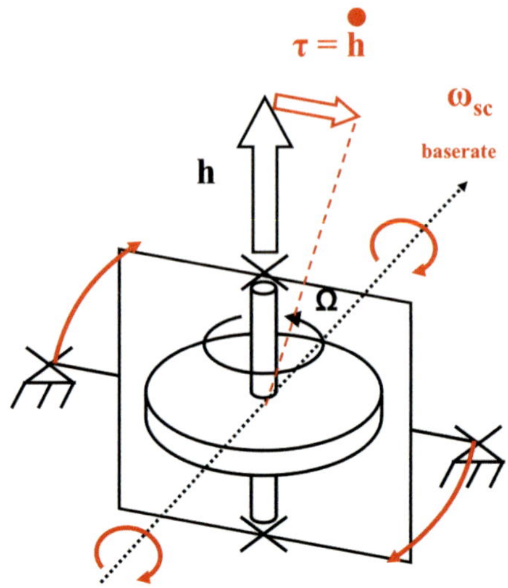

Fig. 3.10 Response to vehicle base rate

As we are assuming the worst case, where the vehicle base rate is purely about the CMG output axis, the gimbal must produce a torque of $\tau_g = \omega_s h$. Remembering that the momentum of the CMG is sized to produce a desired vehicle momentum $H = J_s \omega_s$, we conclude that the requirement for gimbal torque is

$$\tau_g \approx J_s \omega_s^2 \tag{3.5}$$

We use the $\approx$ sign above to acknowledge that this equation does not distinguish between CMG momentum (h) and array momentum (H). However, regardless that the gimbal torque may be somewhat less, the recognition that it varies as the square of vehicle rate is the basis for the note about high vehicle rates in the lower-right of Fig. 3.5.

As a rule of thumb, gimbal torque less than 5–10 Nm can be produced by a direct-drive CMG, i.e., without gears. Above this level, the weight and power of the gimbal usually drives a geared design. Geared CMGs are common on large, agile spacecraft, but they are more complex and expensive to produce.

3.2.6 Torque Amplification Revisited

Now that we have quantified the effect of base rate on the gimbal torque, we can quantify the torque amplification. Earlier, it was suggested that the gimbal motor need only be sized to accelerate the gimbal inertia, but we now understand that it must also be sized to oppose the precession from base rate. So

$$\tau_g = J_g \alpha_g + \omega_s h \tag{3.6}$$

The output torque from the CMG has already been shown to relate to the gimbal rate, ω_g, as $\boldsymbol{\tau}_o = -\boldsymbol{\omega}_g \times \mathbf{h}$. Since these axes are always orthogonal, we have the scalar magnitudes

$$\tau_o = \omega_g h \tag{3.7}$$

Combining these two equations, and assuming the gimbal acceleration torque is small compared to precession, we find that for any momentum:

$$\text{torque amplification} = \frac{\tau_o}{\tau_g} \approx \frac{\omega_g}{\omega_s} \tag{3.8}$$

This is a very powerful observation for spacecraft, whose maximum rate rarely exceeds a few degrees per second. Gimbals are easily designed with rates of 1 rad/s or more, therefore a SGCMG can achieve torque amplification easily over a factor of 10 on most spacecraft, and nearly 100 on some.

Alternatively, Eq. (3.8) also explains why CMGs are not often found in applications such as aircraft, which might be rolled at 30°/s or more. The potential for high base rate reduces the amplification, drives larger torque motors and negates the advantage of the technology.

3.3 Momentum Device Technology Tradeoffs

3.3.1 Momentum and Torque

All of the candidate devices can be designed to store similar amounts of momentum. The design needs only a rotor with a certain combination of inertia and maximum speed. Given the same momentum, all the devices can produce similar spacecraft velocity, and/or absorb similar amounts of stored momentum.

However, distinctions among the various technologies discussed in Sect. 3.2, make it clear that the range of torque available varies dramatically. The discussion of Sect. 3.1, in which Ka is defined as h/τ, provides the key insight. A reaction wheel, which can have the same momentum as other momentum devices but is severely torque-limited, exhibits the highest Ka. The DGCMG is in the middle, and the SGCMG has the lowest value.

Ka defines the time it takes to accelerate the vehicle to maximum rate and thus determines how much longer a real slew requires than the ideal slew (one not constrained by the device physics). The trade, therefore, comes down to agility. Figure 3.11 illustrates the tradespace for the various MCS technologies. The plot shows torque vs. momentum and includes dashed lines for various values of the agility metric, Ka. For convenience, the abscissa has a second scale showing the vehicle inertia that would require the indicated momentum for a slew rate of 3°/s. Reaction wheel designs fall in the region of torque less than 1 Nm, and the torque is

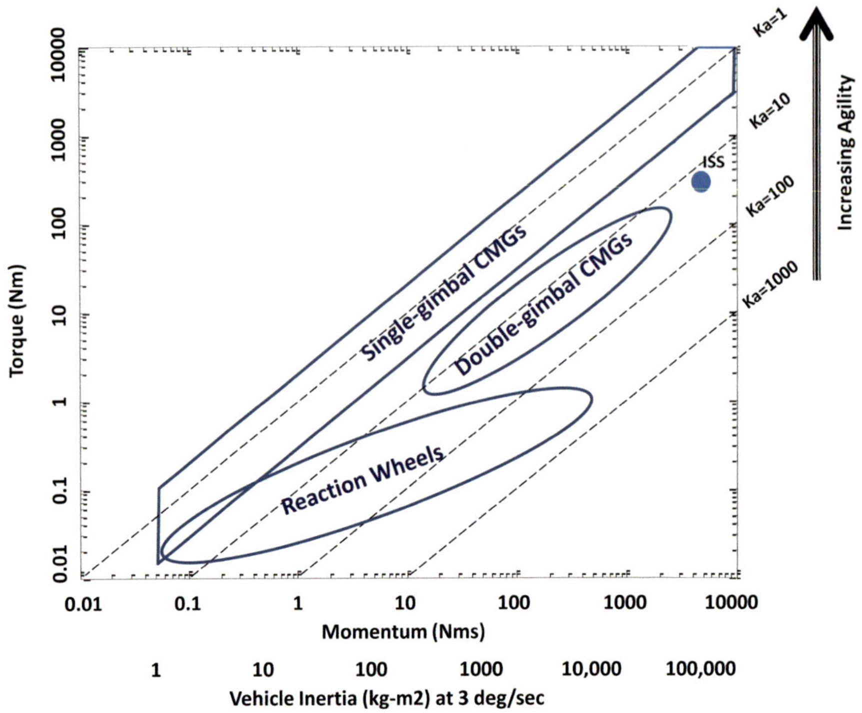

Fig. 3.11 Torque, momentum and agility of MCS technologies

even less for smaller devices with smaller motors. Because the torque does not drop linearly, lower Ka values are achievable using this technology on small spacecraft.

The design range for double-gimbal CMGs is generally less than 100 Nm, although the design mentioned earlier for the international space station is plotted with a single dot, illustrating how far out of family it falls. Because of the torque-amplification principle, devices with more momentum can produce larger torques using similar gimbal rates. This principle results in a linear slope.

Finally, the range suitable for single-gimbal CMGs provides the highest agility regardless of the size. Only SGCMGs can reach Ka values approaching 1, and they can do so over the entire tradespace. Recall that the CMG gimbal torque required is driven mostly by vehicle inertia and vehicle velocity squared, while a gimbal rate of 1 rad/s results in Ka near 1 for any size. This analysis neglects the effects of array steering on Ka, which are discussed in Chap. 9.

The lowest values on the plot presume that we are referring to traditional mechanical systems, fabricated with traditional techniques, and it is understood that by using MEMS technologies, even smaller devices are possible.

3.3.2 Power

As discussed previously, in producing torque the RWA uses mechanical power approximately equal to torque times spin speed. Typical large RWAs have maximum rotor speeds in the range 2000–6000 rpm, with small RWAs going even higher. At 2000 rpm, 1 Nm of torque requires 209 W of power, without even considering the heat power dissipated in the motor, electronics and bearings. However, when not producing torque and at low rotor speeds, RWA "bias" power can be very small.

CMGs, whose output torque is due to a mechanical constraint, require no transfer of energy to the rotor. This feature makes them much more efficient in torque production. One consideration, however, is that because they run continuously at high speed, they draw steady-state power to maintain that speed. This spin motor power (P_M) can be approximated as the product of spin-bearing drag torque (τ_d) and rotor speed, scaled up by a factor s to account for heat dissipated in the motor windings:

$$P_M = s\tau_d\omega_r$$

The factor s varies nonlinearly as a function of motor design and operating point, but can be taken to be roughly 1.1. When P_M is combined with the bias power of the electronics, P_E, the sum is referred to as "quiescent" power, P_Q:

$$P_Q = P_M + P_E$$

The term "quiescent" implies this is power drawn even when the CMG is not being used to produce torque. Driven predominantly by τ_d, P_Q is larger when larger spin bearings are required, and at higher spin speeds. Increased quiescent power impacts not only the power that the spacecraft bus must supply to the CMG, but also increases the heat dissipated in the spin bearings. This dissipation in turn affects the thermal design. These impacts of quiescent power and heat in the cost of optimizing an MCS is why larger CMGs tend not be run at speeds in excess of 10,000 rpm. Smaller CMGs with smaller bearings are able to more readily take advantage of the benefits of high rotor speed in reducing CMG size. Recent work at Honeywell has shown that integrating vibration isolation technology within momentum systems can allow the use of smaller spin bearings by attenuating the launch loads that often drive their design.

3.3.3 Vibration

Like all rotating machines, momentum devices induce vibrations in the structure that supports them. These vibrations have a rich but well understood spectral content. The spectrum is not broadband "white" or "colored" noise of the type that might be examined using power spectral density (PSD) plots. Instead, it consists of a comb of discrete frequencies at various multiples of the wheel speed. On a power spectral density (PSD) plot, a tone at a discrete frequency would produce a peak that is infinitely narrow and infinitely tall. Therefore PSD plots are inappropriate to the study of these vibrations. Instead, we plot the root-mean-square (rms) sinusoidal amplitude of each component frequency, such as would be obtained from a Fourier analysis. We can calculate the total rms noise in a given band as the root-sum-square of the peaks in that band. Let us examine some of the sources of vibration in these devices.

One of the key contributors is rotor balance, both static and dynamic. Static balance measures how far the rotor center of mass is displaced from the spin axis. This displacement is responsible for a centripetal force $F = Me\omega^2$ transmitted to the spacecraft structure, where M is the rotor mass and e is the center-of-mass displacement. As such, the static balance is normally expressed in units of mass times displacement, for example oz-in. Dynamic balance expresses how far the rotor's principal axis of inertia is tilted from the spin axis and is generally measured in oz-in^2. The disturbance that arises is a torque vector, tracing out a cone around the spin axis. Static and dynamic balance effects are generally felt predominantly at the rotor spin frequency, with a small amount of harmonic content.

Another major contributor is bearing eccentricity. The bearings contribute a variety of frequencies because there are multiple rotation rates within them. The inner race of the bearing moves with the shaft at the rotor spin speed. The group of balls in their ball carrier rotates at a slower speed. Individual balls rotate at a higher speed. When we begin to track the frequency that results from dimensional tolerances or manufacturing flaws on various surfaces, we find these bearing frequencies, their harmonics, and beat (difference) frequencies between these tones. At each frequency, the bearing pushes the shaft off the centerline in a circular motion. For certain frequencies, that circular motion can be "retrograde," i.e. in the direction opposite the rotor spin.

There is generally a bearing at each end of a rotor shaft. Further complicating this spectrum is the fact that the tones from these two bearings are never identical. Consider the ball group frequency; it is determined by the diameters of the raceways and balls, all of which are subject to dimensional tolerance. As a result, while the tones on each end of the shaft are close, they are not exactly equal. The beat frequency between these nearly identical tones can have a period of many seconds or minutes. This beat accounts for the slow audible modulation in vibrations that we are accustomed to when listening to a rotating machine. It also makes routine measurements of the vibrations difficult to repeat unless very long time averages are taken and processed. Tones appear to come and go. The relative phase of the spinning vectors at each end of the shaft changes, resulting in excitation of different structural tones, etc.

Despite the stochastic nature of these vibrations, it is possible, with some care, to approximate their effect in computer modeling. When attempting to do so, the most important thing to recognize is that, unlike many disturbances, the "forcing function" arising from bearing rotation is neither a force nor a torque.

The balls are much stiffer than anything else in a typical structure, and their geometric imperfections impose a displacement between the spinning shaft and the surrounding structure. In so doing, they are stiff enough to push everything else out of the way. They can bend shafts, housings and support structure, etc. as necessary to achieve this end. As a result, the amount of vibration force applied by the bearing is determined by the structures to which it is attached. When modeling a bearing disturbance as an input to a finite-element model, we must represent the bearing by a very high-stiffness element (orders of magnitude higher than the rest of the structure), and then apply a force differentially across that stiffness of $F = kx$, where k is the element stiffness, and x is the desired displacement. Typical bearing displacements vary from tenths to tens of microinches, with each tone having its own value. For a given bearing design, these values must be determined empirically, and their statistical nature across a production lot kept in mind.

All of the tones in the spectrum vary linearly with wheel speed, resulting in a comb of spectral content that shifts along the frequency axis with rotor speed. The CMG, whose speed is held constant, exhibits a stable and repeatable set of vibration frequencies. As such it is possible, within reason, to engineer these devices such that their disturbance frequencies do not coincide with the frequencies of certain key spacecraft structural modes. In fact, it is possible to "tune" the system on orbit to minimize such excitations by making small adjustments to the rotor speeds. By comparison, the RWA, whose rotor speed varies continually as momentum is exchanged, imparts vibration tones that sweep up and down the spectrum. In this case, it is not practical to avoid structure excitation by design.

3.3.4 Torque Accuracy

As the method of torque production for RWAs differs completely from that of CMGs, each must be discussed separately.

For a reaction wheel, the net output torque experienced by the spacecraft is equal to the motor torque less the friction torque of the spin bearings and any other source of drag. While the motor torque can be readily estimated from a measurement of the current, the effects of friction are difficult to predict and vary nonlinearly with speed, temperature, lubrication, etc. For a medium-sized RWA, it is not unusual for the friction torque to be 10% of the full scale output torque of the device. Therefore, if the RWA were commanded to 5% of full scale torque, the friction could be twice the control torque. This effect introduces significant error into the attitude control loop, often necessitating the use of a state-estimation strategy to determine the friction. Such strategies do not perform well when the rotor crosses zero speed (reverses direction). This is because friction and stiction variations are largest in this region, and speed measurements are generally poor near zero speed. Many spacecraft attempt to bias the RWAs in an array away from zero speed during normal operation. This can be achieved either by operating the spacecraft with a net momentum bias, or by including more than 3 RWAs and biasing them against each other, using the null space of the array as discussed in Chap. 6.

Like RWA bearings, the bearings of a CMG also exhibit friction. Because CMGs operate at a nearly constant spin speed, the spin-bearing friction is nearly constant. Spin-bearing friction is a consideration in system power but has little effect on performance. Because the CMG output torque is determined by gimbal rate, CMG gimbals usually employ inner loops controlling the gimbal rate. Therefore gimbal bearing friction effects are mitigated by closed loop control to a great extent. Typical gimbal rate-loop control bandwidths are on the order of 10–15 Hz. While the effects of friction on performance are far lower than for an RWA, they still appear as a slowly changing torque command passes through zero, i.e. when the gimbal rate changes direction. While the integral feedback in the rate loop "winds up" to offset the abrupt change in Coulomb friction, the gimbal momentarily stalls at zero rate. Geared CMGs exhibit this effect to a lesser degree than direct-drive ones.

Aside from friction effects, torque accuracy is generally the result of gain and offset errors. In the case of the RWA, the offset torque is low, but gain error can occur as a result of changes in motor Kt and the resistance of the current-sense resistor, both of which vary with temperature. The CMG performance, because of the closed rate loop, is driven by the performance of the tachometer that measures gimbal rate and the tachometer's electronics. This is often an electromagnetic device such as a precision generator. Older CMGs with analog electronics did exhibit a rate-offset error due to electronics offset. Modern digital CMGs can mitigate this offset by monitoring changes in the gimbal position sensor in software and applying a slow correction. Gain error in an electromagnetic tachometer can still occur due to variations in Kt over temperature. Some modern CMGs employ digital encoder technology to measure gimbal angle and rate. Such devices have effectively zero error.

3.3.5 Array Control

In this section, we consider the ease with which a body torque in three dimensions can be produced by an array of momentum devices.

An array of n reaction wheels is by far the simplest to control. The change in net momentum (in 3 axes) that results from changes in individual rotor momentum (n rotors) is a $3\times n$ Jacobian matrix A. Because the RWAs are mounted in fixed orientations on the vehicle, the coefficients of A are constant. The columns of A are the unit vectors in the direction of the RWA spin axes. By taking the inverse of A (for $n = 3$), or a pseudoinverse (for n > 3), we obtain a constant n $\times$ 3 control matrix A^+ such that

$$\tau_{RWA} = A^+ \tau_{cmd} \tag{3.9}$$

This matrix allows the attitude-control system or the momentum control system to calculate the torque to apply to each wheel to obtain the desired body torque.

For DGCMG, this mapping becomes more difficult. The output torques from this device are applied by the gimbals, and the rotor precesses in response. Each device has not only one gimbal fixed in the body frame, but another whose torque axis varies with time. As a result, the Jacobian varies with gimbal angle, which can change quickly. Furthermore, a strategy to avoid gimbal-lock must be incorporated.

For SGCMG, array control is the most difficult. Again, the Jacobian varies with gimbal angle. In this case, the Jacobian relates changes in net momentum to changes in the CMG gimbal angles. Furthermore, arrays of SGCMGs exhibit a rich and complex set of internal singularities, which can make inverting the Jacobian impossible. Singularities and methods for steering arrays of SGCMGs around and through them are addressed at length in Chaps. 5 and 9.

3.4 Guidelines for Selecting Momentum Device Technology

Table 3.1 offers general guidelines summarizing the trades among various momentum system technologies.

Table 3.1 Momentum device selection guidelines

	RWA	DGCMG	SGCMG
Agility (Fig. 3.11)	Best for station keeping, or slews that take minutes	Midrange	Most agile, slews completed in seconds
Power	Lower quiescent power	Higher quiescent power	
	Very inefficient in torque production	Midrange	Very efficient in torque production
Gear train	N/A	Recommended for output torques in excess of 10 Nm	Recommended for gimbal torques in excess of 10 Nm (see Eq. (3.5))
Vibration	Broad spectrum over operating speed range	Adjustable constant speed makes avoiding key modes possible	
Torque accuracy	Open loop torque errors due to friction	Errors mitigated by inner gimbal rate loop Errors mitigated further by gear train	
	Motor Kt, Rsense	Tach performance	
Control	Simplest, with constant control matrix	More complex, avoidance of gimbal lock	Most complex, internal singularity avoidance
Cost	$	$$$	$$

3.5 Chapter Summary

This chapter offers metrics for quantifying the agility of a spacecraft that apply directly to the selection and sizing of momentum devices. The most common momentum device technologies are described, along with various tradeoffs among them. Finally, general guidelines are presented for the selection of a momentum device for a specific application.

3.6 Exercises

1. Shown here are the spin-axis directions for a three-RWA array. Which of them is capable of producing torque in any direction in three dimensions?

   (a) $\hat{a}_1 = \frac{\sqrt{2}}{2}[1\ -1\ 0]\,,\ \hat{a}_2 = \frac{\sqrt{2}}{2}[-1\ 1\ 0]\,,\ \hat{a}_3 = \frac{\sqrt{2}}{2}[0\ 0\ 1]$
   (b) $\hat{a}_1 = \frac{\sqrt{3}}{3}[1\ 1\ 1]\,,\ \hat{a}_2 = \frac{\sqrt{3}}{3}[-1\ -1\ -1]\,,\ \hat{a}_3 = \frac{\sqrt{2}}{2}[1\ -1\ 0]$
   (c) $\hat{a}_1 = \frac{\sqrt{2}}{2}[1\ 1\ 0]\,,\ \hat{a}_2 = \frac{\sqrt{2}}{2}[1\ -1\ 0]\,,\ \hat{a}_3 = \frac{\sqrt{2}}{2}[1\ 0\ 1]$
   (d) $\hat{a}_1 = \frac{\sqrt{2}}{2}[1\ 0\ 1]\,,\ \hat{a}_2 = [0\ 1\ 0]\,,\ \hat{a}_3 = [0\ 0\ 0]$

2. It's 2045, and life has been found on Enceladus, a moon of Saturn. These life-forms are primitive phosphorescent bacteria, colonies of which emit occasional flashes of UV light from just beneath the surface of the ice layer that covers Enceladus's water-ice ocean. The China National Space Administration (CNSA) has designed a spacecraft that is to orbit the moon and photograph the flashes to understand how many of these colonies there are. CNSA's spacecraft is in a very low orbit (50 km from the surface) and must quickly slew from one attitude to another in order to capture images of these flashes. The camera's field of view is circular and 1° in diameter. The bacterial colonies are about 100 m in diameter. It's an expensive and important mission; so, reliability is of great interest

   (a) What kind of attitude sensors should this spacecraft use, and why?
   (b) What kind of attitude-control actuators should it use, and why? There is no need to calculate the required angular-momentum and/or thrust magnitudes explicitly for these actuators.

3. A spacecraft is designed to perform large-angle attitude maneuvers using three reaction wheels as actuators. The inertia dyadic for the spacecraft's mass center, including the wheel inertia, is

$$\mathbf{I}_c = \begin{bmatrix}\mathbf{b}_1\ \mathbf{b}_2\ \mathbf{b}_1\end{bmatrix}\begin{bmatrix}3500 & 0 & -100\\ 0 & 4000 & 50\\ -100 & 50 & 5000\end{bmatrix}\begin{bmatrix}\mathbf{b}_1\\ \mathbf{b}_2\\ \mathbf{b}_1\end{bmatrix}\ \text{kg-m}^2$$

The reaction wheels are oriented such that their spin axes are the following unit vectors:

$$^{B}a_1 = \begin{bmatrix} 2/3 \\ 2/3 \\ -1/3 \end{bmatrix}, \qquad ^{B}a_2 = \begin{bmatrix} -1/3 \\ 2/3 \\ 2/3 \end{bmatrix}, \qquad ^{B}a_2 = \begin{bmatrix} 2/3 \\ -1/3 \\ 2/3 \end{bmatrix}$$

The following parameters describe a specific slew of interest here:

- There is no nutation damping.
- The system angular momentum (sum of body and wheels) just before the start of the slew at t = 0 is

$$\mathbf{H} = \begin{bmatrix} \mathbf{b_1} \ \mathbf{b_2} \ \mathbf{b_1} \end{bmatrix} \begin{bmatrix} 5 \\ -5 \\ 10 \end{bmatrix} \text{Nms}$$

- The body angular velocity at $t = 0$ is $\omega^{B/N} = 0$.
- All wheel and body angular accelerations are zero.
- The attitude motion in terms of a time-varying quaternion is

$$q(t) = \begin{bmatrix} 0 \\ 0 \\ \sin(0.1 - -0.1\cos(0.1t)) \\ \cos(0.1 - -0.1\cos(0.1t)) \end{bmatrix} \text{for } 0 < t < \pi \, \text{s}$$

- Calculate the following for the entire slew. If you use a computer simulation (e.g. MATLAB), show plots of the results.
- The torque applied by each of the three wheels about its spin axis
- The angular momentum in each of the three wheels about its spin axis

4. How many reaction wheels are necessary if a spacecraft design is to meet the following three requirements?

   - (1.0) The reaction-wheel array shall to apply independent torque simultaneously along body-fixed basis vectors $\mathbf{b}_1$, $\mathbf{b}_2$, and $\mathbf{b}_3$.
   - (2.0) The reaction-wheel array shall include a single degree-of-freedom null space for optimizing wheel usage.
   - (3.0) The reaction-wheel array shall provide the torque of Requirement (1.0) and the null space of Requirement (2.0) in the presence of any single wheel failure.

5. A spacecraft features an array of CMGs that offer Ka = 3 and Kc = 0.2. A bang-bang jerk maneuver is performed which reaches a maximum spacecraft velocity of 3°/s. How long will it take for this vehicle to slew 10°?

6. A spacecraft features an array of 4 CMGs, each of whose rotors has 200 Nms of momentum. The spacecraft can rotate about any axis with a maximum rate of 3°/s. To maintain control of the CMGs, what is the absolute minimum amount of torque that the CMG gimbal motors must be designed to produce?
7. A new startup company plans to offer a high-torque reaction wheel to compete with CMGs. They boast a maximum rotor speed of 6000 rpm, and an output torque of 50 Nm over the entire speed range. Why is this a bad idea?

Chapter 4
Dynamics of Momentum Control Systems

It is not enough to treat the momentum system as a black box, simply receiving power and control signals and imparting torque or momentum, as if it were any other element in a feedback-control block diagram. A momentum system offers unique opportunities to achieve robust and lightweight spacecraft designs, but taking advantage of these opportunities requires careful attention to rigid- and flexible-body dynamics. This chapter provides the foundational equations of motion for spacecraft with momentum-control devices and summarizes important flexible effects that drive the design of spacecraft that incorporate these actuators.

4.1 Notation

It may be that dynamics analysis is fundamentally unsuited to being expressed with the simple notation of other areas of physics; but for whatever reason, notation among dynamicists varies widely. There is so little universally accepted notation that we find it necessary to explain the approach taken here.

The following approach is based on vector-dyadic (or tensor) notation, which recognizes the key distinction between vectors and matrices. A matrix is a collection of scalars—some say "measure numbers"—each of which represents the projection of a vector or a tensor onto a basis vector. For purposes of this discussion, the basis vectors are of unit length and always comprise a right-handed coordinate system as suggested in Fig. 4.1. In this figure, the three $\mathbf{n}_i$ are basis vectors ordered so that $\mathbf{n}_1 \times \mathbf{n}_2 = \mathbf{n}_3$.

All vectors we refer to are sometimes known as "free vectors." They are mathematical objects with magnitude and direction but are not associated with a coordinate system and do not involve a coordinate-system origin of some kind. They are not rays emanating from a point. When referring to three numbers in a vertical stack, one often casually says "vector" when the more explicit term would be

© The Author(s), under exclusive license to Springer Nature Switzerland AG 2025

F. A. Leve et al., *Spacecraft Momentum Control Systems*, Space Technology Library 46, https://doi.org/10.1007/978-3-031-96165-6_4

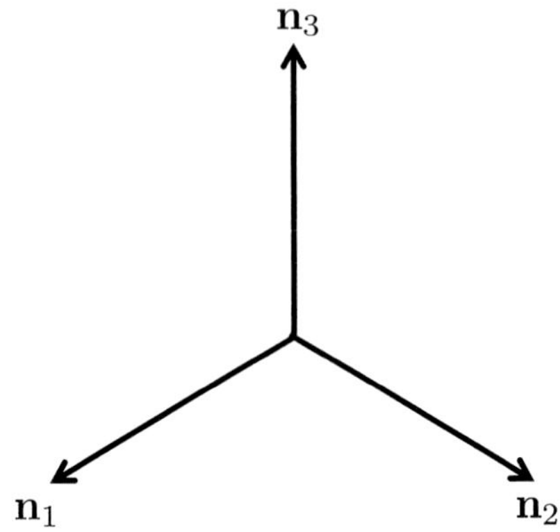

Fig. 4.1 Right-handed coordinate system with three orthonormal basis vectors

Table 4.1 Vector/dyadic notation

Scalar	a
Matrix	v, D
Vector	$\mathbf{v}$
Dyadic	$\mathbf{D}$

"column matrix." Such matrices always presuppose a coordinate system. However, in this chapter we emphasize the coordinate-free world of vectors and tensors so that the derivations are unencumbered by a choice of coordinate system. Only when a coordinate system is helpful do we explicitly define one.

A dyadic is a second-order tensor, which gives rise to a square matrix when it is projected onto the basis vectors of a coordinate system. The notation for a dyadic consists of juxtaposing pairs of vectors—simply writing one vector next to another. This placement does not imply multiplication, although the placement follows the order of operations for multiplication. For example, a dyadic $\mathbf{D}$ might be written in terms of two vectors $\mathbf{c}_1$ and $\mathbf{c}_2$:

$$\mathbf{D} = \mathbf{c}_1\mathbf{c}_2 \tag{4.1}$$

Algebraically manipulating dyadics uses familiar vector operations, such as cross product and dot products, on the individual vectors in the dyadics in a way that is analogous to the rules for matrix algebra, as in the examples provided in Table 4.1.

We denote vectors and dyadics with boldface type. A scalar is denoted by an italic font. A matrix is written the same way except that it may include a left superscript indicating the coordinate system that led to the scalars in the matrix. This indication of a coordinate system is often missing when the choice is implicit or irrelevant. Note that a scalar is a special case of a matrix, one with dimension 1×1, and the choice of coordinate system is therefore irrelevant. For clarity, often an upper-case letter is used to denote a matrix, while a lower-case letter denotes a scalar. Table 4.1 shows some examples.

Table 4.2 Example matrix dimensions

Matrix type	Example
3×1 (Column) Matrix	$\begin{bmatrix} x_1 \\ x_2 \\ x_3 \end{bmatrix}$
1×4 (Row) Matrix	$\begin{bmatrix} x_1 & x_2 & x_3 & x_4 \end{bmatrix}$
2×3 Matrix	$\begin{bmatrix} x_{11} & x_{12} & x_{13} & x_{14} \\ x_{21} & x_{22} & x_{23} & x_{24} \end{bmatrix}$

A matrix is a collection of scalars stacked in an m × n array. Examples include the column matrix, the row matrix, and the m × n matrix, examples of which are shown in Table 4.2.

Here is an example of a vector. Call it **v**. Suppose there are three basis vectors fixed in the body of a spacecraft, $\mathbf{b}_1$, $\mathbf{b}_2$, and $\mathbf{b}_3$. Then some scalars v_i can be defined such that

$$\mathbf{v} = v_1\mathbf{b}_1 + v_2\mathbf{b}_2 + v_3\mathbf{b}_3 \tag{4.2}$$

Furthermore, one can extract the v_i simply by projecting **v** onto each basis vector. For example,

$$\begin{aligned} v_1 &= \mathbf{b}_1 \cdot \mathbf{v} \\ &= \mathbf{b}_1 \cdot (v_1\mathbf{b}_1) + \mathbf{b}_1 \cdot (v_2\mathbf{b}_2) + \mathbf{b}_1 \cdot (v_3\mathbf{b}_3) \\ &= \mathbf{b}_1 \cdot (v_1\mathbf{b}_1) + 0 + 0 \\ &= v_1\mathbf{b}_1 \cdot \mathbf{b}_1 + 0 + 0 \\ &= v_1 \end{aligned} \tag{4.3}$$

Consider another set of basis vectors, $\mathbf{a}_1$, $\mathbf{a}_2$, and $\mathbf{a}_3$. If they differ from the $\mathbf{b}_i$, the scalars associated with the $\mathbf{a_i}$ differ from the scalars associated with the $\mathbf{b_i}$. Since one cannot change a vector by choosing a different basis, then the scalars (which are basis dependent) cannot really be equivalent to the vector (the zero vector is a trivial exception).

P. C. Hughes [25] introduced a convenient notation, the "vectrix." A vectrix is simply a matrix of vectors. For example,

$$\mathbf{b} = \begin{bmatrix} \mathbf{b}_1 & \mathbf{b}_2 & \mathbf{b}_3 \end{bmatrix} \text{or } \mathbf{a} = \begin{bmatrix} \mathbf{a}_1 & \mathbf{a}_2 & \mathbf{a}_3 \end{bmatrix} \tag{4.4}$$

This object has 1 row and 3 columns. It is not a shorthand for a 3×3 matrix of scalars. Vectrices are useful when dealing with vectors and bases. They lead to compact expressions such as

$$\mathbf{v} = \begin{bmatrix}\mathbf{b}_1 \ \mathbf{b}_2 \ \mathbf{b}_3\end{bmatrix} \begin{bmatrix} v_1 \\ v_2 \\ v_3 \end{bmatrix} = \begin{bmatrix}\mathbf{b}_1 \ \mathbf{b}_2 \ \mathbf{b}_3\end{bmatrix} {}^{\mathrm{B}}v \tag{4.5}$$

This matrix multiplication expresses the vector $\mathbf{v}$ in terms of a vectrix of basis vectors and a column matrix of scalars. That column matrix is specific to the choice of basis. So, for clarity we indicate that choice with a left-superscript on the column matrix ${}^{\mathrm{B}}v$.

Vectrices can also be used in vector expressions, such as the dot product and the cross product. The vector operation is performed for each element in the matrix that results from standard matrix algebra. For example,

$$\mathbf{c} = \begin{bmatrix}\mathbf{a}_1 \ \mathbf{a}_2 \ \mathbf{a}_3\end{bmatrix} \cdot \begin{bmatrix} \mathbf{b}_1 \\ \mathbf{b}_2 \\ \mathbf{b}_3 \end{bmatrix} = \mathbf{a}_1 \cdot \mathbf{b}_1 + \mathbf{a}_2 \cdot \mathbf{b}_2 + \mathbf{a}_3 \cdot \mathbf{b}_3$$

$$\mathbf{C} = \begin{bmatrix} \mathbf{b}_1 \\ \mathbf{b}_2 \\ \mathbf{b}_3 \end{bmatrix} \cdot \begin{bmatrix}\mathbf{a}_1 \ \mathbf{a}_2 \ \mathbf{a}_3\end{bmatrix} = \begin{bmatrix} \mathbf{b}_1 \cdot \mathbf{a}_1 & \mathbf{b}_1 \cdot \mathbf{a}_2 & \mathbf{b}_1 \cdot \mathbf{a}_3 \\ \mathbf{b}_2 \cdot \mathbf{a}_1 & \mathbf{b}_2 \cdot \mathbf{a}_2 & \mathbf{b}_2 \cdot \mathbf{a}_3 \\ \mathbf{b}_3 \cdot \mathbf{a}_1 & \mathbf{b}_3 \cdot \mathbf{a}_2 & \mathbf{b}_3 \cdot \mathbf{a}_3 \end{bmatrix} \tag{4.6}$$

Vector operations on a dyadic apply to either the first or the second vector, depending on whether the other vector in the operation is pre- or post-multiplied. Some common operations with the dyadic $\mathbf{D}$ are shown in Table 4.3.

A dyadic can be written in terms of vectrices and a matrix. For example take the tedious-looking dyadic

$$\begin{aligned} \mathbf{J} = {} & J_{11}\mathbf{a}_1\mathbf{a}_1 + J_{12}\mathbf{a}_1\mathbf{a}_2 + J_{21}\mathbf{a}_2\mathbf{a}_1 + J_{13}\mathbf{a}_1\mathbf{a}_3 + +J_{31}\mathbf{a}_3\mathbf{a}_1 \ldots \\ & \ldots J_{22}\mathbf{a}_2\mathbf{a}_2 + J_{23}\mathbf{a}_2\mathbf{a}_3 + J_{33}\mathbf{a}_3\mathbf{a}_3 \end{aligned} \tag{4.7}$$

Table 4.3 Examples of dyadic operations. $\mathbf{D} = \mathbf{a}_1\mathbf{a}_2$; $\mathbf{a}_1$ and $\mathbf{a}_2$ are arbitrary vectors

Operation	Result	Description
$\mathbf{D} \cdot \mathbf{w}$	$= \mathbf{a}_1(\mathbf{a}_2 \cdot \mathbf{w})$	Vector: $\mathbf{a}_1$ multiplied by the scalar result of dotting $\mathbf{w}$ onto $\mathbf{a}_2$
$\mathbf{v} \cdot \mathbf{D}$	$= (\mathbf{v} \cdot \mathbf{a}_1)\mathbf{a}_2$	Vector: $\mathbf{a}_2$ multiplied by the scalar result of dotting $\mathbf{v}$ onto $\mathbf{a}_1$
$\mathbf{v} \cdot \mathbf{D} \cdot \mathbf{w}$	$= (\mathbf{v} \cdot \mathbf{a}_1)(\mathbf{a}_2 \cdot \mathbf{w})$	Scalar: scalar result of dotting $\mathbf{v}$ onto $\mathbf{a}_1$ multiplied by the scalar result of dotting $\mathbf{w}$ onto $\mathbf{a}_2$
$\mathbf{v} \times \mathbf{D}$	$= (\mathbf{v} \times \mathbf{a}_1)\mathbf{a}_2$	Dyadic: on the left, the cross product of $\mathbf{v}$ and $\mathbf{a}_1$; on the right, $\mathbf{a}_2$
$\mathbf{v} \times \mathbf{D} \cdot \mathbf{w}$	$= (\mathbf{v} \times \mathbf{a}_1)(\mathbf{a}_2 \cdot \mathbf{w})$	Vector: the cross product of $\mathbf{v}$ and $\mathbf{a}_1$ multiplied by the scalar result of dotting $\mathbf{w}$ onto $\mathbf{a}_2$

It can be written

$$\mathbf{J} = \begin{bmatrix}\mathbf{a}_1 & \mathbf{a}_2 & \mathbf{a}_3\end{bmatrix} \begin{bmatrix} J_{11} & J_{12} & J_{13} \\ J_{21} & J_{22} & J_{23} \\ J_{31} & J_{32} & J_{33} \end{bmatrix} \begin{bmatrix} \mathbf{a}_1 \\ \mathbf{a}_2 \\ \mathbf{a}_3 \end{bmatrix} \tag{4.8}$$

Notice that there is no dot-product operation here. Carrying out what appears to be matrix multiplication merely places the vectors $\mathbf{a}_i$ and $\mathbf{a}_j$ next to one another in the appropriate combinations. The inertia dyadic takes the form of $\mathbf{J}$ in Eq. (4.7), but with $J_{ij} = J_{ji}$; i.e., the inertia matrix is symmetric.

We also distinguish between "coordinate system" and "frame of reference," although, again, casual usage confuses the two. Texts sometimes say "frame" when "coordinate system" would be less confusing. A coordinate system consists of three orthonormal basis vectors ordered according to the right-hand rule. It is used to extract scalar and matrix values from vector-dyadic expressions and represent these expressions in terms of meaningful directions, such as compass points or the manufacturing coordinate system for a spacecraft. The choice of coordinate system has nothing at all to do with the underlying physics. It's like tilting your head: the choice of coordinates does not affect what you're looking at.

In contrast, a frame of reference is a powerful physical concept that establishes the ground rules for how equations of motion are written. Such frames are also known as "observational frames of reference." Ideas of space, time, and motion are central to this concept. Therefore, a frame is independent of arbitrarily chosen basis vectors and allows one to establish a common three-dimensional space for vectors and tensors considered in that frame.

As one might expect, the attitude motions of spacecraft do not require that we distinguish among frames of reference with relative motion near the speed of light. If that were necessary, the story would be more complicated. Instead, in the cases considered here, a frame of reference may be fixed either in a body or in inertial space. An inertial frame is Newtonian, i.e. Newton's laws can be stated in the familiar way in any such frame. Body-fixed frames are not necessarily Newtonian. One frame of reference may accelerate or rotate relative to another, and this relative motion can be described with kinematics. So, each rigid body considered here has a frame of reference associated with it, and rigid-body kinematics describes the relative motion among these frames. A table of symbols found at the beginning of the book defines key variables in these terms.

4.2 Spacecraft Attitude and Momentum-Device Kinematics

Attitude can be understood as the orientation of one set of basis vectors relative to another. Spacecraft attitude control (see Chap. 8 and Appendix D) seeks to drive this orientation and/or its derivatives toward some commanded value, for example aligning the spacecraft's nadir axis or a payload boresight with the vector from the spacecraft to a target on the ground or in space. In practice, spacecraft attitude is parameterized in a number of ways, the most common of which is the quaternion. It is favored because of its simplicity and compactness. However, direction-cosine matrices are sometimes used and, even more rarely, Euler angles. These and other attitude parameterizations are discussed thoroughly in (Stuelpnagel [26], Hughes [25], Shuster [27], and Crassidis and Markley [28]).

An attitude parameterization stores information about how two coordinate systems are related. For example, the direction-cosine matrix (derived in Eq. (4.6)) is an attitude parameterization that allows one to transform the representation of any vector in one coordinate system to its representation in any other in a conveniently direct way:

$$ {}^{\mathrm{B}}v = {}^{\mathrm{B}}Q^{\mathrm{A}\,\mathrm{A}}v \tag{4.9} $$

This notation places the symbols for the "A" coordinate system–one for the direction-cosine matrix and one for the column matrix–next to one another. This placement helps ensure that coordinate systems are used consistently. Here

$$ {}^{\mathrm{B}}Q^{\mathrm{A}} = \begin{bmatrix} \mathbf{b}_1 \\ \mathbf{b}_2 \\ \mathbf{b}_3 \end{bmatrix} \cdot \begin{bmatrix} \mathbf{a}_1 \; \mathbf{a}_2 \; \mathbf{a}_3 \end{bmatrix} \tag{4.10} $$

This relationship arises because the vector is independent of coordinate system, and it can therefore be written in terms of either basis-vector set:

$$ \mathbf{v} = \begin{bmatrix} \mathbf{b}_1 \; \mathbf{b}_2 \; \mathbf{b}_3 \end{bmatrix} \begin{bmatrix} {}^{\mathrm{B}}v_1 \\ {}^{\mathrm{B}}v_2 \\ {}^{\mathrm{B}}v_3 \end{bmatrix} = \begin{bmatrix} \mathbf{a}_1 \; \mathbf{a}_2 \; \mathbf{a}_3 \end{bmatrix} \begin{bmatrix} {}^{A}v_1 \\ {}^{A}v_2 \\ {}^{A}v_3 \end{bmatrix} \tag{4.11} $$

Projecting onto the B basis vectors leads to

$$ \begin{bmatrix} \mathbf{b}_1 \\ \mathbf{b}_2 \\ \mathbf{b}_3 \end{bmatrix} \cdot \mathbf{v} = \begin{bmatrix} \mathbf{b}_1 \\ \mathbf{b}_2 \\ \mathbf{b}_3 \end{bmatrix} \cdot \begin{bmatrix} \mathbf{b}_1 \; \mathbf{b}_2 \; \mathbf{b}_3 \end{bmatrix} \begin{bmatrix} {}^{\mathrm{B}}v_1 \\ {}^{\mathrm{B}}v_2 \\ {}^{\mathrm{B}}v_3 \end{bmatrix} = \begin{bmatrix} \mathbf{b}_1 \\ \mathbf{b}_2 \\ \mathbf{b}_3 \end{bmatrix} \cdot \begin{bmatrix} \mathbf{a}_1 \; \mathbf{a}_2 \; \mathbf{a}_3 \end{bmatrix} \begin{bmatrix} {}^{A}v_1 \\ {}^{A}v_2 \\ {}^{A}v_3 \end{bmatrix} \tag{4.12} $$

After some simplification,

$$ {}^{\mathrm{B}}v = I_{3\times 3} \begin{bmatrix} {}^{\mathrm{B}}v_1 \\ {}^{\mathrm{B}}v_2 \\ {}^{\mathrm{B}}v_3 \end{bmatrix} = {}^{\mathrm{B}}Q^{\mathrm{A}\,\mathrm{A}}v \tag{4.13} $$

which is merely Eq. (4.6).

While the fundamental physics is independent of coordinate system, certain traditional choices of coordinate systems allow spacecraft engineers to communicate

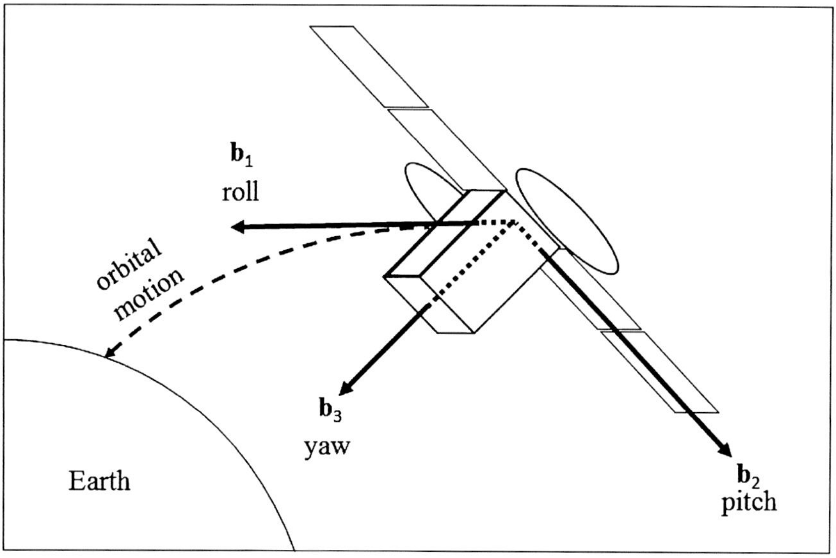

Fig. 4.2 Spacecraft reference coordinate system

with momentum-device engineers about alignment of the hardware and important behaviors of the momentum system. For the spacecraft, canonical reference axes include roll, pitch, and yaw, which arose originally from nautical convention, then were applied to aircraft, then to rockets, and then ultimately to spacecraft. Roll, pitch, and yaw are often represented with **x**, **y**, and **z** respectively. Sometimes $\mathbf{b}_1$, $\mathbf{b}_2$, and $\mathbf{b}_3$ are used in place of **x**, **y**, and **z**, and we adopt this more versatile convention. For many satellites, the pitch axis runs through the solar array (or solar wings), recalling that earlier aircraft use. Yaw is aligned with nadir or antinadir. The combination comprises a right-handed coordinate system, as shown in Fig. 4.2.

The convention for specifying the orientation of a momentum device focuses on its intended angular-momentum axis, i.e. the direction of the angular momentum for a perfectly built device. The actual angular momentum may vary because of dynamic imbalance, bearing runout, mechanical misalignment, and other terms; nevertheless, the coordinate system is based on the intended direction.

For an RWA, that axis is a fixed parameter $\hat{\mathbf{s}}_i$ in the spacecraft body reference frame. Here, i is an index that represents one of n momentum devices in an array. The other axes are orthogonal to this one and merely specify how connectors and fasteners need to be accommodated in the design of the spacecraft bus structure to which the momentum device is mounted. So, an array of RWAs is traditionally specified in terms of the various wheels' $\hat{\mathbf{s}}_i$ directions.

In contrast, a CMG's momentum axis is not fixed in the spacecraft structure. The gimbal motor causes it to precess. So, choosing coordinates is a little more subtle. For a CMG, it is the gimbal axis, $\hat{\mathbf{g}}_i$, that is fixed in the spacecraft reference frame.

The gyroscopic torque from the CMG, acting upon the spacecraft, lies along an output-torque axis $\hat{\mathbf{o}}_i$, related to these other axes by

$$\hat{\mathbf{o}}_i = \hat{\mathbf{s}}_i \times \hat{\mathbf{g}}_i \tag{4.14}$$

As the CMG gimbals, the $\hat{\mathbf{s}}_i$ vector changes direction with the gimbal angle δ. So, in the spacecraft coordinates B, its matrix representation ${}^{\mathrm{B}}\hat{s}_i$ is

$$\begin{aligned} {}^{\mathrm{B}}\hat{s}_i &= {}^{\mathrm{B}}\hat{s}_i(0)\cos(\delta) - {}^{\mathrm{B}}\hat{o}_i(0)\sin(\delta) \\ &= {}^{\mathrm{B}}\hat{s}_i(0)\cos(\delta) + {}^{\mathrm{B}}\hat{g}_i^{\times}\hat{s}_i(0)\sin(\delta) \end{aligned} \tag{4.15}$$

where ${}^{\mathrm{B}}\hat{s}_i(0)$ and ${}^{\mathrm{B}}\hat{o}_i(0)$ indicate the direction of the rotor momentum and output-torque axes respectively for $\delta = 0$. The superscript '$\times$' indicates the matrix equivalent of the cross-product operation, as described in Eq. (4.32). This reference configuration is set by the CMG hardware and software. Figure 4.3 shows this relationship

An array of CMGs is traditionally described with one or two simple parameters, exploiting symmetry for a compact description. For a SGCMG array, the β angle specifies the plane within which each CMG's momentum can lie. A plane is fully described in terms of a normal vector. In the case of a SGCMG, that vector is clearly $\hat{\mathbf{g}}_i$. So, β represents the tilt of $\hat{\mathbf{g}}_i$ away from a common reference, e.g. the vector that identifies the vertex of a pyramid or the centerline of a roof array. Figure 4.4 shows β for a four-CMG pyramid and a roof array.

The single parameter β is sufficient to specify the geometry of a roof array. A collection of scissored pairs does not even require β because their common output axis can be specified in the same way as the torque axis for a RWA. However, for a pyramid, one can also specify an angle γ that describes the cant of each $\hat{\mathbf{g}}_i$ around the axis of symmetry. This cant or "clocking" angle is 90° for a regular four-CMG pyramid, 72° for a five-CMG pyramid, 60° for a six-CMG pyramid, and so on (see Chap. 6). Although these general descriptions apply to contemporary CMG systems,

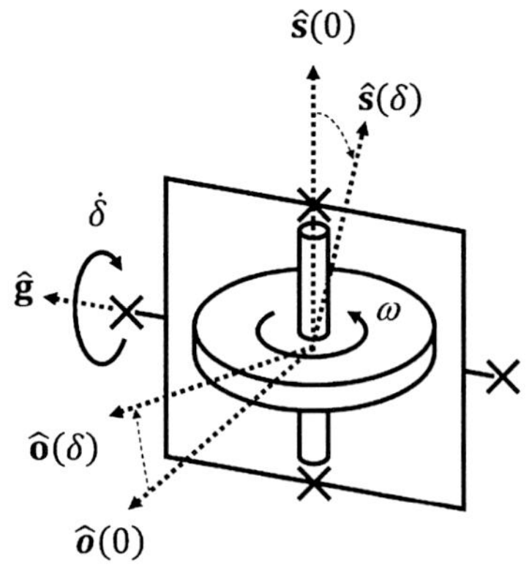

Fig. 4.3 CMG momentum in terms of reference axes for $\delta = 0$

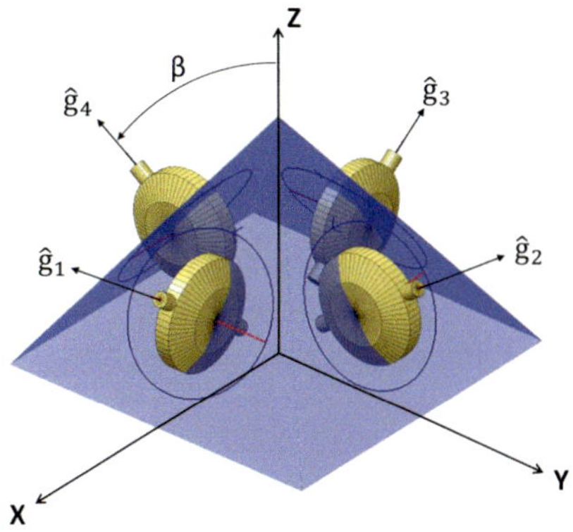

Fig. 4.4 β for a four-CMG pyramid

only regular pyramids are of current interest. So, it is considered sufficient to specify β and the number of CMGs.

Representing the alignment of the overall array in the spacecraft coordinates enables a Jacobian matrix to be constructed. For example, in spacecraft coordinates, the gimbal axis of any of the four CMGs in the pyramid shown above is represented as follows:

$$ {}^{B}\hat{g}_i = \begin{bmatrix} \sin(\beta)\cos(\gamma(i-1)) \\ \sin(\beta)\sin(\gamma(i-1)) \\ \cos(\beta) \end{bmatrix} \tag{4.16} $$

The momentum axis for an arbitrary gimbal angle is

$$ {}^{B}\hat{s}_i(\delta_i) = \begin{bmatrix} -\cos(\delta_i)\sin(\gamma(i-1)) - \sin(\delta_i)\cos(\beta)\cos(\gamma(i-1)) \\ \cos(\delta_i)\cos(\gamma(i-1)) - \sin(\delta_i)\cos(\beta)\sin(\gamma(i-1)) \\ \sin(\delta_i)\sin(\beta) \end{bmatrix} \tag{4.17} $$

Therefore, the output axis is

$$ {}^{B}\hat{o}_i(\delta_i) = \begin{bmatrix} \cos(\delta_i)\cos(\beta)\cos(\gamma(i-1)) - \sin(\delta_i)\sin(\gamma(i-1)) \\ \cos(\delta_i)\cos(\beta)\sin(\gamma(i-1)) + \sin(\delta_i)\cos(\gamma(i-1)) \\ -\cos(\delta_i)\sin(\beta) \end{bmatrix} \tag{4.18} $$

It may be of some historical interest that α, β, γ, δ and ϵ were chosen to describe a general array of CMGs, whether comprised of SGCMGs or DGCMGs. The outer gimbal was specified with β, as is still the case for a SGCMG array. The inner gimbal lay in the same plane in the reference configuration, and its orientation was specified with α. γ specified the cant of this reference plane, and both δ and ϵ specified the respective gimbal rotations to orient $\hat{\mathbf{s}}_i$. This more general convention is no longer in use. Only the β and δ angles remain in contemporary usage.

4.3 Equations of Motion for a Gyrostat with Balanced Rotors

A gyrostat is an assembly of interconnected rigid bodies whose relative motion does not alter the total inertia dyadic of the assembly. This model describes the basic behaviors of many spacecraft of practical interest: rigid-body spinners, the simplest configuration, consisting of a single spinning rigid body; dual-spin satellites, which use a single axially symmetric rotor connected by a motor to a counterspun or non-spinning platform on which generally a payload is mounted; momentum-bias satellites, which incorporate internal wheels to store angular momentum while the central body spins slowly about a single inertial axis or remains fixed in all three axes; zero-momentum satellites, which nominally have no angular momentum, except for what might be stored temporarily in preparation for momentum dumping, and which typically do not spin except to remain earth-pointed; and agile spacecraft, which undergo comparatively fast, large-angle slews as they acquire and scan celestial or terrestrial targets with the help of control-moment gyros (CMGs).

By this definition, the inertia dyadic $\mathbf{J}$ of a gyrostat for its mass center is constant in body axes:

$$\mathbf{J} = \begin{bmatrix} \mathbf{b}_1 & \mathbf{b}_2 & \mathbf{b}_3 \end{bmatrix} \begin{bmatrix} J_{11} & J_{12} & J_{13} \\ J_{21} & J_{22} & J_{23} \\ J_{31} & J_{32} & J_{33} \end{bmatrix} \begin{bmatrix} \mathbf{b}_1 \\ \mathbf{b}_2 \\ \mathbf{b}_3 \end{bmatrix} = \begin{bmatrix} \mathbf{b}_1 & \mathbf{b}_2 & \mathbf{b}_3 \end{bmatrix} {}^{\mathrm{B}}J \begin{bmatrix} \mathbf{b}_1 \\ \mathbf{b}_2 \\ \mathbf{b}_3 \end{bmatrix} \tag{4.19}$$

where the $\mathbf{b}_i$ are mutually orthogonal unit vectors constant in a body-fixed reference frame F_B. The 3×3 matrix in this expression is known as the inertia matrix for the body's mass center, which we denote as ${}^{\mathrm{B}}J$ to indicate that this body-fixed basis is used to express it.

Wheels with perfect static and dynamic balance move in a way that does not change the wheel's contribution to the total inertia dyadic in a spacecraft-fixed reference frame. The same is true for CMGs if the combination of rotor motion and gimbal motion can be correctly described by the motion of a uniformly dense sphere whose mass center is fixed in the spacecraft. For these special (or approximate) cases, the inertia dyadic of the rigid body ($\mathbf{J}_s$) combines with that of its constant-inertia rotors

$$\mathbf{J}_r = \sum_{i=1}^{n} \mathbf{J}_{r,i} \tag{4.20}$$

to produce the sum

$$\mathbf{J} = \mathbf{J}_s + \mathbf{J}_r \tag{4.21}$$

The reason is straightforward and begins with a few definitions. The angular velocity vector of a spacecraft body-fixed reference frame F_B relative to an inertial frame F_N is written $\boldsymbol{\omega}^{\mathrm{B/N}}$. Similarly, F_{R_i} is a reference frame fixed in the ith rotor, and the angular velocity of F_{R_i} relative to F_B is $\boldsymbol{\omega}^{\mathrm{R_i/B}}$. For n wheels of inertia $\mathbf{J}_{r,i}$, the momentum system's angular momentum is the sum

$$\mathbf{h}_r = \sum_{i=1}^{n} \mathbf{J}_{r,i} \cdot (\boldsymbol{\omega}^{\mathrm{B/N}} + \boldsymbol{\omega}^{\mathrm{R_i/B}}) \tag{4.22}$$

The total spacecraft angular momentum, $\mathbf{H}$ can be written in terms of this wheel angular momentum $\mathbf{h}_r$:

$$\mathbf{H} = \mathbf{J}_s \cdot \boldsymbol{\omega}^{\mathrm{B/N}} + \mathbf{h}_r \tag{4.23}$$

i.e.,

$$\mathbf{H} = \mathbf{J}_s \cdot \boldsymbol{\omega}^{\mathrm{B/N}} + \sum_{i=1}^{n} \mathbf{J}_{r,i} \cdot (\boldsymbol{\omega}^{\mathrm{B/N}} + \boldsymbol{\omega}^{\mathrm{R_i/B}}) \tag{4.24}$$

Collecting on $\boldsymbol{\omega}^{\mathrm{B/N}}$ results in

$$\mathbf{H} = \left(\mathbf{J}_s + \sum_{i=1}^{n} \mathbf{J}_{r,i}\right) \cdot \boldsymbol{\omega}^{\mathrm{B/N}} + \sum_{i=1}^{n} \mathbf{J}_{r,i} \cdot \boldsymbol{\omega}^{\mathrm{R_i/B}} \tag{4.25}$$

Defining

$$\mathbf{h} = \sum_{i=1}^{n} \mathbf{J}_{r,i} \cdot \boldsymbol{\omega}^{\mathrm{R_i/B}} \tag{4.26}$$

and making a few substitutions results in a compact expression for the gyrostat's angular momentum:

$$\mathbf{H} = \mathbf{J} \cdot \boldsymbol{\omega}^{\mathrm{B/N}} + \mathbf{h} \tag{4.27}$$

The equations of motion for the gyrostat can be derived simply from the derivative of this momentum in F_N and application of the Transport Theorem:

$$\frac{{}^{\mathrm{N}}\mathrm{d}\mathbf{H}}{\mathrm{d}t} = \mathbf{J}_s \cdot \frac{{}^{\mathrm{B}}\mathrm{d}\boldsymbol{\omega}^{\mathrm{B/N}}}{\mathrm{d}t} + \frac{{}^{\mathrm{B}}\mathrm{d}\mathbf{h}}{\mathrm{d}t} + \boldsymbol{\omega}^{\mathrm{B/N}} \times \mathbf{J} \cdot \boldsymbol{\omega}^{\mathrm{B/N}} \tag{4.28}$$

The left superscript in this vector expression indicates the reference frame with respect to which a vector derivative is taken. It differs from the way in which we indicate the coordinate system used to represent a vector's components. It is unambiguous because this use is exclusive to vectors, while the coordinate-system notation is exclusive to matrices. The fact that a gyrostat's inertia dyadic is constant in a body-fixed frame motivates using the transport theorem to eliminate derivatives of the inertia dyadic that would appear if the derivative were taken with respect to

F_N. The governing equations include both this result and kinematic relationships, which might be written in terms of a quaternion, direction-cosine matrices, and many other parameterizations of attitude [26, 27, 29].

4.4 Relative Equilibria and Stability of Gyrostats

The equations of motion exhibit some simple equilibrium solutions in the absence of applied torque. Of interest here are so-called relative equilibria, for which the derivatives in F_B are zero. Despite that wheel momentum and angular velocity are constant for a gyrostat in a relative equilibrium, the spacecraft may still spin. The motion is known as a simple spin, a constant angular velocity along a body-fixed direction.

In equilibrium (and with no applied moments), the equation of motion is

$$0 = \boldsymbol{\omega}^{\mathrm{B/N}} \times \left(\mathbf{J} \cdot \boldsymbol{\omega}^{\mathrm{B/N}} + \mathbf{h}\right) \tag{4.29}$$

The body-fixed torque due to wheel momentum, $\boldsymbol{\omega}^{\mathrm{B/N}} \times \mathbf{h}$, balances the dynamic torque associated with the rigid body, $\boldsymbol{\omega}^{\mathrm{B/N}} \times \mathbf{J} \cdot \boldsymbol{\omega}^{\mathrm{B/N}}$. The net torque (the sum) is zero in equilibrium, a situation known as dynamic balance. If there were no wheels, the rigid-body torque in equilibrium would be zero on its own. The equation of motion would then represent spin about a principal axis of the rigid inertia matrix. [29]

For a given $\boldsymbol{\omega}^{\mathrm{B/N}}$ this equilibrium condition determines the wheel momentum in the plane perpendicular to $\boldsymbol{\omega}^{\mathrm{B/N}}$ in equilibrium. It does not constrain the wheel momentum parallel to $\boldsymbol{\omega}^{\mathrm{B/N}}$, momentum which can be used to augment the dynamics in ways other than dynamic balance. Let us denote the projection of $\mathbf{h}$ along $\boldsymbol{\omega}^{\mathrm{B/N}}$ as the scalar h_s (spin-axis wheel momentum). The selection of h_s for some spacecraft application is generally a design choice and is independent of dynamic balance. It is often used to augment the effective moment of inertia, stabilizing an otherwise unstable minor-axis spinner. In fact, dual-spin spacecraft, which are minor-axis rigid bodies due to geometry that allows them to fit well inside launch-vehicle fairings, owe their success to this passive approach to attitude stabilization. Such a design choice can be enforced through an additional equation

$$0 = \boldsymbol{\omega}^{\mathrm{B/N}} \cdot \mathbf{h} = \Omega h_s \tag{4.30}$$

where $\Omega = \|\boldsymbol{\omega}^{\mathrm{B/N}}\|$.

The equilibrium system is described by four scalar equations that are linear in the three components of the 3×1 matrix

$$h = \mathbf{h} \cdot \begin{bmatrix} \mathbf{b}_1 \\ \mathbf{b}_2 \\ \mathbf{b}_2 \end{bmatrix}$$

The four scalar equations are

$$Ah = \begin{bmatrix} \omega^T \\ \omega^\times \end{bmatrix} h = \begin{bmatrix} \Omega h_s \\ \omega^\times J \omega \end{bmatrix} \tag{4.31}$$

where ω is the 3×1 matrix of components of $\boldsymbol{\omega}^{\mathrm{B/N}}$ in the **b** basis, and the superscript '$\times$' indicates the matrix equivalent of the cross-product operation, i.e.,

$$\omega^\times = \begin{bmatrix} 0 & -\boldsymbol{\omega}^{\mathrm{B/N}} \cdot \mathbf{b}_3 & \boldsymbol{\omega}^{\mathrm{B/N}} \cdot \mathbf{b}_2 \\ \boldsymbol{\omega}^{\mathrm{B/N}} \cdot \mathbf{b}_3 & 0 & -\boldsymbol{\omega}^{\mathrm{B/N}} \cdot \mathbf{b}_1 \\ -\boldsymbol{\omega}^{\mathrm{B/N}} \cdot \mathbf{b}_2 & \boldsymbol{\omega}^{\mathrm{B/N}} \cdot \mathbf{b}_1 & 0 \end{bmatrix} \tag{4.32}$$

A is a 4×3 matrix. Rank(A)=3 if ω is nonzero. A simple proof of this statement is that $\omega^\times$ is rank 2, and ω^T cannot lie in its range space. Therefore, a matrix comprised of both $\omega^\times$ and ω^T must be rank 3. With this fact about rank(A), h is found from a Moore-Penrose pseudoinverse:

$$h = \left(A^{\mathrm{T}} A\right)^{-1} A^{\mathrm{T}} \begin{bmatrix} \Omega h_s \\ \omega^\times J \omega \end{bmatrix} \tag{4.33}$$

The solution is

$$h = \frac{1}{\Omega^2} \left[\Omega \omega \ \omega^\times \omega^\times\right] \begin{bmatrix} h_s \\ J \omega \end{bmatrix} \tag{4.34}$$

or, in terms of the normalized matrix

$$\hat{\omega} = \frac{\omega}{\Omega} \tag{4.35}$$

the result is

$$h = \hat{\omega} h_s + \Omega \hat{\omega}^\times \left(\hat{\omega}^\times J \hat{\omega}\right) \tag{4.36}$$

This expression for h is a closed-form solution for the total, embedded angular momentum in a gyrostat for a given equilibrium angular velocity and desired superspin. It provides the foundation for passive stabilization of a rigid spacecraft spinning about any axis $\hat{\omega}$.

It is well known that a spinning rigid body—a gyrostat with zero embedded momentum—has six relative equilibria for a given angular momentum, each corresponding to spin about one of the principal axes of the inertia matrix, either in the positive or negative sense. This result arises from the equilibrium condition for a rigid body,

$$0 = \boldsymbol{\omega}^{\mathrm{B/N}} \times \mathbf{J} \cdot \boldsymbol{\omega}^{\mathrm{B/N}} \tag{4.37}$$

which is simply Euler's equation with the angular-velocity derivative set to zero. This condition implies

$$\boldsymbol{\omega}^{\mathrm{B/N}} \parallel \left(\mathbf{J} \cdot \boldsymbol{\omega}^{\mathrm{B/N}}\right) \tag{4.38}$$

i.e.,

$$\lambda \boldsymbol{\omega}^{\mathrm{B/N}} = \mathbf{J} \cdot \boldsymbol{\omega}^{\mathrm{B/N}} \tag{4.39}$$

The three solutions for λ are the three eigenvalues of the inertia matrix ${}^{\mathrm{B}}J$. The eigenvectors are the directions of the angular-velocity vector for each of these equilibria. Since an eigenvector and its negative are both solutions, there are six such directions.

The solution in which λ is the maximum principal moment of inertia represents a minimum-energy state for a given angular-momentum magnitude. Spin about this maximum axis is therefore stable when kinetic energy can dissipate, and this fact is used in many different spacecraft architectures to ensure mission success. Syncom 2, launched in 1963, was the first successful geosynchronous communications satellite. It is an example of a passively stable maximum-axis spinner. Spin about the minor axis is a maximum-energy state and is therefore unstable, diverging from the equilibrium over time, depending on the rate at which kinetic energy is lost. Each of the early Explorer series of spacecraft is an example of an unstable, minor-axis spinner. The intermediate axis is a saddle point that can be considered unstable, but in fact any perturbation away from this equilibrium represents large-angle motion about either the minimum or maximum-inertia axis. No spacecraft intended to be passively stable would be designed to spin about the intermediate-inertia axis.

The situation is more subtle for a gyrostat with nonzero embedded momentum. Hughes [29] provides a detailed assessment of the equilibria for such spacecraft, using a polynomial whose roots can be used to find these equilibria. Here we offer a compact approach to finding these equilibria based on a generalized eigenvalue problem. For small $\|\mathbf{h}\|$ compared to $\|\mathbf{H}\|$, the gyrostat solution resembles that of a single rigid body, as one might expect. Larger values of embedded momentum can reduce the number of real equilibria to four or as few as two.

For nonzero $\mathbf{H}$ and $\boldsymbol{\omega}^{\mathrm{B/N}}$, a gyrostat in equilibrium must exhibit

$$\boldsymbol{\omega}^{\mathrm{B/N}} \parallel \left(\mathbf{J} \cdot \boldsymbol{\omega}^{\mathrm{B/N}} + \mathbf{h}\right) \tag{4.40}$$

Equivalently,

$$\lambda \boldsymbol{\omega}^{\mathrm{B/N}} = \left(\mathbf{J} \cdot \boldsymbol{\omega}^{\mathrm{B/N}} + \mathbf{h}\right) \tag{4.41}$$

λ here plays a role very similar to how it appears in the case of a single rigid body. It is a scalar that is related to the magnitudes $H = \|\mathbf{H}\|$ and Ω by

$$\lambda = \frac{H}{\Omega} \tag{4.42}$$

Projecting $\mathbf{J}$ onto the $\mathbf{b}_i$ bases leads to the 3×3 inertia matrix ${}^{\mathrm{B}}J$ (denoted simply by J here). In terms of these matrices,

$$\lambda\omega = J\omega + h \tag{4.43}$$

This statement is a matrix pencil: almost an eigenvalue problem in λ, but for the presence of h.

An expression for 6 scalars that can be used to compute the corresponding directions of ω for known H, J, and h is

$$\det\left(\Omega I - \begin{bmatrix} 2HJ^{-1} & J^{-2}\left(hh^T - H^2 I\right) \\ I & 0 \end{bmatrix}\right) = 0 \tag{4.44}$$

where 'det' indicates the determinant of a matrix, and I is the 3×3 identity matrix. Each eigenvalue Ω_k is a scalar angular-velocity magnitude for one of the equilibria. Only the real values correspond to physical solutions, and because the energy and momentum are real numbers, the non-physical solutions must be complex-conjugate pairs. Therefore, there may be two, four, or six real solutions. The direction of the kth angular velocity vector is then given in terms of its $\mathbf{b}$ components as

$$\omega_k = \left(\frac{H}{\Omega_k} I - J\right)^{-1} h \tag{4.45}$$

The stability of these equilibria can be assessed in a number of ways, as described in Hughes [29]. A fail-safe rule of thumb is that an equilibrium is stable if its value of λ is the largest of all six, again analogous to the single rigid-body case. This rule applies to any gyrostat and reduces to the maximum-axis rule for the case of a single rigid spinning body, for which there are two solutions (positive and negative spin). For a more general gyrostat, it is possible for the stability of positive and negative spin to differ.

A spacecraft design that is informed by this principle benefits from a guarantee of passive stability as long as the momentum system can maintain constant momentum. Software failures related to the spacecraft attitude-control subsystem do not threaten stability as long as the operation of the momentum system is decoupled in these failure cases. Such a failure response is available in many contemporary RWAs, which are designed to hold the most recent speed command in the event of interrupted commands from the spacecraft's command and data handling subsystem.

4.4.1 RWA Spacecraft

Superspin is a technique that augments the open-loop dynamics to achieve passive stabilization. The term comes originally from dual-spin spacecraft, which typically separate from the launch-vehicle upper stage with a high spin rate (by today's standards) of 30–60 RPM. After separation, a motor between the rotor and the platform causes one part to spin at a different rate from the other. The higher-speed portion is responsible for "super" spin. These two counter-rotating bodies exhibit gyrostat dynamics. Depending on the amount of energy dissipation in the bodies, the faster-spinning portion can stabilize the other, per the Iorillo stability criterion.

Contemporary spacecraft can use superspin to stabilize a spacecraft that spins during some or all of its operations, but they do so with reaction wheels instead of a dual-spin rotor. Typically the flight software is responsible for establishing this embedded momentum by commanding the momentum system. By analogy to the maximum-axis principle, the simplest approach to superspin stabilization is to use the momentum system to make the effective inertia, λ, for that axis the largest among all six possible equilibria. In practice, λ is chosen to be at least 20% greater when possible. For example, to stabilize a rigid body around one of its principal-inertia axes, the spacecraft design sets the effective inertia to be

$$\lambda \geq 1.2 J_{mm} \tag{4.46}$$

where J_{mm} is the maximum principal moment of inertia of the rigid body.

Superspin can be used to stabilize a spacecraft meant to spin around any axis, not exclusively one of these basis vectors. If wheel augmentation does not already provide sufficient effective inertia, the spin-axis momentum to achieve superspin would attempt to exceed that insufficient value λ_{max} by 20% through a new value:

$$\lambda \geq 0.2 \lambda_{max} \Omega \tag{4.47}$$

For the case of a rigid spinner, with λ equal to the rigid inertia for the desired spin axis,

$$h_s \geq (1.2 J_{mm} - \lambda)\, \Omega \tag{4.48}$$

When the desired spin axis is not a passive equilibrium, it can be made so through dynamic balance. This technique consists of placing some of the momentum system's momentum vector perpendicular to the desired spin axis. In closed form, this dynamic-balance momentum is

$$h_d = \hat{\omega}^{\times} \left(\hat{\omega}^{\times} J \omega \right) \tag{4.49}$$

For example, a spacecraft intended to spin about its $\mathbf{b}_3$ axis $\hat{\omega} = [0\ 0\ 1]^T$ can use its momentum system to dynamically balance the J_{13} and J_{23} products of inertia with the following constant angular momentum:

$$h_d = \begin{bmatrix} -J_{13}\Omega \\ -J_{23}\Omega \\ 0 \end{bmatrix} \tag{4.50}$$

Both superspin and dynamic balance require less momentum to be effective when the spacecraft spin rate is low. In the limit, as the spacecraft approaches a non-spinning state, the only gyroscopic stiffness becomes that of the wheel. In that limit, the system is a momentum-bias spacecraft.

4.4.2 CMG Spacecraft

Spacecraft with CMGs exhibit the same equilibria as any other gyrostat when the gimbal angle of each CMG is held constant, whether by active torque from the gimbal motor or by mechanical resistance in the gimbal mechanism. In principle, such spacecraft can also use dynamic balance and superspin to augment the rigid-body dynamics to achieve a variety of architectural benefits. However, unlike a spacecraft with RWAs, a CMG exhibits different behavior in the first few moments after attitude control no longer commands the momentum-control devices. This open-loop behavior depends on the friction and stiction in the CMG gimbals, which is not simple to predict. Nevertheless, the following general principles provide some insight.

As in the reaction-wheel case, the equilibrium condition is

$$0 = \boldsymbol{\omega}^{\mathrm{B/N}} \times \left(\mathbf{J} \cdot \boldsymbol{\omega}^{\mathrm{B/N}} + \mathbf{h}\right) \tag{4.51}$$

but with the added constraint that the gimbal rate is also zero. Furthermore, in equilibrium, the spacecraft's motion does not accelerate any CMG's rotor about that CMG's gimbal axis. For the ith single-gimbal CMG and in the absence of resistance torques along the gimbal axis, that condition is

$$0 = \left(\boldsymbol{\omega}^{\mathrm{B/N}} \times \mathbf{h}_{c,i}\right) \cdot \hat{\mathbf{g}}_i \tag{4.52}$$

The equilibrium solution therefore varies with the array geometry, and there is a larger family of solutions than in the case of the gyrostat with RWAs. A computer simulation would identify the stable configurations of CMG gimbal angles and rigid-body angular velocity that meets these constraints in the presence of stiction.

Motor cogging, gear-tooth meshing, and bearing friction resist free rotation of the rotor about the gimbal axis. In the absence of attitude control, open-loop motion of the gimbals can quickly damp kinetic energy and place the spacecraft in an equilibrium spin. This damping effect resembles how mechanical gyroscopes have been used to resist and damp out rolling and pitching motion of ships at sea. In the space application, this effect can also be beneficial. It is a valuable way to ensure bounded dynamics, resulting in an equilibrium spin, during any one of several failures that leave the spacecraft without attitude control.

4.4.3 Large-Angle Slews

A conceptually simple way to describe a slew from one attitude to another is through Euler's Theorem for rigid-body motion. A consequence of that theorem is that one can always reorient a rigid body by rotating it about a single axis fixed both in the body and in an inertial reference frame. Let the attitude of the rigid body at the start time of the slew t_0 be described by a direction-cosine matrix ${}^{\mathrm{N}}Q^{\mathrm{B}(t_0)}$, and let its attitude at completion time t_f be described by ${}^{\mathrm{N}}Q^{\mathrm{B}(t_f)}$. The change in attitude, again parameterized as a direction-cosine matrix, is

$$
{}^{\mathrm{B}(t_f)}Q^{\mathrm{B}(t_0)} = \left({}^{\mathrm{N}}Q^{\mathrm{B}(t_f)}\right)^{\mathrm{T}}\, {}^{\mathrm{N}}Q^{\mathrm{B}(t_0)} \tag{4.53}
$$

The axis-angle representation of the attitude change is found from an eigenvalue decomposition of ${}^{\mathrm{B}(t_f)}Q^{\mathrm{B}(t_0)}$. It can also be found from the matrix logarithm of ${}^{\mathrm{B}(t_f)}Q^{\mathrm{B}(t_0)}$, and then the components of the rotation axis extracted from the resulting 3×3 skew-symmetric matrix. Specifically, the eigenvector ${}^{\mathrm{B}}c$ of ${}^{\mathrm{B}(t_f)}Q^{\mathrm{B}(t_0)}$ that is associated with the eigenvalue 1 is the axis of rotation such that

$$
{}^{\mathrm{B}(t_f)}Q^{\mathrm{B}(t_0)} = \cos\phi I_{3\times3} + (1 - \cos\phi)\,{}^{\mathrm{B}}c\,{}^{\mathrm{B}}c^{\mathrm{T}} - \sin\phi\,{}^{\mathrm{B}}c^{\times} \tag{4.54}
$$

The superscript B on ${}^{\mathrm{B}}c$ does not specify time simply because the vector $\mathbf{c}$ is constant throughout the motion, per Euler's Theorem. The matrix ${}^{\mathrm{B}}c$ therefore consists of the same three scalars regardless of the fact that the spacecraft's attitude is changing. In the case of an eigenaxis slew, the angle of rotation is $\phi(t)$. It starts at $\phi = 0$ and reaches a commanded (final) value $\phi = \phi_c$ has the following angular velocity:

$$
\boldsymbol{\omega}^{\mathrm{B/N}} = \dot{\phi}(t)\mathbf{c} \tag{4.55}
$$

This behavior is also known in kinematics as a "simple spin" because the unit vector $\mathbf{c}$ does not change.

For purposes of this discussion, the definition of "versine slew" is the following: an eigenaxis slew for which the angle follows a half-wave versine path in time until the end of the maneuver. "Versine" is the name for the following trigonometric function:

$$
\mathrm{versin}\theta = 2\sin^2\frac{\theta}{2} = 1 - \cos\theta \tag{4.56}
$$

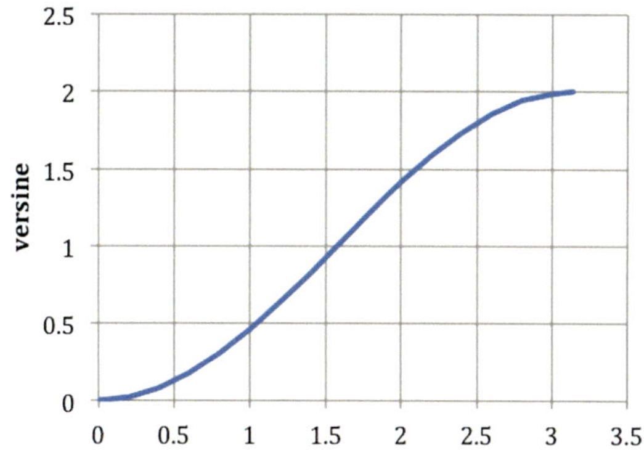

Fig. 4.5 Versine

This function is illustrated in Fig. 4.5.

For single-axis motion, let the initial angle for a versine slew be ϕ_1 and the total angle be $\Delta\phi$. Then the final angle is

$$\phi_2 = \phi_1 + \Delta\phi \tag{4.57}$$

Since the initial angle is arbitrary, let $\phi_1 \equiv 0$. Then

$$\phi_2 = \Delta\phi \tag{4.58}$$

The kinematics of an angle-versine slew are

$$\begin{aligned}
\phi(t) &= \frac{\phi_2}{2}(1 - \cos at) \\
\dot{\phi}(t) &= a\frac{\phi_2}{2}\sin at \\
\ddot{\phi}(t) &= a^2\frac{\phi_2}{2}\cos at \\
\dddot{\phi}(t) &= -\frac{\phi_2}{2}\sin at
\end{aligned} \tag{4.59}$$

where $\frac{\pi}{a}$ is the duration of the slew, and time t is time in seconds, measured from the beginning of the slew such that

$$0 \le t \le \frac{\pi}{a} \tag{4.60}$$

This function is infinitely differentiable and therefore smooth from the perspective of actuation and numerical implementation. The maxima of these functions lead to expressions for the timing parameter a in terms of prescribed kinematic limits:

$$a_1 = \frac{\dot{\phi}_{max}}{2|\phi_2|}, a_2 = \left(\frac{\ddot{\phi}_{max}}{2|\phi_2|}\right)^{1/2}, \text{ or } a_3 = \left(\frac{\dddot{\phi}_{max}}{2|\phi_2|}\right)^{1/3} \tag{4.61}$$

Here, the indices on a denote the degree of the derivative, not the initial or final point in the slew. The best choice of a is the maximum value (therefore the shortest slew time) for which the other two time histories are within their respective limits. I.e.,

$$\begin{aligned}
&\text{for } a_1 = \frac{\dot{\phi}_{max}}{2\,|\phi_2|}, \text{ require } \left|\frac{a_1^2\phi_2}{2}\right| \le \ddot{\phi}_{max} \text{ and } \left|\frac{a_1^3\phi_2}{2}\right| \le \dddot{\phi}_{max} \\
&\text{for } a_2 = \left(\frac{\ddot{\phi}_{max}}{2\,|\phi_2|}\right)^{1/2}, \text{ require } \left|\frac{a_2\phi_2}{2}\right| \le \dot{\phi}_{max} \text{ and } \left|\frac{a_2^3\phi_2}{2}\right| \le \dddot{\phi}_{max} \\
&\text{for } a_3 = \left(\frac{\dddot{\phi}_{max}}{2\,|\phi_2|}\right)^{1/3}, \text{ require } \left|\frac{a_3\phi_2}{2}\right| \le \dot{\phi}_{max} \text{ and } \left|\frac{a_3^2\phi_2}{2}\right| \le \ddot{\phi}_{max}
\end{aligned} \tag{4.62}$$

A solution is always possible because each derivative increases monotonically with a.

A rate versine is suitable for rate control, not necessarily for attitude control. The rate versine begins with an expression for $\dot{\phi}$ that is similar to the angle-versine expression for ϕ. The expression for angle is the integral, with the initial angle as the constant of integration (i.e. 0). The result is

$$\begin{aligned}
\phi(t) &= \frac{1}{2}\left[\left(\dot{\phi}_2 + \dot{\phi}_1\right)t - \frac{1}{a}\left(\dot{\phi}_2 - \dot{\phi}_1\right)\sin at\right] \\
\dot{\phi}(t) &= \frac{1}{2}\left(\dot{\phi}_2 - \dot{\phi}_1\right)(1 - \cos at) + \dot{\phi}_1 \\
\ddot{\phi}(t) &= \frac{1}{2}\left(\dot{\phi}_2 - \dot{\phi}_1\right)a\sin at \\
\dddot{\phi}(t) &= \frac{1}{2}\left(\dot{\phi}_2 - \dot{\phi}_1\right)a^2\cos at
\end{aligned} \tag{4.63}$$

The maximum value of $\dot{\phi}$ occurs at the beginning or the end of the slew. Since these boundaries are given as conditions of the problem, the planning algorithm does not need to implement a means of observing the rate limits during the slew. Some means of rejecting infeasible commands would likely be implemented in ground software, before such commands reach the spacecraft, or in some higher-level onboard autonomy software.

The maximum acceleration and jerk provide conditions for the slew duration:

$$a_1 = \frac{2\dot{\phi}_{max}}{\left|\dot{\phi}_2 - \dot{\phi}_1\right|}, \quad a_2 = \left(\frac{2\dot{\phi}_{max}}{\left|\dot{\phi}_2 - \dot{\phi}_1\right|}\right)^{1/2} \tag{4.64}$$

As in the case of the angle-versine slew, the best choice is the maximum, subject to the requirement that the other derivative not exceed its kinematic limit:

$$\begin{aligned} &\text{for } a_1, \text{ require } \left|a_1 \frac{\dot{\phi}_2 - \dot{\phi}_1}{2}\right| \le \ddot{\phi}_{max} \\ &\text{for } a_2, \text{ require } \left|a_2^2 \frac{\dot{\phi}_2 - \dot{\phi}_1}{2}\right| \le \dddot{\phi}_{max} \end{aligned} \tag{4.65}$$

One may prefer to think of ϕ starting at some nonzero initial condition, ϕ_0. For example, an attitude-control system might autonomously track this angle as a large error and continuously compute the eigenaxis parameters until the angle is small enough for the dynamics to be considered linear. In that case, ϕ_c in these expressions can be replaced with a delta angle, $\delta\phi = \phi_c - \phi_0$. Regardless of how the initial condition is specified, the results provide a smooth slew from the initial value to the final, commanded value regardless of the sign of these angles.

These kinematics provide fundamental insight into the peak momentum, torque, and gimbal acceleration (in the case of CMGs) necessary to achieve a desired agility. Specifically, the axis of rotation determines the scalar inertia J_e that the momentum system must move:

$$J_e = \mathbf{c} \cdot \mathbf{J} \cdot \mathbf{c} \tag{4.66}$$

The actuator's peak performance derives from the slew kinematics according to

$$\begin{aligned} h_{max} &= \dot{\phi}_{max} J_e \\ \tau_{max} &= \ddot{\phi}_{max} J_e \end{aligned} \tag{4.67}$$

The gimbal's acceleration is driven by the jerk limit ($\dddot{\phi}$) and a variety of gimbal rate-loop requirements that were discussed in Chap. 3. These simple results do not take into account contingency, margin, and control-related phenomena such as overshoot, but it's a start.

However, an eigenaxis slew is not necessarily the fastest route nor the most power-efficient. Fully optimizing for time and power would take into account the fact that the inertia varies across the range of possible paths. Slewing around an axis with lower inertia than that of the eigenaxis might speed up the maneuver. The momentum system torque and momentum envelope also comes into play: some corners of the array's performance envelope might be exploited for higher performance if an axis other than this geometrically shortest one is chosen. Finally, other spacecraft-specific constraints on orientation relative to the sun or the Earth, for thermal or communications reasons, might have to be accommodated, regardless of the purely dynamical benefits of a time-optimal slew. Eigenaxis slews provide useful insight, particularly at the early stages of spacecraft design, but a detailed design for an agile spacecraft's operations concept may ultimately demand different kinematics.

4.5 Control Moment Gyroscopes

The dynamics of a CMG includes a wide variety of rigid-body, flexible, and nonlinear effects. This section derives the foundational equations for a CMG that is statically balanced—i.e., the mass centers of each component do not translate in a spacecraft-fixed reference frame as the CMG's rotor tilts. In practice, this static imbalance and dynamic imbalance of the rotor induce vibration in the spacecraft structure, as do gear-tooth meshing and bearing-related disturbances. This induced vibration (IV) represents a high-frequency mechanical disturbance that can affect pointing performance for high-precision payloads. Imprecision in the gimbal motion because of motor cogging and the transmission of gimbal-motor torque through gears (among many effects) is responsible for so-called output-torque ripple (OTR), which is of lower frequency than IV. However, OTR can be of larger magnitude and influences the spacecraft pointing in ways that IV does not, such as excitement of low-frequency spacecraft flexible modes. These effects are taken up in Sect. 4.7.

Although the multiple-gimbal case is the general one from a theoretical perspective, single-gimbal CMGs are preferred in practice because the torque produced by precessing the rotor is reacted by structural components rather than the motors on other gimbal axes, saving weight and power. Acknowledging the widespread use of single-gimbal CMGs, this derivation assumes that **h** is the sum of the rotor angular momentum for any number of internal components with no more than one gimbal axis each. Figure 4.6 shows the components of a single-gimbal CMG.

Consider a spinning rigid spacecraft whose angular-momentum vector (**H**) is given by

$$\mathbf{H} = \mathbf{J} \cdot \boldsymbol{\omega}^{\mathrm{B/N}} + \mathbf{h} \tag{4.68}$$

as discussed above. The assumption that the internal components are statically balanced about their angular-velocity vectors means that the rigid body's mass center does not accelerate, that equations for the rotational dynamics alone completely describe the motion of the system. So, the gyrostat principles apply here.

As in the case of the RWA spacecraft, we lump the momentum of the momentum devices due to the spacecraft's rotation in with that of the rigid body. I.e., the term

$$\mathbf{J}_g \cdot \boldsymbol{\omega}^{\mathrm{B/N}} + \mathbf{J}_r \cdot \boldsymbol{\omega}^{\mathrm{B/N}} \tag{4.69}$$

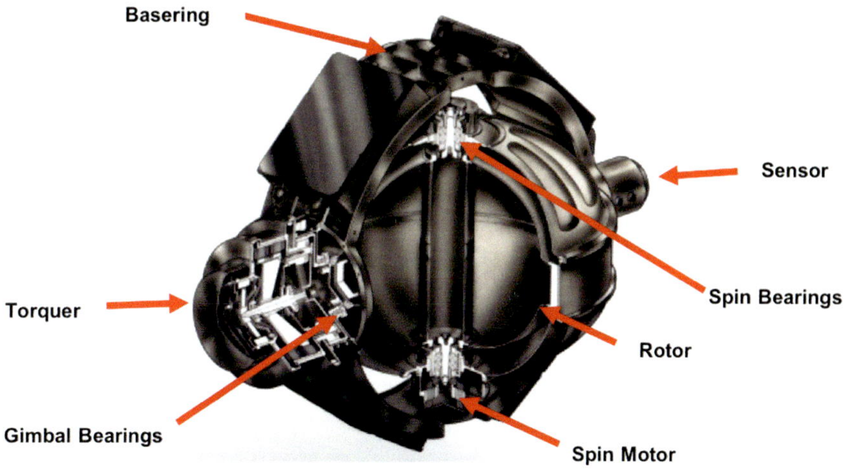

Fig. 4.6 Single-gimbal CMG (Image courtesy of Honeywell, Inc.)

does not appear here because it is accounted for in the spacecraft bus motion. Then, the remaining momentum of all n CMGs is

$$\mathbf{h} = \sum_{i=1}^{n} \mathbf{J'}_{g,i} \cdot \boldsymbol{\omega}^{\mathrm{G}_i/\mathrm{B}} + \mathbf{J}_{r,i} \cdot \boldsymbol{\omega}^{\mathrm{R}_i/\mathrm{B}} \tag{4.70}$$

in which another simplification is possible. Let the gimbal structure's inertia be $\mathbf{J'}_{g,i}$ defined so that an augmented inertia $\mathbf{J}_{g,i}$ can include the rotor's inertia:

$$\mathbf{J}_{g,i} = \mathbf{J'}_{g,i} + \mathbf{J}_{r,i} \tag{4.71}$$

Then

$$\mathbf{h} = \sum_{i=1}^{n} \mathbf{J'}_{g,i} \cdot \boldsymbol{\omega}^{\mathrm{G}_i/\mathrm{B}} + \mathbf{J}_{r,i} \cdot \left(\boldsymbol{\omega}^{\mathrm{G}_i/\mathrm{B}} + \boldsymbol{\omega}^{\mathrm{R}_i/\mathrm{G}_i}\right) \tag{4.72}$$

becomes

$$\mathbf{h} = \sum_{i=1}^{n} \mathbf{J}_{g,i} \boldsymbol{\omega}^{\mathrm{G}_i/\mathrm{B}} + \mathbf{J}_{r,i} \boldsymbol{\omega}^{\mathrm{R}_i/\mathrm{G}_i} \tag{4.73}$$

substituting the scalar gimbal speed $\dot{\delta}_i$, rotor speed $\Omega_{r,i}$, and the unit vectors along the gimbal and rotor-spin directions leads to

$$\mathbf{h} = \sum_{i=1}^{n} \mathbf{J}_{g,i} \cdot (\dot{\delta}_i \hat{\mathbf{g}}_i) + \mathbf{J}_{r,i} \cdot (\Omega_i \hat{\mathbf{s}}_i) \tag{4.74}$$

Because the gimbal and rotor are taken to be dynamically balanced, their respective inertia dyadics include no products of inertia. The inertia for the $\hat{\mathbf{g}}_i$ and $\hat{\mathbf{s}}_i$ directions are $I_{g,i}$ and $I_{r,i}$ respectively. So,

$$\mathbf{h} = \sum_{i=1}^{n} I_{g,i}\dot{\delta}_i\hat{\mathbf{g}}_i + I_{r,i}\Omega_i\hat{\mathbf{s}}_i \tag{4.75}$$

This spacecraft's dynamics resemble those of the RWA spacecraft. Both begin with the inertial derivative of the total angular momentum. However, for the CMG spacecraft, this derivative includes additional terms that result from the gimbals:

$$\begin{aligned}\frac{{}^{\mathrm{N}}\mathrm{d}\mathbf{H}}{\mathrm{d}t} &= \mathbf{J}\cdot\frac{{}^{\mathrm{B}}\mathrm{d}\boldsymbol{\omega}^{\mathrm{B/N}}}{\mathrm{d}t} + \frac{{}^{\mathrm{B}}\mathrm{d}}{\mathrm{d}t}\left(\sum_{i=1}^{n} I_{g,i}\dot{\delta}\hat{\mathbf{g}}_i + I_{r,i}\Omega_i\hat{\mathbf{s}}_i\right)\\ &+ \boldsymbol{\omega}^{\mathrm{B/N}}\times\left(\sum_{i=1}^{n} I_{g,i}\dot{\delta}_i\hat{\mathbf{g}}_i + I_{r,i}\Omega_i\hat{\mathbf{s}}_i\right)\end{aligned} \tag{4.76}$$

The first term in parentheses can be made explicit:

$$\begin{aligned}\frac{{}^{\mathrm{B}}\mathrm{d}}{\mathrm{d}t}\left(\sum_{i=1}^{n} I_{g,i}\dot{\delta}\hat{\mathbf{g}}_i + I_{r,i}\Omega_i\hat{\mathbf{s}}_i\right) &= \left(\sum_{i=1}^{n} I_{g,i}\ddot{\delta}_i\hat{\mathbf{g}}_i + I_{r,i}\dot{\Omega}_i\hat{\mathbf{s}}_i\right)\\ &+ \left(\sum_{i=1}^{n} I_{g,i}\dot{\delta}_i\frac{{}^{\mathrm{B}}\mathrm{d}}{\mathrm{d}t}\hat{\mathbf{g}}_i + I_{r,i}\Omega_i\frac{{}^{\mathrm{B}}\mathrm{d}}{\mathrm{d}t}\hat{\mathbf{s}}_i\right)\end{aligned}$$

The body-frame derivatives of the gimbal axes and the rotor axes result from the kinematics of the CMG. The gimbal axis is fixed in F_B. So,

$$\frac{{}^{\mathrm{B}}\mathrm{d}}{\mathrm{d}t}\hat{\mathbf{g}}_i = 0 \tag{4.77}$$

From the Transport Theorem and our definitions,

$$\frac{{}^{\mathrm{B}}\mathrm{d}}{\mathrm{d}t}\hat{\mathbf{s}}_i = \frac{{}^{\mathrm{G}_i}\mathrm{d}}{\mathrm{d}t}\hat{\mathbf{s}}_i + \boldsymbol{\omega}^{\mathrm{G}_i/\mathrm{B}}\times\hat{\mathbf{s}}_i = 0 + \dot{\delta}\hat{\mathbf{g}}_i\times\hat{\mathbf{s}}_i \tag{4.78}$$

This direction is exactly the opposite of the output torque ($\hat{\mathbf{o}}_i$) as defined in 4.14. So,

$$\frac{{}^{\mathrm{B}}\mathrm{d}}{\mathrm{d}t}\hat{\mathbf{s}}_i = -\dot{\delta}\hat{\mathbf{o}}_i \tag{4.79}$$

Substituting this result into Euler's equation leads to a complete expression for the time-derivative of angular momentum for the CMG spacecraft with rotors that are not necessarily at constant speed:

$$\frac{{}^{\mathrm{N}}\mathrm{d}\mathbf{H}}{\mathrm{d}t} = \mathbf{J} \cdot \frac{{}^{\mathrm{B}}\mathrm{d}\boldsymbol{\omega}^{\mathrm{B/N}}}{\mathrm{d}t} + \left(\sum_{i=1}^{n} I_{g,i} \ddot{\delta}_i \hat{\mathbf{g}}_i + I_{r,i} \left(\dot{\Omega}_i \hat{\mathbf{s}}_i - \dot{\delta}_i \Omega_i \hat{\mathbf{o}}_i \right) \right)$$
$$+ \boldsymbol{\omega}^{\mathrm{B/N}} \times \left(\mathbf{J} \cdot \boldsymbol{\omega}^{\mathrm{B/N}} + \sum_{i=1}^{n} I_{g,i} \dot{\delta}_i \hat{\mathbf{g}}_i + I_{r,i} \Omega_i \hat{\mathbf{s}}_i \right) \tag{4.80}$$

It is instructive to consider these terms individually:

- $\mathbf{J} \cdot \frac{{}^{\mathrm{B}}\mathrm{d}\boldsymbol{\omega}^{\mathrm{B/N}}}{\mathrm{d}t} + \boldsymbol{\omega}^{\mathrm{B/N}} \times \mathbf{J} \cdot \boldsymbol{\omega}^{\mathrm{B/N}}$—This term represents the rigid-body motion due to spacecraft bus angular acceleration and angular velocity, as it would appear in Euler's Equation.
- $\boldsymbol{\omega}^{\mathrm{B/N}} \times \left(\sum_{i=1}^{n} I_{g,i} \dot{\delta}_i \hat{\mathbf{g}}_i\right)$—This term represents the gyroscopic effect on the spacecraft of the moving gimbal. While important in a high-fidelity simulation, the gimbal's momentum is much lower than that of the rotor and may be neglected for first-order torque- and momentum-sizing analyses. Neglecting it would be analogous to setting $I_{g,i} = 0$.
- $\boldsymbol{\omega}^{\mathrm{B/N}} \times \left(\sum_{i=1}^{n} I_{r,i} \Omega_i \hat{\mathbf{s}}_i\right)$—This term represents the gyroscopic effect on the spacecraft of the spinning rotors. This term has a significant effect on the spacecraft dynamics. For each CMG, it is also a torque that the gimbal motor must react, i.e. the "base rate" effect. This effect is one of the design drivers for the gimbal motor.
- $\sum_{i=1}^{n} I_{g,i} \ddot{\delta}_i \hat{\mathbf{g}}_i$—This term represents the torque on the spacecraft due to the gimbal's accelerating everything in the inner gimbal assembly, i.e. the rotor and the gimbal structure that supports it. This term is an important one in sizing the gimbal motor, but it may be neglected in simpler analyses that focus on momentum system torque and momentum capability.
- $\sum_{i=1}^{n} I_{r,i} \dot{\Omega}_i \hat{\mathbf{s}}_i$—This term represents the torque on the spacecraft due to spinning up or down the rotor. For typical single-gimbal CMGs, this event is rare, but it is an important effect for the case of variable-speed CMGs.
- $\sum_{i=1}^{n} -I_{r,i} \dot{\delta}_i \Omega_i \hat{\mathbf{o}}_i$—This term represents the gyroscopic output torque from gimbaling the rotor's angular momentum. This value dominates the output torque of a single-gimbal CMG.

A simplistic model of a spacecraft with constant-speed rotors would include only the output-torque effect. In that case, the spacecraft's dynamics would be

$$\frac{{}^{\mathrm{N}}\mathrm{d}\mathbf{H}}{\mathrm{d}t} = \mathbf{J} \cdot \frac{{}^{\mathrm{B}}\mathrm{d}\boldsymbol{\omega}^{\mathrm{B/N}}}{\mathrm{d}t} - \sum_{i=1}^{n} I_{r,i} \dot{\delta}_i \Omega_i \hat{\mathbf{o}}_i$$
$$+ \boldsymbol{\omega}^{\mathrm{B/N}} \times \left(\mathbf{J} \cdot \boldsymbol{\omega}^{\mathrm{B/N}} + \sum_{i=1}^{n} I_{r,i} \Omega_i \hat{\mathbf{s}}_i \right) \tag{4.81}$$

To construct a model suitable for attitude-control simulations, one can split this equation in such a way that the momentum system appears as a separate subsystem from the spacecraft body:

$$\frac{{}^{\mathrm{N}}\mathrm{d}\mathbf{H}}{\mathrm{d}t} + \boldsymbol{\tau}_o = \mathbf{J} \cdot \frac{{}^{\mathrm{B}}\mathrm{d}\boldsymbol{\omega}^{\mathrm{B/N}}}{\mathrm{d}t} + \boldsymbol{\omega}^{\mathrm{B/N}} \times \left(\mathbf{J} \cdot \boldsymbol{\omega}^{\mathrm{B/N}} + \sum_{i=1}^{n} I_{r,i} \Omega_i \hat{\mathbf{s}}_i \right)$$

$$\boldsymbol{\tau}_o = \sum_{i=1}^{n} I_{r,i} \dot{\delta} \Omega_i \hat{\mathbf{o}}_i \tag{4.82}$$

The torque acting on the spacecraft is the opposite of what acts on the momentum system and is positive as shown on the left-hand side of this equation, according to the conventional definition of output torque.

These equations can be written in matrix form and solved for the angular acceleration. Here, the matrices have been constructed via the body-fixed basis vectors $\mathbf{b}_i$ and are indicated as such with a left superscript B.

$${}^{\mathrm{B}}\dot{\omega} = {}^{\mathrm{B}}J^{-1} \left[{}^{\mathrm{B}}\dot{H} + {}^{\mathrm{B}}\tau_o - {}^{\mathrm{B}}\omega^{\times} \left({}^{\mathrm{B}}J^{\mathrm{B}}\omega + \sum_{i=1}^{n} I_{r,i} \Omega_i {}^{\mathrm{B}}\hat{s}_i \right) \right] \tag{4.83}$$

$${}^{\mathrm{B}}\tau_o = \left(\sum_{i=1}^{n} I_{r,i} \dot{\delta}_i \Omega_i {}^{\mathrm{B}}\hat{o}_i \right) \tag{4.84}$$

4.6 Actuator Jacobian Matrices

A Jacobian matrix A, or simply "Jacobian," is an $n \times m$ matrix formed by taking the partial derivative of an $m \times 1$ real-valued matrix $F(x)$ of torque acting on a spacecraft with respect to an $n \times 1$ column matrix x of inputs to individual momentum devices:

$$A = \frac{\partial F}{\partial x} = \left[\frac{\partial F}{\partial x_1} \cdots \frac{\partial F}{\partial x_n} \right] = \begin{bmatrix} \frac{\partial F_1}{\partial x_1} & \cdots & \frac{\partial F_1}{\partial x_n} \\ \vdots & \ddots & \vdots \\ \frac{\partial F_m}{\partial x_1} & \cdots & \frac{\partial F_m}{\partial x_n} \end{bmatrix} \tag{4.85}$$

where F_i and x_i are the matrix components of the real-valued function and column matrix.

A linear, scalar example offers some insight into the physical significance of a Jacobian. Consider a function for the output torque of a motor $\tau = Ki$, where τ is the output torque, K is a scalar gain known as a torque constant, and i is the input current to the motor. The partial derivative of the output torque $\tau(i)$ with respect to the output current i yields a scalar Jacobian matrix in this case and, because K is a constant and not a function of i, it is the gain K. This gain amplifies the current, producing torque, for any value of i.

The case for an array of RWAs is analogous. Where the gain K characterizes the amplification in the scalar example, the singular values of the Jacobian matrix characterize it in the general case. A zero singular value is roughly analogous to zero gain–at least for some directions, if not all. The angular momentum of the RWAs is the sum of all of the rotor speeds multiplied by their respective inertias. In the likely case where all RWAs have the same rotor inertia, the array angular momentum is

$$h = I_r \sum_{i=1}^{n} \Omega_i {}^{\mathrm{B}}\hat{s}_i \tag{4.86}$$

The output torque is the time derivative of the spacecraft body's array angular momentum, the negative of the RWAs' momentum. The chain rule yields

$$\frac{\mathrm{d}h}{\mathrm{d}t} = -{}^{\mathrm{B}}\tau_o = \frac{\partial h}{\partial \Omega}\dot{\Omega} = I_r \begin{bmatrix} {}^{\mathrm{B}}\hat{s}_1 & \cdots & {}^{\mathrm{B}}\hat{s}_n \end{bmatrix} \begin{bmatrix} \dot{\Omega}_1 \\ \vdots \\ \dot{\Omega}_n \end{bmatrix} \tag{4.87}$$

and therefore, the Jacobian matrix for the RWA angular momentum with respect to the RWA rotor rates is

$$A = I_r \begin{bmatrix} {}^{\mathrm{B}}\hat{s}_1 & \cdots & {}^{\mathrm{B}}\hat{s}_n \end{bmatrix} \tag{4.88}$$

Like the linear, scalar example for the output torque of a motor, the Jacobian matrix for an array of RWAs is constant. Therefore, if a momentum system includes three or more RWAs with linearly independent spin axes, its Jacobian matrix cannot be singular.

The case of a Jacobian matrix for an array of CMGs is more complicated because the Jacobian matrix is not constant. It is a function of the gimbal-axis orientation, gimbal angle, and rotor speed for each CMG. The Jacobian matrix for an array of CMGs is

$$A = \frac{\partial h}{\partial \Delta} = \begin{bmatrix} \frac{\partial h}{\partial \delta_i} & \cdots & \frac{\partial h}{\partial \delta_n} \end{bmatrix} \tag{4.89}$$

where Δ is a column matrix of gimbal angles. For the likely case of uniform rotor momentum for each CMG in the array, the Jacobian matrix is a concatenation of all of the output-torque directions in the spacecraft body coordinates, multiplied by the single-CMG rotor angular momentum:

$$A = \frac{\partial h}{\partial \Delta} = -h_r \begin{bmatrix} {}^{\mathrm{B}}\hat{o}_1 & \cdots & {}^{\mathrm{B}}\hat{o}_n \end{bmatrix} \tag{4.90}$$

Therefore, singularities arise when the output-torque directions for the CMGs lie in a plane. Mathematically, this condition corresponds to a situation in which the matrix $\left[{}^{\mathrm{B}}\hat{o}_1 \cdots {}^{\mathrm{B}}\hat{o}_n\right]$ has rank 2.

For an array of VSCMGs, the Jacobian matrix for the CMG gimbal rates and rotors-as-RWAs are concatenated but without the assumption that the rotor angular momenta are uniform (because VSCMG rotor rates vary). The VSCMG array Jacobian matrix is therefore

$$A = \left[\frac{\partial h}{\partial \Delta} \ \frac{\partial h}{\partial \Omega}\right] \tag{4.91}$$

and the total output torque from the VSCMG array is

$${}^{\mathrm{B}}\tau_o = -A \begin{bmatrix} \dot{\Delta} \\ \dot{\Omega} \end{bmatrix} \tag{4.92}$$

This Jacobian matrix is always full rank for nonzero rotor velocity if there are at least two VSCMGs in the array with linearly independent gimbal axes. However, rank alone is not the full story; there are deeper issues with VSCMG arrays, including loss of gyroscopic torque, that are discussed in later chapters.

For an array of DGCMGs, the Jacobian matrix is similar to that for an array of SGCMGs but includes additional columns because of the additional gimbals. This Jacobian matrix is

$$A = \left[\frac{\partial h}{\partial \Delta_1} \ \frac{\partial h}{\partial \Delta_2}\right] \tag{4.93}$$

where Δ_1 and Δ_2 are the column matrices of gimbal angles for the inner and outer gimbals. Even with constant-magnitude and uniform angular momentum among all the CMGs, the outer-gimbal Jacobian matrix contribution is not a concatenation of unit vectors (Kurokawa [30]).

It is often useful to measure the distance from singularity by the rank properties of the Jacobian matrix for an array of momentum devices. In such a calculation, the Jacobian matrix should be normalized (scaled) by rotor momentum before it is used in a "metric" such as the manipulability, $m = \sqrt{\det(AA^T)}$ equal to the product of the singular values of the Jacobian matrix (as discussed in Chap. 7). For example, with constant uniform angular momentum across all across all CMGs, the metric is $m = \sqrt{\det(\hat{A}\hat{A}^T)}$, where $A = h_r \hat{A}$. The normalization of the Jacobian matrix provides not only a measure of singularity that is independent of CMG size but also serves as a numerically stable value in studies of singularity avoidance.

4.7 Rotor and Gimbal Structural Dynamics

The analysis of flexible dynamics on rotating bodies is a specialty discipline. The equations are difficult to formulate, and all but the most trivial cases must be solved numerically. This subtlety motivates some straightforward design principles: momentum devices are designed for high stiffness so that their flexible dynamics are decoupled from fundamental vibration modes of the spacecraft structure and from attitude-control loops. Separating these frequencies simplifies control design and improves the predictability of the spacecraft's flexible response in the context of launch-vehicle coupled-loads analysis, among many advantages. This stiffness also is consistent with the need for flight hardware to survive high loads associated with launch-vehicle dynamics, including acoustic loads and rocket-engine vibrations. However, it is generally stiffness, not material-stress considerations, that drives the mechanical design of the actuators.

RWAs and CMGs achieve stiffness in part by maximizing bearing span—that is, the distance between multiple bearings that support the rotor and (in the case of CMGs) the gimbal. The more widely separated the bearings, the larger the moment they can react, all things being equal. These devices also incorporate high-stiffness rotors, structural elements, and bracketry while trying to minimize the mass of these elements. Details of the design are specific to the devices and the application, and they are beyond the scope of this discussion. Nevertheless, two general behaviors are evident for a wide variety of actuators: (1) the vibration mode of the rotor, where gyroscopic effects couple rigid-body nutation to structural deflection of the rotor, and (2) a CMG specific effect in which the so-called output-axis stiffness influences the torque necessary from the gimbal motor in the form of an augmented (effective) inertia.

This first effect is evident in a simple model of a spinning symmetric body on a two degree-of-freedom torsional spring. The subtle flexible behaviors of the entire gimbal structure to which the rotor is mounted is captured in a single spring rate k, which is assumed to be the same in any direction normal to the gimbal axis. A body-fixed reference frame F_B is fixed to this rotor, and a coordinate system consisting of the $\mathbf{b}_i$ basis vectors are also fixed in F_B. In this model, the body represents a rotor spinning at high speed about the axis $\mathbf{b}_1$ shown in Fig. 4.7. The compliant base permits the rotor to deflect about any axis in the $\mathbf{b}_2$-$\mathbf{b}_3$ plane with uniform stiffness.

This simplified representation includes an embedded angular momentum vector, h_r about the $\mathbf{b}_1$ axis and assumes small-angle motion of the rotor in the transverse directions. For sufficiently small motion, the angular-velocity vector is the derivative of a three-parameter attitude representation consisting of a small rotation in each of three axes:

$$^{B}\begin{bmatrix} \theta_1 \\ \theta_2 \\ \theta_3 \end{bmatrix} \approx {}^{B}\begin{bmatrix} \int \omega_1 dt \\ \int \omega_2 dt \\ \int \omega_3 dt \end{bmatrix} \tag{4.94}$$

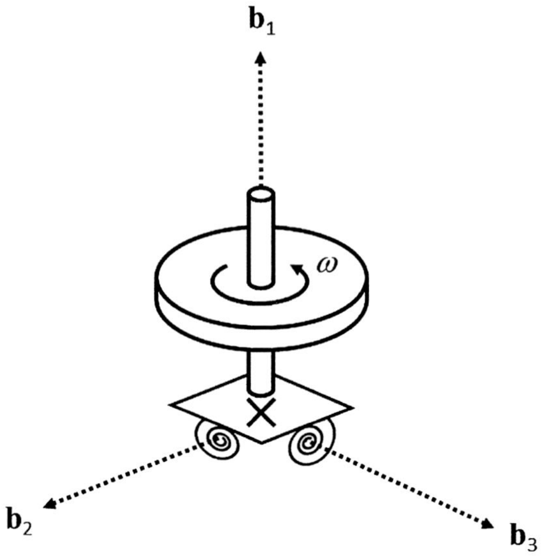

Fig. 4.7 Rotor on a compliant base

the left-superscript B indicates that the matrix of components is associated with the B coordinate system.

However, the rotation in $\mathbf{b}_1$ is taken to be zero because the springs deflect only in the transverse direction. If the inertia dyadic for the rotor's mass center is $\mathbf{J}_r = I_r\mathbf{b}_1\mathbf{b}_1 + I_t(\mathbf{b}_2\mathbf{b}_2 + \mathbf{b}_3\mathbf{b}_3)$, the equations of motion in matrix form are

$$ {}^{\mathrm{B}}J_r\,{}^{\mathrm{B}}\dot{\omega}^{\mathrm{B/N}} + {}^{\mathrm{B}}\omega^{\mathrm{B/N}\times}\left({}^{\mathrm{B}}J_r\,{}^{\mathrm{B}}\omega^{\mathrm{B/N}} + {}^{\mathrm{B}}h\right) = {}^{\mathrm{B}}\tau_K \tag{4.95} $$

where again, the superscript $\times$ indicates the matrix equivalent of the vector cross product, and τ_K is the torque that results from the transverse deflection of the rotor on its flexible base. Substituting $\dot{\theta}$ for ${}^{\mathrm{B}}\omega^{\mathrm{B/N}}$ leads to

$$ {}^{\mathrm{B}}\begin{bmatrix} I_r & 0 & 0 \\ 0 & I_t & 0 \\ 0 & 0 & I_t \end{bmatrix} {}^{\mathrm{B}}\begin{bmatrix} \ddot{\theta}_1 \\ \ddot{\theta}_2 \\ \ddot{\theta}_3 \end{bmatrix} + {}^{\mathrm{B}}\begin{bmatrix} 0 & -\dot{\theta}_3 & \dot{\theta}_2 \\ \dot{\theta}_3 & 0 & -\dot{\theta}_1 \\ -\dot{\theta}_2 & \dot{\theta}_1 & 0 \end{bmatrix} \left({}^{\mathrm{B}}\begin{bmatrix} I_r & 0 & 0 \\ 0 & I_t & 0 \\ 0 & 0 & I_t \end{bmatrix} {}^{\mathrm{B}}\begin{bmatrix} \dot{\theta}_1 \\ \dot{\theta}_2 \\ \dot{\theta}_3 \end{bmatrix} + \begin{bmatrix} h_r \\ 0 \\ 0 \end{bmatrix} \right) = \begin{bmatrix} 0 \\ -k\theta_2 \\ -k\theta_3 \end{bmatrix} \tag{4.96} $$

Because the small-angle assumption eliminates coupling of this spring torque into the spin axis, the solution is

$$ \begin{aligned} \ddot{\theta}_2 &= -\frac{1}{I_t}(h_r\dot{\theta}_3 + k\dot{\theta}_2) \\ \ddot{\theta}_3 &= -\frac{1}{I_t}(h_r\dot{\theta}_2 - k\dot{\theta}_3) \end{aligned} \tag{4.97} $$

Converting this second-order, coupled, ordinary differential equation into first-order form provides the state-space dynamics matrix

$$A = \begin{bmatrix} 0 & -\frac{h_r}{I_t} & -\frac{k}{I_t} & 0 \\ \frac{h_r}{I_t} & 0 & 0 & -\frac{k}{I_t} \\ 1 & 0 & 0 & 0 \\ 0 & 1 & 0 & 0 \end{bmatrix} \tag{4.98}$$

with eigenvalues (natural frequencies)

$$\lambda = \left(\left(\frac{h_r}{2I_t} \right)^2 + \frac{k}{I_t} \right)^{1/2} \pm \frac{h_r}{2I_t} \tag{4.99}$$

With no rotor momentum, these roots are simply the natural frequencies of an inertia on a torsional stiffness:

$$\lim_{h_r \to 0} \lambda = \pm \left(\frac{k}{I_t} \right)^{1/2} \tag{4.100}$$

One of these vibrational modes is rotational oscillation about one transverse axis, and the second is about the other transverse axis, as shown in Fig. 4.8.

However for nonzero rotor momentum, these two modes couple and separate. In the limit of infinitely high rotor momentum (or zero stiffness), the eigenvalues approach:

$$\lim_{h_r \to \infty} \lambda = 0, \frac{h_r}{I_t} \tag{4.101}$$

The "precession" frequency asymptotes to zero with increasing rotor momentum, and the "nutation" frequency asymptotes to the rigid-body nutation frequency. These gyroscopically coupled modes exhibit "coning" motion, with the nutation mode coning in the same direction as the rotor spin, and the precession mode in the opposite direction, i.e., retrograde, as shown in Fig. 4.9.

Fig. 4.8 Modes with no rotor speed

Fig. 4.9 Modes with rotor spinning: nutation (left) and precession (right). The rotor angular velocity vector is directed upward here. The red cone indicates the path that the angular-velocity vector carves out during these motions

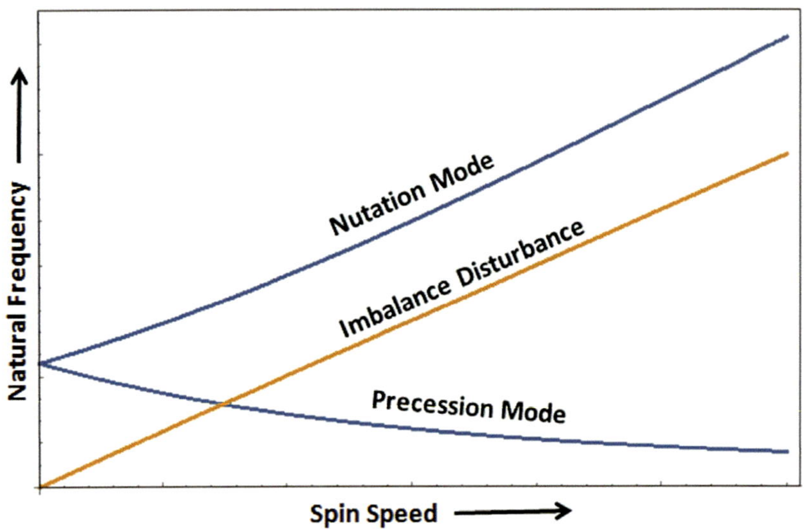

Fig. 4.10 Variations in natural frequency with spin speed

If one of these mode frequencies coincides with the spin speed, excitation can result. Variations of natural frequency with speed can be seen in Fig. 4.10. The precession mode always intersects at the frequency $\sqrt{K/(I_r + I_t)}$. The nutation mode never crosses so long as $(I_r > I_t)$ i.e., the rotor is oblate (a disc), not prolate (a skinny cylinder).

The precession mode, which can coincide with the spin frequency, cones in the retrograde direction. Retrograde modes exhibit an interesting phenomenon directly traceable to that of pole-zero cancellation. From a simple mathematical model, one might conclude that spin vibrations cannot excite retrograde modes because their eigenvectors are orthogonal. However, in a practical context these modes are excited through subtle effects that include imperfections, nonlinearities, and other unmodeled features. Rotor spin vibrations can transmit energy through surrounding structure.

The discussion above applies to stationary, spinning rotors. However, gimbal motion can result in further complexity: in particular, the second flexible effect of general interest considered here is that of augmented inertia. This phenomenon arises because of compliance in all structural elements that are meant to constrain the CMG's momentum vector to move in a plane perpendicular to the gimbal axes. These elements include the rotor shell or web, the rotor shaft, rotor bearings, gimbal structure, gimbal bearings, and possibly elements of the gimbal drive train. The mechanical fixture between the spacecraft bus and the gimbal may also play a role. The combined stiffness of all of these elements, whether in series or in parallel, comprises the "output-axis stiffness." The phrase suggests that the stiffness is associated with the output axis of the CMG. Specifically, it is a stiffness that is responsible for keeping the output axis perpendicular to the gimbal axis.

One way to estimate the participation of these elements in the output-axis stiffness is to begin with an analytical model or a finite-element representation of this subsystem. Applying a couple M at the rotor rim in a direction parallel to the output axis deflects this model by an angle, ϕ, as shown in Fig. 4.11. The stiffness is the ratio of the couple to the angle.

The more compliant this subsystem, the more that the rotor can tilt toward or away from the gimbal. As its momentum $\mathbf{h}_r$ projects along the gimbal axis $\hat{\mathbf{g}}_i$, the resulting change in momentum represents a torque that the gimbal must react. Here, the various parameters are shown as scalars because this analysis is much simpler than the general case that would be analyzed with a detailed model. For a given scalar gimbal rate δ and output-axis stiffness, K_{OA}, the output torque is

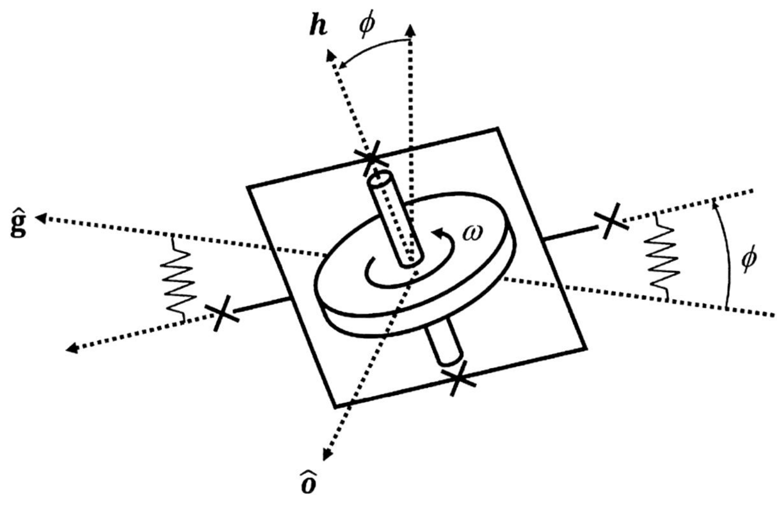

Fig. 4.11 Conceptual model of output-axis stiffness

approximately

$$\tau_o = \dot{\delta} h_r$$
$$\tau_o = -K_{OA}\phi \tag{4.102}$$

or

$$\phi = -\frac{\dot{\delta} h_r}{K_{OA}} \tag{4.103}$$

$\mathbf{h}_r$ deflects by an angle ϕ in proportion to the CMG output torque $\boldsymbol{\tau}_o$ causing angular momentum h_g to appear along the gimbal axis:

$$h_g = h_r \sin(-\phi) \tag{4.104}$$

For small angles,

$$h_g = -h_r \phi \tag{4.105}$$

Some substitutions lead to

$$h_g = \frac{\dot{\delta} h_r^2}{K_{OA}} \tag{4.106}$$

The time derivative of this momentum is the torque on the gimbal due to the output-axis stiffness effect:

$$\tau_{g,K} = \frac{\ddot{\delta} h_r^2}{K_{OA}} \tag{4.107}$$

The gimbal must accelerate an effective inertia $J_{g,eff}$ that combines the rigid inertia of the IGA and this output-axis stiffness effect:

$$\tau_g = \tau_{g,r} + \tau_{g,K} = J_{g,eff}\ddot{\delta} \tag{4.108}$$

Since the rigid IGA accelerates due to a torque

$$\tau_{g,r} = J_g \ddot{\delta} \tag{4.109}$$

the combined effect is the effective inertia

$$J_{g,eff} = J_g + \frac{h_r^2}{K_{OA}} \tag{4.110}$$

This simple expression conveniently represents how compliance in the gimbal structure degrades the gimbal's ability to accelerate the IGA. There is also a dynamic effect beyond this quasistatic behavior: an initial rotation of the gimbal imparts an oscillatory disturbance that degrades spacecraft-pointing performance somewhat. The key insight is that low stiffness costs power and ultimately increases the weight of the gear train and the gimbal. So, in designing a CMG, adding mass to stiffen this structure may save mass in the gimbal motor, but the best combination of stiffness and motor power depends on the specifics of the application. Regardless of the stiffness of a specific design, this effective inertia must be accounted for in estimates of the gimbal torque required to achieve the attitude-control requirements at the spacecraft level.

4.8 Effects of Scaling CMG Actuators

The rise of small satellites, or "smallsats," since the late 1990s has led to attitude actuators that meet volume, mass, and power requirements for precision pointing of these new spacecraft architectures. Some miniature RWAs for slow-moving spacecraft have been available for many years. Many of the earliest smallsats, such as NASA's Phonesat, simply incorporated brushless DC motors, using the motor's own inertia as a rotor. Some small RWAs are packaged with attitude determination sensors as turnkey attitude-control subsystems–for example, Blue Canyon Technologies' XACT unit. Even miniature CMGs have been developed to provide CubeSats with agile capabilities. The University of Florida's IMPACT CMGs and Honeybee Robotics' Microsat CMGs are examples. They are shown in Figs. 4.12 and 4.13 and respectively.

However, the performance of these momentum systems does not necessarily scale well. For small rotors, the overhead associated with motors, sensors, bearings, structure, slip rings, electronics, etc. can dominate the size and weight of the device. This overhead, in turn, makes the momentum devices a disproportionately larger fraction of the total spacecraft than is the case for large satellites. So, dynamics

Fig. 4.12 IMPACT single CMG (left) and IMPACT 4-CMG pyramid array (right) (Courtesy of Prof. Norman Fitz-Coy, University of Florida)

Fig. 4.13 Honeybee robotics Microsat CMGs (Image courtesy of Honeybee Robotics)

effects considered to be negligible for larger spacecraft must be reconsidered for these smaller ones. Some of the detrimental effects of small scale include:

- Lack of availability of small, high-performance components
- Lack of manufacturing procedures to meet alignment and other mechanical tolerances
- Less accurate position and rate sensors lead to less precise inner-loop control, necessitating more robust control architectures, unfamiliar operations concepts, or simply degraded performance
- Actuators may be on the same size scale as the spacecraft bus and therefore apply more disturbance (e.g., induced vibration and output-torque ripple) for smallsats
- Smallsats very rarely incorporate precise vibration isolation and have less mass with which to absorb vibration energy, leading to worse pointing performance
- Dynamics of the gimbal and other components that were once considered to be negligible are now more apparent. Without new, more detailed models and control approaches, they may lead to unexpected performance degradation

Miniature RWAs are by far the most common momentum control actuators for smallsats. The primary drawback of RWAs, their low torque, becomes less important for small spacecraft. Companies such as Microsat Systems Canada Inc. (MSCI, formerly Dynacon), Sinclair Interplanetary by RocketLab, AAC Clyde Space, Adcole Maryland Aerospace Incorporated, and Blue Canyon Technologies in addition to some universities have built such actuators. The promise of higher agility through RWAs at the small scale is a benefit.

Gimbal dynamics of small CMGs must be considered in modeling and control. Neglecting these dynamics produces inaccuracies when an inexact steering algorithm is used to find a solution to the gimbal rates to track a desired torque. Also, with the increase in gimbal inertia and the higher possible base rate of the smallsat using an array of oversized miniature CMGs, the CMG gimbal motors must offset the additional dynamics with high power. These effects lower accuracy and reduce torque amplification at the same time as shortening lifetime and increasing thermal challenges. For an in-depth derivation of the equations of motion for a spacecraft containing momentum systems with offset mass, as well as the effects of additional dynamics on the performance in terms of precision and torque amplification, see Appendix A.

The performance of small CMGs for smallsat applications suffers from these scaling issues. This is due mostly to the lack of high-performance components at this scale in the industry base. Disproportionately higher bearing friction in the rotors and gimbals, lower torque-per-input-power motors, and higher base rates of the smallsats all threaten to limit the performance of smallsats in the near future. Consider the assumptions from Appendix A corresponding to $|\Omega_i| >> |\dot{\delta}_i|, |\ddot{\delta}_i| >> |\omega^{\text{B/N}}|, |\dot{\omega}^{\text{B/N}}|$.

Consider the output torque of a SGCMG with constant rotor speed ($\dot{\Omega}_i = 0$):

$$\tau_o = -I_{r,i}\hat{\mathbf{o}}_i\dot{\delta}_i\Omega_i + I_{g,i}\ddot{\delta}_i \tag{4.111}$$

where $I_{r,i}\hat{\mathbf{o}}_i\dot{\delta}_i\Omega_i$ is the gyroscopic output torque,and $I_{g,i}\ddot{\delta}_i$ the gimbal shaft torque. This simple model represents a case in which the gimbal axis is perfectly aligned and in which a negligible rotor torque maintains the rotor speed. Defining an input torque (see Appendix A with assumptions for (A.25))

$$\tau_i = \tau_{Gf_i} + I_{g,i}\ddot{\delta}_i + I_{r,i}\Omega_i\hat{\mathbf{g}}_i \cdot (\omega^{\text{B/N}} \times \hat{\mathbf{s}}_i) \tag{4.112}$$

leads to an equation for the torque amplification of a smaller CMG:

$$\tau_a = \frac{\tau_o}{\tau_i} \approx \frac{|I_{r,i}\hat{\mathbf{o}}_i\dot{\delta}_i\Omega_i| + |I_{g,i}\ddot{\delta}_i|}{|\tau_{Gf_i}| + |I_{g,i}\ddot{\delta}_i| + |I_{r,i}\Omega_i\hat{\mathbf{g}}_i \cdot (\omega^{\text{B/N}} \times \hat{\mathbf{s}}_i)|} \tag{4.113}$$

Equation (4.112) may be even further simplified if the gimbal friction is negligible. With these assumptions in place,

$$\tau_a = \frac{|h_{r_i}\dot{\delta}_i| + |I_{g,i}\ddot{\delta}_i|}{|I_{g,i}\ddot{\delta}_i| + \|h_{r_i}\boldsymbol{\omega}^{\text{B/N}}\|} = \frac{|\dot{\delta}_i| + \frac{I_{g,i}}{h_{r_i}}|\ddot{\delta}_i|}{\|\boldsymbol{\omega}^{\text{B/N}}\| + \frac{I_{g,i}}{h_{r_i}}|\ddot{\delta}_i|} = \frac{|\dot{\delta}_i| + \varepsilon|\ddot{\delta}_i|}{\|\boldsymbol{\omega}^{\text{B/N}}\| + \varepsilon|\ddot{\delta}_i|} \tag{4.114}$$

The value ε is a performance parameter that indicates how closely the CMG's gimbal torque resembles RWA torque (but about the gimbal axis): $\varepsilon \rightarrow \infty$ if $h_{r_i} \rightarrow 0$ when the gimbal behaves as a RWA, and $\varepsilon \rightarrow 0$ if the rotor angular momentum approaches infinity or the gimbal inertia is negligible, approaching an ideal SGCMG.

As gimbal friction increases, torque amplification drops. Similarly, it is reduced when the magnitude of rotor angular momentum is small in comparison to the size of the gimbal inertia. In addition to these many difficulties, machining tolerances introduce disproportionately high errors in alignments and runout for rotating devices such as these, with associated degradation in spacecraft pointing. Challenges in fabricating small CMGs that meet the size, mass, power, and precision requirements of smallsats make it likely that the power benefits of torque amplification for such devices will be severely limited.

4.9 Chapter Summary

A momentum system offers unique opportunities to achieve robust and lightweight spacecraft designs, but taking advantage of these opportunities requires careful attention to rigid- and flexible-body dynamics. This chapter provides the physical basis for the motion of reaction wheels and CMGs in the context of spacecraft that use them for attitude control. Rigid-body dynamics captures most of the behaviors of interest in these momentum devices at a level that permits sizing and preliminary assessments of space-vehicle performance. Accordingly, the chapter focuses on rigid-body models of them. It does so through a very general vector/dyadic representation of the physics. This coordinate-free approach offers compact and unambiguous derivations. Common coordinate systems are also presented. Some flexible effects are considered, specifically the mechanical compliance in the rotor and related structure, which produces coupled flexible/gyroscopic modes and can retard the gimbal acceleration in the form of an effective inertia. While these effects are well understood for large-scale devices, CMG technology for smallsats has not yet matured to the point where these many issues are straightforward.

4.10 Exercises

1. A rigid spacecraft has one single-gimbal CMG with a constant rotor angular momentum magnitude $h > 0$. The CMG's momentum vector depends on the gimbal angle δ such that $\mathbf{h} = h\,(\sin(\delta)\mathbf{p}_1 + \cos(\delta)\mathbf{p}_2)$ where the $\mathbf{p}_i$ are principal axes for the spacecraft's inertia dyadic, $\mathbf{I}$, taken to be spherical (same moment of inertia for any axis).
   Assume the following:
   - The CMG's rotor is spherical (in the sense of moments of inertia). Its center of mass is constant in the $\mathbf{p}_i$ axes.
   - The CMG's gimbal is massless and has no inertia.
   - The CMG's inertia is already accounted for in $\mathbf{I}$.
   - Let $\Omega = \left\|\omega^{\mathrm{B/N}}\right\|$.

- $\omega^{B/N}$ is nonzero.

For the case of a constant gimbal angle δ, write a general expression for $\omega^{B/N}$ when this system is in a relative equilibrium. Write this expression in terms of Ω, $\omega^{B/N}$, the $\mathbf{p}_i$, and δ.

2. The XSS-15 spacecraft's inertia matrix is

$$^{B}I = \begin{bmatrix} 400 & 0 & 0 \\ 0 & 440 & 0 \\ 0 & 0 & 600 \end{bmatrix} \text{kg-m}^2$$

Its angular velocity in body-fixed basis vectors is

$$\omega^{B/N} = \begin{bmatrix} 0.1 \\ -0.1 \\ 0.5 \end{bmatrix} \text{rad/s}$$

It houses a single momentum wheel whose angular momentum in body-fixed basis vectors is

$$^{B}h = \begin{bmatrix} 0 \\ 0 \\ 30 \end{bmatrix} \text{Nms}$$

There is no nutation damping.

(a) What is its nutation angle at this instant?
(b) What is its wobble angle at this instant?
(c) What is its cone angle at this instant?
(d) What is the magnitude of the system angular-momentum vector?
(e) What is the maximum cone angle the spacecraft will ever experience?
(f) What is the minimum cone angle the spacecraft will ever experience?
(g) What is the nutation frequency?
(h) Is the relative equilibrium nearest to $\mathbf{b}_3$ for $-\,^{B}h$ stable?

3. (Include a picture of a spacecraft showing coordinate axes) This spacecraft performs a versine slew about the $\mathbf{b}_3$ axis. The total angular motion about this axis is 12°. The maneuver lasts 10 s. Which of the following is an expression for the total angle vs. time for the duration of this maneuver, neglecting any errors in sensing, actuation, or mass-properties knowledge?

(a) $\phi(t) = 10\left(1 - \cos(\frac{\pi}{5}t)\right)$
(b) $\phi(t) = 6\left(1 - \cos(\frac{\pi}{10}t)\right)$
(c) $\phi(t) = 12\left(1 - \cos(\pi t)\right)$
(d) $\phi(t) = 3\left(1 - \cos(\frac{\pi}{12}t)\right)$

4. (Include a diagram showing vectors involved) What is the minimum, constant, wheel angular momentum ${}^{\mathrm{B}}h$ that enables the following spacecraft to spin about the $\mathbf{b}_3$ axis without nutation at 10 RPM?

$$^{\mathrm{B}}I = \begin{bmatrix} 300 & 2 & 5 \\ 2 & 450 & -50 \\ 5 & -50 & 500 \end{bmatrix} \text{kg-m}^2$$

5. (Include a diagram showing vectors involved) What is the minimum, constant, wheel angular momentum ${}^{\mathrm{B}}h$ that enables the following spacecraft to spin about the $\mathbf{b}_3$ axis with an effective inertia ratio of 1.2?

$$^{\mathrm{B}}I = \begin{bmatrix} 200 & 0 & 0 \\ 0 & 200 & 0 \\ 5 & 0 & 210 \end{bmatrix} \text{kg-m}^2$$

6. (Include a diagram showing vectors involved) What is the minimum, constant, wheel angular momentum ${}^{B}h$ that enables a 100 kg, 1 m radius, uniform, spherical spacecraft to spin stably with an angular-velocity vector $\omega^{\mathrm{B/N}} = 1\mathbf{b}_1 + 2\mathbf{b}_2$ rad/s?
7. (Include a diagram showing vectors involved) A spacecraft stably spinning about its $\mathbf{b}_2$ axis at $\Omega = 2$ RPM without active control does so thanks to constant wheel angular momentum $\mathbf{h} = 3\mathbf{b}_1$. What are I_{12} and I_{23}?
8. In the following problem, the basis vectors $\mathbf{b}_1$, $\mathbf{b}_2$, and $\mathbf{b}_3$ are unit vectors fixed in the spacecraft (or, equivalently, fixed in a frame B that rotates with the spacecraft). You may use an engineering reference of some sort to look up the moments of inertia of the rigid bodies described. Unless stated otherwise, each body is assumed to be rotating freely (i.e. about its center of mass) with no externally applied torque.
   Consider a box-shaped spacecraft of uniform density spinning with an angular velocity of $\omega^{\mathrm{B/N}} = 0.1\mathbf{b}_1 + 0.2\mathbf{b}_2 + 0.3\mathbf{b}_3$ rad/s. This rectangular prism's length, measured along $\mathbf{b}_1$, is 1 m. Its width, measured along $\mathbf{b}_2$, is 1.5 m. It is 2 m high. Its mass is 1500 kg. What is the angular momentum vector of this spacecraft in terms of the $\mathbf{b}_i$ basis vectors?
9. Consider a cylindrical spacecraft of uniform density spinning with an angular velocity of $\omega^{\mathrm{B/N}} = 0.1\mathbf{b}_1 + 0.2\mathbf{b}_2 + 0.3\mathbf{b}_3$ rad/s. The cylinder's axis of symmetry is parallel to the $\mathbf{b}_3$ axis. Its height is 1 m; its diameter is 1 m, and its mass is 200 kg. What is the angular momentum vector of this spacecraft in terms of the $\mathbf{b}_i$ basis vectors?
10. Consider a 1 m diameter sphere of uniform density $\rho = 1000\,\text{kg/m}^3$ with an angular velocity of $\omega^{\mathrm{B/N}} = 0.1\mathbf{b}_1 + 0.2\mathbf{b}_2 + 0.3\mathbf{b}_3$ rad/s. What is the angular momentum vector of this spacecraft in terms of the $\mathbf{b}_i$ basis vectors?

11. The magnitude of a uniformly dense sphere's angular velocity vector is 1 rad/s. The magnitude of its angular momentum vector is 2 rad/s. What is the inertia matrix of this sphere?
12. Consider a cylindrical spacecraft of uniform density. The cylinder's axis of symmetry is parallel to the $\mathbf{b}_3$ axis. Its height is 1 m; its diameter is 1 m, and its mass is 200 kg. What is the nutation frequency of the cylinder if its angular velocity vector at a certain instant is rad/s? What is the nutation frequency of the cylinder if its angular velocity vector at a certain instant is $\omega^{B/N} = 0.1\mathbf{b}_1 + 0.1\mathbf{b}_2 + 0.5\mathbf{b}_3$ rad/s?
13. What is the nutation frequency of a uniformly dense sphere?
14. Consider a cylindrical spacecraft whose axis of symmetry is aligned with $\mathbf{b}_3$. Telemetry indicates that its spin rate is $\Omega = \omega^{B/N} \cdot \mathbf{b}_3 = 0.75$ rad/s. Its nutation frequency is measured to be 0.1 rad/s. Its spin moment of inertia is known to be 1200 kg-m^2. What is its transverse moment of inertia?
15. The magnitude of a uniformly dense sphere's angular velocity vector is 1 rad/s. The magnitude of its angular momentum vector is 2 rad/s. What is the angular acceleration vector of this sphere?
16. Compute the angular acceleration vector of a spacecraft whose angular velocity is $\omega^{B/N} = 0.2\mathbf{p}_1 - 0.2\mathbf{p}_2 - 0.2\mathbf{p}_3$ rad/s and whose total angular-momentum vector is $\mathbf{h} = 9\mathbf{p}_1 - 12\mathbf{p}_2 - 15\mathbf{p}_3$ Nms, where the $\mathbf{p}_i$ represent the principal axes of the inertia matrix.
17. Compute the angular velocity vector of a spacecraft whose total angular momentum is $\mathbf{H} = 200\mathbf{b}_1 + 300\mathbf{b}_2 + 1000\mathbf{b}_3$ Nms and whose inertia matrix is

$$^{B}I = \begin{bmatrix} 3000 & -40 & 50 \\ -40 & 4000 & -60 \\ 50 & -60 & 5000 \end{bmatrix} \text{kg-m}^2$$

There's no need to perform this calculation by hand, but at least show the steps if you use a computer.
18. Write down the principal axes of inertia $\mathbf{p}_1$, $\mathbf{p}_2$, $\mathbf{p}_3$ and the principal inertia matrix ^{P}I for

$$^{B}I = \begin{bmatrix} 300 & 0 & 0 \\ 0 & 450 & -50 \\ 0 & -50 & 450 \end{bmatrix} \text{kg-m}^2$$

19. Compute the nutation frequency of a spacecraft in stable, equilibrium spin at $\Omega = 1$ RPM and whose inertia matrix is

$$^{B}I = \begin{bmatrix} 55 & 0 & 5\sqrt{3} \\ 0 & 50 & 0 \\ 5\sqrt{3} & 0 & 65 \end{bmatrix} \text{kg-m}^2$$

20. (Include a figure here) Compute the mass, center of mass (in terms of the $\mathbf{b}_i$ basis vectors), and centroidal inertia matrix (in terms of the bi basis vectors) of a spacecraft consisting of the following two components: The spacecraft bus, represented by a uniformly dense 200 kg sphere of radius R = 0.2 m whose mass center is located at $\mathbf{r}_s = 0.01\mathbf{b}_1 + 0.02\mathbf{b}_2 + 0.2\mathbf{b}_3$ m from the center of the spacecraft/launch-vehicle interface plane.
    The spacecraft payload, represented by a point mass of 50 kg located at $\mathbf{r}_p = -0.04\mathbf{b}_1 - 0.08\mathbf{b}_2 + 0.4\mathbf{b}_3$ m from the center of the spacecraft/launch-vehicle interface plane.
21. Which one of the following matrices can be a direction-cosine matrix (i.e. it meets all the requirements for being a member of the algebraic group SO(3), the right-handed, orthonormal 3×3 matrices)?

    (a) $Q = [1\ 0\ 0;\ 0\ 1\ 0;\ 0\ 0\ -1]$
    (b) $Q = [1/3\ 1/3\ 8/9;\ 2/3\ 2/3\ 0;\ 2/3\ -2/3\ 4/9]$
    (c) $Q = [-1\ 0\ 0;\ 0\ 1\ 0;\ 0\ 0\ -1]$
    (d) $Q = [1\ 0\ 0;\ 0\ \sqrt{3}/2\ 1/2;\ 0\ 1/2\ \sqrt{3}/2]$

Chapter 5
Singularities of Control Moment Gyroscopes

A book on spacecraft momentum control systems would not be complete without a discussion of kinematic singularities inherent in CMGs. Therefore, this chapter starts with a discussion of common singularities associated with DGCMGs (e.g., gimbal lock) and then moves on to the more troublesome singularities associated SGCMGs. The mathematical structure of these singularities and how it relates to difficulties in avoiding them is also discussed in a way made more accessible to the reader not familiar with the subject matter. Also discussed is the location of singularities associated with an array of CMGs by way of a three-dimensional surface in the momentum space. This chapter concludes with some brief discussion on techniques to perform zero-momentum spin up of an array of CMGs in the presence of singularities at zero momentum.

5.1 Coordinate Singularities Versus Kinematic Singularities

There exist purely mathematical coordinate singularities i.e., those associated with a possibly poor choice of parameterization. For example, a parameterization that involves tangent, secant, or cosecant functions of the gimbal angles becomes undefined at certain values of gimbal angle regardless of the physics. There are also kinematic singularities that are fundamentally physical and must show up in any mathematical representation of the array's behavior. The differences between these two situations are discussed here.

© The Author(s), under exclusive license to Springer Nature Switzerland AG 2025
F. A. Leve et al., *Spacecraft Momentum Control Systems*, Space Technology Library 46, https://doi.org/10.1007/978-3-031-96165-6_5

5.1.1 Coordinate Singularities

Coordinates describe the state or configuration of a dynamical system. For example, angles and positions serve as coordinates that specify the configuration of many mechanical devices. Gimbal angles typically serve this purpose for an array of CMGs. The angular coordinate associated with rotation of the rotor is often ignored as a cyclic variable, a variable that constantly changes in a way that can be captured with a static parameter, such as scalar angular momentum h_r. Coordinate singularities can occur as the configuration changes and reaches a point where a mathematical function of the coordinates changes infinitely fast or is otherwise undefined. Such singularities arise because of the choice of coordinates. An analyst or a systems engineer makes this choice, for reasons related to convenience or legacy definitions of the hardware. Therefore, the choice is arbitrary with regard to the physics of the problem and can be altered. Specifically, singular sets of parameters may be replaced with non-singular parameter sets that still describe the same system.

Coordinate singularities can be found in the kinematic differential equations that propagate the system's evolution as a function of time or in expressions that relate these coordinates to others (see Chap. 4). Some specific examples follow.

Example 5.1 Consider the case of a three-coordinate attitude representation known as Euler angles that describes the attitude of one set of basis vectors relative to another. This attitude can be described by a sequence of three rotations through non-coincident axes. An example is the popular 3-2-1, i.e. yaw-pitch-roll rotation sequence. The sequence begins with a rotation ψ about the Z (or 3) axis, followed by a rotation θ about y"', and finally by a rotation ϕ about an axis x" to convert from a coordinate basis consisting of the X,Y,Z axes to a coordinate basis in x',y',z' as shown in Fig. 5.1.

The three direction-cosine matrices (DCMs) that make up the rotation sequence in Fig. 5.1 are

$$D_3(\psi) = \begin{bmatrix} c(\psi) & s(\psi) & 0 \\ -s(\psi) & c(\psi) & 0 \\ 0 & 0 & 1 \end{bmatrix},$$

$$D_2(\theta) = \begin{bmatrix} c(\theta) & 0 & -s(\theta) \\ 0 & 1 & 0 \\ s(\theta) & 0 & c(\theta) \end{bmatrix},$$

(continued)

$$D_1(\phi) = \begin{bmatrix} 1 & 0 & 0 \\ 0 & c(\phi) & s(\psi) \\ 0 & -s(\phi) & c(\phi) \end{bmatrix},$$

where c(·)= cos(·) and s(·) = sin(·). The product of three DCMs is itself a DCM, because of the algebraic properties of the group of rotation matrices:

$$D_{321} = D_1(\phi)D_2(\theta)D_3(\psi),$$

or expanded as

$$D_{321} = \begin{bmatrix} c(\theta)c(\psi) & c(\theta)s(\psi) & -s(\theta) \\ s(\phi)s(\theta)c(\psi) - c(\phi)s(\psi) & s(\phi)s(\theta)s(\psi) + c(\phi)c(\psi) & s(\phi)c(\theta) \\ c(\phi)s(\theta)c(\psi) + s(\phi)s(\psi) & c(\phi)s(\theta)s(\psi) - s(\phi)c(\psi) & c(\phi)c(\theta) \end{bmatrix},$$

where this 3 × 3 matrix relates a matrix of the final ' axes to a matrix of the initial axes as follows:

$$[\text{x' y' z'}] = D_{321}[\text{X Y Z}].$$

Solving for the angles from this matrix is where the singularity can occur. The coordinate ψ is determined from the arctangent (inverse tangent) of the ratio $D_{321}(1,2)/D_{321}(1,1)$:

$$\psi = \tan^{-1}\left(\frac{D_{321}(1,2)}{D_{321}(1,1)}\right), \tag{5.1}$$

and

$$\phi = \tan^{-1}\left(\frac{D_{321}(2,3)}{-D_{321}(3,3)}\right). \tag{5.2}$$

If either of the divisors in Eqs. (5.1) and (5.2) is zero, the value of ψ and/or ϕ is/are undefined. Therefore, singularities exist in the extraction of Euler angles from other parameter sets, in this case from DCMs. Stuelpnagel [26] explains that any three-parameter attitude representation must be singular, non-global or ambiguous. Any use of a three-parameter attitude representation risks introducing singularities.

Consider the kinematic differential equations of the 3-2-1 Euler sequence

(continued)

$$\begin{bmatrix} \dot{\phi} \\ \dot{\theta} \\ \dot{\psi} \end{bmatrix} = \frac{1}{\mathrm{c}(\theta)} \begin{bmatrix} 0 & -\mathrm{s}(\psi) & \mathrm{c}(\phi) \\ 0 & \mathrm{c}(\phi)\mathrm{c}(\theta) & -\mathrm{s}(\phi)\mathrm{c}(\theta) \\ \mathrm{c}(\theta) & \mathrm{s}(\phi)\mathrm{s}(\theta) & \mathrm{c}(\phi)\mathrm{s}(\theta) \end{bmatrix} \begin{bmatrix} \omega_1 \\ \omega_2 \\ \omega_3 \end{bmatrix}, \tag{5.3}$$

where ω_i are the components of the angular velocity ω of one reference frame relative to another. Clearly, the kinematic differential equation in Eq. (5.3) is singular when $\theta = (2n+1)\frac{\pi}{2}, n = 0, 1, \ldots, N$. Therefore, coordinate singularities may be singular not only in their extraction from other coordinates but also in their kinematic differential equations used for propagating attitude.

Other examples of singular sets of attitude parameterizations are the Gibbs vector and Modified Rodrigues parameters. At least four coordinates are needed for an attitude description to be singularity free (Shuster [27], Stuelpnagel [26]).

This example illustrates that coordinate singularities can arise from how we chose to describe a system's kinematics. Furthermore, coordinate singularities can be avoided with a different choice of parameters. The next section shows that singularities directly associated with the system's physical behavior, not merely coordinate singularities, cannot be avoided with a change of parameters.

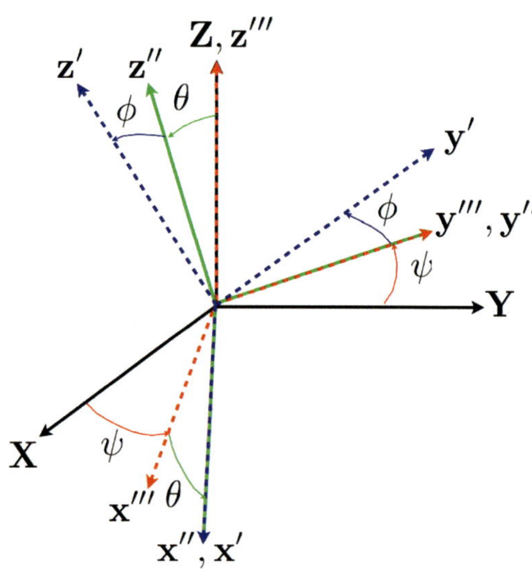

Fig. 5.1 Euler rotation sequence

5.1.2 Singularities Associated with Kinematic Constraints

Singularities associated with geometric constraints (e.g., limited joint motion) are independent of the choice of coordinates. These singularities are fundamental products of the dynamical system's physics and cannot be avoided through mere mathematics. Caution should be exercised when dealing with systems that exhibit these singularities because coordinates that seem to describe a singularity-free system must be incorrect; the underlying physics of this system is lost in such an attempt, and any such description of its dynamics is incorrect.

Some of the earliest literature on kinematic singularities comes from that associated with robotic manipulators and their forward and inverse kinematics.

Example 5.2 A robotic manipulator system shown in Fig. 5.2 is represented as

$$\dot{x} = \frac{\partial x}{\partial q}\dot{q} = A(q)\dot{q},$$

where x is a column matrix of coordinates for the end-effector position and orientation relative to a fixed frame, $\dot{x}$ is a column matrix of the rates (or coordinate derivatives), $A(q)$ is the Jacobian matrix as a function of the configuration space, q is a column matrix consisting of joint angles, and $\dot{q}$ is a column matrix of joint velocities. Typically, A is nonsquare and therefore noninvertible. The joint velocities $\dot{q}$ corresponding to a desired $\dot{x}$ may be found through the Moore-Penrose pseudoinverse. In what follows, the rank of A is assumed to be the min $\{\dim(x), \dim(q)\}$.

This inverse provides the minimum two-norm solution, i.e. the solution that lies entirely in the range space of A. This solution minimizes the joint rates, which is likely a preferred outcome for reasons of power and mechanical design. Motion in the null space of A does not influence $\dot{x}$; therefore, null-space motion carries electromechanical cost with no performance benefits. A solution found via this pseudoinverse can also be interpreted as the minimum-error solution in the case of a least-squares estimate.

The case of $\dim(x) < \dim(q)$ corresponds to an over-actuated robotic system. For A full-rank, Moore-Penrose inverse solution for this case is

$$\dot{q} = A^{\mathrm{T}}(AA^{\mathrm{T}})^{-1}\dot{x}.$$

Otherwise, if dim(x) > dim(q), i.e. the system is underactuated,

$$\dot{q} = (A^{\mathrm{T}}A)^{-1}A^{\mathrm{T}}\dot{x}.$$

(continued)

For A full rank, the special case where dim(x) = dim(q), the familiar inverse of a square matrix provides the solution:

$$\dot{q} = A^{-1}\dot{x}$$

A is rank deficient when its rank is less than dim(x). Specifically, rank$(A) < 6$ for translation and rotation of a rigid end effector in a typical robotic arm. In such cases, the robotic system contains kinematic singularities. In such cases, the inverse that appears in the Moore-Penrose pseudoinverse is singular. The Moore-Penrose inverse always exists, but it can only be put in the matrix form above when the matrix A is of full rank.Therefore, the shown version of the Moore-Penrose inverse is invalid. However, in this case Moore-Penrose inverse can still be formed by methods such as singular value decomposition (SVD). Please see the Appendix C for more details. An example of such a singularity is shown in Fig. 5.2 where the robot manipulator is maximally extended.

Consider the three joint angles of the planar robotic manipulator Fig. 5.2, θ_1, θ_2, and θ_3 relative to the horizontal. Assume that its link lengths are equal to 1 and that only the position and velocity of the end effector are known. Then the position of the end effector is related to the joint angles by

$$r = \begin{bmatrix} \cos(\theta_1) + \cos(\theta_2) + \cos(\theta_3) \\ \sin(\theta_1) + \sin(\theta_2) + \sin(\theta_3) \end{bmatrix},$$

and the velocity of the end effector is

$$\dot{r} = \begin{bmatrix} -\sin(\theta_1) & -\sin(\theta_2) & -\sin(\theta_3) \\ \cos(\theta_1) & \cos(\theta_2) & \cos(\theta_3) \end{bmatrix} \begin{bmatrix} \dot{\theta}_1 \\ \dot{\theta}_2 \\ \dot{\theta}_3 \end{bmatrix} = A\dot{\Theta}.$$

With the manipulator maximally extended, $\theta_1 = \theta_2 = \theta_3 = 0$, and

$$\dot{r} = \begin{bmatrix} 0 \; 0 \; 0 \\ 1 \; 1 \; 1 \end{bmatrix} \dot{\Theta}.$$

Clearly, this configuration is singular, i.e., where rank$(A) = 1$, of the robotic manipulator with singular direction, $u = \pm[1 \; 0]^{\mathrm{T}}$. This singularity is analogous to the external or saturation singularities for CMGs discussed in subsequent sections.

A spacecraft with CMGs is a special case of a robotic system of this type, as observed by Bedrossian et al. [31], who relates the singularities associated with

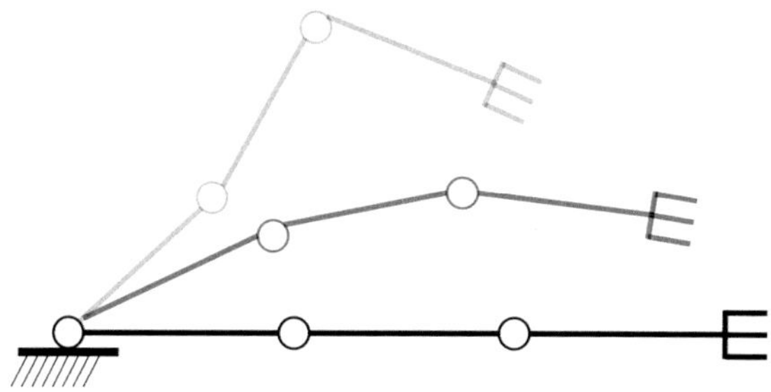

Fig. 5.2 Three-link robot manipulator at singular configuration of maximum extension

robotic manipulators to those associated with SGCMGs. Kinematic singularities associated with CMGs are discussed below.

5.2 Control Moment Gyroscope Singularities

The Jacobian matrix of an array of RWAs is constant in the spacecraft body. In fact its geometry is decided well before launch. The CMG Jacobian matrix, in contrast, varies with the gimbal angles of the CMGs and therefore can become singular as the CMG gimbals rotate in flight (see Chap. 4). When the CMG Jacobian matrix is singular, one or more directions exist where no torque is available at that instant. The singular direction consists of a single vector when the Jacobian matrix is rank 2—a so-called rank 2 singularity. It consists of a plane of singular directions when the Jacobian matrix is of rank 1—a rank 1 singularity. However, redundant arrays of CMGs offer additional null solutions or null motion that may provide a means to skirt the singularity while maintaining the required angular momentum.

As the Jacobian matrix approaches a singularity, the gimbal rates necessary to produce a finite torque approach infinity. Near the singularity, the finite capability of the hardware makes the array incapable of producing torque in one or more directions. A loss of control of the vehicle would follow. So, not only is a CMG array singular when the Jacobian matrix is rank deficient, the array is also effectively unusable for a finite region around these points.

As an example, consider the momentum vectors of a planar array of three SGCMGs as shown in Fig. 5.3. All three gimbal axes are parallel and point out of the page. This array is not of practical interest on its own, but it helps illustrate the principle of interest here. The array is shown in a zero-momentum state. By rotating the three gimbals the spacecraft can redirect the net **h** vector anywhere in this plane. Figure 5.4 shows how the gimbals can be rotated to produce maximum momentum in either the X or Y directions.

The three CMGs combine to produce at most $3h$ momentum in the desired direction. When the array reaches this $3h$ state, it is impossible to increase the net $|h|$ any further. This condition is known as a "saturation" singularity. Like any physically realizable mechanical system, the hardware comprising the array

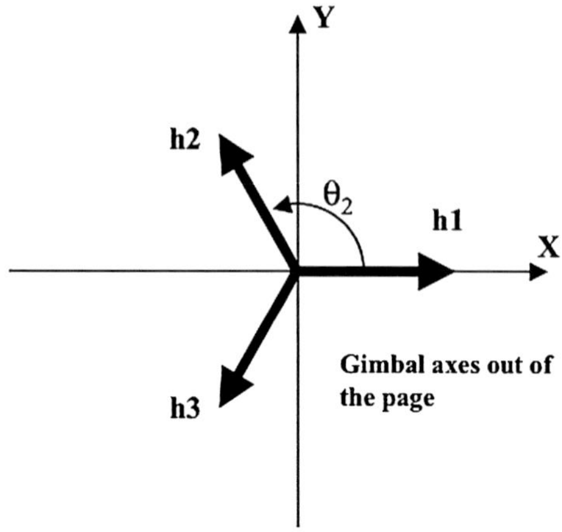

Fig. 5.3 Momentum vectors of 3 CMGs with parallel gimbal axes

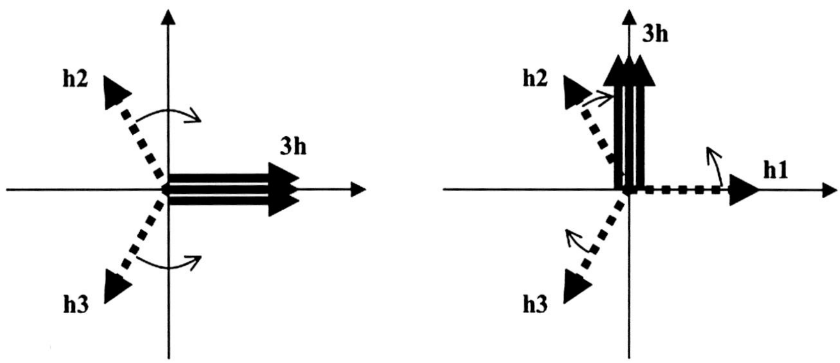

Fig. 5.4 Maximum momentum states of the array (saturation)

has limits, specifically a maximum gimbal rate. To avoid unexpected consequences of the reduction in torque authority (or analogously loss of feedback-control gain) that arises as the array approaches saturation, the attitude-control design must incorporate margin between the required momentum and the array's limited capability (see Chap. 8).

A more subtle situation occurs when the array is commanded to produce net momentum in the -X direction, h_X (see Fig. 5.5). A simple steering algorithm would likely seek to change the gimbal angles δ_2 and δ_3 in the correct direction, but since

$$\frac{\partial h_X}{\partial \delta_1} = 0 \tag{5.4}$$

at this point, the algorithm would deduce that there is no value in changing δ_1. The array momentum is now only $1h$ in magnitude, but there is no way to produce any more without producing error in the Y-axis. Specifically, no incremental change in gimbal angles increases h in the desired direction. This situation is called an

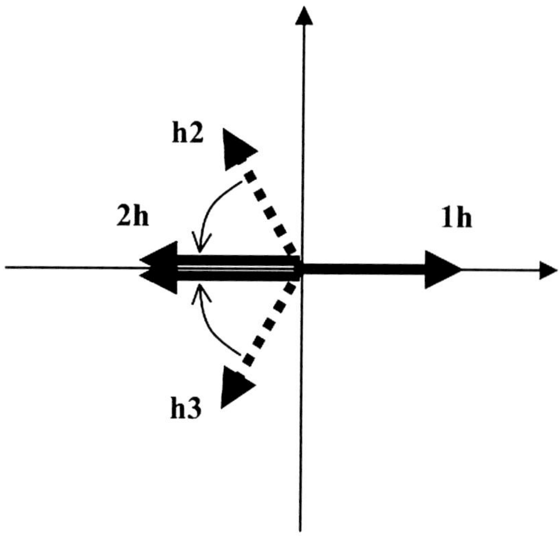

Fig. 5.5 A singular state of the array (internal)

Table 5.1 Singularity regions associated with the number of CMG in an array

CMG number in array	Singularity region
2	$0h$,$2h$
3	$1h$,$3h$
4	$0h$,$2h$,$4h$
5	$1h$,$3h$,$5h$
6	$0h$,$2h$,$4h$,$6h$

"internal singularity" because it occurs within, not on, the saturation boundary of the available momentum volume. When one of the CMGs opposes the rest of the array in this fashion, the problematic CMG is known as a "hung gyro." The avoidance of such singularities has been the subject of considerable research over the years.

It has been shown that an array of three SGCMGs has singularities at $1h$ and $3h$, an array of four has singularities at $0h$, $2h$ and $4h$, an array of five at $1h$, $3h$, and $5h$, and so on. This is discussed further in Sect. 6.5.2, and Meffe and Stocking [32]. (Unlike this simple planar example, the so-called "nh" singular surface of most CMG arrays does not have magnitude nh). Table 5.1 lists the combinations of singular surface regions for the number of CMGs and their associated regions.

Conventional wisdom has established a safe strategy: use a three-SGCMG array to avoid the singularity at $0h$ and size the momentum system such that attitude control never requires more than $1h$ from the array (see 3/4 box array in Chap. 6). In this way, the entire singularity problem is neatly sidestepped. However, this book argues against restricting the design to a $1h$ sphere of performance. Singularity-avoidance algorithms enable arrays of four or CMGs to achieve considerably larger momentum envelopes and/or reduce the size, weight, and power of the momentum system (see Chap. 7).

5.3 Double-Gimbal Control Moment Gyroscope Singularities

Double-gimbal control moment gyroscopes (DGCMGs), shown in Fig. 5.6, are among the most mechanically complex CMGs.

This complexity is due to their multiple gimbals and the additional drive electronics, slip rings, and sensors needed for their use. Severe performance limitations or mass penalties arise from the physics of these nested gimbals. There is also mathematical complexity with implications for flight-software development. The columns of the Jacobian matrix for DGCMGs contain n vectors that are not in general unit-norm. In addition, parts of the Jacobian matrix associated with the inner and outer gimbal are coupled functions of the inner and outer gimbal angles (see Chap. 4). Nevertheless, these mathematical complexities do not compromise the usability of such an array.

The kinematic singularities associated with an array of DGCMGs are much like those for a SGCMGs array with one exception—an array with three or more DGCMGs includes only passable internal singularities (Kurokawa [30], Leve [33]). Passable singularities are easily handled by one of various algorithms that provide null motion for singularity avoidance (see Chap. 9). Therefore, the full momentum volume is available in a singularity-free manner (i.e., modulo the points on the boundary that contribute zero volume).

The most familiar of the singularities associated with DGCMGs is gimbal lock. Figure 5.7 shows this configuration. Gimbal lock for DGCMG occurs when the rotor spin axis aligns with the outer gimbal. With the gimbals in this configuration, the outer gimbal can no longer redirect the momentum vector. Mathematically, gimbal lock represents a loss of one control degree of freedom. In this configuration, the behavior of the DGCMG degrades to that of a SGCMG using the inner gimbal.

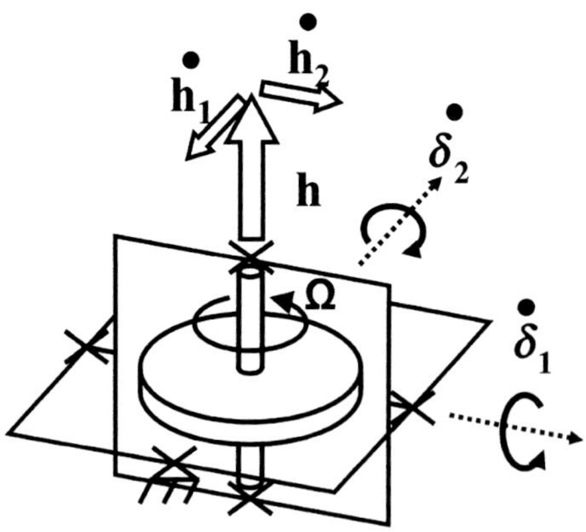

Fig. 5.6 Double gimbal control moment gyroscope

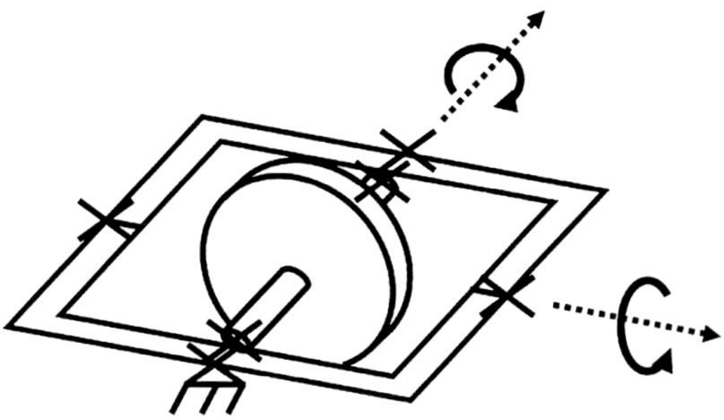

Fig. 5.7 Gimbal lock in a DGCMG

5.4 Single-Gimbal Control Moment Gyroscope Singularities

Typically four or more devices (RWAs or SGCMGs) are used for attitude control of a spacecraft. Such an array is known as "redundant." In the case of RWAs, the redundancy is usually added for fault tolerance. The choice of a redundant SGCMG array offers an opportunity to avoid internal singularities, thereby increasing the momentum and torque capability of the array in a mass-efficient way. For all redundant arrays, the Jacobian matrix A, is nonsquare and thus noninvertible.

Unlike coordinate singularities, kinematic singularities inherent in CMG arrays are not only mathematical but also physical phenomena—there is physically no torque available in a singular direction or plane of singular directions. In addition, external singularities correspond to the maximum allowable angular momentum in any single direction. It is important to recognize that a CMG array's singularities are physical because merely preventing a numerical divide-by-zero error is insufficient to address the problem. Instead, null motion is required.

The effectiveness of avoiding internal singularities through null motion depends on the form of the singularity. This chapter classifies them as hyperbolic or elliptic. However, in the analysis of CMG singularities, the direction of the singularity is of the utmost importance: if the desired torque has components orthogonal to the singular direction, a proper steering algorithm could exploit this for singularity avoidance.

The case where all components of requested torque are along the singular direction is also known as gimbal lock. It is impassable when a null motion solution does not exist at the singularity. Unlike gimbal lock for a DGCMG, gimbal-lock for an array of SGCMGs occurs at any singularity for which

$$\dot{\Delta} = A^{+}\dot{h} = 0, \;\dot{h} \neq 0,$$

where A^{+} is the SVD computation of the Moore-Penrose inverse since A is assumed rank deficient, and Δ is a matrix of the gimbal rates $\dot{\delta}_i$. When gimbal lock occurs, the system is trapped in this singular configuration. Both hyperbolic and elliptic singularities can encounter gimbal lock depending on the commanded torque.

However, unlike elliptic singularities, hyperbolic singularities offer the possibility of avoiding gimbal lock.

5.5 Classification of Singularities

Two distinct types of singularities exist, with different classes within them. As discussed previously, singularities for SGCMGs can be classified into the groups/subgroups shown in Fig. 5.8.

As discussed above, saturation singularities at the limit of the array capability are common to any array. Such singularities, also known as *external* singularities, are accommodated in the sizing of the control system and are not germane to the discussion of singularity avoidance.

Another class of singularities called *internal* singularities are of much greater importance. The array in Fig. 5.9 offers a practical example. For purposes of discussion, the "shared direction" for this roof array (as discussed earlier) is aligned with the Y axis of a reference coordinate system.

Suppose that the gimbals in each plane were positioned so the momentum vectors (red) were opposite one another, one pointing in the +Y direction, the other in −Y, as shown. All the momentum vectors cancel. The array momentum is zero. No incremental motion of any of the four gimbals can change momentum along the Y axis. Therefore, no collection of gimbal rates can produce a Y torque from

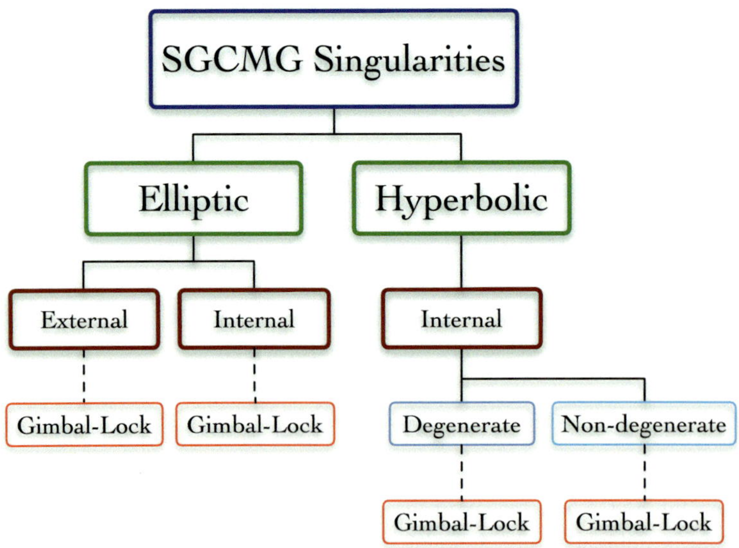

Fig. 5.8 Classification of SGCMG singularities

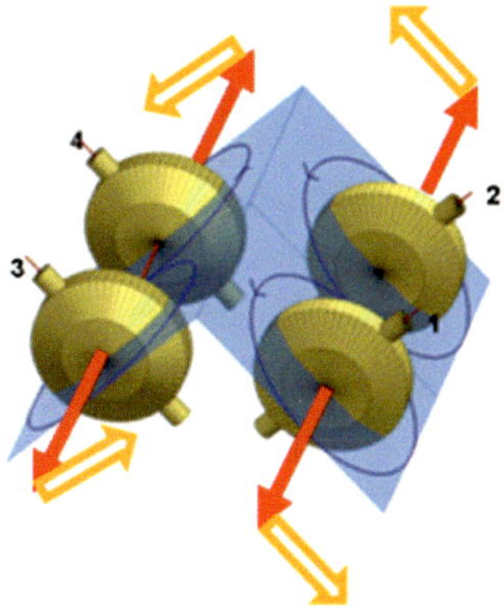

Fig. 5.9 Shared direction singularity configuration

this configuration. This is an internal singularity—no torque available, and yet the momentum state is far from the limit of the array's capability.

If only three CMGs are operating, there is a 1:1 map between gimbal angle sets and momentum states on $\mathbb{R}^3$. Therefore, any internal singularity corresponds to a single momentum state, and the only way to avoid them is not to go there. This absence of alternatives severely restricts how much of the momentum envelope can be used in practice. Thus a three-CMG array often uses its capability inefficiently.

However, if four or more CMGs are operating, the same momentum state can be achieved through infinite combinations of gimbal angles, some of which may be non-singular. This opens the door to creativity in choosing how to steer the gimbals to avoid singularities.

5.6 Singularity Conditions Defined Mathematically

Typically, differential geometry is used to represent hyperbolic and elliptic internal singularities as surfaces or manifolds. Margulies and Aurbrun [34] and Tokar and Platonov [35] were the first to use differential geometry for analysis of kinematic singularities associated with SGCMGs. Margulies and Aubrun derived singularity type conditions from a differential-kinematic singularity condition in the form of a quadratic product. Their results gave the names elliptic and hyperbolic to the kinematic singularities associated with SGCMGs. Kurokawa [30] and Bedrossian et al. [36] compared the kinematic singularities of SGCMGs to those of robotic manipulators, and set up the null motion condition from Margulies [34] to consider the possibility of null motion at singularities. Bedrossian et al. derived a method of determining if a hyperbolic singularity is or isn't degenerate for a specific set of gimbal angles. Wie [37] extended the results of Margulies-Aubrun and Bedrossian et al. to show an example of a degenerate hyperbolic singularity contained at the zero-momentum state of a single scissored pair of SGCMGs. Kurokawa [30] investigated the singularities for two classes of arrays—those with parallel gimbal axes and

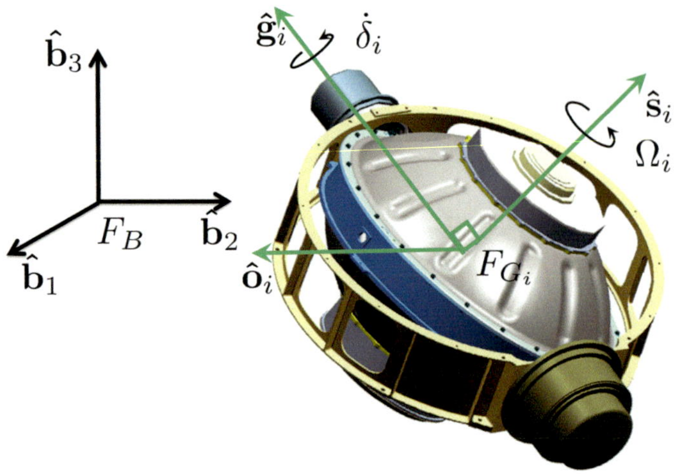

Fig. 5.10 CMG coordinate system fixed in the spacecraft body

those without. Through differential geometry and principles of curvature, he showed the properties of singularities for parallel and independent type arrays of up to six SGCMGs. Leve [33], defined different categories of passable and impassable singularities based off the topology calculated in the form of homological invariants of regions of null-space solutions. Rather than rely on the sophisticated mathematical tools of differential geometry, and algebraic topology, we can use linear algebra to provide insight into the behavior of internal singularities. We use the methods presented in reference [36] to define singularity mathematically.

It is customary to define an orthonormal basis $\{\hat{\mathbf{s}}_i, \hat{\mathbf{o}}_i, \hat{\mathbf{g}}_i\}$ shown in Fig. 5.10, where $\hat{\mathbf{s}}_i$ is the spin axis of the flywheel, $\hat{\mathbf{o}}_i$ is the SGCMG torque direction, and $\hat{\mathbf{g}}_i$ is the gimbal axis direction of a SGCMG.

A Taylor series expansion of the SGCMG angular momentum about a singular configuration, to second order, gives

$$h(\Delta) - h(\Delta^S) = \sum_{i=1}^{n} \frac{\partial h_i}{\partial \delta_i}\bigg|_{\delta_i^S} \partial\delta_i + \frac{1}{2}\frac{\partial^2 h_i}{\partial \delta_i^2}\bigg|_{\delta_i^S} \partial\delta_i^2 + \text{H.O.T.} \tag{5.5}$$

where $h(\Delta^S)$ is the angular momentum of a SGCMG array at a singular set of gimbal angles Δ^S, $\partial\delta_i = \delta_i - \delta_i^S$, h_i is the angular momentum of the ith SGCMG, n is the number of SGCMGs in the system, and H.O.T. are the higher-order terms taken to be negligible because only admissible null solutions are considered. Therefore, the Taylor series expansion in Eq. (5.5) remains valid, i.e., convergent for the present case.

The first term on the right-hand side (RHS) of Eq. (5.5) contains the ith column of the Jacobian matrix:

$$\hat{o}_i = \frac{\partial h_i}{\partial \delta_i}\bigg|_{\delta_i^S} \tag{5.6}$$

associated with the ith SGCMG's torque direction. The second term on the RHS of Eq. (5.5) is a Hessian and contains the partial derivative of the Jacobian matrix's ith column with respect to the ith gimbal angle

$$\frac{\partial^2 h_i}{\partial \delta_i^2} = \frac{\partial \hat{o}_i}{\partial \delta_i} = -h_{r_i}\hat{s}_i = -h_i. \tag{5.7}$$

Next, Eq. (5.7) is substituted into Eq. (5.5). The inner product of the result with the singular direction u, obtained from null(A^{T}), yields

$$u^{\mathrm{T}}[h(\Delta) - h(\Delta^S)] = \frac{1}{2}\sum_{i=1}^{n} u^{\mathrm{T}} h_i \partial \delta_i^2, \tag{5.8}$$

since the first term on the RHS of Eq. (5.5), $u^{\mathrm{T}}\hat{o}_i$ has zero contribution because of the definition of the singular direction. Equation (5.8) can be written more compactly in matrix form as

$$u^{\mathrm{T}}[h(\Delta) - h(\Delta^S)] = \frac{1}{2}\partial \Delta^{\mathrm{T}} P \partial \Delta = 0, \tag{5.9}$$

where P is the singularity projection matrix defined as $P = \mathrm{diag}(u^{\mathrm{T}} h_i)$.

By definition, null motion does not affect the total system angular momentum, which requires that $h(\Delta) = h(\Delta^S)$. Consequently, the left-hand side (LHS) of Eq. (5.9) is zero, i.e., $\partial \Delta^{\mathrm{T}} P \partial \Delta = 0$. As in reference [36], null motion is expressed in terms of the basis $N = \mathrm{null}(A)$, concatenated in matrix form as

$$\partial \Delta = \sum_{i=1}^{n} \lambda_i \nu_i = N\lambda, \tag{5.10}$$

where λ is a column matrix of the scaling components of the null-space basis vectors ν_i. In this concatenation step, the ν_i are columns representing the null-space basis. So, $N \in \mathbb{R}^{n\times(n-\mathrm{rank}(A))}$ is a $n \times (n - \mathrm{rank}(A))$ real matrix for any system of SGCMGs. Substituting Eq. (5.10) into Eq. (5.9) while observing the null motion constraint, we get

$$\lambda^{\mathrm{T}} S \lambda = 0. \tag{5.11}$$

As a result of this analysis, the singularity-definition matrix S is defined as

$$S = N^{\mathrm{T}} P N. \tag{5.12}$$

Therefore, it is always true that $S \in \mathbb{R}^{n-\text{rank}(A))\times(n-\text{rank}(A))}$. The eigenvalues of S determine whether a singularity is hyperbolic or elliptic. If S is definite, i.e., has all positive or negative eigenvalues, it contains only a non-trivial, i.e., zero null space since no nonzero null vector λ satisfies Eq. (5.11). Therefore, situations where the S is definite constitute elliptic singularities. They are so named because of the ellipsoidal nature of the quadratic in Eq. (5.11). For example, it can be rewritten as the equation of an ellipse for a rank 2 singularity:

$$au_1^2 + bu_2^2 = r^2, \tag{5.13}$$

where u_1 and u_2 are components that are a function of the eigenvectors of S and null-space coefficients, λ, a and b, are the eigenvalues of S or semimajor axes of the ellipse/ellipsoid/hyperellipsoid, and r^2 is the value of the result in Eq. (5.11) for arbitrary λ. Clearly where we desire $r = 0$.

When S is semi-definite, i.e., it has at least one $\lambda \neq 0$ at the point Δ^S, then the null space is non-trivial. That value of λ satisfies Eq. (5.11). Singularities that are semi-definite are known as hyperbolic in the literature even though they produce quadratic products that are paraboloids in Eq. (5.11) with one or more of the eigenvalues zero.

If S is indefinite, the eigenvalues are positive, negative, and/or zero. In that case, Eq. (5.11) may equal zero even though S itself may not have a null space. For example, consider the case

$$\lambda^{\mathrm{T}} S \lambda = \begin{bmatrix} 1 & 1 \end{bmatrix} \begin{bmatrix} -1 & 0 \\ 0 & 1 \end{bmatrix} \begin{bmatrix} 1 \\ 1 \end{bmatrix}.$$

In this example the solution happens to equal zero, but that fact is unrelated to the rank of S or P. By analogy, nontrivial null motion or at least a null solution to the gimbal angles is possible near singularity and the possibility of singularity avoidance may hold for all S, semi-definite or indefinite. It should be noted that "possibility" is stressed because additional null solutions do not constitute a sufficient condition for the singularity to be passable, or avoidable through null motion. An example, which is discussed below, is the case of degenerate-hyperbolic singularities discussed in (Bedrossian et al. [36]).

The authors in reference [36] were the first to include the null-space basis as part of determining whether a specific point of angular momentum offers null solutions, as shown in the preceding analysis. They also provided an analytic way to determine if all of such null solutions are singular, i.e., that the singularity is a degenerate hyperbolic singularity. When the Jacobian matrix is rank 1, more terms are needed from the Taylor series expansion in Eq. (5.10) to determine if null solutions exist.

Degenerate solutions of null motion in the vicinity of singularity lie on a local curve along which the Jacobian matrix is less than full rank. Therefore, null motion at a degenerate hyperbolic singularity satisfies the null motion condition in Eq. (5.11) and does not change the array angular momentum. Although the individual output-torque directions of the CMGs, (the columns of the Jacobian matrix) may change, the Jacobian matrix instantaneously remains less than full rank.

To define a hyperbolic singularity as degenerate we first define the commonly used "metric" for the distance from singularity. This singularity metric is $m = \sqrt{\det(AA^{\mathrm{T}})}$ which is the product of the Jacobian matrix singular values (Yoshikawa [38]). The objective is to find the hyperbolic-singularity null solutions for which $m(\Delta) > 0$, thereby showing that there may exist null motion that can move the system away from singularity. To accomplish this objective we first need to find the non-zero, i.e., non-trivial admissible null solutions λ that satisfy Eq. (5.11). Having found λ, we then determine if any of these null solutions meet the constraint $m(\Delta) > 0$.

The analysis begins with a Taylor series expansion of m about a specific δ:

$$m(\Delta^S + \Delta) = m(\Delta^S) + \sum_{i=1}^{n} \frac{\partial m}{\partial \delta_i}\bigg|_{\delta_i^S} \partial \delta_i + \frac{1}{2} \sum_{j=1}^{n} \sum_{i=1}^{n} \frac{\partial^2 m}{\partial \delta_j \partial \delta_i} \partial \delta_j \partial \delta_i \bigg|_{\delta_i^S} + \text{H.O.T.} \tag{5.14}$$

It is clear that $m(\Delta^S) = 0$ by the definition of singularity and

$$\frac{\partial m}{\partial \delta_i}\bigg|_{\delta_i^S} = 0$$

because it is an unconstrained stationary point of m. Therefore, neglecting the $H.O.T.$, Eq. (5.14) in matrix form is

$$m(\Delta^S + \Delta) = \frac{1}{2} \partial \Delta^{\mathrm{T}} \frac{\partial^2 m}{\partial \Delta \partial \Delta^{\mathrm{T}}} \partial \Delta, \tag{5.15}$$

where the null solutions λ are incorporated as a linear combination of null-space basis vectors, and

$$W = N^{\mathrm{T}} \frac{\partial^2 m}{\partial \Delta \partial \Delta^{\mathrm{T}}} N = N^{\mathrm{T}} M N, \tag{5.16}$$

where M is the Hessian of m.

The fact that we require $m > 0 \; \forall \; \delta$ implies that M and W are also positive definite. Therefore, a sufficient condition for null motion to permit the array to escape singularity, i.e., the hyperbolic singularity is not degenerate if $M > 0$. However, if W is singular and thus contains a null-space, solutions λ exist to keep the quadratic form in Eq. (5.15) zero. If these are the same solutions to the quadratic result in Eq. (5.11), the test for degeneracy is inconclusive. When this occurs, either no null solutions exist to move the system away from singularity or the $H.O.T$ in the quadratic forms in Eqs. (5.11) and (5.15) must be considered.

Here, degeneracy of a hyperbolic singularity is found either through a local analysis and searching the entire set or through globally searching all of the null-space trajectories that produce the same angular momentum. Neither is a straightforward process. Some cases may arise, such as the four-CMG roof array, where null motion may be available at a singularity but where a connected set of null motions can reach a non-singular angular momentum in finite time (see Leve [33]).

5.7 Hyperbolic Singularities

The previous section defined singularities mathematically by the ability of null motion to pass through or avoid them. All such singularities lie inside the angular-momentum envelope shown in Fig. 5.19, and all points on the internal singular surface shown in Fig. 5.20 that correspond to a hyperbolic singularity have null solutions or null motion that have the possibility of being used for singularity avoidance. This section offers some examples of hyperbolic singularities.

5.7.1 Non-degenerate Hyperbolic Singularities

Non-degenerate hyperbolic singularities have a continuum of non-singular null solutions at singularity, where null motion is possible for singularity avoidance. The four CMG pyramid array has both hyperbolic and elliptic singularities (see Chap. 6). The following example considers a four CMG pyramid array and uses the tools developed in the previous section to classify singular configurations for an array of CMGs. Consider the example for a four CMG pyramid array at a skew angle β taken from reference (Bedrossian [39]). The "skew" angle is the angle of the plane of angular momentum of the rotor, spanned by rotating the gimbal, with respect to the horizontal axis of the SGCMG array. The skew angle is therefore the complement of the "inclination" angle (see Sect. 4.2).

Example 5.3 The angular momentum of this array is

$$h = h_r \begin{bmatrix} -\mathrm{c}(\beta)\mathrm{s}(\delta_1) \\ \mathrm{c}(\delta_1) \\ \mathrm{s}(\beta)\mathrm{s}(\delta_1) \end{bmatrix} + h_r \begin{bmatrix} -\mathrm{c}(\delta_2) \\ -\mathrm{c}(\beta)\mathrm{s}(\delta_2) \\ \mathrm{s}(\beta)\mathrm{s}(\delta_2) \end{bmatrix} + h_r \begin{bmatrix} \mathrm{c}(\beta)\mathrm{s}(\delta_3) \\ -\mathrm{c}(\delta_3) \\ \mathrm{s}(\beta)\mathrm{s}(\delta_3) \end{bmatrix} + h_r \begin{bmatrix} \mathrm{c}(\delta_4) \\ \mathrm{c}(\beta)\mathrm{s}(\delta_4) \\ \mathrm{s}(\beta)\mathrm{s}(\delta_4) \end{bmatrix},$$

(continued)

$$A = h_r \begin{bmatrix} -\mathrm{c}(\beta))\mathrm{c}(\delta_1) & \mathrm{s}(\delta_2) & \mathrm{c}(\beta))\mathrm{c}(\delta_3) & -\mathrm{s}(\delta_4) \\ -\mathrm{s}(\delta_1) & \mathrm{c}(\beta)\mathrm{c}(\delta_2) & \mathrm{s}(\delta_3) & \mathrm{c}(\beta)\mathrm{c}(\delta_4) \\ \mathrm{s}(\beta)\mathrm{c}(\delta_1) & -\mathrm{s}(\beta)\mathrm{c}(\delta_2) & \mathrm{s}(\beta)\mathrm{c}(\delta_3) & \mathrm{s}(\beta)\mathrm{c}(\delta_4) \end{bmatrix}.$$

For a four CMG pyramid array with zero angular momentum, a set of gimbal angles $\Delta = [\frac{\pi}{2},\ \pi,\ -\frac{\pi}{2},\ 0]^{\mathrm{T}}$ deg is a hyperbolic singularity that has the following Jacobian matrix

$$A = h_r \begin{bmatrix} 0 & 0 & 0 & 0 \\ -1 & -\mathrm{c}(\beta) & -1 & \mathrm{c}(\beta) \\ 0 & -\mathrm{s}(\beta) & 0 & \mathrm{s}(\beta) \end{bmatrix},$$

and singular direction

$$u = \begin{bmatrix} 1 \\ 0 \\ 0 \end{bmatrix}.$$

The resultant projection matrix is

$$P = h_r \mathrm{diag}(-\mathrm{c}(\beta), 1, -\mathrm{c}(\beta), 1)$$

and the null space of the Jacobian matrix, concatenated in matrix form, is

$$N = \begin{bmatrix} -1 & 2\mathrm{c}(\beta) \\ 0 & 1 \\ 1 & 0 \\ 0 & 1 \end{bmatrix}.$$

The resulting singularity-definition matrix is

$$S = 2h_r \begin{bmatrix} -\mathrm{c}(\beta) & \mathrm{c}(\beta)^2 \\ \mathrm{c}(\beta)^2 & 1 - 2\mathrm{c}(\beta)^3 \end{bmatrix}.$$

The determinant of the singularity definitions matrix is $2h_r\mathrm{c}(\beta)(\mathrm{c}(\beta)^3 - 1)$. If $0 \le \beta \le \pi/2$, $\mathrm{c}(\beta) > 0$, the term $2h_r\mathrm{c}(\beta)$ can be neglected. Also, $\mathrm{c}(\beta)^3 \le 1$. So, S is either indefinite or semi-definite, and therefore this singularity is a hyperbolic singularity.

5.7.2 *Degenerate Hyperbolic Singularities*

Degenerate hyperbolic internal singularities occur when all of the null-space solutions for a particular point of angular momentum correspond to a singular Jacobian matrix, leaving no room for singularity avoidance solely using null-motion. Hyperbolic singularities that are degenerate are also considered impassable and therefore may be handled in a manner similar to elliptic singularities. Consider the example of a degenerate hyperbolic singularity from reference Wie [37] for a pair of 2 SGCMG with parallel gimbal axes.

Example 5.4 The angular momentum and Jacobian matrix for a set of two SGCMGs with parallel gimbal axes is

$$h = h_r \begin{bmatrix} s(\delta_1) \\ c(\delta_1) \end{bmatrix} + h_r \begin{bmatrix} s(\delta_2) \\ c(\delta_2) \end{bmatrix},$$

$$A = h_r \begin{bmatrix} c(\delta_1) & c(\delta_2) \\ -s(\delta_1) & -s(\delta_2) \end{bmatrix},$$

For this array, the singularity measure m is

$$m = s(\delta_1 - \delta_2)$$

Now, there are two singular configurations for this array: one in which $\delta_1 - \delta_2 = 0$, which is the saturation singularity at $2h$, and one that occurs internally at $h = 0$, when $\delta_1 - \delta_2 = \pi$. Because all hyperbolic singularities are internal, the latter case is of interest for degenerate hyperbolic singularities. By inspection, all $0h$ configurations must have $\delta_1 - \delta_2 = \pi$ and thus are singular. This singular configuration is shown in Fig. 5.11. In addition, the singular condition continues to be in force when both gimbals are moving using null motion. Therefore, this point of angular momentum has only null motion solutions that are always singular. Furthermore, any set of gimbal angles at this point of angular momentum is a degenerate hyperbolic singularity.

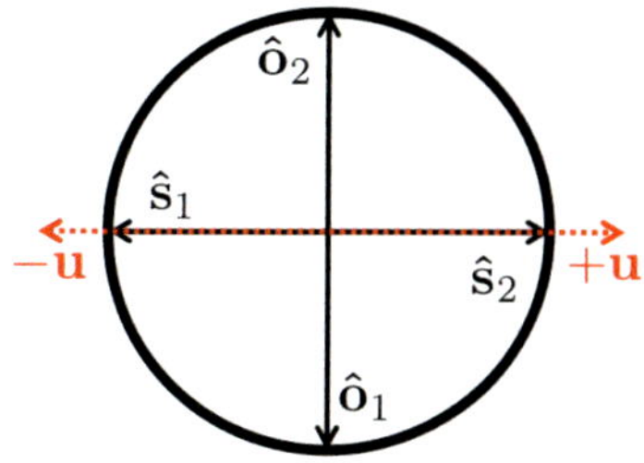

Fig. 5.11 Degenerate hyperbolic singularity

5.8 Elliptic Singularities

Elliptic singularities are those in which null motion does not exist for a specific value of array angular momentum. Since elliptic singularities do not permit null motion, the only way to escape singularity is for the array's angular momentum to be perturbed. This perturbation induces error in the spacecraft's motion relative to what the attitude-control subsystem attempts to achieve.

5.8.1 *External Elliptic Singularities*

Elliptic singularities comprise both external and internal singularities. As explained above, external singularities occur on the surface of the array angular-momentum envelope as shown in Fig. 5.20. Consider again the example of a four CMG pyramid array with skew angle β.

Example 5.5 For a four CMG pyramid array, the set of gimbal angles $\Delta^S = [\frac{\pi}{2}, \frac{\pi}{2}, \frac{\pi}{2}, \frac{\pi}{2}]^{\mathrm{T}}$ is an external elliptic singularity with angular momentum and Jacobian matrix

$$A = h_r \begin{bmatrix} 0 & 1 & 0 & -1 \\ -1 & 0 & 1 & 0 \\ 0 & 0 & 0 & 0 \end{bmatrix},$$

and singular direction

$$u = \begin{bmatrix} 0 \\ 0 \\ 1 \end{bmatrix}.$$

(continued)

The resultant projection matrix is

$$P = h_r \text{diag}(\text{s}(\beta), \text{s}(\beta), \text{s}(\beta), \text{s}(\beta)),$$

and the null space of the Jacobian matrix concatenated in matrix form is

$$N = \begin{bmatrix} 0 & 1 \\ 1 & 0 \\ 0 & 1 \\ 1 & 0 \end{bmatrix}.$$

The resulting singularity definition matrix is

$$S = 2h_r \begin{bmatrix} \text{s}(\beta) & 0 \\ 0 & \text{s}(\beta) \end{bmatrix}.$$

The determinant of the singularity definitions matrix is $2h_r \text{s}(\beta)^2$ which is obviously positive meaning that S is definite and corresponds to an elliptic singularity. Therefore, no null motion can exist in the vicinity of Δ^S, and this singularity is elliptic and impassable. Also, note that a definite P proves the singularity to be an external singularity, i.e., maximum projection of angular momentum. S is always definite when P is definite, proving that external singularities are all elliptic.

5.8.2 *Internal Elliptic singularities*

Elliptic singularities that lie on the internal singular surfaces, such as those shown for the four CMG pyramid array in Fig. 5.20, are referred to as elliptic internal singularities.

Unlike external singularities, these singularities are not a simple matter of sizing a CMG array; furthermore, since they occur instantaneously, they cannot be avoided by a naive algorithm on the fly. Consider the example for an internal singularity of a four CMG pyramid array.

Example 5.6 For a four CMG pyramid array, the set of gimbal angles $\Delta^S = [-\frac{\pi}{2},\ 0,\ \frac{\pi}{2},\ 0]^{\mathrm{T}}$ is an internal elliptic singularity with angular momentum and Jacobian matrix

$$h = h_r \begin{bmatrix} \mathrm{c}(\beta) \\ 0 \\ -\mathrm{s}(\beta) \end{bmatrix} + h_r \begin{bmatrix} -1 \\ 0 \\ 0 \end{bmatrix} + h_r \begin{bmatrix} \mathrm{c}(\beta) \\ 0 \\ \mathrm{s}(\beta) \end{bmatrix} + h_r \begin{bmatrix} 1 \\ 0 \\ 0 \end{bmatrix},$$

$$A = h_r \begin{bmatrix} 0 & 0 & 0 & 0 \\ 1 & -\mathrm{c}(\beta) & 1 & \mathrm{c}(\beta) \\ 0 & \mathrm{s}(\beta) & 0 & \mathrm{s}(\beta) \end{bmatrix},$$

and singular direction

$$u = \begin{bmatrix} 1 \\ 0 \\ 0 \end{bmatrix}.$$

The resultant projection matrix is

$$P = h_r \mathrm{diag}(\mathrm{c}(\beta), -1, \mathrm{c}(\beta), 1)$$

and the null space of the Jacobian matrix concatenated in matrix form is

$$N = \begin{bmatrix} -1 & 2\mathrm{c}(\beta) \\ 0 & -1 \\ 1 & 0 \\ 0 & 1 \end{bmatrix}.$$

The resulting singularity definition matrix is

$$S = 2h_r \begin{bmatrix} \mathrm{c}(\beta) & \mathrm{c}(\beta)^2 \\ \mathrm{c}(\beta)^2 & 2\mathrm{c}(\beta)^3 \end{bmatrix}.$$

The determinant of the singularity definitions matrix is $2h_r\mathrm{c}(\beta)^4$ which is positive and therefore, the singularity is defined as elliptic. Therefore, no null motion can exist in the vicinity of Δ^S, and this singularity is elliptic and impassable. Also, P is indefinite and therefore, it is an internal singularity.

5.9 Passability and Impassability of Singular Points

Having defined and categorized the types of singularities encountered in CMG arrays, we can now discuss an important property of singular points which has direct practical applicability to the choice of array architectures and controls. The concept of passability for a singular point is neither consistent nor clearly explained in the literature. This probably has to do with the fact that there has been no set standard in defining passability in terms that are understood by more readers than the mathematically inclined. Therefore, here we will define in our own terms the passability of points of angular momentum and their corresponding singularities. Defining passability of points of angular momentum, provides a means for the user of a CMG-based spacecraft momentum control system to map out troublesome parts of the angular momentum workspace (see Sect. 5.11).

In this discussion we must distinguish between a point in gimbal space (a specific set of n gimbal angles), and the corresponding point in momentum space (a vector in 3 axes). Any array of greater than 3 CMGs has many possible combinations of gimbal angles that map to the same point in momentum space. Some of these combinations of gimbal angles may be singular, others not. Passability is a characteristic of a point in momentum space, and it derives from an examination of the entire family of gimbal angle sets that map to that point. Classically, this problem is analogous to inverse kinematics in robotics.The concepts discussed in this section are not based on local analysis and therefore are different from those shown in references [36] and [40].

A point in momentum space is called passable if it is possible to change the point from singular to nonsingular using only null motion of the gimbals. Sub-categories of impassable and passable singularities are shown in Fig. 5.12. Note that we cannot simply extend the categories of singularities presented in Fig. 5.8. This is because passability is a property of the momentum point, and the singularities of Fig. 5.8 are properties of a specific gimbal angle set.

In the rest of this section, we shall examine each of the different types described in Fig. 5.12. To illustrate, examples will be provided using simple planar arrays of CMGs with parallel gimbal axes. These arrays offer a convenient way to visualize the distinctions among the various categories of singularity.

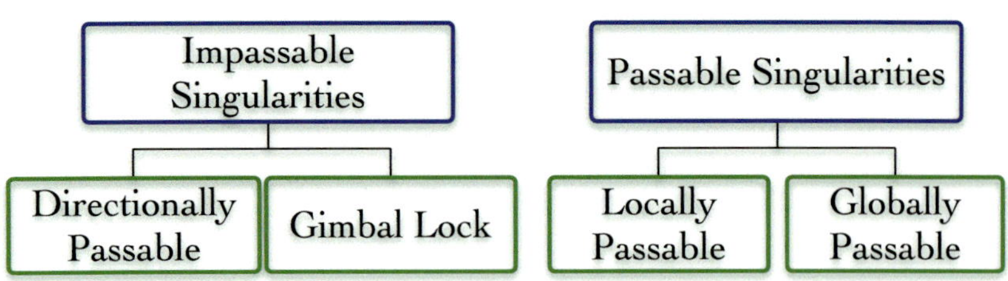

Fig. 5.12 Impassability and passability types

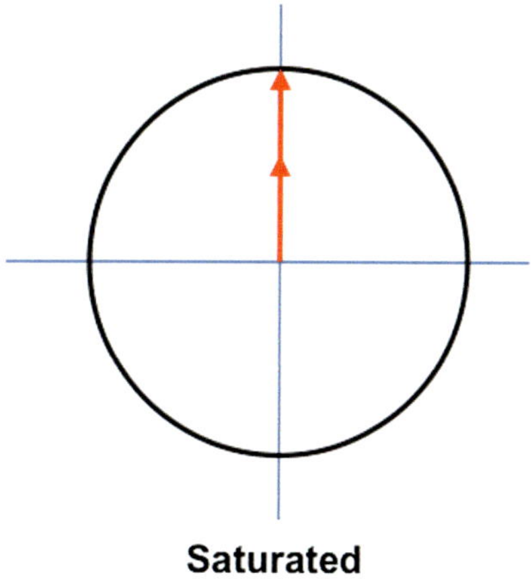

Fig. 5.13 Saturation singularity (Elliptic)

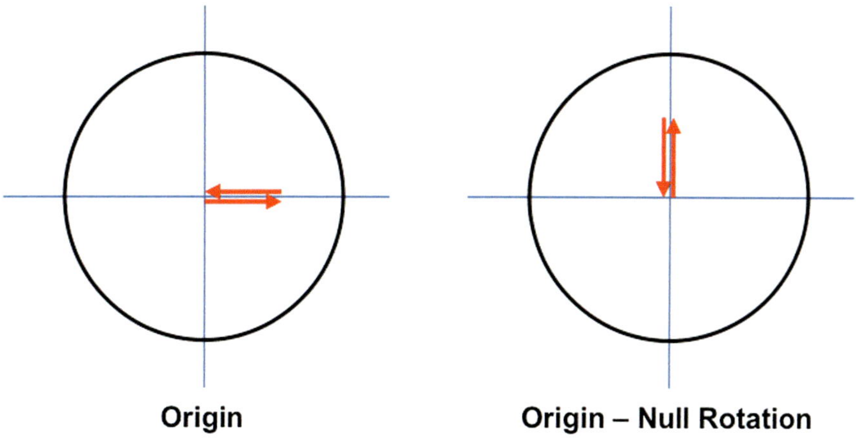

Fig. 5.14 Degenerate hyperbolic singularity

5.9.1 Impassable Singular Points

First, consider a point in momentum space characterized as an elliptic singularity (see Sect. 5.9). As this point has no null space, it must be impassable by definition. An example would be any point on the saturation boundary, such as shown in Fig. 5.13. Only one set of gimbal angles maps to this point in momentum space.

Next, we consider a point in momentum space characterized as a degenerate hyperbolic singularity (see Sect. 5.8.2). This point has a null space, but all possible gimbal angle sets are singular. Therefore this point must also be impassable by definition. An example is the origin of a 2-disc, as shown in Fig. 5.14.

The fact that these points are impassable does not imply that they are inescapable. However, traversing these points will, in general, result in torque error, i.e., the net momentum of the array must deviate from the intended momentum trajectory.

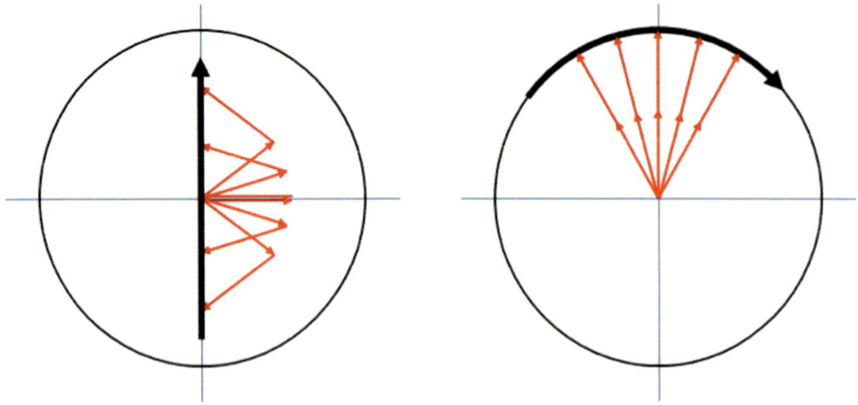

Fig. 5.15 "Directional" passability

Whether a torque error results depends upon the instantaneous direction of the commanded torque, and the singular direction of the gimbal set. Consider the following possibilities.

Torque command is in exactly the singular direction: This case is referred to as "gimbal lock". It is impossible to move in the intended direction.

Torque command has some components in the singular direction: In this case, called "directional passability", some of the desired torque can be provided, but there will be error. Nevertheless, the act of providing this torque moves us to a different momentum point, and may facilitate singularity escape. The practice of employing trajectories in the range space for singularity escape is discussed by Yamada and Jikuya [40].

Torque command has no component in the singular direction: In this case, despite the fact that the point is singular, we can traverse it with no torque error. This unusual case, illustrated in Fig. 5.15, is also referred to as "directional passability" and is possible but unlikely for any singularity. Note how we can follow the indicated momentum trajectories exactly despite the singularities.

The odds of the requested torque happening to fall precisely in the non-singular subspace of the Jacobian matrix where no torque error is accumulated are vanishingly small. No practical system would be designed in the hope that this situation would occur; any torque required along the singular direction leads to attitude-control error. Therefore, a spacecraft would likely be designed to operate only in a situation with passable singularities.

5.9.2 Passable Singular Points

A point in momentum space may have gimbal angle sets characterized by non-degenerate hyperbolic singularity (Sect. 5.8.1). This point has a null space, and it is possible, using only null motion, to move from a singular to a non-singular set of gimbal angles. It is therefore passable by definition.

Passable singular points fall into two subcategories, referred to as local or global. Unlike the subcategories of impassable points discussed previously, which depended upon the instantaneous torque command, local or global passability is a characteristic of the momentum point (see Leve [33]). Categorizing the types of passability of "local" and "global" has some symmetry with the controls community's viewpoint of locally and globally stable.

> **Local passability:** Consider the four-CMG roof array. This array consists of two of the two-CMG discs discussed earlier. The internal singular surfaces for the four-CMG roof array are shown in Fig. 6.8 as blue-green discs. These discs correspond to the origin of one pair of CMGs being swept by the $2h$ disc of the other pair.

We can visualize the motion of the 4 CMGs in this array by drawing the two disks. In Fig. 5.16, the top row shows the disk in the XY plane, and the bottom shows the disk in the YZ plane. The shared axis of the array is "Y" in this example. Figure 5.16 illustrates the null space at the origin of the four-CMG roof array. Examination of snapshots 1–5 verifies that all of these gimbal angle sets map to the net momentum $[0,\ 0,\ 0]^{T}$, therefore motion between them occurs in null space.

Snapshot 1 is singular, unable to produce torque in the Y direction. Snapshots 2 and 3 are nonsingular. Snapshot 4 is singular again, and it is interesting to note that this is a "rank 1" singularity—the array can produce torque ONLY in the Y direction. Since 4 CMGs have rank 1, 3 degrees of freedom of null space are

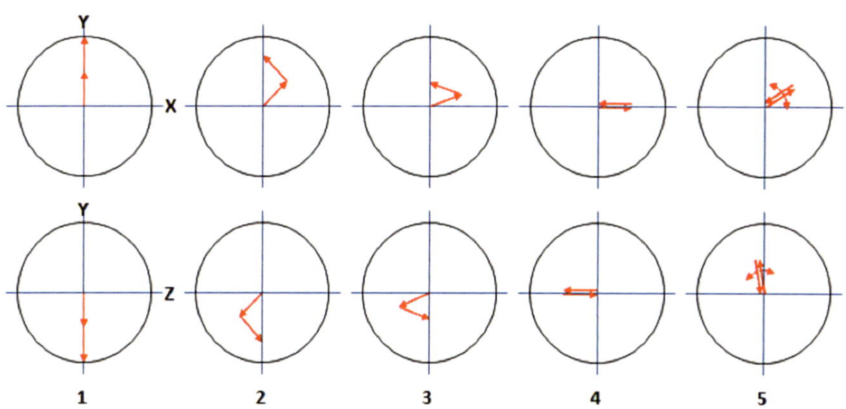

Fig. 5.16 Null space at the origin of the four-CMG roof array

available. They could move to return to snapshot 3, or either of the disks could rotate in pairs to arrangements such as those shown in snapshot 5.

Once the array has reached any of the arrangements depicted by snapshot 5, it is again rank 2—each disc can produce torque in one direction. Thus it has 2 degrees of freedom of null space, and the CMGs in each disc can rotate in pairs, independently. However, note that there is no way (using only null space motion) to get from one of these configurations back to the (nonsingular) snapshot 3 without first passing through the arrangement of snapshot 4.

Such difficulties earn this point in momentum space the characteristic of "local" passability. Specifically, while it is possible using only null space to return to a non-singular state (passable), it cannot be done instantaneously. It takes a finite amount of time to navigate the CMGs to arrangement 4, then 3.

Of course, singularities such as those shown in snapshot 5 can be escaped instantaneously by using range space, but doing so results in torque error, as discussed earlier. This is why steering through the entire momentum envelope of a four-CMG roof array may require torque error at some points. The steering law design can influence where these errors occur, for example, moving them well away from the origin.

> **Global passability:** If all singular combinations of gimbal angles at a point in momentum space can instantaneously move in null space to a nonsingular arrangement, the point is globally passable. An example of global passability is the $1h$ singular circle for 3 parallel CMGs, as illustrated in Fig. 5.17.

From symmetry, we can see that any point on the $1h$ internal singular surface above is globally passable. It can then be shown that a six-CMG roof array, comprised of two sets of three parallel CMGs, has all of its internal points of angular momentum globally passable (see references [30, 35]). Such an array can avoid all internal singularities with proper steering.

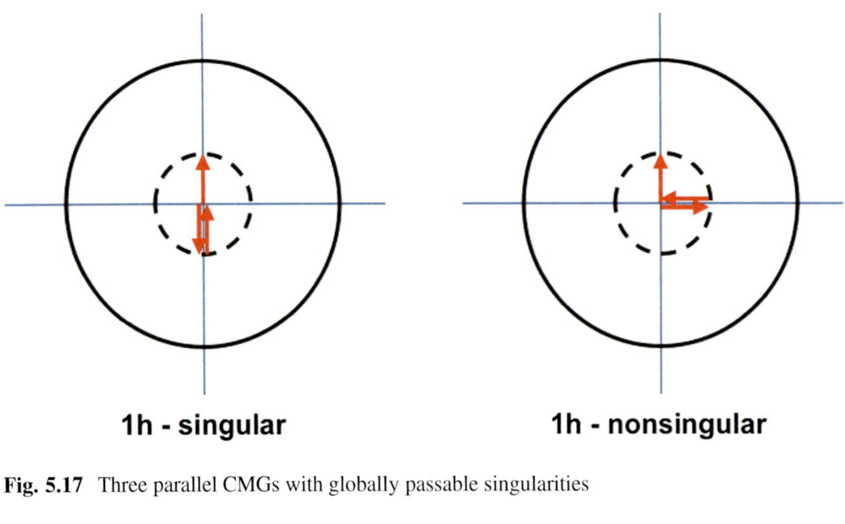

Fig. 5.17 Three parallel CMGs with globally passable singularities

5.10 Singular Surfaces for SGCMG

It is beneficial to understand where in the angular momentum space singularities arise. The satellite attitude control designer may wish to know all angular momenta that risk a singular Jacobian matrix a-priori so that he or she may design an algorithm to escape or avoid such points.

Typically, arrays of SGCMGs are redundant for the purpose of dealing with singularities by use of null motion. This redundancy and the fact that the Jacobian matrix elements are nonlinear functions of the SGCMG gimbal angles make it difficult to visualize a unique relationship between the solution of gimbal angles and the angular momentum. However, it is still possible to construct three-dimensional surfaces that represents the full extent of singular surfaces for the n-dimensional case.

This is done by first constructing a basis for the singular directions $\mathbf{u} \in \mathbb{R}^3$ in terms of angular momentum. In terms of the orthonormal basis shown in Fig. 5.10, the singular direction is

$$\mathbf{u} \in \mathbb{R}^3 : \mathbf{u} \cdot \hat{\mathbf{o}}_i = 0, \; i = 1, 2, \ldots, n. \tag{5.17}$$

In this basis, all torque vectors lie in a plane, and $\mathbf{u}$ is orthogonal to that plane (or along a line). This constraint constitutes a maximum (or minimum) projection of $\hat{\mathbf{s}}_i$ onto $\mathbf{u}$. For simplicity, this development is restricted to the common case where the magnitude of angular momentum is the same for all SGCMGs in the array. For a given singular direction $\mathbf{u} \neq \hat{\mathbf{g}}_i$ where $\mathbf{u} = \hat{\mathbf{g}}_i$ occurs for DGCMGs and for roof arrays (see Chap. 6). The conditions for singularity are

$$\mathbf{u} \cdot \hat{\mathbf{s}}_i > 0 \text{ and } \mathbf{u} \cdot \hat{\mathbf{s}}_i < 0, \tag{5.18}$$

where the definition of the singular direction $\mathbf{u} \cdot \hat{\mathbf{o}}_i = 0$ must also hold, i.e., a singular direction must exist for singularity. From the orthonormal basis in Fig. 5.10,

$$\begin{aligned}
\hat{\mathbf{s}}_i &= \hat{\mathbf{g}}_i \times \hat{\mathbf{o}}_i \\
\hat{\mathbf{o}}_i &= \hat{\mathbf{s}}_i \times \hat{\mathbf{g}}_i \\
\hat{\mathbf{g}}_i &= \hat{\mathbf{o}}_i \times \hat{\mathbf{s}}_i, \; i = 1, 2, \; \ldots, n.
\end{aligned} \tag{5.19}$$

Define the projections $\varepsilon_i \triangleq \mathbf{u} \cdot \hat{\mathbf{s}}_i$. Then the torque axis is

$$\hat{\mathbf{o}}_i = \varepsilon_i \frac{\mathbf{u} \times \hat{\mathbf{g}}_i}{||\mathbf{u} \times \hat{\mathbf{g}}_i||}, \; \mathbf{u} \neq \hat{\mathbf{g}}_i, i = 1, 2, \ldots, n, \tag{5.20}$$

which satisfies Eqs. (5.17) and (5.19), for all torque vectors. The resulting spin axis is

$$\hat{\mathbf{s}}_i = \varepsilon_i \frac{\hat{\mathbf{g}}_i \times (\mathbf{u} \times \hat{\mathbf{g}}_i)}{||\mathbf{u} \times \hat{\mathbf{g}}_i||}, \; \mathbf{u} \neq \hat{\mathbf{g}}_i, i = 1, 2, \ldots, n. \tag{5.21}$$

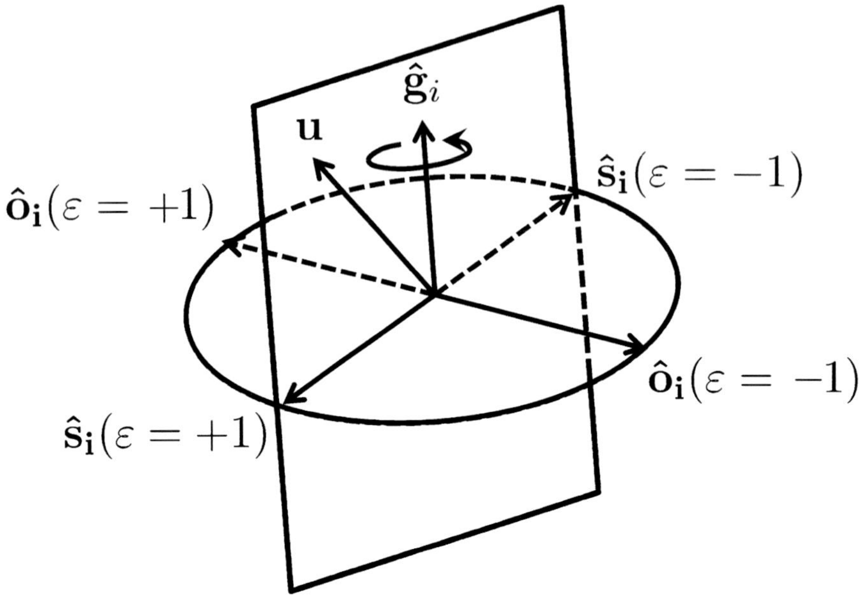

Fig. 5.18 Example of the singular direction in terms of CMG coordinate frame reconstructed from (Yoon and Tsiotras [41])

Because there are two singular directions—along the positive and negative directions, i.e., $\varepsilon_i \pm 1$—at each point of singular momentum, there are 2^n singular configurations of gimbal angles for n SGCMGs, for any singular direction. Therefore, adding more SGCMGs does not necessarily provide better singularity avoidance. An example of a singular direction projected onto a CMG's axes is shown in Fig. 5.18 in the coordinate system of a single SGCMG.

In Fig. 5.18, the singular direction shares a plane with the ith gimbal axis, and the ith torque vector is always orthogonal to the ith spin axis. This geometry is consistent with the definitions in Eqs. (5.17) through (5.21). Summing Eqs. (5.21) results in the total normalized angular momentum from the SGCMGs:

$$\bar{\mathbf{h}} = \sum_{i=1}^{n} \hat{\mathbf{s}}_i = \sum_{i=1}^{n} \varepsilon_i \frac{\hat{\mathbf{g}}_i \times (\mathbf{u} \times \hat{\mathbf{g}}_i)}{||\mathbf{u} \times \hat{\mathbf{g}}_i||}, \mathbf{u} \neq \hat{\mathbf{g}}_i, \ i = 1, 2, \ldots, n. \tag{5.22}$$

When $\mathbf{u} = \hat{\mathbf{g}}_i$ for any of the gimbal axes, Eqs. (5.20) through (5.22) corresponding to those gimbal axes are indeterminate.

The locus of total normalized angular momentum $\bar{\mathbf{h}}$, from Eq. (5.22), for all $\mathbf{u} \in \mathbb{R}^3$ and all $\varepsilon_i \neq 0$, i.e., $\mathbf{u}$ not collinear to $\hat{\mathbf{g}}_i$ produces the external singular surface known as the angular momentum envelope and the internal singular surfaces. Typically, the sequence of signs of ε_i corresponds to the singularity region of the associated singular surface. For example, for a four CMG array with $\varepsilon =$

Fig. 5.19 External (saturation) and internal singular surfaces for four CMG pyramid array (External surface shown in red)

$\{+, +, +, +\}$ or $\varepsilon = \{-, -, -, -\}$ corresponds to a $4h$ saturation-singularity region. $\varepsilon = \{+, +, +, -\}$, $\varepsilon = \{+, -, -, -\}$ or any other of the 16 combinations of ε with just one sign different, correspond to $2h$ singularity regions.

The angular momentum envelope in Fig. 5.19 represents the case where the set of all $\varepsilon_i \neq 0$ of the array have equal sign. Similarly, each of the four internal singular surfaces shown in Fig. 5.20 for a four CMG pyramid array can be found by setting the sign of one of the ε_i to be different from the others. This process yields points inside the momentum envelope that may be singular. The subset of these points for which the rank of the Jacobian matrix is less than 3 are the true singularities.

5.11 Characteristics of Singular Surfaces

The internal and external singular surfaces consist of the locus of points derived from all the singular angular-momentum points identified via Eq. (5.22). The external surface is the set of all singular momentum states corresponding to the maximum array angular momentum, i.e., $\varepsilon = \{+, +, +, +\}$ or $\varepsilon = \{-, -, -, -\}$. This limit makes sense physically: null motion cannot provide angular momentum beyond the array's capability.

Within the saturation limit lie the internal singular surfaces. They are shown in Fig. 5.20 for the case of a four CMG pyramid. All points on this surface correspond to the possibility of internal elliptic or hyperbolic singularities. These surfaces are connected smoothly to the external singular surface in Fig. 5.19 by trumpet-like dimples. They are also sometimes referred to as "lily pads.". These dimples are tangent to the plane orthogonal to a CMG's gimbal axis, along which none of that

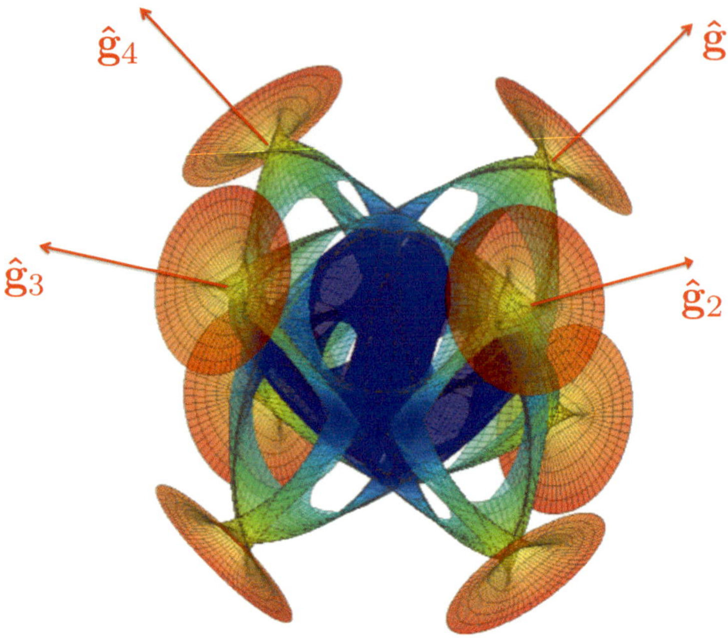

Fig. 5.20 Internal singular surfaces for a four CMG pyramid array

CMG's momentum can lie. There are four of these surfaces, each corresponding to one of the four CMGs in this array.

Note that the surface of the angular momentum envelope in Fig. 5.20, is almost a spheroid but for the dimples and the flat parts of the angular momentum envelope near them. A commonly cited choice of skew angle is $\beta = 54.74^\circ$ for this array because this angle produces the symmetry needed for the array to be as close to a spheroid as possible. However, only in the rare case where a spherical angular momentum envelope is of practical use would this angle be implemented: for example, an unlikely spacecraft with identical inertia and identical agility requirements in all axes. More commonly, the momentum envelope would be biased to maximize agility in the one or two directions where it has operational value (see Sect. 6.8).

As more SGCMGs are added, the external singular surface becomes more spherical. However, adding more SGCMGs to most arrays does not eliminate the singularity problem. Actually, the number of singular combinations of gimbal angles for any singular direction goes up as 2^n. Therefore, it is incorrect to assume that more SGCMGs always offer more effective singularity avoidance. In addition, research has shown that it is beneficial to optimize the shape of the singular surface as well as to maximize the usable angular momentum. Such an approach can maximize the use of the angular momentum for attitude control instead of devoting gimbal activity to avoiding singularities (see Chaps. 6 and 7). Leve et al. [42] and

Sands et al. [43] choose the orientation of SGCMG gimbal axes in the context of spacecraft inertia properties to maximize overall attitude-control performance.

The singular surfaces discussed and those in general for SGCMGs are found by projecting the angular momentum from an n-dimensional set of gimbal angles to 3-D space. Therefore, the surfaces are usually subtle and hard to visualize. Techniques by Dominguez and Wie [44] and Meffe and Stocking [32] construct cross sections of the singular surfaces to show attachments of the internal to external singular surfaces as well as the singularity free angular momentum.

5.12 Numerical Sensitivity in the Vicinity of a Singularity

According to Eq. (5.17), a SGCMG array singularity occurs when a singular direction $\mathbf{u}$ exists that is orthogonal to all output-torque directions $\hat{\mathbf{o}}_i$. However, this result is theoretical and neglects hardware limitations and imprecision. In reality, rather than points of singularities, singularity regions exist where the array cannot produce the torque requested due to finite gimbal-rate limits.

Accommodating these limits requires revisiting Eq. (5.17) with the following modification

$$\mathbf{u} \in \mathbb{R}^3 : |\mathbf{u} \cdot \hat{\mathbf{o}}_i| \leq \alpha(\dot{\delta}_{\max}), \, i = 1, 2, \ldots, n, \tag{5.23}$$

where $\alpha(\dot{\delta}_{\max})$ is a parameter that limits the distance to singularity based on the maximum gimbal rates achievable. If the condition in Eq. (5.23) for all CMG torque vectors is less than $\alpha(\dot{\delta}_{\max})$, the Jacobian matrix is close enough to singularity that at least one of the CMG cannot produce the gimbal rate needed to match the requested torque. The condition in Eq. (5.23) depends to some extent upon the steering law.

A simple planar case describes what is meant here by region of singularity. Consider the scissored pair of SGCMGs in Example 5.4. This pair of SGCMGs has singularities at $0h$ and $2h$, where the singularity at 0h is just a point. The Jacobian matrix is

$$A = h_r \begin{bmatrix} \mathrm{c}(\delta_1) & \mathrm{c}(\delta_2) \\ -\mathrm{s}(\delta_1) & -\mathrm{s}(\delta_2) \end{bmatrix}.$$

At the $0h$ singularity, the Jacobian matrix and singular direction are

$$A = h_r \begin{bmatrix} 1 & -1 \\ 0 & 0 \end{bmatrix},$$

$$u = \begin{bmatrix} 0 \\ 1 \end{bmatrix}.$$

For this system, let $h_r = 1$ and assume a 1° departure of the torque and angular momentum vectors from their original angle along the gimbal axis. The Jacobian matrix for this case is

$$A = \begin{bmatrix} 0.9998 & -1 \\ -0.0175 & 0 \end{bmatrix}.$$

Therefore, the requested torque is achieved by

$$\Delta = A^{-1}\dot{h} = \frac{1}{\det(A)}\mathrm{adj}(A)\dot{h},$$

where adj(A) is the adjoint of A or the cofactor transpose of the A matrix, and

$$\frac{1}{\det(A)} = -57.2987. \tag{5.24}$$

Equation (5.24) shows that a gain of approximately 57 to the gimbal rates needed to achieve the requested torque $\dot{h}$. It is clear that as the array approaches singularity, even without encountering it, det(A) becomes smaller. Therefore, the gain that multiplies the torque may lead to unachievable gimbal rates.

5.13 Variable-Speed Control Moment Gyroscope Singularities

A variable-speed control moment gyro (VSCMG) differs from a SGCMG in that it offers an extra control degree of freedom in the form of the rotor torque. This speed variation can be used for singularity avoidance, in principle. So, it is possible for a set of two VSCMGs with linearly independent gimbal axes to provide full three-axis control to a spacecraft without a singular Jacobian matrix (Yoon and Tsiotras [41]).

However, the torque does not scale linearly to power. RWAs, like VSCMGs, also have variable-speed rotors that produce torque through rotor accelerations. They offer at best a 1:1 ratio of output to input torque. SGCMGs produce torque through a gyroscopic effect that multiplies the torque input to the gimbal axis to produce a much higher output torque. In practice, the VSCMG rotor torque is considerably lower than the CMG output torque—sometimes one or more orders of magnitude. Beyond about three orders of magnitude (as is the case in typical SGCMGs), this difference between rotor-torque capability and CMG output-torque capability makes any benefits of variable speed largely irrelevant. Enabling a large VSCMG to provide comparable torque through its rotor requires a comparatively heavy spin motor, introduces thermal challenges, stresses bearing design, and ultimately creates a less-than optimal system design. It may be that very small CubeSat CMGs

(CSCMGs)—of a scale suitable for CubeSats—exhibit a more successful balance of rotor torque and gyroscopic output torque.

To reduce the amount of RWA accelerations used for singularity avoidance, and therefore the input power used by VSCMGs, Schaub and Junkins [45] developed a weighted Moore-Penrose pseudoinverse approach that used a local gradient to steer away from singularities and as well as tracking a desired rotor rate. Schaub's and Junkins' algorithm shows promise in converging to a set of desired rates or accelerations, thereby reducing the rotor accelerations needed for singularity avoidance. However, such approaches do not address the problem of the SGCMG Jacobian matrix contribution that depends on the speed of the rotors. As the VSCMG rotor speed decreases, the contribution of the gyroscopic torque from the SGCMG is reduced linearly. Therefore, the fact that the rotor speed is often lower sacrifices output-torque capability. Even worse, if the CMGs' rotor momenta vary too widely, voids can appear in the momentum envelope that resemble saturation singularities but within the interior. Such voids are not necessarily reflected in a singular Jacobian matrix, but nevertheless these configurations are impassable by gimbal motions alone and represent angular-momentum values that cannot be achieved even for nonsingular gimbal-angle configurations. These many issues explain why VSCMGs are not widely adopted in practice.

5.14 Zero-Momentum Spin Up

During launch, momentum devices are turned off. There are many reasons, including that the launch vehicle typically prohibits spacecraft actuators from operating until after separation. Also, many rotors are tied down to manage launch loads. Those tiedowns are released only after spacecraft separation. Thus, the momentum devices have zero gimbal and rotor speed, i.e., zero momentum at the beginning of mission operations. Specific procedures prevent the spacecraft from tumbling during on-orbit checkout and spin-up of the RWA or SGCMG. These procedures spin up CMG rotors to maximum momentum capacity within power constraints while staying at zero momentum and keeping the array far from singularities.

These requirements demand that zero-momentum spin up is performed in the null space of the momentum control system. There may be an infinite continuum of spin-up profiles that can be used to spin up the momentum control systems while keeping the array at zero momentum. Also, the generally low-torque rotor spin motors and power limitations within the momentum device mean that a spin-up maneuver in reality can take many hours. For example, the ISS CMGs spin up at approximately 17 RPM per minute, requiring about 6.5 hours to reach their full speed of 6600 RPM. In addition, uncertainties in actuator parameters such as alignments, gimbal and rotor inertias, and friction may impact spacecraft attitude during spinup. Kim et al. [46] addressed these parametric uncertainties and provided an asymptotic result to de-tumbling and zero-momentum spin up after launch tip-off of a four CMG pyramid array in their presence.

5.14.1 RWA Zero Momentum Spin-Up

RWAs do not typically spin up to full speed during on-orbit checkout. Unlike SGCMGs, they have the ability to spin down to lower speeds, where viscous friction effects are less and therefore quiescent power is lower (see Chap. 3). Nevertheless, some RWAs may normally operate at a speed well away from zero in order to avoid imprecise torque that occurs at zero speed. For such RWAs, the initial spin-up maneuver is quite straightforward. They do not contain an extra degree of freedom in the gimbals, and RWA arrays involve no singular configurations as long as the array contains at least three independent RWA rotor axes. At least four RWAs are required if a RWA array is to be spun up in the null space to a set of speeds that minimizes zero crossings. Such a spin-up maneuver is typically quick: on the scale of minutes because the bias speeds of the rotors are usually a small fraction of full speed.

5.14.2 Four-CMG Roof Zero Momentum Spin-Up

The strategy employed for spinning up the CMGs in a roof array involves biasing the momentum of one pane of the roof against the other along the axis they share. This allows both panes of the roof to be well separated from their $0h$ singularity at the origin. This singularity is discussed in Sect. 9.4.3.4, and illustrated in Fig. 9.8. If the 4 CMGs do not spin up with equal torque, whether due to spin motor tolerances or power-saving measures, the gimbal angles are adjusted as necessary to keep the net momentum from each pane of the roof equal in magnitude and opposite in direction

5.14.3 Four CMG Pyramid Zero Momentum Spin-Up

There are two popular, mirror-image configurations for the four CMG pyramid that can be used for zero momentum and are far from singularity. These configurations place all of the angular momentum vectors in a plane (torque vectors not coplanar) as shown in Fig. 5.21. It is possible that the power needed to spin up all four CMGs simultaneously while maintaining zero momentum is infeasible. Therefore, the CMG rotors can spin up at least two at a time, with the anti-parallel spin axes, so as to minimize attitude perturbations and achieve the maneuver in a timely manner.

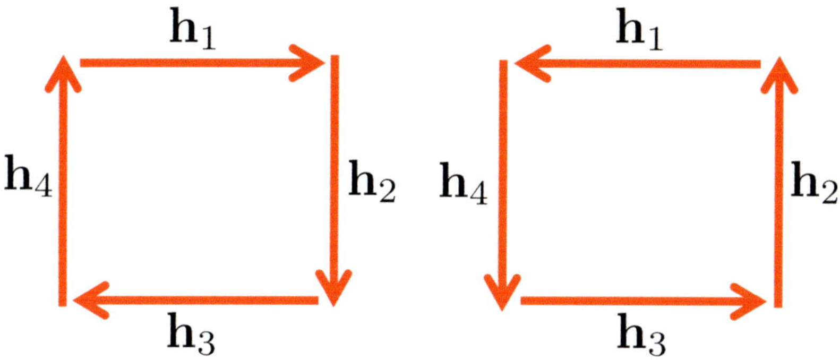

Fig. 5.21 Non-singular zero-momentum configurations for a four CMG pyramid array

5.15 Chapter Summary

This chapter discusses one of the major issues inherent in control-moment gyroscopes, known as singularities. What may appear as a purely mathematical phenomenon in the literature turns out to have important physical consequences for spacecraft design. Classifying singularities helps identify options for attitude-control architectures, such as the choice of arrays that lend themselves to singularity avoidance. The thorough understanding that these classifications make possible may help new momentum-control systems evolve more successfully into viable technology.

This systems-engineering activity can benefit from creating visualizations of the singular surfaces in angular-momentum space. They help establish a common vocabulary for communicating the risks and opportunities that momentum systems offer. Balancing the optimistic enthusiasm of rigorously mathematical academic literature with the practical implications of dealing with singularities can provide insight that is sometimes lacking in new space-system concepts.

5.16 Exercises

1. Which is true of a spacecraft with four operating control-moment gyroscopes (CMGs) aligned so that their gimbal axes are directed along the vertices of a regular tetrahedron?

   (a) The CMG array has internal singularities
   (b) The CMG array has no internal singularities
   (c) The CMG array requires at least one more CMG to impart three-axis torque for attitude control in every situation.

2. Consider the following Jacobian matrix for a four CMG box arrangement

$$A = h_r \begin{bmatrix} c(\delta_1) & c(\delta_2) & s(\delta_3) & s(\delta_4) \\ -s(\delta_1) & -s(\delta_2) & 0 & 0 \\ 0 & 0 & c(\delta_3) & c(\delta_4) \end{bmatrix}.$$

   (a) What are the singular directions at $\Delta^S = [0,\ 0,\ \frac{\pi}{2},\ \frac{\pi}{2}]^T$ (hint it is a set of higher dimension than one)
   (b) What type of singularity is at this point?
   (c) Does a four CMG pyramid arrangement have these type of singularities (see Chap. 6?

3. Consider the following Jacobian matrix for a four CMG box arrangement

$$A = h_r \begin{bmatrix} c(\delta_1) & c(\delta_2) & s(\delta_3) & s(\delta_4) \\ -s(\delta_1) & -s(\delta_2) & 0 & 0 \\ 0 & 0 & c(\delta_3) & c(\delta_4) \end{bmatrix}.$$

   What is the singular value decomposition of this matrix at $\Delta^S = [0,\ 0,\ \frac{\pi}{2},\ \frac{\pi}{2}]^T$ and $\Delta^S = [\frac{\pi}{2},\ \frac{\pi}{2},\ \frac{\pi}{2},\ \frac{\pi}{2}]^T$
4. (Scissor Pair) Consider the Jacobian matrix of a scissor pair with the constraint $\delta_1 = -\delta_2$. How many scissor pairs are needed for full three axis control and where do the singularities occur for the entire array of three orthogonal scissor pairs?

$$A = h_r \begin{bmatrix} c(\delta_1) & c(\delta_2) \\ -s(\delta_1) & -s(\delta_2) \\ 0 & 0 \end{bmatrix}.$$

Chapter 6
Momentum Control System Array Architectures

This chapter provides the analysis tools and fundamental theory for the design of an array architecture consisting of momentum devices. First, the properties of the actuator alignments and their effect on shaping the performance envelope of the momentum control system are discussed. A survey of common array types for RWA, CMG, and mixed arrays follows. A discussion of performance metrics and methods used to optimize the array architecture concludes the chapter.

6.1 The Nature of Momentum Devices

Thrusters apply external forces to a spacecraft at a point where the thruster expels propellant. The combination of the thruster force vector and position relative to the center of mass results in an external torque on the spacecraft. Like thrusters, the performance envelope for an array of momentum devices depends explicitly on their alignment with respect to the spacecraft body. However, unlike thrusters, the location of momentum devices with respect to the spacecraft's center of mass is unrelated to its ability to influence the attitude. This is entirely due to rigid-body dynamics: all points on a rigid body experience the same angular velocity and angular acceleration. So, a torque applied to the rigid body accelerates all of its inertia in the same way, regardless of what mechanical means applies that torque. This must be the case—otherwise, the body would twist and warp in a non-rigid fashion. The fact that momentum devices apply torque, not force, to the spacecraft allows them to be placed anywhere on the body.

Momentum devices impart internal torque, which conserves angular momentum of the combined spacecraft and momentum-control system. The torque is transferred through the structure of the actuator to the spacecraft body. Along with mechanical design considerations (e.g. bearing and motor choices, rotor shapes and size), actuator alignment for an array of momentum devices is critical to system-level

© The Author(s), under exclusive license to Springer Nature Switzerland AG 2025
F. A. Leve et al., *Spacecraft Momentum Control Systems*, Space Technology Library 46, https://doi.org/10.1007/978-3-031-96165-6_6

performance. The following sections describe best practices in choosing alignments for momentum-control actuators in their array. Sources of these requirements are described in Chap. 3, and the dynamics of these actuators is discussed in Chap. 4.

6.2 Momentum and Torque Capability of an Array of RWAs

The momentum and torque capability of an array of RWAs depends on the actuators' alignment with respect to the spacecraft body as well as the rate and acceleration limits of the rotors. It is customary to assume that the RWAs in an array are identical; therefore, the rate and acceleration limits are uniform across all RWAs in the array. This assumption provides a straightforward means to visualize the performance capability of the array. Consider an array of three orthogonal RWAs. The momentum and torque envelopes for this array are box-shaped. For the momentum capacity, the sides of this box are of length $2h_{max}$, which results from the positive and negative momentum limits for each of the three wheels. For torque, the sides are of length $2\tau_{max}$. Therefore, the maximum angular momentum and torque for this array lies along the direction $h = \frac{\sqrt{3}}{3}[1\ 1\ 1]^{\mathrm{T}}$, with a maximum momentum of $\sqrt{3}h_{max}$ and torque of $\sqrt{3}\tau_{max}$ along one of the vertices, as shown in Fig. 6.1. For a general array of three RWAs, not necessarily mutually orthogonal, the torque and momentum envelopes are parallelepipeds.

One of the two commonly implemented RWA arrays, an early favorite at NASA, is the 3-orthogonal and 1-skew (ortho-skew) RWA array shown in Fig. 6.2. In this array, the skew RWA can be used in one of two ways. First, it can act as a dormant spare which, because it has components in all three axes, can allow three-axis control to be retained in the event that any other single RWA fails. Second, it can be used to as part of a four RWA array. The simultaneous use of all four introduces a degree of freedom of null space, which allows operations such as avoiding zero wheel speeds (where friction can be troublesome), as discussed in Chap. 3.

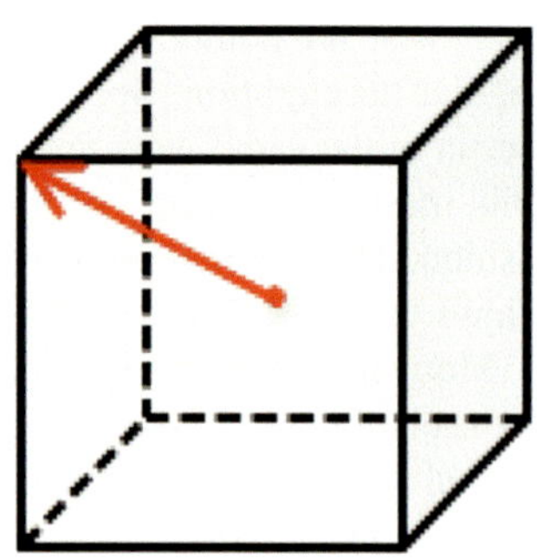

Fig. 6.1 Angular momentum envelope for a Three-Orthoskew RWA array

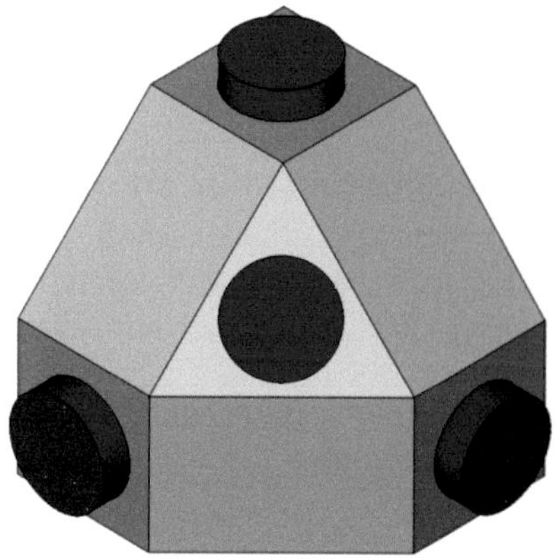

Fig. 6.2 Ortho-skew RWA array

Fig. 6.3 Four-RWA pyramid array

This array of RWAs has one disadvantage. If any wheel other than the skew rotor fails, the resulting torque authority will be unequal among the three axes, with very little capability in the direction of the lost RWA.

The other common use of four RWAs is the four-pyramid array shown in Fig. 6.3. In this array, any single RWA failure leads to the same sort of angular momentum and torque envelope. This array still does not have equal control authority about every axis following a failure, but the capability is far better in any axis than the ortho-skew array, no matter which actuator fails. Markley et al. [47] offers more insight into optimizing momentum via the choice of an RWA array.

6.3 Momentum and Torque Capability of an Array of CMGs

Unlike an RWA, a CMG typically has constant-magnitude rotor angular momentum. The maximum angular momentum, or performance envelope, of CMG arrays consists of a surface where all points correspond to a singular Jacobian (see Chap. 5).

The envelope for an array of SGCMGs is determined by the alignment of all CMG gimbal axes. The locus of all possible angular-momentum states of any single SGCMG is a circle (but not the points inside, i.e., not a disc) in the plane normal to its gimbal axis. The combination of these one-dimensional loci comprises the range of possible configurations. Therefore, a minimum of three SGCMGs is required for linear, three-axis attitude control.

The locus of angular momentum states of a single DGCMG is a sphere (but not the points inside, i.e., not a ball). Therefore, a minimum of two DGCMGs is required for full three-axis control, representing four independent degrees of freedom. The single redundant degree of freedom may be used for optimization or failure tolerance.

6.4 Double-Gimbal CMG Arrays

DGCMGs were used on the first U.S. space station, known as Skylab.

> *Skylab was the first manned spacecraft to utilize large CMGs for momentum storage and attitude control, the first to utilize vehicle maneuvers for CMG momentum desaturation, the first to utilize a fully digital control system with in-orbit reprogramming capability and extensive automatic redundancy management, and the first to utilize an attitude reference system based on a four-parameter strapdown computation which allowed utilization of an all-attitude eigenaxis maneuvering scheme. [48]*

Skylab flew three orthogonal DGCMGs, shown in Fig. 6.4, with a momentum capability of approximately 3000 Nm-s (2200 ft-lb-s). Two modes of CMG control existed, two- and three-CMG control. After the initial sizing of the CMG momentum, the vehicle moments of inertia increased tenfold. Consequently, the two CMG control mode became hopelessly ineffectual, unable to handle the secular torques and maneuvers in addition to the cyclic torques. On the ninth day into the third manned mission, CMG no. 1 showed evidence of failure, and the mode was changed to two-CMG control. After the failure of CMG no. 1, CMG no. 2 began to show signs of distress as well, and the CMG control allocation and maneuver choice for Skylab was altered to prolong the life of CMG no. 2 until the end of the mission.

The International Space Station also uses DGCMGs. These CMGs were built by Bendix Corp, now part of L3 Space and Navigation. Each provides 4760 Nms of angular momentum with stainless-steel rotors that spin at 6600 RPM. The four-CMG array can store about 19,000 Nms in any direction. With two attitude-control degrees of freedom available each DGCMG, at least two CMGs are necessary to provide attitude control for the ISS. NASA's ISS program has six such CMGs: four on ISS and two spares. Another CMG was shut down because of high vibrations, suggesting bearing failure.

While there are points in a DGCMG momentum envelope where higher gimbal rates are required for a certain torque to be achieved, a DGCMG array is not used for high torque. Its advantage over SGCMG arrays is its ability to distribute momentum efficiently in any direction. This distribution is achieved at low gimbal rates, with

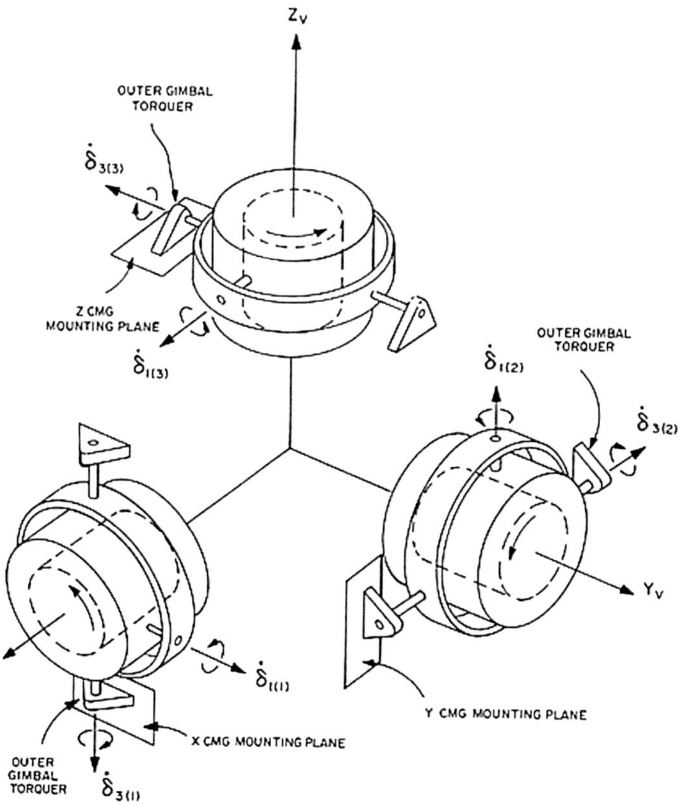

Fig. 6.4 Skylab's array of three orthogonal DGCMGs (Image courtesy of NASA)

attention paid to avoiding gimbal lock on any single CMG. This steering is a form of singularity-avoidance, but the challenges for DGCMGs in this regard are generally less demanding than in the case of SGCMGs.

6.5 Single-Gimbal CMG Arrays

This section discusses typical arrays of multiple CMGs and heterogeneous arrays of CMGs and RWAs.

6.5.1 Scissored Pairs of CMGs

Perhaps the most intuitive array of CMGs for three-axis attitude control is an array of three orthogonal scissored pairs (Carpenter and Peck [22]). A scissored pair consists of two CMGs with parallel gimbal axes and a linear constraint that keeps the gimbal angles equal in magnitude. Without this constraint, these two CMGs could produce momentum anywhere within a disc of radius $2h$. However, for the scissored pair, the two gimbals are moved in such a way that the locus of net momentum forms a line. This configuration is illustrated in Fig. 6.5, in which the gimbal axes are out of the page. Since both CMGs sweep through the same angle, δ, at all times, the net momentum along the X-axis in the figure is always zero. The net momentum along Y is $2h_r \sin(\delta)$. The torque produced about the Y axis for a gimbal rate, $\dot{\delta}$, is $2\dot{\delta}h_r \cos(\delta)$.

This constraint is, perhaps, the simplest form of steering law. It is so simple that it can be imposed by a gear train or mechanical linkage in place of algorithms involving feedback control. The 1970s era Astronaut Maneuvering Unit, which incorporated three scissored pairs, adopted this mechanical solution. Mathematically, the constraint requires gimbal rates of $\tau/(2h\cos(\delta))$ in opposite directions to produce output torque τ. A single scissored pair offers full access to every momentum state on the line and encounters no singularities (aside from, of course, saturation at $2h$). Without the constraint, a singularity exists at the origin of the disc, complicating the steering of even these two simple devices.

Using a mechanical linkage to impose the constraint is of particular value in the presence of base motion. For example, consider the case in which the array of Fig. 6.5 is nearly saturated (δ is near 90°), and the base to which the CMGs are mounted is instantaneously rotating about the X axis. Gyroscopic precession tries to rotate both gimbals in the same direction. If the gimbals are controlled separately, each gimbal motor would have to dissipate power to oppose this motion. However, if a mechanism forces the gimbals to rotate only in opposite directions, this mechanism maintains the constraint without mechanical work or power dissipation.

Because of their inherent simplicity, scissored pairs have been used in many applications such as robotic manipulators, astronaut maneuvering, and automobiles

Fig. 6.5 A scissored pair of CMGs

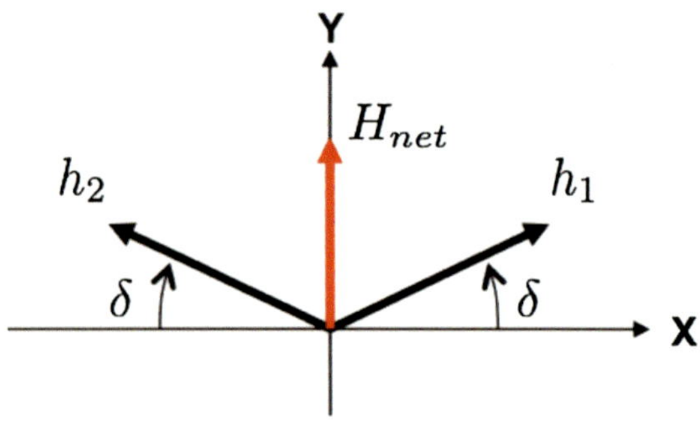

(Carpenter and Peck [22], Cunningham and Driskill [49], and Litmotors [50]). In one example, a scissored pair is aligned to produce roll torque to stabilize a ship at sea. Because only one axis of torque is required, this application does not carry any penalty from the absence of transverse momentum that the constraint imposes.

This simplicity extends to the general case of three-axis attitude control. A set of three orthogonal scissored pairs of CMG has a maximum angular momentum envelope similar to that of the RWA array in Fig. 6.1 except that the sides of the cube-shaped envelope have length $4h$. However, the ease and simplicity of use of three orthogonal sets of scissor pairs comes with a cost: the spacecraft must operate six CMGs at all times for a $\pm 2h$ cube of angular momentum. A single failed CMG halves the capability in one axis and couples some transverse torques into the others, further reducing the capability in those directions. Therefore, if this array of CMGs is used, the satellite designer has chosen to trade complex steering algorithms for additional volume, mass, and power.

6.5.2 Collinear (Multiple-Type) Arrays

Scissored pairs are only one of a family of array types characterized by the use of two or more CMGs with common gimbal axes. The most popular of these so-called collinear arrays in the literature are the roof arrays. These arrays have two common gimbal-axis directions. The angular momentum and torque for CMGs with a common gimbal axis reside in a plane ("roof side") shown as blue surfaces in Fig. 6.6 for the six CMG and four-CMG roof arrays. Roof arrays are also known as "multiple-type" arrays in the literature (Kurokawa [30]). An older term for the two-triplet architecture shown in the figure is "single-gimbal double-trident," which dates from 1980 or earlier (Rodden [51]).

Because roof arrays have only two planes of angular momentum and torque, they may experience rank-1 singularities. At a rank-1 singularity, torque is available

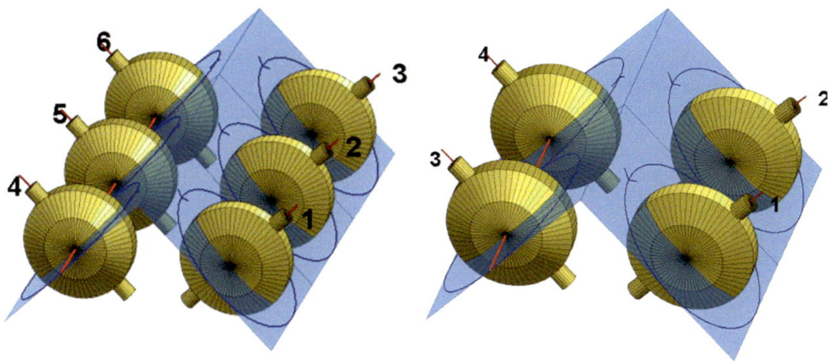

Fig. 6.6 Six-CMG roof array (left) and four-CMG roof array (right)

in only a single direction. Nevertheless, roof arrays of four or more CMGs are preferred over other CMG array architectures such as skew, pyramid, or symmetric type arrays because they have no elliptic internal singularities (See Chap. 5) and reference [35]. Therefore, all roof arrays can provide a non-singular gimbal-angle solution at every point of angular momentum internal to the momentum envelope. This feature enables roof-array steering laws to achieve nonsingular angular momentum close to the edges of the envelope. However, the fact that the internal singularities are hyperbolic does not guarantee that null motion is sufficient to pass a singularity for all combinations of gimbal angles. The null-space motion may have to be preplanned so that certain gimbal-angle combinations are avoided. For such cases, gradient methods are ineffective (Kurokawa [52]).

At this juncture, we digress briefly to introduce the subject of characterizing and plotting singular surfaces. Chapter 5 discussed singularities in detail and construction of singular surfaces, but here we introduce a common convention for describing the various singular structures and how to visualize them.

In general, arrays of CMGs are intended to provide momentum in a three-dimensional envelope. Because the direction of output torque from a SGCMG varies with gimbal angle, sets of gimbal angles may exist for which torque is not possible in all 3 axes. When this occurs, the array is said to be singular. Consider the four-CMG roof array of Fig. 6.6. With the gimbals positioned as shown, all spin axes, i.e., momentum vectors are parallel with the peak of the roof, which we call the shared axis. A small change to any gimbal angle in the array produces dH/dt in a direction orthogonal to the shared axis, and there is no small change of gimbal angles that can produce a dH/dt, i.e., a torque along the shared axis. Thus this set of gimbal angles is singular.

Once a set of gimbal angles like this is encountered, rotating any CMG gimbal angle 180° also results in a singular state. This gives rise to a convention for describing which of these singular states is present. In Fig. 6.7, if all four of the momentum vectors were not only parallel, but pointing in the same direction, we would refer to this as a $4h$ singularity. For an n-CMG array, the nh singularities will always be on the saturation surface of the array. Next, if one of the CMGs was pointed oppositely from the other three, this would be referred to as a $2h$ singularity. The names derive from the "net" momentum in a simple configuration like this, but the reader is cautioned that the "nh" singular surfaces do not always have momentum magnitude of nh. Finally, if two CMGs point opposite the other two, the condition is known as a $0h$ singularity. These combinations are illustrated in

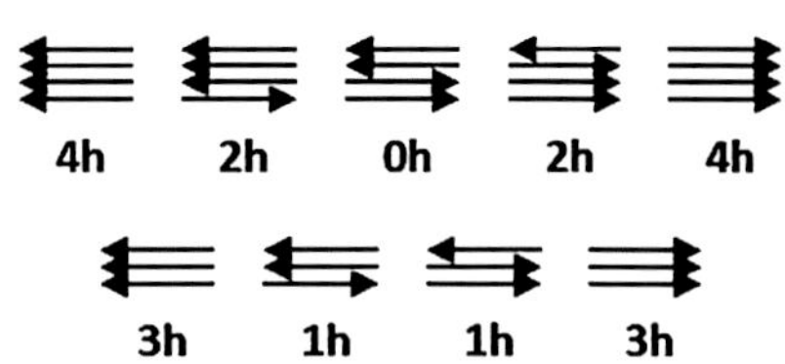

Fig. 6.7 Singularity designations

Fig. 6.7. In general, we can determine this designation by projecting the momentum vectors onto the singular direction and examining their sign.

It follows from this definition that an array with an even number of CMGs can have $0h$, $2h$, $\ldots nh$ singularities, and an array with an odd number can have $1h$, $3h$, $\ldots nh$ singularities (see Table 5.1 for locations of singularities for arrays with typical numbers of CMGs). All the array momentum states corresponding to all of the singular gimbal sets are referred to as the singular surface or the set of singular surfaces. Examination of these surfaces can be useful in understanding the nature of singularities for a given array architecture.

The singular surfaces for the four-CMG and six-CMG roof arrays are shown in Figs. 6.8 and 6.9. Blue and cyan indicates internal singular surfaces, which for roof arrays of 6 or more CMGs are passable by null motion at all singularity points within the angular momentum envelope. Recall from Sect. 5.10, that passability is a local property; despite that null motion may be unable to escape a singularity associated with a specific set of gimbal angles instantaneously, such solutions may still be reached in finite time. In orange and red, the outer surfaces represent the external or saturation singular surfaces that are never passable through null motion, regardless of how much gimbal-angle motion occurs.

In general, these $0h$, $2h$, $\ldots Nh$ surfaces, when plotted together, form a continuous connection between all possible singular momentum states. It is important to

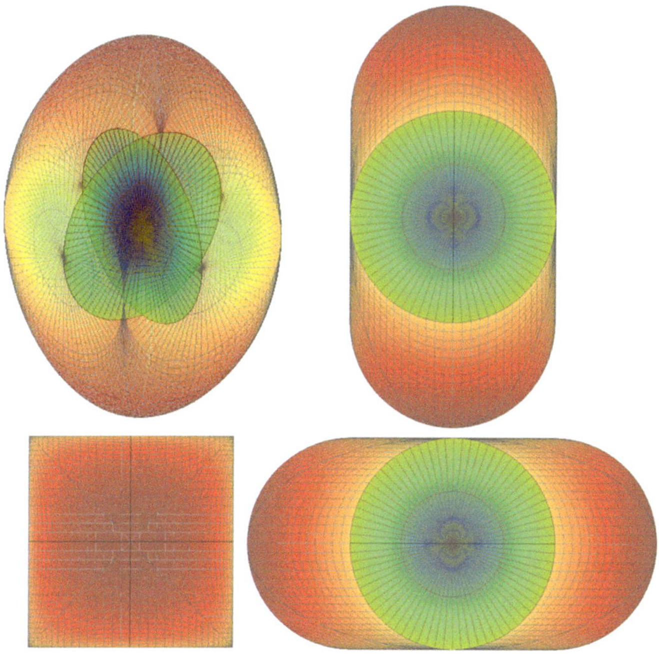

Fig. 6.8 Four-CMG roof array singular surfaces

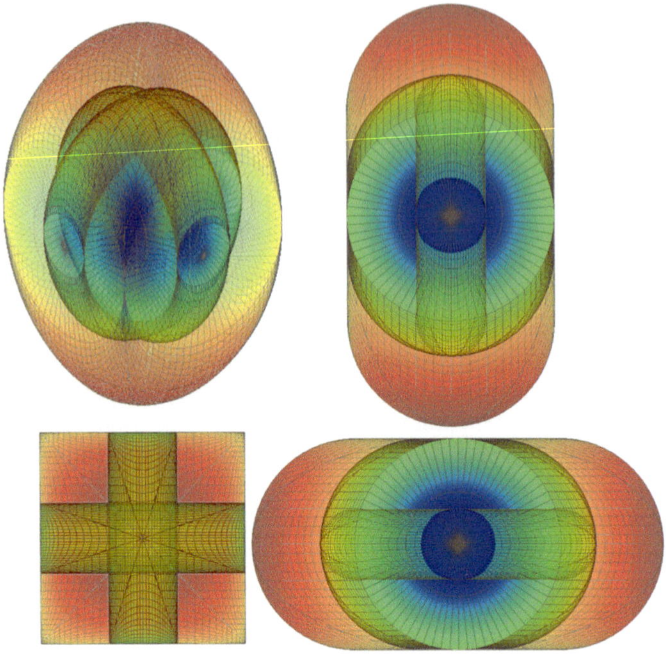

Fig. 6.9 Six-CMG roof array singular surfaces

recognize, however, that points on the internal surfaces are shown in terms of three-dimensional angular momentum, not n-dimensional gimbal angles. So, such a point is not necessarily singular: it may be that some gimbal-angle sets are singular for that momentum, and some are not. Many of these points can be reached by combinations of gimbal angles that keep the array well conditioned. Choosing these combinations is the central challenge of steering algorithms that enable spacecraft to utilize an array effectively.

An example of the four-CMG roof array that has a 90° angle between the two sets of gimbal axes is also known as the box-90 array. The box-90 array is shown in Fig. 6.10.

6.5.3 3/4 (3 of 4) Box Array

Whenever a single CMG fails in a box-90 array, the consequence is the 3/4 box array. Therefore, in the likely case that fault tolerance is required, the performance requirements must be met by a 3/4 box. Consequently, during typical operations, only 3 CMG out of 4 are active, with a dormant spare. The singular surfaces for the 3/4-CMG box array are shown in Fig. 6.11.

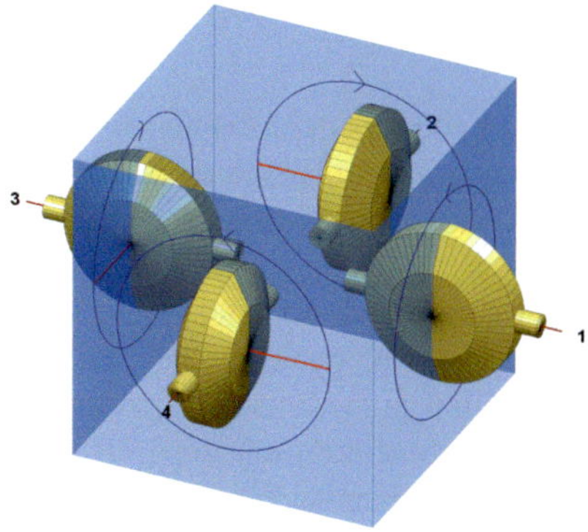

Fig. 6.10 Box-90 CMG array

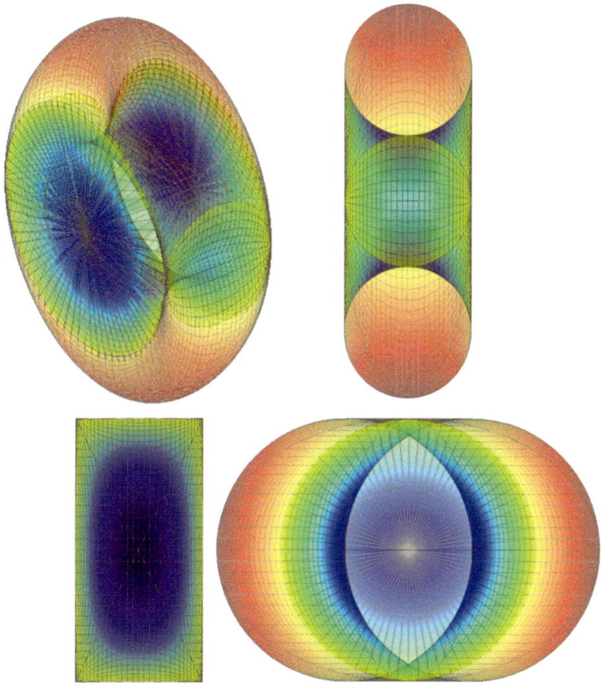

Fig. 6.11 3/4-CMG box array singular surfaces

In Fig. 6.11, there are no singularities within a $1h$ sphere about zero angular momentum (Sands et al. [43]). The use of this $1h$ singularity-free sphere is one of the reasons this array is one of the first to have been used and has been the array of choice for many spacecraft. With three CMGs, there is no way of avoiding singularities through null motion because there is no null space (see Chap. 5). Therefore, any spacecraft-level performance requirements must be met within the singularity-free region of the momentum envelope.

6.5.4 Pyramid Arrays

Perhaps the most commonly discussed array architecture is the pyramid array. It frequently appears in the academic literature. It is also one of the most complex. The four-CMG pyramid array shown in Fig. 6.12 has independent gimbal axes, unlike collinear roof arrays. The figure shows that pyramid arrays of 4 or more CMGs never have a rank-1 singularity.

A four-CMG pyramid array at a skew angle, i.e., angle from horizontal to angular momentum plane of 54.74° produces a somewhat spherical three-axis symmetric angular momentum envelope. The envelope is indicated in orange and red, with internal singular surfaces in cyan and blue, in Fig. 6.13. The internal singular surfaces are smoothly connected to the angular momentum envelope. They contain trumpet-like funnels or tubes in which no angular momentum is available. These tube structures are smoothly connected to the angular momentum envelope through surfaces that resemble lily pads (orange). The circular outer edge of each lily pad is formed by sweeping the momentum vector of the CMG whose gimbal axis passes through it (Yoon and Tsiotras [41]). We recall that a SGCMG cannot access the points inside this circular locus.

This somewhat spherical momentum envelope provides roughly equal control authority in angular momentum and torque about any axis and provides a large volume of accessible momentum. Perhaps this uniformity in angular momentum and torque is the reason the four-CMG pyramid array initially became an object of academic study. The nature of academic publication encouraged many follow-on studies of this particular architecture. However, from a practical perspective, the angular jerk, acceleration, and velocity are of primary importance for many agile spacecraft, not this uniform distribution of momentum and torque. These agility metrics are calculated by inverting the spacecraft inertia onto the control authority (see Chap. 4). Therefore, equal momentum and torque about three axes does not

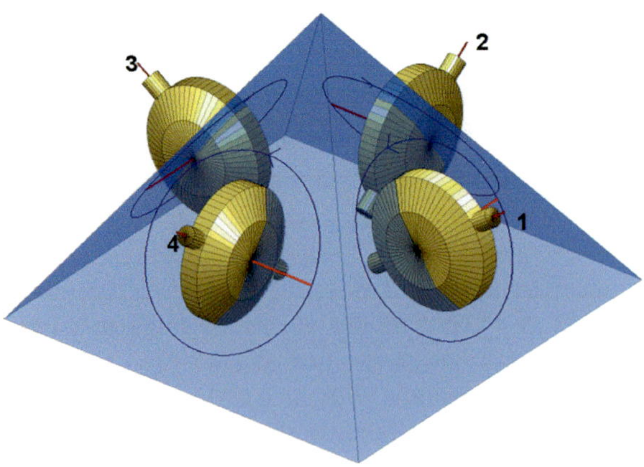

Fig. 6.12 Four-CMG pyramid array

Fig. 6.13 Four-CMG pyramid array singular surfaces

correspond to equal agility about these axes. Moreover, many Earth-observing spacecraft, for example those that perform push-broom imaging (sweeping the imager back and forth through small angles throughout the orbit). Such satellites require high torque and momentum about one or two axes, with less capability needed about the boresight axis of the imaging payload. An example of an application that takes advantage of agility in the third degree of freedom is described by Karpenko et al. [53] and Ross et al. [54]. They seek optimal slew trajectories that can exploit the reduced inertia axis to reduce overall slew time when compared with an eigenaxis slew.

The uniformity of the angular-momentum envelope comes at a cost: the very complex internal singular surfaces shown in Fig. 6.13, which contain both passable non-degenerate hyperbolic singularities and impassable elliptic singularities. Recall from Chap. 5, in the case of impassable elliptic singularities, no null motion exists for singularity avoidance; therefore, passing through points of angular momentum near an elliptic singularity sacrifices accuracy in torque for attitude control. Also, unlike collinear arrays, the internal singular surfaces for a pyramid array are scattered throughout the angular momentum envelope. Therefore, the pyramid array has a smaller non-singular (or singular-but-passable) angular momentum envelope to offer the attitude control system.

6.5.5 *Dynamic Arrays*

Some have proposed making the gimbals of a CMG dynamic , in the sense that reorienting them can change the angular momentum envelope and corresponding singular surfaces during operations. Kojima [55] suggests frequent reconfiguration as a singularity-avoidance approach. However, approaches that involve frequent reconfiguration during maneuvers ignore the influence of additional torque during reconfiguration. Essentially, such methods make the array of CMGs act as a semi-static DGCMG array and therefore come with a similar increase in volume, mass, power, and mechanical complexity as well as the loss in performance (see Chap. 3). This disadvantage can be overcome in the case where the CMG gimbals can be reconfigured with the CMG despun, and the new configuration locked into place with a mechanical constraint, as suggested by Bonn et al. [56], where it is a response to failures or infrequent changes in mission requirements

6.6 Blended Arrays and Other Designs

The early days of CMG technology saw efforts to create heterogeneous arrays of momentum devices, or "blended arrays". These arrays combine reaction wheels of various sizes, single- and double-gimbal CMGs, and some less-familiar devices. None have caught on, the most likely reason being that parts count, reliability, and related cost penalties motivate arrays of identical devices. Nevertheless, these early innovations deserve some recognition. And there still may come a day when one of them proves useful in a spacecraft architecture.

6.6.1 *Double-Gimbal/Single-Gimbal Scissored Pairs*

A pair of DGCMGs provides four independent control degrees of freedom, one for each gimbal axis. This number is sufficient for three-axis control of attitude, within the limits imposed by gimbal-lock singularities. In fact, it is more than enough (by one), but certain failures of one of the DGCMG's components prevents the entire array from serving the needs of a three-axis attitude-control system. Adding a scissored pair as shown in Fig. 6.14 adds a degree of freedom–only one, because the gimbal-angle constraint subtracts one from what would otherwise be two independent SGCMGs. The result is an array that is single-fault tolerant, any single CMG failure leading to a momentum system capable of providing torque in three axes, again within the limits imposed by singularities. The steering law is straightforward, but this momentum system is considered to have high weight and volume for its performance (Rodden [51]). It is also suited primarily to a system with single-axis agility, since the scissored pair might apply torque along its axis that is

Fig. 6.14 Two DGCMGs and one scissored pair of SGCMGs

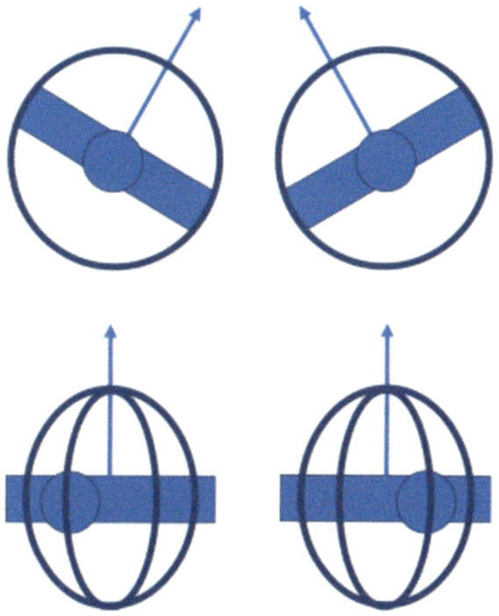

Fig. 6.15 Three DGCMGs uniformly distributed in a plane

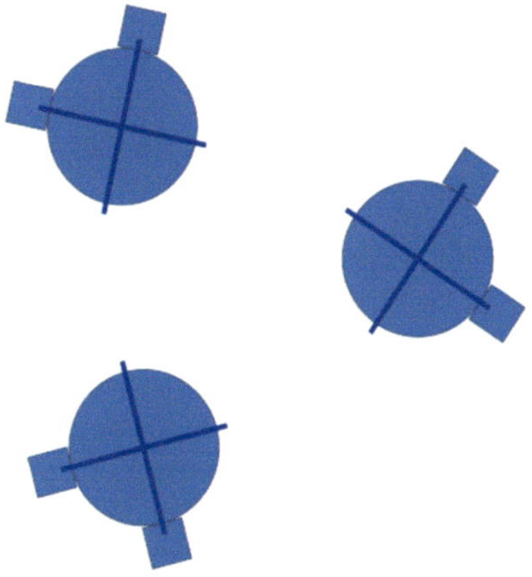

hundreds of times as much as torque from one of the DGCMGs in the transverse axes.

6.6.2 Langley "Six-Pac"

A combination of three DGCMGs provides similar failure tolerance. This architecture has been referred to as the "Langley Six-Pac" see Rodden [51] , in the specific case where the momentum vectors for zero gimbal angle lie in a plane and are separated by 120°, as shown in Fig. 6.15. A failure of any DGCMG results in an array with four remaining degrees of freedom, but the complex control law and high weight and volume has made it an unpopular choice.

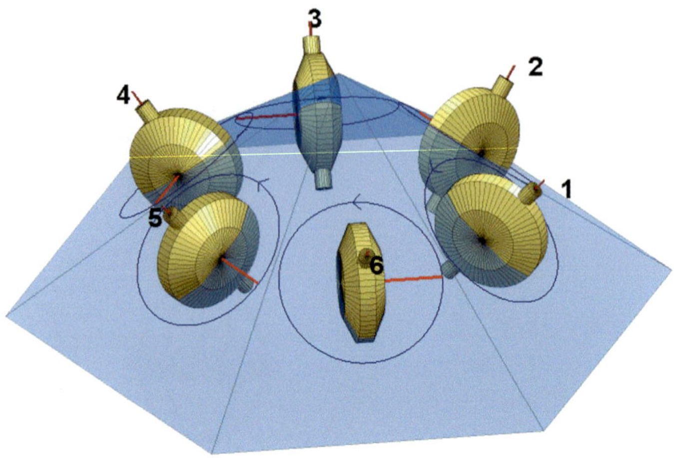

Fig. 6.16 Six GAMS pyramid array

6.6.3 Single-Gimbal 6 GAMS (Six-CMG Pyramid Array)

Six SGCMGs can be arranged so that each CMG momentum vector sweeps out a plane that forms one of the faces of a six-faceted pyramid. This so-called gimbaled-axis momentum system (GAMS) is known as the 6-GAMS shown in Fig. 6.16, as described by Cornick [57] and Rodden [51]. Like any six-CMG array, this one offers active, multiple redundancy but represents lower weight and volume than DGCMG arrays for a given torque requirement.

The singular surfaces of the six-CMG pyramid array in Fig. 6.17 differ from the four-CMG pyramid array Fig. 6.13. The addition of two CMGs provides a more nearly spherical momentum envelope, with a significantly larger momentum capacity at the expense of a more complex set of internal singular surfaces. These internal singular surfaces appear throughout the volume of the momentum envelope and also have more lily pads that connect internal structures to the edges of the envelope. Therefore, unlike the six-CMG roof array, the six-pyramid's increased momentum capacity and spherical symmetry alone does not eliminate geometric singularities. Nevertheless, Cornick describes this array as one in which an internal envelope can be exploited through an appropriate steering law.

6.6.4 Scissored Pair with High-Torque Reaction Wheels

A single scissored pair of variable-speed control-moment gyroscopes (VSCMGs) represents a three degree-of-freedom system with minimal hardware. Such an array may be suitable for applications in which single-axis agility is required, with limited

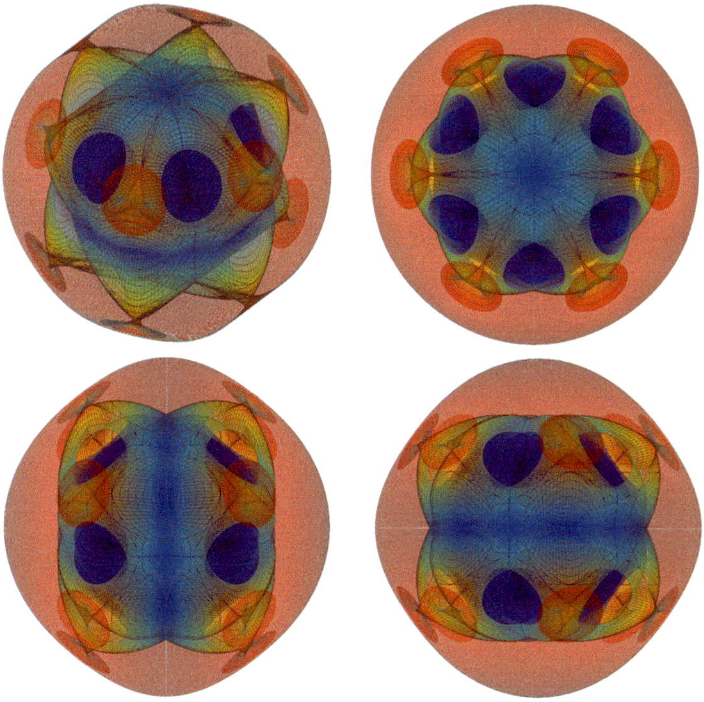

Fig. 6.17 Six-CMG pyramid array singular surfaces

control authority in the transverse directions. Such a design suggests severe volume constraints. In such circumstances, further simplifications may be motivated. For example, the gimbal angle might be limited to a finite range, eliminating the need for continuous rotation and thereby saving the mass and complexity of slip rings and related electronics.

Limiting the gimbal angle can be based on some straightforward sizing principles, as suggested in the following example. High-torque reaction wheels can provide about 2 Nm of torque about their spin axes. With 16.6 Nms of angular momentum and a 1 rad/s gimbal rate, this array can produce about 1/10 the torque in the transverse axes as it can along the spin axis, up to a gimbal angle of about 45°. The momentum and torque performance of this example are shown in Fig. 6.18.

6.7 Variable-Speed CMG Arrays

The perennial search for new options for SGCMG singularity avoidance motivates the case of an array of VSCMGs. Basically, at the values of angular momentum where an array of SGCMGs would be singular, the VSCMG relies on rotor accel-

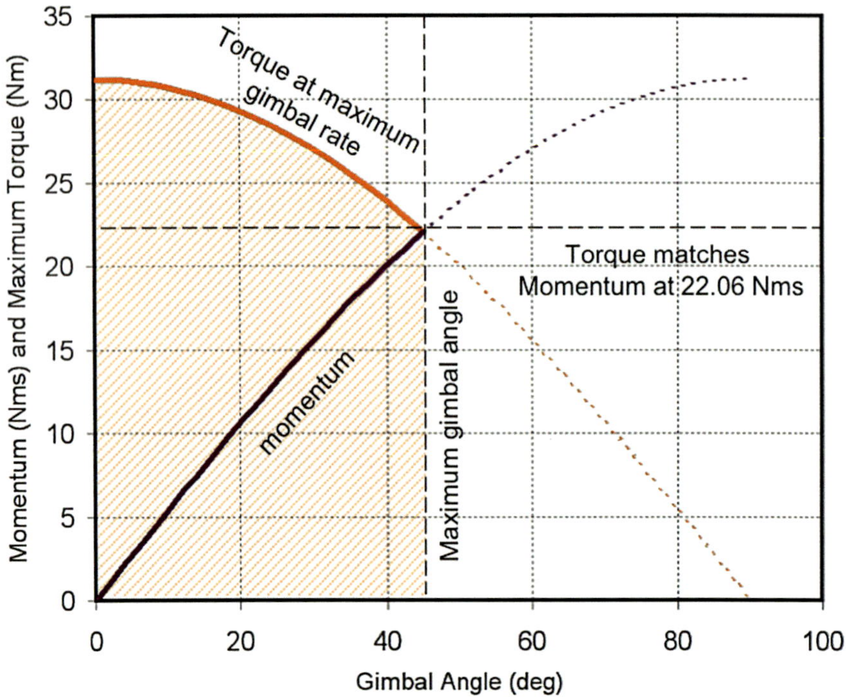

Fig. 6.18 6 Agile-axis torque and momentum performance of a scissored pair with limited gimbal angles, for a factor of 10 greater momentum in one axis than in the axes transverse to it

eration as a control DOF to provide the needed torque along the singular direction. As a consequence, an array of two or more VSCMGs with linearly independent gimbal axes sufficiently far from rotor-speed saturation is non-singular,[1] i.e., has the non-singular Jacobian shown in Eq. (6.1).

$$\dot{h} = [A(\Delta, \Omega)\ B(\Delta)] \begin{bmatrix} \dot{\Delta} \\ \dot{\Omega} \end{bmatrix} = D\dot{X} \tag{6.1}$$

As explained in Chap. 4, one of the major disadvantages of RWAs also impacts VSCMGs: their ratio of input power to torque is high. They require direct shaft power because their torque is proportional to the acceleration of the rotor. A few successful algorithms for minimizing rotor accelerations while avoiding CMG

[1] The absence of singularities for an array of two or more VSCMGs with linearly independent gimbal axes assumes that the rotors do not reach their maximum speed or zero. With sufficiently many rotors at maximum speed, the array can no longer provide rotor accelerations to aid in SGCMG singularity avoidance. Similarly, when enough rotors have very little momentum, the gyroscopic torque vanishes, worsening the singularities.

singularities can be found in Yoon [41] and Schaub [58]. Even neglecting the hardware and systems-engineering issues, the singularity-avoidance algorithms alone dramatically compromise the performance of a VSCMG array.

The chief drawback of VSCMGs is that they sacrifice rotor speed to avoid singularities or for power management. Reducing rotor speed reduces torque amplification, and yet that amplification is the reason for using a CMG in the first place. Equation (6.1) shows that the VSCMG array Jacobian concatenates a Jacobian for CMG torque and one for RWA style torque.

There are several other reasons why VSCMG arrays have not been used in attitude control on-board contemporary satellites. Some of these reasons arise from system-level concerns:

1. The spacecraft's jitter response varies with frequency. A constant rotor speed (as in a SGCMG) may permit the spacecraft to be designed to minimize structural amplification of the rotor-frequency disturbance. In contrast, variable rotor speeds make this approach infeasible.
2. Larger rotor motors are needed for rotor accelerations. This mass increases inner-gimbal inertia, thereby requiring a larger gimbal motor. The mass, volume, and power is therefore increased in such systems.
3. These systems demand more powerful motor electronics because the rotors require higher-bandwidth speed control than SGCMGs.
4. Singularity avoidance for arrays of SGCMGs is becoming well understood and proven on many systems, eliminating the motivation for more exotic solutions.
5. Higher bearing wear and shorter lifetime results from the heat introduced by higher rotor power.

These many reasons motivate contemporary spacecraft designs to maintain constant rotor speeds.

6.8 Energy Storage

An additional benefit of an array of VSCMGs is its ability to use the flywheels in RWA mode to provide not only CMG singularity avoidance, but also electrical power. Some of the earliest literature on using flywheels to store energy on spacecraft in the form of kinetic energy appeared as far back as the early 1960s (Roes [59]). Batteries alone cannot generally provide large current to their connected loads; some form of capacitance usually complements batteries for this reason. Spacecraft that require very high, instantaneous power—perhaps satellites whose payloads involve directed energy—may realize mass savings by implementing flywheels whose kinetic energy can be converted to electrical energy (Babuska et al. [60]). Contemporary technology for flywheel energy storage is roughly competitive with satellite batteries, with power density around 100 W/kg. However, to achieve this energy storage and extraction efficiently, the flywheels typically require very high speeds, along with additional electrical circuitry for the

Fig. 6.19 Honeywell FACETS CMG

conversion of kinetic energy into electrical energy. Magnetic bearings can reduce frictional losses and extend lifetime in the presence of these high speeds for such systems but also consume considerable power on their own in maintaining a stiff constraint. And they are unstable if power to the bearings is lost. Furthermore, high rotor speed comes with high mechanical stress in the flywheels. Therefore, an array of VSCMGs for energy storage and attitude control would have to trade the additional mass, electronics, and complexity for these modest power benefits. More traditional power subsystems are likely still the best choice.

NASA and the Air Force Research Laboratory (AFRL) have performed considerable research and development in the area of integrated power and attitude control (IPACS) (Notti et al. [61]). A popular program that looked at the use of flywheel energy storage and energy control is the Flywheel Attitude Control and Energy Transmission System (FACETS) formerly at AFRL in Kirtland AFB New Mexico (Fausz et al. [62]). In this program, three VSCMG were designed by Honeywell for AFRL (see Fig. 6.19) and were tested at AFRL on the ASTREX air-bearing satellite simulator by AFRL, Honeywell, and its contractors.

During this study, algorithms were developed to be validated on FACETS and published in the literature (Tsiotras et al. [63] and Yoon and Tsiotras [64]). NASA continued to study IPACS at their Glenn Research Center (McLallin et al. [65]). The programs were eventually cut, and there has been little applied research in this area since.

6.9 Optimization and Arbitrary Array Configurations

If the spacecraft's inertia and the statistics of its expected pointing time history are known a priori, then optimizing the CMG array geometry can take advantage of this knowledge. Such optimization seeks to increase the amount of non-singular angular

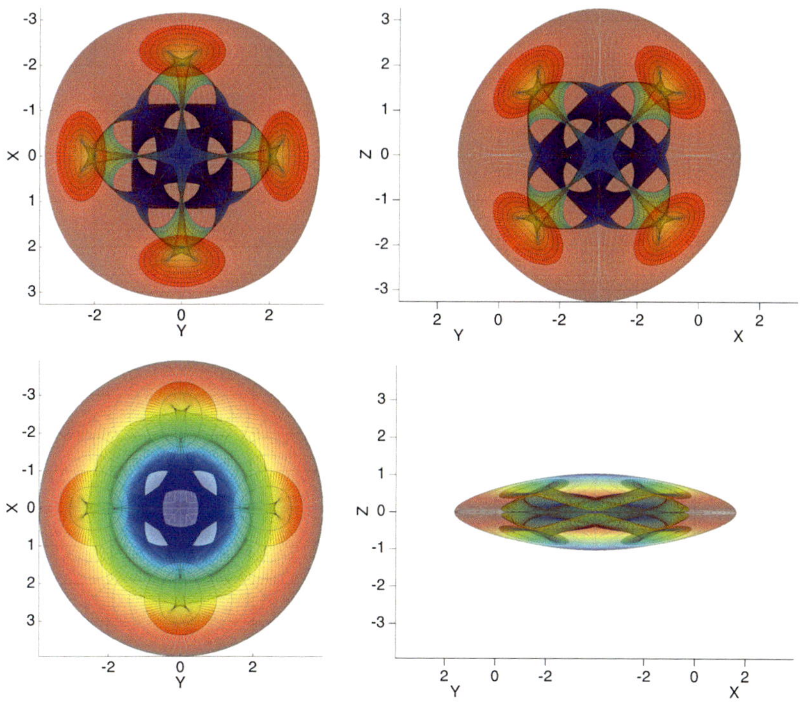

Fig. 6.20 Singular surfaces for a four-CMG pyramid array $\beta = 54.74°$ (top) and $\beta = 30°$ (bottom)

momentum, the size or shape of the angular momentum,or some combination of the two. Examples of optimization for these mission-specific parameters may be found in Leve [42] and Sands [43].

A fairly intuitive example of a optimizing the shape of the angular-momentum envelope is shown in Figs. 6.20 and 6.21. The objective of this optimization is to maximize the angular momentum about two of the three axes, i.e. to make the envelope more oblate. Reducing the CMG array skew angle β (See Chap. 4) has this effect, providing more angular momentum in the pitch/yaw plane. However, Fig. 6.20 shows that the internal singularities appear throughout the angular momentum envelope and are both elliptic and hyperbolic for the four-CMG pyramid array. In contrast, the internal singularities for the 6 CMG roof array in Fig. 6.21 are merely hyperbolic, offering the possibility of a nonsingular solution for angular momentum value within the envelope.

As Figs. 6.20 and 6.21 show, the internal singular surfaces for both arrays differ, and those for the four-CMG pyramid array are spread throughout the entire momentum envelope including passable and impassable singularities. Therefore, even when the spacecraft inertia and its jerk, acceleration, and rate requirements are not considered in the optimization process, the location of internal singularities and their type is of higher importance than the volume of useable angular momentum.

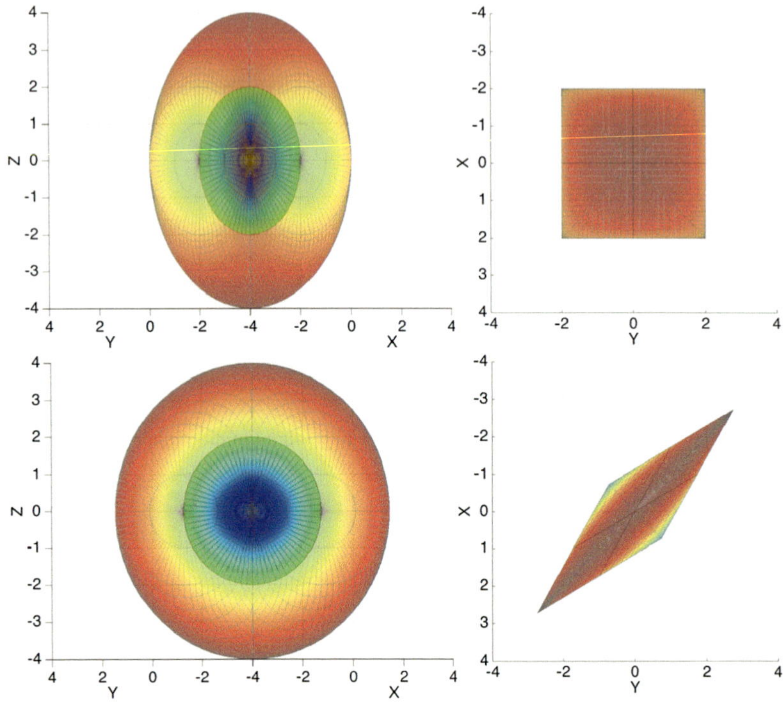

Fig. 6.21 Singular surfaces for a four-CMG roof array $\beta = 45°$ (top) and $\beta = 15°$ (bottom)

This example makes it clear that the shape and size of the angular momentum envelope is not the only driver of CMG array architecture. The architecture must also offer a workable singularity structure.

6.10 Chapter Summary

This chapter discussed how the performance of RWA and CMG arrays depend on the momentum devices' orientation within the spacecraft. Performance envelopes were generated for RWA and CMG arrays in terms of maximum angular momentum surfaces and, for the case of CMG arrays, internal singular surfaces. The geometry and the physical significance of these surfaces was discussed. Some common arrays used in practice were also introduced. It was shown that although it may seem beneficial to optimize an array of CMG for the shape of the angular momentum envelope, the more important requirements are vehicle rate and acceleration, which are driven by vehicle inertia. The types of internal singularities limit the array momentum that can be used to meet these agility requirements.

6.11 Exercises

1. How many CMGs, operated as scissored pairs (two CMGs per pair), are required to achieve instantaneous three-axis attitude control?

   (a) 1
   (b) 2
   (c) 4
   (d) 6

2. Which is true of a spacecraft with three operating reaction wheels and no other ACS actuators?

   (a) It needs no attitude sensors to estimate attitude.
   (b) It is capable of three-axis attitude control if the wheels' spin axes are along linearly independent directions
   (c) It can achieve high-agility motions with low electrical power.

3. An array of four reaction wheels has as its Jacobian matrix

$$A = \frac{\sqrt{2}}{2}\begin{bmatrix} 1 & 0 & -1 & 0 \\ 0 & -1 & 0 & 1 \\ 1 & 1 & 1 & 1 \end{bmatrix}$$

such that

$${}^{\mathrm{B}}h = A\frac{\sqrt{2}}{2}\begin{bmatrix} h_1 \\ h_2 \\ h_3 \\ h_4 \end{bmatrix} \text{Nms}$$

where ${}^{\mathrm{B}}h$ is the total wheel momentum in a spacecraft body-fixed rotating frame B in terms of some body-fixed basis vectors. The maximum angular momentum possible in a single wheel is 100 Nms.

$$h_\alpha = \begin{bmatrix} h_1 \\ h_2 \\ h_3 \\ h_4 \end{bmatrix} = \begin{bmatrix} 15 \\ 30 \\ -20 \\ -25 \end{bmatrix} \text{Nms} \qquad h_\beta = \begin{bmatrix} h_1 \\ h_2 \\ h_3 \\ h_4 \end{bmatrix} = \begin{bmatrix} 17.5 \\ 27.5 \\ -17.5 \\ -27.5 \end{bmatrix} \text{Nms}$$

   (a) Does $h_\alpha - h_\beta$ lie entirely in the null space of A?
   (b) Find h_1, h_2, h_3 and h_4 such that Nms and the kinetic energy in the wheels is minimized.
   (c) In what situations is this array singular?

Chapter 7
In-Flight Estimation of Spacecraft and Momentum-Control System Parameters

The term mass properties refers to the set of parameters that describes the distribution of mass in a rigid body. These parameters include moments of inertia, products of inertia, mass, and center of mass. It is common for some of these parameters to be measured relative to the spacecraft's manufacturing coordinate system before a spacecraft launched and for all to be predicted for the on-orbit lifetime. However, not all are measured, and the predictions may include assumptions (for example, about propellant usage) that do not match the real system to some extent. Such information can be valuable, for example, in verifying the success of a deployment maneuver which is predicted to change the inertia matrix in a certain way. It can also be indispensable in diagnosing anomalous performance, characterizing broken appendages, or recovering from an unanticipated loss of attitude control. Knowledge of the mass properties can help operators select optimized control-system gains to improve performance or recover the spacecraft.

Knowing the alignment and other parameters of an attitude control system's actuators and sensors has similar benefits. The problem of estimating sensor alignments relative to some preferred coordinate system is beyond the scope of this book, and it has been treated successfully in work by Schuster [66, 67], Pittelkau [68], and Crassidis [69, 70], among many authors. This chapter focuses on estimating the alignment of momentum-control system parameters, i.e. RWA and CMG alignments, including approaches to estimate mass properties and MCS parameters simultaneously. They are presented as simple batch or recursive least-squares solutions, although Kalman filters, and even more sophisticated schemes involving machine learning certainly can be integrated into these fundamental results.

The environmental torques τ applied to spacecraft with an MCS–a gyrostat–are small enough not to disturb its angular momentum significantly over a time interval on the scale of the spin period. This principle is the basis for momentum-bias satellites, which largely hold the spin-axis attitude despite the presence of such disturbances. In the case of negligible applied torque, the equations of

© The Author(s), under exclusive license to Springer Nature Switzerland AG 2025

F. A. Leve et al., *Spacecraft Momentum Control Systems*, Space Technology Library 46, https://doi.org/10.1007/978-3-031-96165-6_7

motion contain a cyclic coordinate, a momentum integral. The angular-momentum magnitude and direction are constant in an inertial frame. Because practical implementations of an inertia-estimation technique must accommodate digital, or discretized, measurements of the angular velocity and wheel speeds, the following derives a discrete-measurement version of the equations of motion. This derivation takes advantage of the presence of this momentum integral.

The following derivation follows Peck's work in this area, [71, 71]. It begins with the assumption that τ is small or that some other least-squares filtering can remove it. Then, for two measurements instances 1 and 2,

$$\mathbf{H}_1 = \mathbf{H}_2 \ldots \mathbf{H}_n \tag{7.1}$$

which holds for any pair of the n measurements. For two different vectors ω and h, this result becomes

$$\mathbf{I} \cdot \omega_1 + \mathbf{h}_1 = \mathbf{I} \cdot \omega_2 + \mathbf{h}_2 \tag{7.2}$$

Projecting Eq. (7.2) onto some basis vectors needs to accommodate the estimation scheme's assumption that the six unique quantities in I remain constant. Doing so uses a transformation that accounts for the relative attitude change during the time between the two measurements. This transformation can be computed (for example) from a quaternion propagated from measurement time 1 to time 2 or it can be taken directly from some on-board attitude determination algorithm. Using this transformation is analogous to integrating (or somehow filtering) ω in Eq. (7.2). In matrix form,

$$^2_1Q(I\omega_1 + h_1) = I\omega_2 + h_2 \tag{7.3}$$

in the form $A\Theta = B + \nu$ (convenient for identifying the parameter array Θ), equation is

$$\left(^2_1Q\tilde{\omega}_1 - \tilde{\omega}_2\right)\tilde{I} = h_2 - {}^2_1Qh_1 + \nu \tag{7.4}$$

for some stochastic noise process ν, where

$$\tilde{\omega} = \begin{bmatrix} \omega_1 & 0 & 0 & \omega_2 & \omega_3 & 0 \\ 0 & \omega_2 & 0 & \omega_1 & 0 & \omega_3 \\ 0 & 0 & \omega_3 & 0 & \omega_1 & \omega_2 \end{bmatrix} \tag{7.5}$$

$$\tilde{I} = \left[I_{11}\ I_{22}\ I_{33}\ I_{12}\ I_{13}\ I_{23}\right]^T \tag{7.6}$$

Note that

$$I\omega = \tilde{\omega}\tilde{I} \tag{7.7}$$

The indices on the scalar components in Eq. (7.5) and (7.6) are those of the angular-rate measure numbers in Eq. (7.3), not to be confused with the measurement-time indices 1 and 2. With many measurements,

$$A = \begin{bmatrix} {}^{m}_{1}Q\tilde{\omega}_1 - \tilde{\omega}_m \\ {}^{m}_{2}Q\tilde{\omega}_2 - \tilde{\omega}_m \\ \vdots \\ {}^{m}_{n}Q\tilde{\omega}_n - \tilde{\omega}_m \end{bmatrix} \tag{7.8}$$

and

$$B = \begin{bmatrix} h_m - {}^{m}_{1}Qh_1 \\ h_m - {}^{m}_{2}Qh_2 \\ \vdots \\ h_m - {}^{m}_{n}Qh_n \end{bmatrix} \tag{7.9}$$

Then the weighted least-squares solution, assuming a known measurement covariance V, is

$$\tilde{I} = \left(A^T V^{-1} A\right)^{-1} A^T V^{-1} B \tag{7.10}$$

if A is full rank. The rank-deficient case is readily addressed by inverting only the nonzero singular values. An advantage of this method over one that requires continuous, streaming measurements is that the index m can be chosen arbitrarily for each row in A and B. This feature leads to a convenient form for batch least-squares estimation: the index m is selected randomly, and the least-squares solution for I is computed recursively until it converges to within some user-defined tolerance. This recursive least-squares approach has been implemented successfully for existing spacecraft, such as Messenger [72].

In the event that the full transformation ${}^{m}_{n}Q$ is unknown, but that observations of a single vector (such as the sun unit vector in the body coordinates) are known, the momentum-integral equations can be projected onto this single vector. Again, for simplicity, we assume that H is constant in an inertial frame, as is the body-coordinates measurement of the inertial vector s. The angle between them is constant:

$$s_1^T \left(I\omega_1 + h_1\right) = s_2^T \left(I\omega_2 + h_2\right) \tag{7.11}$$

Then the $A\Theta = B$ relation for a single measurement becomes

$$\left(s_1^T \tilde{\omega}_1 - s_2^T \tilde{\omega}_2\right) \tilde{I} = s_2^T h_2 - s_1^T h_1 \tag{7.12}$$

and for multiple measurements,

$$A = \begin{bmatrix} \left(s_1^T \tilde{\omega}_1 - s_m^T \tilde{\omega}_m\right) \\ \left(s_2^T \tilde{\omega}_2 - s_m^T \tilde{\omega}_m\right) \\ \vdots \\ \left(s_n^T \tilde{\omega}_n - s_m^T \tilde{\omega}_m\right) \end{bmatrix} \tag{7.13}$$

and

$$B = \begin{bmatrix} s_m^T h_m - s_1^T h_1 \\ s_m^T h_m - s_2^T h_2 \\ \vdots \\ s_m^T h_m - s_n^T h_n \end{bmatrix} \tag{7.14}$$

where m is an index chosen by the user, as before, and which may differ for each row in A and B.

7.1 General Solution

The general solution accounts for external torques to the extent they are known. This section focuses on environmental torques because flight data for vehicle motions in the presence of these effects can be extensive and rich, while thruster torques are much less well known: unexpected variation may exceed 10%. That variation translates directly into error in the mass-properties estimates. Still, these principles can be adapted to the application of thruster torque if needed. It begins with

$$\frac{{}^{\mathrm{N}}\mathrm{d}}{\mathrm{dt}}\mathbf{H} = \tau \tag{7.15}$$

Integration leads to

$$\mathbf{H}_2 - \mathbf{H}_1 = \int_{t_1}^{t_2} \tau \mathrm{dt} \tag{7.16}$$

In terms of the B and N axes **b** and **n** and the 3x1 matrix ${}^{\mathrm{N}}\tau$ of τ components in N axes,

$$\mathbf{b}(t_2)^{\mathrm{B}} H_2 - \mathbf{b}(t_1)^{\mathrm{B}} H_1 = \int_{t_1}^{t_2} \mathbf{n}^{\mathrm{N}} \tau \mathrm{dt} \tag{7.17}$$

Where, of course, the **n** axes are constant but the **b** axes evolve with the rigid-body kinematics. Premultiplying by $\mathbf{b}(t_1)$ yields

$$\mathbf{b}(t_1)^T\mathbf{b}(t_2)^{\mathrm{B}}H_2 - {}^{\mathrm{B}}H_1 = \int_{t_1}^{t_2} \mathbf{b}(t_1)^T\mathbf{n}^{\mathrm{N}}\tau \mathrm{dt} \tag{7.18}$$

Representing **n** in terms of **b** results in

$$\mathbf{b}(t_1)^T\mathbf{b}(t_2)^{\mathrm{B}}H_2 - {}^{\mathrm{B}}H_1 = \int_{t_1}^{t_2} \mathbf{b}(t_1)^T\mathbf{b}\,{}^{\mathrm{B(t)}}_{\mathrm{N}}Q(t)^{\mathrm{N}}\tau \mathrm{dt} = {}^{\mathrm{B}(t_1)}_{\mathrm{N}}Q\int_{t_1}^{t_2} {}^{\mathrm{N}}\tau \mathrm{dt} \tag{7.19}$$

which results in

$${}^{\mathrm{B}(t_1)}_{\mathrm{B}(t_2)}Q^{\mathrm{B}}H_2 - {}^{\mathrm{B}}H_1 = {}^{\mathrm{B}(t_1)}_{\mathrm{N}}Q\int_{t_1}^{t_2} {}^{\mathrm{N}}\tau \mathrm{dt} \tag{7.20}$$

In terms of the measured quantities ω and h,

$$\left({}^{\mathrm{B}(t_2)}_{\mathrm{B}(t_1)}Q^{\mathrm{B}}\tilde{\omega}_2 - \tilde{\omega}_1\right)\tilde{I} = h_2 - {}^{2}_{1}Q + {}^{\mathrm{B}(t_1)}_{\mathrm{N}}Q\int_{t_1}^{t_2} {}^{\mathrm{N}}\tau \mathrm{dt} \tag{7.21}$$

As before, the $A\Theta = B + \nu$ relation can be found from this result, where

$$A = \begin{bmatrix} {}^{m}_{1}Q\tilde{\omega}_1 - \tilde{\omega}_m \\ {}^{m}_{2}Q\tilde{\omega}_2 - \tilde{\omega}_m \\ \vdots \\ {}^{m}_{n}Q\tilde{\omega}_n - \tilde{\omega}_m \end{bmatrix} \tag{7.22}$$

and

$$B = \begin{bmatrix} h_m - {}^{m}_{1}Qh_1 + {}^{\mathrm{B}(t_1)}_{\mathrm{N}}Q\int_{t_1}^{t_2} {}^{\mathrm{N}}\tau \mathrm{dt} \\ h_m - {}^{m}_{2}Qh_2 + {}^{\mathrm{B}(t_1)}_{\mathrm{N}}Q\int_{t_1}^{t_2} {}^{\mathrm{N}}\tau \mathrm{dt} \\ \vdots \\ h_m - {}^{m}_{n}Qh_n + {}^{\mathrm{B}(t_1)}_{\mathrm{N}}Q\int_{t_1}^{t_2} {}^{\mathrm{N}}\tau \mathrm{dt} \end{bmatrix} \tag{7.23}$$

An important conclusion to draw from this exercise is that incorporating inertially applied torques requires some knowledge of the attitude in inertial reference axes—i.e., the attitude of B in N must be known explicitly, where in Eq. (7.4) it is only the relative motion of B that matters. Also, the integral in Eq. (7.21) presupposes some means of obtaining continuous samples of τ. These features of the applied-torque solution make it less appealing than the torque-free solution because the estimation now requires continuous data and absolute attitude knowledge. Thus, one would prefer to use the torque-free solution in practice if possible. If the external torques are known in body coordinates, they can instead be incorporated via

$$H_1 + \int_1^2 {}^1_t Q\tau(t)\mathrm{dt} = H_2 \tag{7.24}$$

which requires relative-attitude tracking from some arbitrary initial orientation.

7.2 Gravity Gradient

The gravity-gradient torque can be modeled as an explicit, linear function of the mass properties. The fact that it is linear in **I** allows this torque to be used to augment A. A derivation is provided here. The effect can even be used in the absence of rich angular-velocity data to improve mass-properties estimates.

The gravitation field, of course, varies with altitude (an inverse-square dependence). This variation is manifested in the force on a satellite, which varies across the satellite's physical extent. Differential elements of the spacecraft's mass closer to the planet experience slightly stronger gravitational force than elements farther away. The result is that gravity acts on the satellite through a point that is offset from the mass center, producing the gravity-gradient torque about the mass center. So, the gravity-gradient torque depends on the distribution of mass in the satellite. For a satellite that is small enough that the gradient of the gravitational field is approximately linear across the satellite, the gravity-gradient torque $\boldsymbol{\tau}_g$ is linear in **I**:

$$\boldsymbol{\tau}_g = \frac{3\mu}{|\mathbf{r}|^3}\hat{\mathbf{r}} \times \mathbf{I} \cdot \hat{\mathbf{r}} \tag{7.25}$$

where μ is the gravitational constant ($3.986012106\,\mathrm{km}^3/\mathrm{s}^2$ for Earth), and **r** is the vector from the spacecraft to the Earth's center (more precisely, the barycenter of the Earth-Moon system). The vector $\hat{\mathbf{r}}$ is a unit vector parallel to **r**. Since $\hat{\mathbf{r}}$ appears quadratically, it might be defined equivalently as the vector from the barycenter to the satellite; the difference in sign from the previous definition would be canceled.

If $\boldsymbol{\tau}_g$ is taken to be the only torque acting on the gyrostat, the equation of motion becomes

$$\mathbf{I} \cdot \dot{\omega} + \dot{\mathbf{h}} + \omega \times (\mathbf{I} \cdot \omega + \mathbf{h}) = \frac{3\mu}{|\mathbf{r}|^3}\hat{\mathbf{r}} \times \mathbf{I} \cdot \hat{\mathbf{r}} \tag{7.26}$$

In terms of components in body axes and after implementing the tilde operator defined in Eq. (7.5), Eq. (7.36) can be written as a linear relation in I:

$$\left(\tilde{\dot{\omega}} + \omega^\times \tilde{\omega} - \frac{3\mu}{|r|^3}\hat{r}^\times \tilde{\hat{r}}\right)\tilde{I} = -\left(\dot{h} - \omega^\times h\right) \tag{7.27}$$

The form in which the derivatives do not appear would make this useful for a wide variety of estimators. However, to arrive at that form, we must modify the earlier torque-free result to accommodate the external torque. To this end, we note that

$$\frac{{}^{\mathrm{N}}\mathrm{d}}{\mathrm{dt}}\mathbf{H} = \frac{3\mu}{|\mathbf{r}|^3}\hat{\mathbf{r}} \times \mathbf{I} \cdot \hat{\mathbf{r}} \tag{7.28}$$

resolved in body axes can be written

$$\frac{{}^{\mathrm{N}}\mathrm{d}}{\mathrm{dt}}\mathbf{b}H = \frac{3\mu}{|\mathbf{r}|^3}(\mathbf{b}\hat{r}) \times \mathbf{b}I\mathbf{b}^{\mathbf{T}} \cdot \mathbf{b}\hat{r} \tag{7.29}$$

Premultiplying by $\mathbf{n}^T$ and canceling some terms leads to

$$\frac{{}^{\mathrm{N}}\mathrm{d}}{\mathrm{dt}}\mathbf{n}^T\mathbf{b}H = \mathbf{n}^T\mathbf{b}\frac{3\mu}{|\mathbf{r}|^3}\hat{r}^{\times}I\hat{r} \tag{7.30}$$

We recognize that $\mathbf{n}^T\mathbf{b}(t) = {}^N_B Q(t)$. Substituting this term and integrating leads to

$${}^{\mathrm{N}}_{\mathrm{B}}Q(t_2)H(t_2) - {}^{\mathrm{N}}_{\mathrm{B}}Q(t_1)H(t_1) = \int_{t_1}^{t_2} {}^{\mathrm{N}}_{\mathrm{B}}Q(t)\left(\frac{3\mu}{|\mathbf{r}|^3}\hat{r}^{\times}I\hat{r}\right)\mathrm{dt} \tag{7.31}$$

Having to know the attitude of B in N can be bypassed if we premultiply by ${}^{\mathrm{N}}_{\mathrm{B}}Q(t_1)^T$ as before, where t_1 is an arbitrary reference time that can be chosen from among the available data. Also, we must assume that r is known in B axes (which would be the case for an Earth-pointed satellite). Substituting $H = I\omega + h$ and using the tilde operator on ω, r, and I leads to

$$\left({}^{\mathrm{B}(t_1)}_{\mathrm{B}(t_2)}QI\tilde{\omega}(t_2) - \tilde{\omega}(t_1) - 3\mu\int_{t_1}^{t} {}^{\mathrm{B}(t_1)}_{\mathrm{B(t)}}Q\frac{\hat{r}^{\times}\tilde{\hat{r}}}{|\mathbf{r}|^3}\mathrm{dt}\right)\tilde{I} = h(t_1) - {}^{\mathrm{B}(t_1)}_{\mathrm{B}(t_2)}Qh(t_2) \tag{7.32}$$

Equation (7.32) is the general form of the linear relation in I that provides the basis for estimating the inertia in the presence of gravity-gradient torques.

7.3 Earth-Pointed Gyrostats with Known Ephemerides

A practical reason for preferring the form of this equation without derivatives is that it allows for gaps in data. However, performing the integration that appears in Eq. (7.32) requires continuous data. In the following development we examine a special case for which this term can be solved explicitly without numerical integration: an Earth-pointed satellite in an elliptical, restricted two-body orbit.

For this case, the angular velocity of the body in N is equal to the orbital angular rate (the derivative of the true anomaly, ν). This angular velocity keeps a given reference vector fixed in the B frame pointed at the Earth as the satellite orbits it. Typically, an onboard active attitude-control system achieves this goal. Thus, the angular velocity of B in N is perpendicular to the orbital plane, whose normal we denote with $\hat{\mathbf{n}}$:

$$\omega^{\mathrm{B/N}}(t) = \Omega(t)\hat{\mathbf{n}} \tag{7.33}$$

The direction of r in B is therefore constant, although for an elliptical orbit the magnitude of r varies with ν. For similar reasons, Ω is constant for a circular orbit but varies for an elliptical one. The change in body attitude Q is therefore such that it is unaffected. Therefore, can be taken outside the integral:

$$3\mu \int_{t_1}^{t_2} {}^{\mathrm{B}(t_1)}_{\mathrm{B}(t)} Q \frac{\hat{r}^{\times}\tilde{\hat{r}}}{|\mathbf{r}|^3} \mathrm{dt} = 3\mu\hat{r}^{\times}\tilde{\hat{r}} \int_{t_1}^{t_2} \frac{1}{|\mathbf{r}|^3} \mathrm{dt} \tag{7.34}$$

It is one of the awkward features of the two-body problem that there is no explicit function for $r(t)$. Therefore, in order to deal with the integral, we perform a change of variables and use the true anomaly ν in place of r:

$$r = \frac{a(1-e^2)}{1+e\cos(\nu)} \tag{7.35}$$

where a is the orbital semimajor axis, and e is the eccentricity [73]. The differential $\mathrm{d}\nu$ is found

$$\begin{aligned} r\dot{\nu} &= \left(\frac{\mu}{a(1-e^2)}\right)^{\frac{1}{2}} (1+e\cos(\nu)) \\ \Rightarrow \mathrm{d}t &= \frac{\left(a(1-e^2)\right)^{\frac{3}{2}}}{\mu^{\frac{1}{2}}(1+e\cos(\nu))^2} \mathrm{d}\nu \end{aligned} \tag{7.36}$$

The integral is then

$$\begin{aligned} &3\mu\hat{r}^{\times}\tilde{\hat{r}} \int_{\nu_1}^{\nu_2} \left(\frac{1+e\cos(\nu)}{a(1-e^2)}\right)^3 \frac{\left(a(1-e^2)\right)^{\frac{3}{2}}}{\mu^{\frac{1}{2}}(1+e\cos(\nu))^2} \mathrm{dt} \\ &= \frac{3\mu^{\frac{1}{2}}}{\left(a(1-e^2)\right)^{\frac{3}{2}}} \hat{r}^{\times}\tilde{\hat{r}} \int_{\nu_1}^{\nu_2} (1+e\cos(\nu))\,\mathrm{dt} \end{aligned} \tag{7.37}$$

The result can be found explicitly:

$$\frac{-3\mu^{\frac{1}{2}}}{\left(a(1-e^2)\right)^{\frac{3}{2}}}\hat{r}^{\times}\tilde{\hat{r}}\left[(\sin(\nu_2)-\sin(\nu_1)e)+\nu_2-\nu_1\right] \tag{7.38}$$

For a circular orbit, where $|r| = a$

$$-3\mu\hat{r}^{\times}\tilde{\hat{r}}\int_{\nu_1}^{\nu_2}\frac{1}{|r|^3}\mathrm{dt} = \frac{3\mu}{|r|^3}\hat{r}^{\times}\tilde{\hat{r}}(\nu_2-\nu_1) \tag{7.39}$$

Thus, the scale of this term's contribution is determined merely by the time (or true anomaly) between the measurements 1 and 2. For this Earth-pointed case the linear relation in I is therefore

$$= \frac{3\mu^{\frac{1}{2}}}{\left(a(1-e^2)\right)^{\frac{3}{2}}}\left[(\sin(\nu_2)-\sin(\nu_1))\,e+\nu_2-\nu_1\right](\hat{r}^{\times}\tilde{\hat{r}})\tilde{I} = h(t_1) - {}^{\mathrm{B}(t_1)}_{\mathrm{B}(t_2)}Qh(t_1) \tag{7.40}$$

As before, a single pair of measurements is insufficient to determine all of I because I has six unique values, while $(\hat{r}^{\times}\tilde{\hat{r}})$ is not rank 6. In fact, it is clear that its rank is 2 (at most) because rank$(\tilde{r})=2$. Furthermore, if (as we supposed) r is constant in B, stacking many of these matrices will not increase the rank. We require at least three linearly independent r for this special case of all of I is to be estimated.

7.4 Estimation via Equilibrium Motions–Observability

Now we examine an even more special case: a gyrostat with zero wheel momentum (i.e., a rigid body). We have shown that the orbital ephemeris becomes a scalar term in the estimation equation. Without wheel momentum it is

$$s(\hat{r}^{\times}\tilde{\hat{r}})\tilde{I} = 0 \tag{7.41}$$

where we have used s to represent the scalar term for notational simplicity. It is also well known that the relative equilibria of a rigid body subject to gravity-gradient torques are such that one principal axis of the inertia matrix is parallel to $\mathbf{r}$, and one is parallel to $\mathbf{n}$. Thus, the principal coordinates are completely determined if we can observe $\mathbf{r}$ and $\mathbf{n}$. In fact, the principal coordinate system can be estimated optimally via attitude-determination methods using many measurements of $\mathbf{r}$ and $\mathbf{n}$ from satellite range data and from on-board attitude sensors. These many measurements are combined to establish the orientation ${}^{\mathrm{B}}_{\mathrm{P}}Q$ of the principal coordinates relative to

B. And if the principal moments of inertia Ip can be estimated, the inertia matrix in B axes is simply

$$I = {}^{\mathrm{B}}_{\mathrm{P}}Q I_P \tag{7.42}$$

By our earlier arguments, knowing the principal coordinates leaves three unknown inertia parameters, the principal moments of inertia. Since, by definition, the products of inertia are zero (and therefore known) in this frame, we can truncate the array of parameters to be estimated so that

$$s \begin{bmatrix} 0 & -r_3 & r_2 \\ r_3 & 0 & -r_1 \\ -r_2 & r_1 & 0 \end{bmatrix} \begin{bmatrix} r_1 & 0 & 0 \\ 0 & r_2 & 0 \\ 0 & 0 & r_3 \end{bmatrix} \begin{bmatrix} I_1 \\ I_2 \\ I_3 \end{bmatrix} \tag{7.43}$$

where I_i are the principal moments of inertia. Thus,

$$\begin{bmatrix} I_1 \\ I_2 \\ I_3 \end{bmatrix} \in \mathrm{Null}\left\{ \begin{bmatrix} 0 & -r_3 r_2 & r_3 r_2 \\ r_3 r_1 & 0 & -r_3 r_1 \\ -r_2 r_1 & r_2 r_1 & 0 \end{bmatrix} \right\} \tag{7.44}$$

It is a simple matter to determine that the null space is parallel to $[1\ 1\ 1]^T$. We conclude that estimating the three moments of inertia requires enforcing a guess for one of the three. For example, if I_1 is assumed to be known by some other means, then we can add an arbitrary vector in the null space to the parameter vector:

$$\begin{bmatrix} 0 & -r_3 r_2 & r_3 r_2 \\ r_3 r_1 & 0 & -r_3 r_1 \\ -r_2 r_1 & r_2 r_1 & 0 \end{bmatrix} \begin{bmatrix} I_1 - I_1 \\ I_2 - I_1 \\ I_3 - I_1 \end{bmatrix} = 0 \tag{7.45}$$

which illustrates that the parameter vector can consist only of two unique values: in this example, the difference between I_1 and the other two moments of inertia. Furthermore, the scale of this difference is not observable–only the ratio $(I_2 - I_1)/(I_3 - I_1)$ is unambiguous.

In the event that the wheel angular momentum is nonzero, the situation is more subtle. Beck and Hall (and others) have revisited the problem of relative equilibria for circular orbits [74]. Here, rather than use some of the explicit results of this and earlier studies, we assume that the satellite remains earth pointed and that the dynamic balance condition described earlier holds in the presence of the gravity-gradient torque. I.e.,

$$\omega \times (\mathbf{I} \cdot \omega + \mathbf{h}) = \tau \tag{7.46}$$

or

$$\omega \times (\mathbf{I} \cdot \omega + \mathbf{h}) - \tau = 0$$

Projecting this result onto the angular-velocity vector, we find

$$\begin{aligned} &\omega \cdot [\omega \times (\mathbf{I} \cdot \omega] + \mathbf{h}) - \omega \cdot \tau = 0 \\ &0 - \omega \cdot \tau = 0 \end{aligned} \tag{7.47}$$

which implies that $\omega \perp \tau$. So, equivalently

$$\omega \times (\mathbf{I} \cdot \omega + \mathbf{h} + \frac{1}{\Omega^2} \omega \times \tau) = 0 \tag{7.48}$$

Now, using the procedure for torque-free gyrostats derived in Eqs. (7.33) through-(7.40), we arrive at the condition for a relative equilibrium in the presence of a body-fixed external torque, gravity gradient or otherwise:

$$h = \hat{\omega} h_s \hat{\omega}^\times \hat{\omega}^\times I \omega - \hat{\omega}^\times \frac{\tau}{\Omega}) = 0 \tag{7.49}$$

To construct an estimator from this result, we again collect on the components of the inertia matrix so that

$$(\hat{\omega}^\times \; \hat{\omega}^\times \; \tilde{\omega}) \tilde{I} = h + \hat{\omega}^\times \frac{\tau}{\Omega} - \hat{\omega} h_s \tag{7.50}$$

The presence of $\hat{\omega}^\times$ ensures that the rank of the measurement matrix is at most 2. In fact, its null space is spanned by

$$\left\{ \begin{matrix} \frac{\omega_1^2 - \omega_3^2}{\omega_1 \omega_3} \\ \frac{\omega_1}{\omega_3} \\ 0 \\ 0 \\ 1 \\ 0 \end{matrix} \right\}, \left\{ \begin{matrix} 1 \\ 1 \\ 1 \\ 0 \\ 0 \\ 0 \end{matrix} \right\}, \left\{ \begin{matrix} -\frac{\omega_2}{\omega_1} \\ -\frac{\omega_1}{\omega_2} \\ 0 \\ 1 \\ 0 \\ 0 \end{matrix} \right\}, \left\{ \begin{matrix} \frac{\omega_2}{\omega_3} \\ \frac{\omega_2^2 - \omega_3^2}{\omega_2 \omega_3} \\ 0 \\ 0 \\ 0 \\ 1 \end{matrix} \right\} \tag{7.51}$$

Thus, for a single equilibrium measurement, only two parameters are observable; furthermore, nothing with a component in the null space defined above is estimated accurately. Any guesses made to shore up the estimated must consist of vectors spanning this null space. Once again, at least three discrete equilibrium measurements of this type must be made (and the method of Eqs. (7.22) and (7.23) used) if all of the inertia-matrix elements are to be estimated without a priori knowledge.

7.5 Estimation of RWA and CMG Alignments

In developing this method, we will keep in mind the environment in which it is likely to be used: an operational setting, where possibly intermittent telemetry is recorded for later processing. Accordingly, this method is based on measurements that are directly available from typical spacecraft: attitude estimates, angular-velocity measurements, and estimated actuator outputs. Angular acceleration is rarely measured, and there is no simple sensor that directly measures the angular momentum of a spacecraft, desirable as that might be. Furthermore, this method must be able to accommodate sporadic information, for which the sampling rate is not constant and which may contain large periods of unusable data.

Recall that a spinning rigid spacecraft's momentum vector (**H**) is given by

$$\mathbf{H} = \mathbf{I} \cdot \omega^{B/N} + \mathbf{h} \tag{7.52}$$

where

- **I** is the inertia dyadic of the rigid spacecraft.
- B is a reference frame fixed in the spacecraft body.
- N is an inertial, or Newtonian, reference frame.
- $\omega^{B/N}$ is the angular velocity of B in N.
- **h** is the sum of the angular-momentum of internal components, such as momentum wheels or control-moment gyros (CMGs), that rotate about axes through their respective mass centers.

The assumption that the internal components are statically balanced about their angular-velocity vectors implies that the rigid body's mass center does not accelerate–that equations for the rotational dynamics completely describe the motion of the system. The ith CMG is characterized by the following:

- $\hat{\mathbf{g}}_i$ is the gimbal axis of the ith CMG and is fixed in B.
- $\omega^{G_i/B} = \dot{\delta}_i \hat{\mathbf{g}}_i$ is the angular velocity in B of a frame G_i that is fixed in the ith CMG in such a way that it rotates with the gimbal. δ_i is the gimbal rate.
- $\omega^{G_i/N} = \omega^{G_i/B} + \omega^{B/N} = \dot{\delta}_i \hat{\mathbf{g}}_i + \omega^{B/N}$ is the angular velocity of G_i in N.
- $\hat{\mathbf{s}}_i$ is the spin axis of the ith CMG's rotor.
- ω^{R_i/G_i} is the angular velocity in G_i of a frame R_i that is fixed in the CMG in such a way that it rotates with the wheel. Ω_i is the ith rotors spin rate.
- $\omega^{R_i/N} = \omega^{R_i/G_i} + \omega^{G_i/B} + \omega^{B/N} = \Omega_i \hat{\mathbf{s}}_i + \dot{\delta}_i \hat{\mathbf{g}}_i + \omega^{B/N}$ is the angular velocity of R_i in N.
- $\mathbf{I}_{gi}$ is the inertia dyadic of the rigid ith gimbal. $\mathbf{I}_{gi}$ is constant in G_i, but it is not constant in B unless the device under consideration is a momentum wheel or a reaction wheel.
- $\mathbf{I}_{ri}$ is the inertia dyadic of the rigid ith rotor. $\mathbf{I}_{ri}$ is constant in R_i and G_i.

Based on these definitions, the internal angular momentum of all k CMGs is

$$\begin{aligned}\mathbf{h} &= \sum_{i=1}^{k} \mathbf{h}_i \\ &= \sum_{i=1}^{k} \mathbf{I}_{gi} \cdot \omega^{\mathrm{G}_i/\mathrm{N}} + \mathbf{I}_{ri} \cdot \omega^{\mathrm{R}_i/\mathrm{N}} \\ &= \sum_{i=1}^{k} \mathbf{I}_{gi} \cdot (\dot{\delta}_i \hat{\mathbf{g}}_i + \omega^{\mathrm{B/N}}) + \mathbf{I}_{ri} \cdot (\Omega_i \hat{\mathbf{s}}_i + \dot{\delta}_i \hat{\mathbf{g}}_i + \omega^{\mathrm{B/N}})\end{aligned} \tag{7.53}$$

In convenient basis vectors, the attitude of the spacecraft and the alignment of the CMG components can be summarized via the following list of direction-cosine matrices:

- ${}^{\mathrm{N}}_{\mathrm{B}}Q$ is the attitude of B in N. This direction-cosine matrix is a parameterization of the spacecraft attitude. A numerically robust way to track ${}^{\mathrm{N}}_{\mathrm{B}}Q$ is by propagating Euler parameters (a quaternion), whose derivative is a simple function of $\omega^{\mathrm{B/N}}$. Alternatively, if a star-tracker or some other device can compute ${}^{\mathrm{N}}_{\mathrm{B}}Q$ from vector observations, the quaternions need not be propagated except for periods when, for example, no stars can be identified.
- ${}^{\mathrm{B}}_{\mathrm{G}_i}Q$ is the attitude, or alignment, of G_i in B. This quantity is of particular interest. Spacecraft that employ CMGs generally do so because they are required to slew quickly and accurately. Often, such slewing maneuvers incorporate a feed-forward torque: based on estimates of the rigid-body mass properties and the gimbal alignment, an open-loop torque profile can be constructed that comes close to an optimal closed-loop sequence. Naturally, closed-loop control is also incorporated to eliminate hang-off and inevitable imprecision in knowledge of the system parameters. The objective of this study is to estimate ${}^{\mathrm{B}}_{\mathrm{G}_i}Q$.
- ${}^{\mathrm{G}_i}_{\mathrm{R}_i}Q$ is the attitude of R_i in G_i. Let us assume that each rotor rotates about an axis of symmetry. I.e., there are no products of inertia with respect to $\hat{\mathbf{s}}_i$, and the transverse moments of inertia (those associated with axes perpendicular to $\hat{\mathbf{s}}_i$) are equal. Therefore, the wheel's inertia in G_i axes is constant. For some definition of R_i, its representation in G_i is equal to its representation in R_i. So, without loss of generality, ${}^{\mathrm{G}_i}_{\mathrm{R}_i}Q$ can be taken as the identity. A more rigorous way to account for the axisymmetry would be to define yet another frame between G_i and R_i, but this added rigor does not enhance the solution.

7.6 Parameterization of CMG Alignment

In order to parameterize the misalignment of the gimbals, we define a rotation ${}_iE$ that describes the error in the ith alignment such that

$${}^{\mathrm{B}}_{\mathrm{G}_i}Q = {}_iE^{\mathrm{B}}_{\mathrm{G}_i}\overline{Q} \tag{7.54}$$

where ${}^{\mathrm{B}}_{\mathrm{G}_i}\overline{Q}$ is the best guess of the alignment, or what the alignment is meant to be. Similarly, a best guess of ${}^{\mathrm{G}_i}\hat{\mathbf{g}}_i$ can be defined:

$$
{}^{\mathrm{B}}\hat{\mathbf{g}}_i = {}^{\mathrm{B}}_{\mathrm{G}_i}Q\,{}^{\mathrm{G}_i}\hat{\mathbf{g}}_i = {}_iE\,{}^{\mathrm{B}}_{\mathrm{G}_i}\overline{Q}\,{}^{\mathrm{G}_i}\hat{\mathbf{g}}_i = {}_iE\,{}^{\mathrm{B}}\overline{\hat{\mathbf{g}}}_i \tag{7.55}
$$

Since $\hat{\mathbf{s}}_i$ is fixed in G_i

$$
{}^{\mathrm{B}}\hat{\mathbf{s}}_i = {}^{\mathrm{B}}_{\mathrm{G}_i}Q\,{}^{\mathrm{G}_i}\hat{\mathbf{s}}_i = {}_iE\,{}^{\mathrm{B}}_{\mathrm{G}_i}\overline{Q}\,{}^{\mathrm{G}_i}\hat{\mathbf{s}}_i = {}_iE\,{}^{\mathrm{G}_i}\overline{\hat{\mathbf{s}}}_i \tag{7.56}
$$

We need a relation ${}_iE$ that allows to be estimated. Beginning with the matrix equation

$$
{}^{\mathrm{B}}h_i = {}^{\mathrm{B}}_{\mathrm{G}_i}Q\,{}^{\mathrm{G}_i}h_i \tag{7.57}
$$

substituting the above results, and collecting some terms ultimately leads to

$$
{}^{\mathrm{B}}h_i = {}_iE\,{}^{\mathrm{B}}_{\mathrm{G}_i}\overline{Q}\left[\left({}^{\mathrm{G}_i}I_{gi} + {}^{\mathrm{G}_i}I_{ri}\right){}^{\mathrm{B}}_{\mathrm{G}_i}\overline{Q}(\dot{\delta}\,{}^{\mathrm{B}}\overline{\hat{\mathbf{g}}}_i + {}^{\mathrm{B}}\omega^{\mathrm{B/N}}) + {}^{\mathrm{G}_i}I_{ri}\Omega_i\right] \tag{7.58}
$$

an expression in which the quantities to be estimated, the contents of ${}_iE$, have been factored out so that they appear only once. Since everything that postmultiplies ${}_iE$ is known, let this quantity be represented by

$$
A_i = {}^{\mathrm{B}}_{\mathrm{G}_i}\overline{Q}\left[({}^{\mathrm{G}_i}I_{gi} + {}^{\mathrm{G}_i}I_{ri}){}^{\mathrm{B}}_{\mathrm{G}_i}\overline{Q}(\dot{\delta}\,{}^{\mathrm{B}}\overline{\hat{\mathbf{g}}}_i + {}^{\mathrm{B}}\omega^{\mathrm{B/N}}) + {}^{\mathrm{G}_i}I_{ri}\Omega_i\right] \tag{7.59}
$$

for notational simplicity. Thus,

$$
{}^{\mathrm{B}}h_i = {}_iE\,A_i \tag{7.60}
$$

7.7 Mass Properties with Small Angle Misalignments

Although a useful intermediate step, Eq. (7.60) does not immediately provide a solution for ${}_iE$ because ${}^{\mathrm{B}}h_i$ is still unknown. To estimate it, we turn to an expression for the total system angular momentum and recognize that the B axes are in general time-varying in N:

$$
{}^{\mathrm{N}}H = {}^{\mathrm{N}}_{\mathrm{B}}Q(t)\,{}^{\mathrm{B}}H(t) \tag{7.61}
$$

Often one can successfully assume that H is fixed in N. In other words, the torques acting on the spacecraft are generally negligible compared to the angular momentum in the wheels and/or in the body. It will be shown later that this assumption can be relaxed in some cases of practical interest. For now, taking H to be constant in N leads to a useful form of conservation of angular momentum across two times t_0 and t_n:

$$ {}^{\mathrm{N}}_{\mathrm{B}}Q(t_0)^{\mathrm{B}}H(t_0) = {}^{\mathrm{N}}_{\mathrm{B}}Q(t)^{\mathrm{B}}H(t) \tag{7.62} $$

Equation (7.62) can be rewritten in a way that parameterizes the rigid-body inertia dyadic as a 6x1 matrix [72]:

$$ {}^{\mathrm{N}}_{\mathrm{B}}Q(t_0)^{\mathrm{B}}\left[\tilde{\omega}(t_0)\tilde{I} + h(t_0)\right] = {}^{\mathrm{N}}_{\mathrm{B}}Q(t)^{\mathrm{B}}H(t)\left[\tilde{\omega}(t)\tilde{I} + h(t)\right] \tag{7.63} $$

Collecting $\tilde{I}$ results in

$$ \left[{}^{\mathrm{N}}_{\mathrm{B}}Q(t_0)^{\mathrm{B}}\tilde{\omega}(t_0) - {}^{\mathrm{N}}_{\mathrm{B}}Q(t)^{\mathrm{B}}\tilde{\omega}(t)\right]\tilde{I} + {}^{\mathrm{N}}_{\mathrm{B}}Q(t)^{\mathrm{B}}h(t) - {}^{\mathrm{N}}_{\mathrm{B}}Q(t_0)^{\mathrm{B}}h(t_0) \tag{7.64} $$

The representation of the B axes in N changes from t_0 to t_n. This relative rotation is

$$ {}^{\mathrm{B}(t)}_{\mathrm{B}(t_0)}Q = {}^{\mathrm{N}}_{\mathrm{B}}Q(t)^{T}\,{}^{\mathrm{N}}_{\mathrm{B}}Q(t_0) \tag{7.65} $$

Multiplying Eq. (7.64) through by ${}^{\mathrm{N}}_{\mathrm{B}}Q(t_0)^{T}$ and using this identity in Eq. (7.65) yields

$$ \left[{}^{\mathrm{B}(t)}_{\mathrm{B}(t_0)}Q^{\mathrm{B}}\tilde{\omega}(t_0) - {}^{\mathrm{B}}\tilde{\omega}(t)\right]\tilde{I} = {}^{\mathrm{B}}h(t) - {}^{\mathrm{B}(t)}_{\mathrm{B}(t_0)}Q^{\mathrm{B}}h(t_0) \tag{7.66} $$

For notational simplicity let

$$ W(t_0, t) = \left[{}^{\mathrm{B}(t)}_{\mathrm{B}(t_0)}Q^{\mathrm{B}}\tilde{\omega}(t_0) - {}^{\mathrm{B}}\tilde{\omega}(t)\right] \tag{7.67} $$

Substituting the expression for CMG momentum in terms of the alignment error leads to

$$ W(t_0, t)\tilde{I} = \sum_{i=1}^{k} {}_{e}i\,QA_i(t) - {}^{\mathrm{B}(t)}_{\mathrm{B}(t_0)}Q^{\mathrm{B}}{}_{i}EA_i(t_0) \tag{7.68} $$

At this point the alignment error is taken to be small so that the rotation ${}_{i}E$ is approximately equal to the first-order terms of the Taylor-series expansion of $e^{-[{}^{\mathrm{B}}\theta_i]^{\times}}$ where ${}^{\mathrm{B}}\theta_i = [\theta_1\ \theta_2\ \theta_3]^T$ are infinitesimal rotations about the B axes and

$$ [{}^{\mathrm{B}}\theta_i]^{\times} = \begin{bmatrix} 0 & -\theta_3 & \theta_2 \\ \theta_3 & 0 & -\theta_1 \\ -\theta_2 & \theta_1 & 0 \end{bmatrix} \tag{7.69} $$

Then,

$$ {}_iE \approx 1 - [{}^{\mathrm{B}}\theta_i]^\times \tag{7.70} $$

Implementing the identity

$$ u^\times v = -v^\times u \tag{7.71} $$

leads to

$$ W(t_0, t)\tilde{I} = \sum_{i=1}^{k} \left[A_i(t) - {}^{\mathrm{B}(t)}_{\mathrm{B}(t_0)}Q^{\mathrm{B}} A_i(t_0) - \left([A_i(t)]^\times - {}^{\mathrm{B}(t)}_{\mathrm{B}(t_0)}Q[A_i(t_0)]^\times \right) {}^{\mathrm{B}}\theta_i \right] \tag{7.72} $$

The result is an expression that is linear in column matrices of the unknown parameters, $\tilde{I}$ and ${}^{\mathrm{B}}\theta_i$, from which follows

$$ \left\{ W(t_0, t) \; \left({}^{\mathrm{B}(t)}_{\mathrm{B}(t_0)}Q[A_i(t_0) - [A_i(t)]^\times -]^\times \right) \cdots \left({}^{\mathrm{B}(t)}_{\mathrm{B}(t_0)}Q[A_k(t_0) - [A_k(t)]^\times -]^\times \right) \right\} $$
$$ \times \begin{bmatrix} \tilde{I} \\ {}^{\mathrm{B}}\theta_1 \\ \vdots \\ {}^{\mathrm{B}}\theta_k \end{bmatrix} = \sum_{i=1}^{k} \left([A_i(t)]^\times - {}^{\mathrm{B}(t)}_{\mathrm{B}(t_0)}Q[A_i(t_0)]^\times \right) \tag{7.73} $$

Again, for the sake of simplicity, we represent the matrices of known quantities as $X(0, n)$ and $Y(0, n)$ and the column matrix of unknown parameters as Θ so that this unwieldy expression becomes

$$ Y(0, n) = X(0, n)\Theta \tag{7.74} $$

7.8 Least-Squares Solution

$X(0, n)$ is a $3 \times (3\mathrm{k} \times 6)$ matrix, and $Y(0, n)$ is $(3\mathrm{k} + 6) \times 1$. For a series of m measurement events we can cast this problem as a least-squares one with a loss function

$$ L(\Theta) = \sum_{n=1}^{m} [Y(0, n) - X(0, n)\Theta]^T [Y(0, n) - X(0, n)\Theta] \tag{7.75} $$

and a corresponding least-squares solution $(\ )^{\dagger}$

$$\Theta = \begin{bmatrix} X(0,1) \\ \vdots \\ X(0,n) \end{bmatrix}^{\dagger} \begin{bmatrix} Y(0,1) \\ \vdots \\ Y(0,n) \end{bmatrix} \tag{7.76}$$

The pseudoinverse shown here can be robustly computed by way of a singular-value decomposition (see Appendix C.5) or simply by

$$\begin{bmatrix} X(0,1) \\ \vdots \\ X(0,n) \end{bmatrix}^{\dagger} = \left\{ \begin{bmatrix} X(0,1) \\ \vdots \\ X(0,n) \end{bmatrix}^{T} \begin{bmatrix} X(0,1) \\ \vdots \\ X(0,n) \end{bmatrix} \right\}^{-1} \begin{bmatrix} X(0,1) \\ \vdots \\ X(0,n) \end{bmatrix}^{T} \tag{7.77}$$

if the measurement matrix has nonzero singular values. The solution is a least-squares estimate of the inertia dyadic in B axes and a least-squares fit of a small-angle representation of the alignments. The equations of motion have not been linearized in any way. Only the alignments have been represented approximately.

7.9 Large Angle Misalignments

A natural objection to this formulation is that it accommodates only small misalignments. Furthermore, even those have some error associated with the Taylor series approximation of the misalignment matrix. To remedy this situation we turn to a venerable body of work on the subject of estimating orthonormal matrices like ${}_iE$: optimal attitude determination.

There are very successful solutions to the problem, first posed by Wahba [75], of how to find the components of the orthonormal matrix Q that minimize the loss function

$$L(Q) = \sum_{j=1}^{n} a_j |{}^{B}v_j - Q^{A}v_j|^2, \tag{7.78}$$

where the v_j are known vector components in the two frames A and B whose relationship (the attitude Q) is to be found. The a_j are weighting factors often taken to be unity.

The solutions are legion, and Markley has summarized them quite effectively [69]. This section does not go into them in detail. Using an attitude-determination method with the CMG model described above requires that everything but the alignments be represented as pairs of vectors. For this, we return to the expression for the total angular momentum for two discrete measurement events and isolate the pth CMG, whose alignments are to be refined:

$$ {}^{\mathrm{N}}H = {}^{\mathrm{N}}_{\mathrm{B}}Q(t)\left[{}^{\mathrm{B}}\tilde{\omega}(t)\tilde{I} + \sum_{i=1,i\neq p}^{k} {}^{\mathrm{B}}h_i(t) + {}^{\mathrm{B}}h_p(t)\right] \tag{7.79} $$

Using some of the earlier definitions yields

$$ {}^{\mathrm{N}}_{\mathrm{B}}Q(t)^{T\mathrm{N}}H - {}^{\mathrm{B}}\tilde{\omega}(t)\tilde{I} - \sum_{i=1,i\neq p}^{k} {}^{\mathrm{B}}h_i(t) = {}_pEA_p(t) \tag{7.80} $$

Although perhaps difficult to recognize, this expression is in the form required for attitude determination:

$$ {}^{\mathrm{B}}\nu_j = Q^{\mathrm{A}}\nu_j \tag{7.81} $$

where the formulation of the CMG alignment problem maps to the attitude-determination one as

$$ \begin{aligned} &\frac{{}^{\mathrm{N}}_{\mathrm{B}}Q(t)^{\mathrm{B}^T\mathrm{N}}H - {}^{\mathrm{B}}\tilde{\omega}(t)\tilde{I} - \sum_{i=1,i\neq p}^{k} {}^{\mathrm{B}}h_i(t)}{|A_p(t)|} \rightarrow {}^{\mathrm{B}}\nu \\ &{}_iE \rightarrow Q \\ &\frac{A_p(t)}{|A_p(t)|} \rightarrow {}^{\mathrm{A}}\nu \\ &|A_p(t)|^2 \rightarrow a \end{aligned} \tag{7.82} $$

In practice there is no need to divide by $|A_p(t)|$ because the weighting factor a rescales the vector measurements anyway. It makes sense that the wheel-momentum measurements with the largest magnitude be weighted the most heavily–presumably the noise-to-signal ratio is smaller for such measurements and they are that much more trustworthy. In the absence of more explicit statistics, we use the (unconditional) mean of all n measurements to compute

$$ {}^{\mathrm{N}}H = \frac{1}{n}\sum_{j=1}^{n} {}^{\mathrm{N}}_{\mathrm{B}}Q(t_j)\left[{}^{\mathrm{B}}\tilde{\omega}(t_j)\tilde{I} + \sum_{i=1}^{k} {}^{\mathrm{B}}h_i(t_j)\right] \tag{7.83} $$

Using the measurements and the previously estimated I and ${}_iE, i \neq p$ to assemble H and A creates a set of n vector pairs for use with the q-method [69] or whatever attitude-determination routine is preferred. The attitude-determination step is repeated for each of the k CMGs. It is worth noting that the transformation from B to N may not be known. However, all that really matters is that the measurements be referenced to some common orientation of B, such as the 0th one, in whatever reference frame is used to express ${}^{\mathrm{B}(t_n)}_{\mathrm{B}(t_0)}Q$.

Equation (7.83) assumes that the attitude of B in N is available for each measurement so that a smoothed ${}^{\mathrm{N}}H$ can be used. If that is not the case, one can incorporate the measurement at each time step, sacrificing the benefits of the smoothed ${}^{\mathrm{N}}H$:

$$ {}^{\mathrm{N}}_{\mathrm{B}}Q(t_0)\left[{}^{\mathrm{B}}\tilde{\omega}(t_0)\tilde{I} - \sum_{i=1}^{k}{}^{\mathrm{B}}h_i(t_0)\right]\tilde{I} - {}^{\mathrm{B}}\tilde{\omega}(t)\tilde{I} - \sum_{i=1}^{k}{}^{\mathrm{B}}h_i(t) = {}_pEA_p(t) \qquad (7.84) $$

If further refinement is required, the following substitutions are made:

$$ \begin{aligned} {}_iE^{\mathrm{B}}_{\mathrm{G}_i}Q &\rightarrow {}^{\mathrm{B}}_{\mathrm{G}_i}\overline{Q} \\ {}_iE^{\mathrm{B}}\hat{\mathbf{g}} &\rightarrow {}^{\mathrm{B}}\overline{\hat{\mathbf{g}}} \end{aligned} \qquad (7.85) $$

Then the two-step sequence—linear parameter estimation and nonlinear attitude estimation—is repeated until ${}_iE$ converges to the identity within some tolerance. A useful check for such convergence is $\mathrm{tr}({}_iE)$, which approaches 3 as the rotation approaches identity.

Known applied torques can certainly be implemented, but often models of these are inaccurate and the torques themselves tend to be very small (and likely negligible) for practical spacecraft. An important exception for some Earth satellites is the gravity-gradient torque addressed above. Conveniently, this torque is accurately modeled as a linear function of the inertia matrix [25], suggesting that this torque can play a useful role in the linear estimation problem. The torque ${}^{B}\tau$ is given in vector-dyadic form by

$$ \boldsymbol{\tau} = 3\left(\frac{\mu}{R_c}\right)^3 \hat{\mathbf{o}} \times \mathbf{I}\cdot\hat{\mathbf{o}} \qquad (7.86) $$

where $\frac{\mu}{R_c}$ is related to the radial distance of the satellite from the barycenter and $\hat{\mathbf{o}}$ is the nadir direction (the vector from the spacecraft mass center to the Earth). Converting Eq. (7.86) to matrix form and using some of the operations in Eqs. (7.5) and (7.6) leads to

$$ {}^{\mathrm{N}}_{\mathrm{B}}Q(t){}^{\mathrm{B}}\tau = {}^{\mathrm{N}}_{\mathrm{B}}Q(t_j)3\left(\frac{\mu}{R_c}\right)^3\left(\hat{o}\times\tilde{\hat{o}}\right)\tilde{I} \qquad (7.87) $$

Generally everything on the RHS can vary with time except $\tilde{I}$. Integration provides the change in angular momentum in N from one measurement to another:

$$ {}^{\mathrm{N}}H(t) - {}^{\mathrm{N}}H(t_0) = \left[\int_{t_0}^{t}{}^{\mathrm{N}}_{\mathrm{B}}Q(t)3\left(\frac{\mu}{R_c}\right)^3\left(\hat{o}\times\tilde{\hat{o}}\right)\mathrm{dt}\right]\tilde{I} \qquad (7.88) $$

This effect can be included with $W(t_0, t)$ if the added precision warrants it:

$$\left\{W(t_0,t) + 3{}^{\rm N}_{\rm B}Q(t)\left[\int_{t_0}^{t} {}^{\rm N}_{\rm B}Q(\sigma)\left(\frac{\mu}{R_c}\right)^3\left(\hat{o}\times\tilde{\hat{o}}\right)\mathrm{d}\sigma\right]\right\}\tilde{I}$$
$$= \sum_{i=1}^{k}\left(A_i(t) - {}^{\rm B(t)}_{\rm B(t_0)}Q^{\rm B}A_i(t_0)\right) \tag{7.89}$$
$$+ \sum_{i=1}^{k} - \left([A_i(t)]^\times - {}^{\rm B(t)}_{\rm B(t_0)}Q - \left\{[A_i(t)]^\times - {}^{\rm B(t)}_{\rm B(t_0)}Q[A_i(t_0)]^\times\right\}{}^{\rm B}\theta_i\right){}^{\rm B}\theta_i$$

Then the least-squares solution and subsequent iterations proceed as before.

These linear relations in the inertia-matrix elements can be used to for estimation with batch filters, recursive least-squares filters, or Kalman filters. They are based on momentum integrals in the equations of motion and augmented with explicitly represented gravity-gradient torques. Equilibrium motions can also be used for this purpose, but the null space (the unobservable parameter subspace) of the measurement matrices suggest that only a few parameters can be observed independently in such cases.

7.10 Messenger Spacecraft

One of the first operational uses of momentum systems for parameter estimation is documented by Tanygin and Williams, who successfully tested a simplified version of the algorithms described above on flight data from the Simplified Aid for EVA Rescue (SAFER) system [76]. In this algorithm, gyroscopic effects are ignored by projecting the full equations onto the angular-velocity components. Lee and Wertz successfully used another simplified subset of the full equations to estimate Cassini mass properties [77].

However, the most general, published use of this estimation approach has been for the Messenger spacecraft. Messenger launched on August 3, 2004 and entered Mercury's orbit in March 2011. The spacecraft is the first to orbit the planet, mapping 100% of Mercury's surface during its primary mission. Its two extended missions ended in 2015 when the spacecraft impacted Mercury on April 30, 2015 [78]. O'Shaughnessy et al. report the results of inertia estimation that used flight data from maneuvers designed to improve star tracker and gyro alignment knowledge [79]. For each of the four maneuvers they consider, a pre-flight estimate of the inertia parameters and actuator alignments is available. Those pre-flight estimates are compared to the values extracted via a recursive estimator that they document in detail, and whose performance they compare to some of these earlier, simplified estimation schemes. They conclude that the approach documented in this chapter produces the lowest error, with residuals of a fraction of a percent. Shown

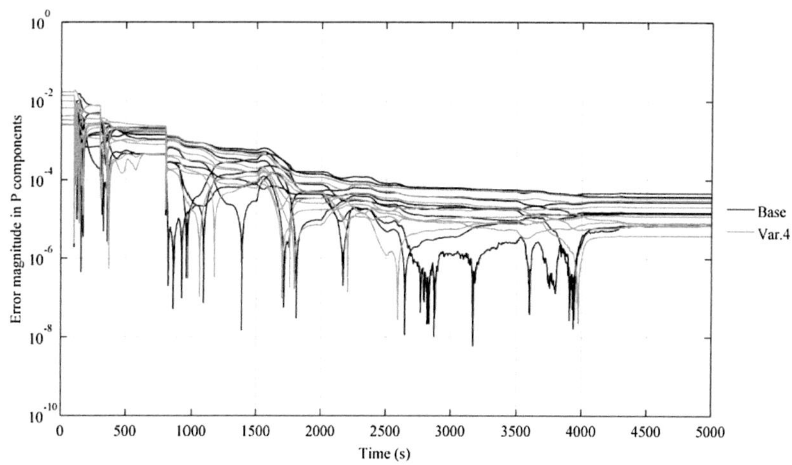

Fig. 7.1 Estimation alignment parameters for simulated maneuvers. The "P components" refer to misalignment values, and "Var. 4" indicates the preferred estimation approach

in Fig. 7.1 is a simulated dataset that resembles the flight data; that raw telemetry data is not available for publication, but the results in Tables 7.1 and 7.2 are based on the flight data itself.

They consider four variants of the estimation approach described above, including or excluding actuator-alignment estimates. Including actuator alignment produces better results, and Variant 4 as shown in the table is the preferred approach because it includes both momentum conservation and these alignment parameters, capturing the details of the true system.

7.11 Chapter Summary

The equations of motion can be rewritten to solve for parameters like inertia-matrix entries and momentum-control device alignments. In this form, the inertia matrices appear linearly, as do the alignments when the angular parameters are small enough that their coupling to each other and to other quantities is negligible. These linear equations enable numerous estimators to be devised–least-squares batch estimators and recursive (Kalman filter) estimators, among them. When the misalignments are large enough that attitude nonlinearities matter, attitude-estimation strategies based on Wahba's loss function can provide simultaneous solutions for the inertia parameters and misalignments.

Momentum-control activity in the form of wheel momentum and torque, or even external torques due to thrusters and gravity-gradient effects, can establish the scale of the inertia-matrix parameters that are estimated in this way. Without those known quantities, scale cannot be determined. Instead, dimensionless quantities such as the

Table 7.1 Inertia-parameter estimates produced by filters from Messenger data.

Calibration Maneuver date and filter configurations													
		2007 Day 262						2008 Day 182					
		Mech.	Base	Var.1	Var.2	Var.3	Var.4	Mech.	Base	Var.1	Var.2	Var.3	Var.4
I_{xx}	$(\mathrm{kg}\cdot\mathrm{m}^2)$	429.0	435.9	409.3	436.5	436.5	435.9	399.5	406.7	369.4	405.9	405.9	406.7
I_{yy}		427.2	438.5	420.0	438.2	438.2	438.5	404.8	430.4	409.1	430.4	430.4	430.4
I_{zz}		537.6	536.4	510.5	534.8	534.8	536.4	527.9	507.0	487.8	511.0	511.0	507.0
I_{xy}		−0.1	−3.9	8.2	−5.2	−5.2	−3.9	−0.1	−4.3	1.9	−3.3	−3.3	−4.3
I_{xz}		0.0	−0.2	−1.1	0.2	0.2	−0.2	0.0	0.2	−0.5	0.3	0.3	0.1
I_{yz}		4.2	5.0	−4.7	10.0	10.0	5.0	4.5	6.2	−1.0	8.7	8.7	6.2
σI_{xx}	$(\mathrm{kg}\cdot\mathrm{m}^2)$	–	0.106	9.321	0.097	0.097	0.106	–	0.116	8.305	0.093	0.093	0.116
σI_{yy}		–	0.134	8.379	0.064	0.064	0.134	–	0.171	8.429	0.064	0.064	0.171
σI_{zz}		–	0.149	15.62	0.112	0.112	0.149	–	0.231	14.69	0.110	0.110	0.231
σI_{xy}		–	0.070	9.584	0.053	0.053	0.070	–	0.063	9.252	0.053	0.053	0.063
σI_{xz}		–	0.088	9.966	0.073	0.073	0.088	–	0.088	9.969	0.071	0.071	0.088
σI_{yz}		–	0.100	9.886	0.056	0.056	0.100	–	0.100	9.983	0.055	0.055	0.100

		2008 Day 184						2009 Day 222					
		Mech.	Base	Var.1	Var.2	Var.3	Var.4	Mech.	Base	Var.1	Var.2	Var.3	Var.4
I_{xx}	$(\mathrm{kg}\cdot\mathrm{m}^2)$	399.5	407.4	393.1	406.7	406.7	407.4	308.5	389.3	355.8	389.1	389.1	389.3
I_{yy}		404.8	432.3	423.9	431.2	431.2	432.3	402.1	428.1	410.4	428.5	428.5	428.1
I_{zz}		527.9	506.9	500.0	512.0	512.0	506.9	508.8	494.8	469.6	496.0	496.0	494.8
I_{xy}		−0.1	−4.0	−2.5	−2.9	−2.9	−4.0	−0.1	−2.8	0.7	−2.7	−2.7	−2.8
I_{xz}		0.0	−0.0	0.7	0.2	0.2	−0.0	0.0	1.4	1.3	0.3	0.3	1.4
I_{yz}		4.5	4.7	0.5	6.9	6.9	4.7	4.5	6.1	0.4	7.1	7.1	6.1
σI_{xx}	$(\mathrm{kg}\cdot\mathrm{m}^2)$	–	0.110	8.312	0.093	0.093	0.110	–	0.116	9.060	0.097	0.097	0.116
σI_{yy}		–	0.165	8.499	0.064	0.064	0.165	–	0.176	7.905	0.065	0.065	0.176
σI_{zz}		–	0.191	14.63	0.111	0.111	0.191	–	0.200	14.97	0.111	0.111	0.200
σI_{xy}		–	0.064	9.253	0.053	0.053	0.064	–	0.066	9.990	0.054	0.054	0.066
σI_{xz}		–	0.087	9.957	0.071	0.071	0.087	–	0.091	9.992	0.073	0.073	0.091
σI_{yz}		–	0.105	9.968	0.055	0.055	0.105	–	0.104	9.990	0.056	0.056	0.104

Table 7.2 Actuator-axis estimates produced by filters from Messenger data.

			Calibration Maneuver date and filter configurations							
			2007 Day 262		2008 Day 182		2008 Day 184		2009 Day 2009	
		Mech.	Base	Var.4	Base	Var.4	Base	Var.4	Base	Var.4
a_{1x}		0.7500	0.7535	0.7535	0.7545	0.7545	0.7561	0.7561	0.7582	0.7582
a_{1y}		0.5000	0.5013	0.5013	0.5004	0.5004	0.4988	0.4988	0.4885	0.4885
a_{1z}		0.4330	0.4254	0.4254	0.4246	0.4246	0.4237	0.4237	0.4318	0.4318
a_{2x}		−0.7500	−0.7510	−0.7510	−0.7559	−0.7559	−0.7565	−0.7565	−0.7522	−0.7522
a_{2y}		0.5000	0.5068	0.5068	0.4996	0.4996	0.4940	0.4940	0.5023	0.5023
a_{2z}		0.4330	0.4232	0.4232	0.4231	0.4231	0.4287	0.4287	0.4264	0.4264
a_{3x}		0.7500	0.7406	0.7406	0.7520	0.7520	0.7523	0.7523	0.7498	0.7498
a_{3y}		−0.5000	−0.5033	−0.5033	−0.5003	−0.5003	−0.5046	−0.5046	−0.5020	−0.5020
a_{3z}		0.4330	0.4452	0.4452	0.4291	00.4291	0.4236	0.4236	0.4310	0.4310
a_{4x}		−0.7500	−0.7492	−0.7492	−0.7440	−0.7440	−0.7401	−0.7401	−0.7455	−0.7455
a_{4y}		−0.5000	−0.4922	−0.4922	−0.5006	−0.5006	−0.5089	−0.5089	−0.5033	−0.5033
a_{4z}		0.4330	0.4433	0.443	0.4425	0.4425	0.4396	0.4396	0.4369	0.4369
σ_{p11}	$(\times 10^{-4})$	–	0.7798	0.7798	0.9733	0.9733	0.9315	0.9315	0.8147	0.8147
σ_{p12}		–	0.9307	0.9307	0.9876	0.9876	0.9824	0.9824	0.8564	0.8564
σ_{p21}		–	0.5634	0.5634	0.8675	0.8675	0.8257	0.8257	0.7513	0.7513
σ_{p22}		–	0.7453	0.7453	0.8720	0.8720	0.8756	0.8756	0.7977	0.7977
σ_{p31}		–	0.7293	0.7293	0.7887	0.7887	0.7255	0.7255	0.8625	0.8625
σ_{p32}		–	0.7515	0.7515	0.8425	0.8424	0.7598	0.7598	0.8896	0.8896
σ_{p41}		–	0.7597	0.7597	0.7892	0.7892	0.7120	0.7120	1.0499	1.0499
σ_{p42}		–	0.6562	0.6562	0.8488	0.8488	0.7418	0.7418	1.0934	1.0934

inertia ratio and wobble-related quantities due to products of inertia are the only observables. So, momentum actuation is an indispensable element of any attempts to estimate complete mass properties.

7.12 Exercises

1. Which of the following is true of a gyroscope?
   (a) A gyroscope measures attitude.
   (b) A gyroscope measures angular velocity and/or incremental attitude change.
   (c) Measurements from a gyroscope and measurements from a sun sensor, used together, provide a full three-axis attitude estimate.
2. Which of the following is true of a star tracker?
   (a) A star tracker provides only single-axis attitude measurements
   (b) A star tracker provides a full three-axis attitude measurement from multiple star observations
   (c) A star tracker works well on fast-spinning spacecraft.
3. For the following problem, use the direction-cosine matrix

$$ {}^{\mathrm{N}}_{\mathrm{B}}Q = \begin{bmatrix} 1 & 0 & 0 \\ 0 & \sqrt{3}/2 & -1/2 \\ 0 & 1/2 & \sqrt{3}/2 \end{bmatrix} $$

   Consistent with Euler's Theorem, the B axes are related to the N axes by a rotation about some single axis by some angle.
   (a) What is that axis?
   (b) What is that angle?
4. For same direction-cosine matrix above given ${}^{B}v = [0\ -1/2\ \sqrt{3}/2]^T$, what is ${}^{N}v$?
5. For same direction-cosine matrix above given ${}^{B}v = [0\ -1\ 0]^T$, what is ${}^{N}v$?
6. For same direction-cosine matrix above given ${}^{B}v = [0\ -1\ 0]^T$, what is ${}^{\mathrm{B}}_{\mathrm{N}}Q$
7. For same direction-cosine matrix above given

$$ {}^{\mathrm{N}}_{\mathrm{A}}Q = \begin{bmatrix} \sqrt{2}/2 & \sqrt{2}/2 & 0 \\ -\sqrt{2}/2 & \sqrt{2}/2 & 0 \\ 0 & 0 & 1 \end{bmatrix} $$

   what is ${}^{\mathrm{N}}_{\mathrm{B}}Q$?
8. In the following problem, the basis vectors $\mathbf{b}_1$, $\mathbf{b}_2$, and $\mathbf{b}_3$ are unit vectors fixed in the spacecraft (or, equivalently, fixed in a reference frame B that rotates

with the spacecraft). The basis vectors $\mathbf{n}_1$, $\mathbf{n}_2$, and $\mathbf{n}_3$ are unit vectors fixed in an inertial reference frame N. The angular-velocity vector $\omega^{B/N}$ represents the rotation of frame B with respect to an inertial reference frame N. Use the direction-cosine matrix provided below to represent the vector $\mathbf{c} = 3\mathbf{b}_1 + 2\mathbf{b}_2 + 1\mathbf{b}_3$ in terms of $\mathbf{n}_1$, $\mathbf{n}_2$, and $\mathbf{n}_3$.

$$ {}^{\mathrm{N}}_{\mathrm{B}}Q = \begin{bmatrix} 1 & 0 & 0 \\ \cos(\frac{\pi}{6}) & \sin(\frac{\pi}{6}) & 0 \\ 0 & -\sin(\frac{\pi}{6}) & \cos(\frac{\pi}{6}) \end{bmatrix} $$

9. Which combination of spacecraft sensors would provide enough information to determine a full (3D) attitude?

    (a) One sun sensor and one rate gyroscope.
    (b) One magnetometer and one rate gyroscope.
    (c) Two magnetometers.
    (d) One star tracker.

10. Without known external torques which of the following cannot be estimated using the approaches summarized in this chapter?

    (a) Trace of the inertia matrix
    (b) Spin-to-transverse inertia ratio
    (c) Wobble angle
    (d) Ratio of the two transverse principal moments of inertia

11. Two measurements of spacecraft attitude, rate, and momentum-actuator state can be sufficient to estimate all 6 inertia parameters in which of the following situations?

    (a) The spacecraft is quiescent, with constant angular velocity.
    (b) One measurement is taken before deployment of the solar array, and the second after.
    (c) The spacecraft has no momentum-control devices.
    (d) The spacecraft is nutating with nonzero wheel angular momentum.

12. The spacecraft angular velocity and attitude are measured at two different times, between which angular-momentum is conserved. Those measurements are the following, where the subscript indicates the time of the measurement: either time 1 or time 2. The time between measurements is irrelevant. What are the six inertia parameters?

$$ {}^{\mathrm{N}}_{\mathrm{B}}Qt_1 = \begin{bmatrix} 1 & 0 & 0 \\ 0 & 0.154251 & -0.988032 \\ 0 & -0.988032 & 0.154251 \end{bmatrix} $$

$$ {}^{\mathrm{N}}_{\mathrm{B}}Qt_2 = \begin{bmatrix} 0.408082 & 0 & 0.912945 \\ 0 & 1 & 0 \\ -0.912945 & 0 & 0.408082 \end{bmatrix} $$

$$ h_1 = \begin{bmatrix} 1 \\ 2 \\ 3 \end{bmatrix} \text{Nms} $$

$$ h_2 = \begin{bmatrix} 3 \\ 1 \\ 2 \end{bmatrix} \text{Nms} $$

$$ \omega_1 = \begin{bmatrix} 0.02 \\ 0.04 \\ -0.01 \end{bmatrix} \text{Nms} $$

$$ \omega_1 = \begin{bmatrix} 0.00548064 \\ -0.000692166 \\ 0.0107317 \end{bmatrix} \text{Nms} $$

Chapter 8
Spacecraft Attitude Control

A book on spacecraft momentum control systems would not be complete without a discussion on spacecraft attitude control and momentum management specific to momentum control actuators. Though there are multiple publications that provide a wide swath of ideas for the theory behind spacecraft attitude control strategies, it behooves the reader to understand the differences between what is practically possible with momentum control actuators versus actuators that provide external torques. These differences will be addressed in the discussions that follow at the intersection of control theory and control engineering as it relates to spacecraft momentum control systems.

The majority of heritage control methods for practical spacecraft attitude control are either for small angle attitude control, feedback linearization, or methods to track prescribed reference attitude, momentum, or angular rate trajectories with respect to small displacements. These reference trajectories paired with their associated open-loop control usually hail from an open-loop optimization problem as part of the guidance system. Since uncertainties and disturbances are present, there are always errors not accounted for by the open-loop control. These errors are typically assumed to be small, and closed-loop attitude control law is computed to address these errors about the linearized system with respect to the reference trajectory. This is possible by linearizing the dynamics about a prescribed attitude and angular rate trajectory as a function of time, providing a linear time-varying set of closed-loop dynamics. Therefore, this chapter will only focus on linear control in terms of linearized error attitude, angular rate, and momentum. See Appendices B and D for a deeper discussion on nonlinear attitude control and momentum management, controllability, stability, safety, and robustness.

In Fig. 8.1, we portray spacecraft momentum control systems within the intersection of control theory and control engineering. Definitions of control theory and control engineering that will be considered are:

© The Author(s), under exclusive license to Springer Nature Switzerland AG 2025
F. A. Leve et al., *Spacecraft Momentum Control Systems*, Space Technology Library 46, https://doi.org/10.1007/978-3-031-96165-6_8

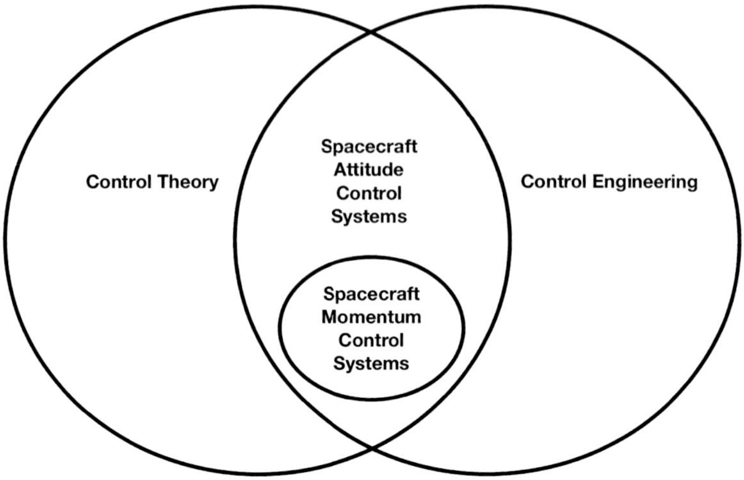

Fig. 8.1 Subarea of spacecraft momentum control systems

Control theory is an area of theoretical research within applied mathematics that is concerned with the analysis and synthesis of control systems for general classes of systems. Control synthesis, better known as control system design, follows from control theory as a subfield.

Control engineering is the practical application of control theory on real systems.

For some good books discussing the theory of general spacecraft attitude control, see references [80–83].

There are many input-output loops between subsystems and within the guidance, navigation, and control (GN&C) subsystem of a spacecraft. Block diagrams are a good visual aid to explain how these loops fit together. The GN&C system shown in Fig. 8.2 of a spacecraft is made up of three important blocks designed together along with the block for the plant provided by the physical model of the spacecraft system.

Guidance systems may take mission objectives and constraints and generate attitude and position way points or trajectories for the spacecraft to track and sends them to the control systems. Modes (e.g., safe mode) are also sent through the guidance system. Navigation systems combine information from various sensors to determine the current position, attitude and velocities of the spacecraft. Control systems are provided that determine the torques and forces required to track these way points or trajectories. Within control systems, there are multiple loops that require attention (See Fig. 8.3).

In this chapter, we will assume way points or trajectories have already been provided via the guidance system and only discuss the attitude, angular rate, and momentum control of spacecraft. We will avoid the discussion of the inner-loop for control allocation of the momentum control actuators known as steering algorithms

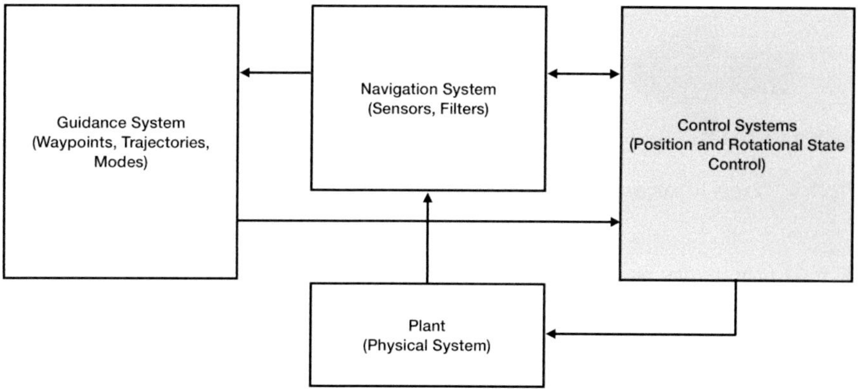

Fig. 8.2 Guidance navigation and control system for a spacecraft

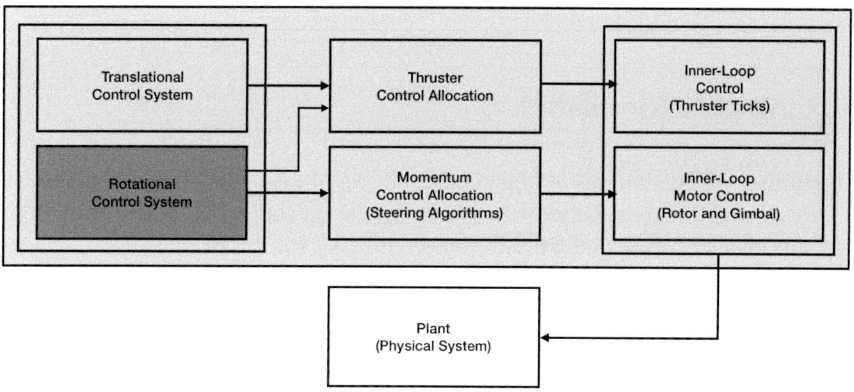

Fig. 8.3 Entire control system of a spacecraft

and the loop inner to this loop for rotor and gimbal motor control since they are discussed in other chapters (see Chaps. 9 and 10). Finally, this chapter will explain how a torque profile is provided by attitude control systems to effect the spacecraft rotation to match attitude, angular velocity or momentum profiles. Note, the block diagrams in Figs. 8.2 and 8.3 abstract away many of the details associated with implementation on a digital processor (e.g., digital to analog and analog to digital converters, noise, disturbances).

For our purposes we will provide our own differentiation between the steps of progression from theory to software (see Fig. 8.4). This book will primarily discuss theory. Algorithms, code, and software will not be discussed, though they are all critical for the future use of anything described here. Algorithms provide a set of steps for how the attitude control law will be implemented (e.g., pseudocode). Code could be scripts in different languages (e.g., Python, C, C++, Java, Matlab, etc.) that are available open source and possibly available on FTP sites such as GitHub

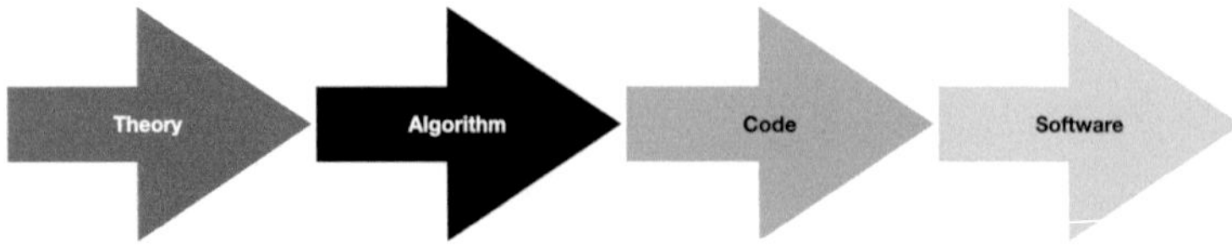

Fig. 8.4 Theory to software progression

or held proprietary by a company. Many sources do not distinguish between code or software. However, we define software as being more sophisticated than code in that it deals with how the actual code is implemented on the real system. This may incorporate available memory, function call frequency, data precision, etc. Also, software may require a certification that it will work correctly. All but theory may be held as proprietary information and will be avoided in this chapter and book in general.

8.1 Attitude Kinematics

Disclaimer All definitions of kinematics and closed-loop equations of motion will be done in matrix form rather than vector dyadic notation. This is due to the fact that this chapter focuses on attitude control utilizing spacecraft momentum control systems for specific sets of coordinates.

Before we can actually define the states to be controlled, we must first understand the different choices of coordinates used to describe attitude and their kinematic differential equations. Spacecraft attitude control design is tightly correlated to the choice of coordinate representations of attitude states. There are a number of popular sets of local coordinates that are considered for spacecraft attitude control synthesis. Most of these coordinates, with the exception of direction cosine matrices, are either local and/or nonunique. Furthermore, many of these coordinate representations also contain coordinate singularities where they are not easily defined. Stuelpnagel, in reference [26], proved that at least 4 coordinates are needed to avoid coordinate singularities, and at least 5 coordinates are required for the representation to be unique in representing attitude. For example, attitude is naturally and globally and represented uniquely (i.e., single set of parameters for each attitude) by the nine distinct values of direction cosine matrices (DCMs), also known as rotation matrices or points of the special orthogonal Lie group manifold SO(3) (see Appendix B). Additionally, Euler parameters (unit quaternions), though nonunique, provide a global set of coordinates to parameterize attitude. It is known that any attitude coordinate parameterization can be related to an axis, c and angle, ϕ through Euler's rotation theorem. All attitude coordinates must also have a mapping from itself to DCMs and from DCMs to itself since DCMs are the natural and global coordinates to describe attitude.

Table 8.1 List of attitude coordinates and their properties

Coordinates	Elements	n	Unique	Singular
Rotation matrix	$Q = \{q_{ij}\}$	9	Yes	No
Euler angles	$\{\theta_1, \theta_2, \theta_3\}$	3	No	Yes
Euler parameters	$\{q_1, q_2, q_3, q_4\}$	4	No	No
ERPs	$\{\rho_1, \rho_2, \rho_3\}$	3	No	Yes
MRPs	$\{\sigma_1, \sigma_2, \sigma_3\}$	3	No	Yes

This section is intended to discuss the problems with specific choices of coordinates on their kinematic differential equations and the evolution from these equations relevant to the succeeding control discussions. Such problems are directly linked to the performance and behavior of practical spacecraft attitude control with or without spacecraft momentum control systems. Some coordinate representations for attitude are shown in Table 8.1.

All attitude coordinate representations provide attitude kinematics of the form

$$\dot{Q} = S(Q)\omega, \tag{8.1}$$

where Q is any of the discussed attitude coordinate representations in Table 8.1. Sometimes $S(Q)$ is linear in the elements of Q and sometimes it consists of nonlinear functions of Q elements. With these considerations, the general closed-loop rotational equations of motion are those in Eq. (8.11).

For the discussion of attitude kinematics, we will assume that attitude, Q, represents a rotation from the body frame to the inertial frame, and that the angular velocity of the body with respect to this inertial frame is ω as a matrix in body coordinates.

8.1.1 Direction Cosine Matrices

Direction cosine matrices have the following kinematics

$$\dot{Q} = Q\omega^{\times}, \tag{8.2}$$

where Q is the direction cosine from one frame to the other, and $\omega^{\times}$ is the skew-symmetric matrix of the angular velocity between the frames in the body coordinates. Note that as stated in Table 8.1, these equations contain no singularities, and provide unique (global) representations for every attitude. Even though nine differential equations are provided to integrate Eq. (8.2), it should be noted that the direction cosine matrix also has six constraints associated with its unitarity (i.e., $Q^T Q = Q Q^T = 1$ are symmetric matrices), which means that only three numbers are required to track the attitude between frames. However, we pay the price of more differential equations to allow for a unique, global, and nonsingular set of coordinates to represent attitude and its evolution.

8.1.2 Euler Angles

Direction cosine matrices can be formed by 3 sequential rotations about different consecutive noncolinear axes by Euler angles. Therefore, Euler angles are intimately related to direction cosine matrices as rotation angles about axes of intermediate frames. There are many different rotation sequences (sets of Euler angles) for any attitude (i.e., a many-to-one map from Euler angles to DCMs) and therefore, they are not unique representations for attitude. However, they are a minimal set of parameters to describe attitude and therefore useful for attitude control when used for parameterizing attitudes away from coordinate singularities. It is a general rule that Euler angles for symmetric sequences (e.g., 3-1-3, 1-2-1, 2-3-2, etc.) have coordinate singularities from the second angle of the sequence, $\theta_2 = \pm\pi$ leaving one of the angles indeterminate via extraction from the associated DCM. Asymmetric rotations (e.g., 3-2-1, 1-2-3, 1-3-2, etc.) have coordinate singularities at $\theta_2 = \pm\pi/2$. All Euler angle sequences are either symmetric or asymmetric and do not have two consecutive angles about the same axes or else they would just be added and collapse to sequence of two rotations. Different sets of Euler angle sequences (i.e., local coordinates or charts) can be switched between to avoid coordinate singularities. However, this is very problematic numerically, and is not good practice for large angle slews.

The attitude kinematics for Euler angles have the following form

$$\begin{bmatrix} \omega_1 \\ \omega_2 \\ \omega_3 \end{bmatrix} = S(\theta_2, \theta_3) \begin{bmatrix} \dot{\theta}_1 \\ \dot{\theta}_2 \\ \dot{\theta}_3 \end{bmatrix}, \tag{8.3}$$

or

$$\begin{bmatrix} \dot{\theta}_1 \\ \dot{\theta}_2 \\ \dot{\theta}_3 \end{bmatrix} = S(\theta_2, \theta_3)^{-1} \begin{bmatrix} \omega_1 \\ \omega_2 \\ \omega_3 \end{bmatrix}, \tag{8.4}$$

As an example, consider the kinematics of a 3-1-3 sequence

$$\begin{bmatrix} \dot{\theta}_1 \\ \dot{\theta}_2 \\ \dot{\theta}_3 \end{bmatrix} = \frac{1}{s(\theta_2)} \begin{bmatrix} s(\theta_3) & c(\theta_3) & 0 \\ c(\theta_3)s(\theta_2) & -s(\theta_3)s(\theta_2) & 0 \\ -s(\theta_3)c(\theta_2) & -c(\theta_3)c(\theta_2) & s(\theta_2) \end{bmatrix} \begin{bmatrix} \omega_1 \\ \omega_2 \\ \omega_3 \end{bmatrix}. \tag{8.5}$$

Clearly, $S(\theta_2, \theta_3)$ in Eq. (8.5) is singular at $s(\theta_2) = 0$ which occurs at $\theta_2 = \pm\pi$.

8.1.3 Euler Parameters (Unit Quaternions)

Euler parameters have the representation in terms of an axis c and an angle ϕ and bilinear kinematics in both directions

$$Q = \begin{bmatrix} q_1 \\ q_2 \\ q_3 \\ q_4 \end{bmatrix} = \begin{bmatrix} \sin(\frac{1}{2}\phi)c \\ \cos(\frac{1}{2}\phi) \end{bmatrix},$$

$$\begin{bmatrix} \dot{\hat{q}} \\ \dot{q}_4 \end{bmatrix} = \frac{1}{2} \begin{bmatrix} (\hat{q}^\times + q_4 1) \\ -\hat{q}^T \end{bmatrix} \omega = S(Q)\omega,$$

or

$$\begin{bmatrix} \dot{\hat{q}} \\ \dot{q}_4 \end{bmatrix} = \frac{1}{2} \begin{bmatrix} -\omega^\times & \omega \\ \omega^T & 0 \end{bmatrix} \begin{bmatrix} \hat{q} \\ q_4 \end{bmatrix} = F(\omega)Q, \tag{8.6}$$

where $Q = [\hat{q}, q_4]^T$ and $\hat{q} = [q_1, q_2, q_3]^T = \sin(\frac{1}{2}\phi)c$ are the eigen-axis components of the Euler parameters, and $q_4 = \cos(\frac{1}{2}\phi)$ is the scalar element.

8.1.4 Euler and Modified Rodrigues Parameters

Euler Rodrigues parameters are a three parameter (minimal) set for attitude that have singularities at $\pm\pi$ and have a representation in terms of an axis c and an angle ϕ about that axis

$$\boldsymbol{\rho} = \tan\frac{\phi}{2}c, \tag{8.7}$$

and highly nonlinear kinematics

$$\dot{\boldsymbol{\rho}} = \frac{1}{2}\left(\boldsymbol{\rho}\boldsymbol{\rho}^T + 1 + \boldsymbol{\rho}^\times\right)\omega = S(\rho)\omega, \tag{8.8}$$

where $\boldsymbol{\rho} = [\rho_1, \rho_2, \rho_3]^T$. Modified Rodrigues parameters (MRPs) have a representation in terms of an axis c and an angle ϕ about that axis

$$\sigma = \tan\frac{\phi}{4}c, \tag{8.9}$$

and highly nonlinear kinematics

$$\dot{\boldsymbol{\sigma}} = \frac{1}{4}\left((1 - |\sigma|^2)1 + 2\sigma\sigma^T + 2\boldsymbol{\sigma}^\times\right)\omega = S(\sigma)\omega, \tag{8.10}$$

and have almost the exactly same form (See Eqs. (8.9) and (8.10)), but have their coordinate singularities shifted to $\pm 2\pi$. Shifting of the coordinate singularity to $\pm 2\pi$ allows for large angle maneuvers past $\pm\pi$ but will encounter such singularities when a satellite is tumbling or rotating multiple revolutions.

8.2 Pointing Precision, Accuracy, Stability

Minimizing spacecraft pointing error when tracking a particular target is a broad topic that affects multiple disciplines. To manage this, the systems engineer uses error budgeting to break the total error down to its constituent components. In the case of pointing error, these components are generally defined as precision, accuracy and stability as illustrated in Fig. 8.5. In the paragraphs below, we shall assume that the target is a fixed attitude.

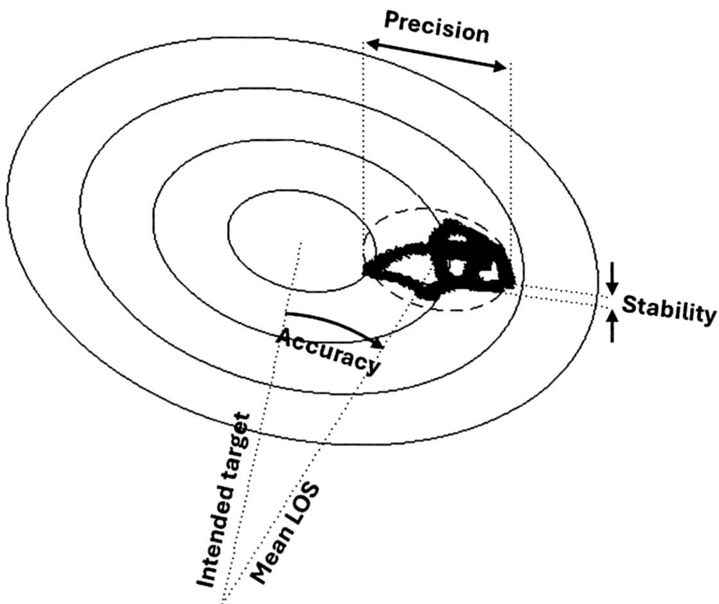

Fig. 8.5 Pointing accuracy, precision, and stability

Pointing is a 3-axis problem, but in discussing these concepts we often visualize using the notion of line of sight (LOS) in two axes. The line of sight is the direction in which the telescope, camera, etc. is pointed.

We also consider the spectral content of LOS motion. The "DC" or "steady state" is the "centroid", or mean LOS over time. "In-band" motion is motion of the LOS with frequency content lower than the bandwidth of the attitude control system (ACS). "Out-of-band" motion has frequency content higher than the ACS bandwidth, and thus is not able to be attenuated by the action of the ACS.

Pointing accuracy is the error between the actual target and the mean LOS. For most well-designed ACS, the mean LOS tracks well with where we think the target is. Thus, accuracy is directly related to errors in the sensors and attitude estimation. We will not focus on pointing accuracy here as there are plenty of good references that consider estimation and filtering for inferring an accurate estimate of the attitude state (e.g., see reference [81]).

Pointing precision describes error between the in-band LOS variations and its mean, usually as a standard deviation. Precision is generally driven by the design of the ACS, and choice of actuators. For example, a "bang-bang" ACS based on thrusters will limit-cycle about the reference attitude, whereas a momentum system can potentially exhibit far greater precision. This chapter discusses ACS design for MCS-based spacecraft.

Pointing stability refers to out-of-band LOS motion. This arises from perturbative sources such as vibrations from cryogenic pumping systems used to cool payloads, jitter from momentum control systems, thermal snap from going in and out of eclipse and other disturbances with high frequency content (see reference [83] for models of these disturbances). Because they are out-of-band, these errors cannot be addressed through control, but depend on passive isolation mechanisms, etc.

For practical implementation of control systems, additional properties beyond those associated with pointing errors are required. These properties are directly associated with the dynamical system and paired control system in question. Some of the most relevant of these properties are controllability, stability, safety, and robustness. Note, these properties are a minimal set of properties that should be considered for spacecraft attitude control.

Controllability is defined as given a set of admissible control solutions, there exists within that set at least one solution that enables a system to get from one point to any another point in the entire configuration space in a finite amount of time. There are different variants of controllability. One variant includes local controllability which limits controllability to an open neighborhood around a point. This form of controllability allows us to focus on heritage methods for linear attitude control or linear systems.

Stability is in the sense of a point is to remain close to a neighborhood of that point when starting at another neighborhood of that point. It is critical for control systems to have this property as instability is dangerous and detrimental to the system itself. There are different kinds of stability which are described in more detail in Appendix B.

Safety is the ability of a control systems to avoid certain regions within the state- and control-space. These regions could may correspond to avoidance cones for imagers, velocity and acceleration constraints associated with IMU saturation and structural limits, etc. See Appendix B for tools on how to ensure and verify safety in control systems.

Robustness is a property usually measured with respect to something undesirable occurring in the system. In control theory, robustness is usually measured as the ability for a system to remain close to nominal when perturbed by small disturbances or under small uncertainties. Since the majority of all systems are estimated from incomplete and imperfect measurements from sensors, and there are external disturbances and model uncertainty always present, control systems require robustness to properly function in implementation.

8.3 Maneuvers

There are many different types of maneuver profiles associated with attitude control. For the purpose of this chapter, we will briefly discuss only a few as they relate to precision pointing and state stability, not to be confused with pointing stability. These maneuvers are numbered below.

1. Small angle attitude control,
2. Point-to-point maneuvers,
3. Attitude tracking,
4. Angular rate and momentum tracking.

Small angle attitude control is concerned with very small angle adjustments to the orientation of the spacecraft. Such adjustments are usually in response to attitude drift from environmental or on-board disturbance torques. Attitude point set control is a special case of this type of maneuver and it corresponds to keeping the satellite at a fixed attitude. Such maneuvers are typically used for space science missions (e.g., Hubble Space Telescope, James Webb Space Telescope, Kepler Space Telescope) that are held at an inertially-fixed orientation for a minimum amount of integration time where light from the very distant past can be collected and focused into the pretty images we see and use to infer our understanding of the universe. Furthermore, these types of maneuvers allow us to linearize the angular rate rotational and attitude kinematics equations of motion. Good isolation and vibration suppression techniques for the payload of the spacecraft are of the utmost importance to mitigate errors in pointing stability. In addition, although momentum control devices may add to disturbances from static and dynamic imbalances (see Sect. 4.7), choosing arrays of CMGs over RWA or vice versa will allow the designer different isolation options. For example, RWA tend to have a large band of jitter disturbance frequencies to avoid, while CMGs with rotors at constant speed, can be isolated fairly easily at high frequencies with minimal effect to the attitude control of the spacecraft.

Point-to-point and even more specifically, rest-to-rest angle slews are the ones shown most often in the literature. They are actually less practical for spacecraft attitude control. Such maneuvers are usually associated with a spacecraft rotating a large angle from an initial attitude and angular velocity to a terminal attitude and angular velocity. Nevertheless, it still makes sense to bring up this type of maneuver. Some of the literature (see references [53]) has been concerned with optimal rest-to-rest slews between two fixed attitudes with zero initial and terminal angular velocities. One objective function for optimality that is considered is minimum time. The goal here is to pick the right set of torques and gimbal rates for those torques, for each time to produce the quickest path between two attitudes. It has been shown that for large angle slews, and for spacecraft that have an inertia tensor with unequal principal moments, the optimal slew is not an eigen-axis slew (i.e., a slew that can parameterized by a single axis and angle). Some great work in optimal control as in reference [54]. Reference [84] has been published and some even demonstrated [53] on this topic for RWA and CMGs. It should be noted though that optimal control or the ability to pick the attitude or angular velocity trajectory is not a free option for attitude tracking. Moreover, such solutions rely on complex numerical solvers for nonconvex optimization problems which may have trouble converging to solutions without a good initial guess, and/or encountering coordinate or kinematic singularities which lead to numerical difficulties in convergence.

Small angle attitude, as well as point-to-point maneuvers allow some wiggle room for the control torque to enable convergence to the zero error state. In contrast, the tracking maneuvers discussed next require exact matching of the attitude, angular rate, or momentum states.

Both agile maneuvers and attitude tracking may require the spacecraft to follow commands whose frequency content greatly exceeds the bandwidth of traditional ACS designs. ACS bandwidth is usually in the range of tenths of Hertz or less. Restrictions on ACS bandwidth typically arise from either actuator capability limits (see Sect. 10.2.3) or structural modes. Spacecraft, designed for minimum mass, often exhibit global or appendage modes at 1 Hz or less, and traditional controls design places closed loop bandwidth an order of magnitude below these modes. As a result, following agile commands requires feedforward architecture, wherein the needed vehicle accelerations are passed directly to the steering laws and implemented by the momentum devices. If the steering laws and momentum devices are accurate, little error remains for the ACS feedback to attenuate. This underscores the importance of high performance steering and actuators for agile vehicles.

Attitude tracking in theory requires the spacecraft to track an attitude trajectory as a function of time exactly. To do this, the filter estimates of angular velocity and attitude states as well as the actuators precision in control must be acceptable. What this also entails is a need for an accurate (i.e., exact) steering algorithm for control allocation. There is no room for error from the control allocation or else trajectory tracking would be error-prone, thus negating the benefits of using the most accurate actuators for attitude control (e.g., RWAs and CMGs) in the first place. Therefore, mathematically, attitude control is a geometric problem, where control allocation coming from inverse kinematics is an algebraic problem (i.e., does not depend on time or distances). Attitude tracking can be at specific points (e.g., way points), disconnected slices of trajectories (e.g., arcs), or over a long connected trajectory. This form of attitude control is practical for Earth imaging spacecraft. Sometimes the tracking trajectory could be a simple arc where the angular velocity remains constant in direction and/or magnitude and the spacecraft just coasts without control allocation of the RWA or CMG array required.

Angular rate and momentum tracking are useful for momentum dumping as well as for modern spacecraft missions such as on-orbit servicing and orbit debris removal (see Chap. 2). It may be the case that you need the spacecraft momentum control system to have a specific angular momentum (e.g., zero momentum control system array angular momentum), but the buildup up of angular momentum from external environmental or control induced torques (e.g., thrusters) needs to be dissipated by external control torques or into the environment by rotating the spacecraft through different orientations exploiting the environmental torques. It is expected that the total angular momentum can never be exactly zero. However, the goal is to get as close to zero as possible to maximize the momentum and thus control capacity of the momentum control system. A great example of such a maneuver to desaturate an array of DGCMGs has been done on the International Space Station (ISS) and has been dubbed the zero-propellant maneuver (ZPM) [84]. Also, it may be the case that you need to match the angular velocity or momentum of another spacecraft or piece of debris you are docking to for servicing or debris removal. This case prioritizes angular rate and momentum tracking. Other examples of angular momentum or angular rate tracking are spin and dual-spin stabilization of spacecraft.

8.4 Closed-Loop Equations of Motion

Assume for the moment the total closed-loop matrix equations in body coordinates for a rigid body in space are

$$\begin{bmatrix} \dot{Q} \\ \dot{\omega} \end{bmatrix} = \begin{bmatrix} 0 & S(Q) \\ 0 & J^{-1}(J\omega)^{\times} \end{bmatrix} \begin{bmatrix} Q \\ \omega \end{bmatrix} + \begin{bmatrix} 0 \\ J^{-1} \end{bmatrix} (\tau + \varepsilon), \tag{8.11}$$

where $(\cdot)^{\times}$ refers to the skew-symmetric operation of turning a column matrix into a skew-symmetric matrix, τ is input torque, and ε is the possible external torques present from the environment or from on-board spacecraft elements, $S(Q)$ is the matrix that comes from the kinematic differential equations with Q being one of the chosen attitude coordinate representations from Table 8.1. For general spacecraft attitude and angular rate control, we will use Eq. (8.11) assuming $\varepsilon = 0$. When we consider spacecraft momentum control systems as the device providing the (internal) control torque, we will use the following closed-loop matrix equations in body coordinates

$$\begin{bmatrix} \dot{Q} \\ \dot{\omega} \\ \dot{h} \end{bmatrix} = \begin{bmatrix} 0 & S(Q) & 0 \\ 0 & J^{-1}(J\omega)^{\times} & -J^{-1}(\omega)^{\times} \\ 0 & 0 & 0 \end{bmatrix} \begin{bmatrix} Q \\ \omega \\ h \end{bmatrix} + \begin{bmatrix} 0 & 0 \\ J^{-1} & J^{-1} \\ 0 & -1 \end{bmatrix} \begin{bmatrix} (\tau + \varepsilon) \\ -\dot{h} \end{bmatrix}, \tag{8.12}$$

or

$$\begin{bmatrix} \dot{Q} \\ \dot{\omega} \\ \dot{h} \end{bmatrix} = \begin{bmatrix} 0 & S(Q) & 0 \\ 0 & J^{-1}(H)^{\times} & 0 \\ 0 & 0 & 0 \end{bmatrix} \begin{bmatrix} Q \\ \omega \\ h \end{bmatrix} + \begin{bmatrix} 0 & 0 \\ J^{-1} & J^{-1} \\ 0 & -1 \end{bmatrix} \begin{bmatrix} (\tau + \varepsilon) \\ -\dot{h} \end{bmatrix} \tag{8.13}$$

where $H = J\omega + h$ in body coordinates, h is the angular momentum contribution from spacecraft momentum control system array where the control is produced. Note, that $\dot{Q} = S(Q)\omega = F(\omega)Q$ for coordinate representations that have bilinear kinematics (e.g., DCMs and Euler parameters). Also, h itself is linear in control inputs for the case of RWA, but it is nonlinearly dependent on gimbal angles for the case of CMGs. This will be discussed more below. Finally, Eqs. (8.11)–(8.13) are presented in a state-space form that will help provide insight for the attitude control problems to be discussed later.

In the literature, the closed-loop equations of motion have $\tau = -\dot{h} - \omega^{\times}h$ as the internal control torque. This is physically accurate, and is most likely used to be most aligned with attitude control of a rigid body with external torques. However, by representing the control variable for the closed-loop dynamical system this way, we may confuse the reader into thinking that τ can be treated as an external torque. Furthermore, this representation does not allow for simplification of control schemes by bringing $\omega^{\times}h$ into the dynamics when $H = J\omega + h = 0$. Moreover, since internal torques are limited by angular momentum capacity of the spacecraft momentum control system array, you cannot always achieve the τ you need due to the limits on h. Therefore, the array control $\dot{h}$, must prohibit h from exceeding the momentum capacity of the array or moment control actuators.

8.5 Linearizations

Typically, attitude and angular rate control is with respect to an equilibrium point or a predefined trajectory. When an open-loop solution to the control for this predefined trajectory is provided by the guidance system, there will always be some uncertainty in the measurements, model, and disturbances that can cause the actual attitude and angular rate to be perturbed and displaced from the desired trajectories (equilibrium point). When that happens, a second controller may be added to the open-loop controller of the reference trajectory to mitigate these small displacements. This motivates the use of linearizations of the closed-loop equations of motion for spacecraft rotation.

There are multiple ways to perform linearizations of differential equations. For the case where the desired trajectory is just an equilibrium point, one such method is to use the Taylor series expansion of Eq. (8.12) or (8.13) about this equilibrium point, keeping only the terms up to order one. The reason we expand about an equilibrium point is it allows us to make statements about local stability of those points. Consider $\dot{x} = f(x)$ with $f(\overline{x}) = 0$ and $x = \overline{x} + \delta x$. Then the Taylor series expansion is

$$\begin{aligned} \dot{x} = \delta\dot{x} &= f(\overline{x}) + \left.\frac{\partial f}{\partial x}\right|_{x=\overline{x}} \delta x + \text{H.O.T} \\ &= 0 + \left.\frac{\partial f}{\partial x}\right|_{x=\overline{x}} \delta x + \text{H.O.T} \\ &\approx \left.\frac{\partial f}{\partial x}\right|_{x=\overline{x}} \delta x \end{aligned} \tag{8.14}$$

When we ignore the higher order terms in δx, this amounts to linearization using the Jacobian matrices of the system. For a controlled dynamical system, $\dot{x} = f(x, u)$ and the linearization for $x = \overline{x} + \delta x$ and $u = \overline{u} + \delta u$ is

$$\begin{aligned} \delta\dot{x} &\approx \left.\frac{\partial f}{\partial x}\right|_{x=\overline{x}} \delta x + \left.\frac{\partial f}{\partial u}\right|_{u=\overline{u}} \delta u \\ &= A\delta x + B\delta u \end{aligned} \tag{8.15}$$

Another method of performing linearizations about an equilibrium is to substitute in $x = \overline{x} + \delta x$ and $u = \overline{u} + \delta u$ to both sides of $\dot{x} = f(x, u)$ and keep on terms of zeroth and first order in $\delta x, \delta u$. Sometimes it is important to not just linearize about an equilibrium but to linearize about a time-based trajectory and control pair $(x^*(t), u^*(t))$. This time-based trajectory control pair is usually provided by an attitude and orbit guidance system as shown in Fig. 8.2. Assume now that $\dot{x}(t) = f(x(t), u(t))$ and its linearization is

$$\begin{aligned} \delta\dot{x} &\approx \left.\frac{\partial f}{\partial x}\right|_{x=x^*(t)} \delta x + \left.\frac{\partial f}{\partial u}\right|_{u=u^*(t)} \delta u \\ &= A(t)\delta x + B(t)\delta u, \end{aligned} \tag{8.16}$$

where the pair $(A(t), B(t))$ is now explicitly time-varying since it is with respect to small deviations of a given trajectory pair $(x^*(t), u^*(t))$. Note, these methods of linearization assume that the state x resides inside a vector space (e.g., Euclidian space) or some other algebraic structure (e.g., ring, module, algebra) where addition and subtraction are allowed.

8.6 Small Angle Attitude Control

In general, the total system linear differential equations of motion have the form,

$$\delta \dot{x}(t) = A\delta x(t) + B\delta u(t). \tag{8.17}$$

For a rigid spacecraft bus with a spacecraft momentum control system, the matrices A and B not only depend on the coordinates chosen for attitude as example from Table 8.1, but they also depend on the number and arrangement of spacecraft momentum actuators. All closed-loop full-state feedback linear systems (see Appendix B) have their nominal dynamics and control properties described by A and B. Therefore, for this section, we will only discuss the effect of control on the dynamical systems for small angle attitude control by deriving the dynamics toward finding A and B (or $A(t)$, $B(t)$ when they are explicitly time varying). Local control problems such as these allow for linearizations from a variety of coordinate representations far away from their associated coordinate singularities and large control inputs.

It should be noted that coordinate singularities associated with the chosen attitude coordinate representation and kinematic singularities for CMG arrays associated with gimbal configurations both need to be considered in the linearization of the closed-loop equations of motion. In practice, the attitude control designer would pick a coordinate representation for attitude and an initial gimbal angle configuration that is far from coordinate and kinematic singularities.

The majority of coordinates far away from coordinate singularities are all acceptable for use in linearization. A typical set of coordinates chosen for this linearization are Euler angles for attitude kinematics. Since we are assuming small angular displacements, we only need the Euler sequence chosen to remain far away from singularity around zero Euler angles which is guaranteed for Euler angle sequences since symmetric and asymmetric sequences have their coordinate singularities about $\pm\pi$ and $\pm\pi/2$.

Let us start with a rigid body and assume for the moment that attitude between two frames is described by a column matrix of Euler angles, $\theta = [\theta_1, \theta_2, \theta_3]^T$. There is always an equation providing a relationship between the time change of Euler angles (i.e., Euler rates) and angular velocity. This equation is of the form $\dot{\theta} = S(\theta)\omega$ where $S(\theta)$ is a function of sines and cosines of θ_i elements. Take for example the roll, pitch, yaw (1-2-3) sequence of rotations for θ. Then the kinematics equations are

$$\begin{bmatrix} \dot{\theta}_1 \\ \dot{\theta}_2 \\ \dot{\theta}_3 \end{bmatrix} = \begin{bmatrix} 1 & \sin(\theta_1)\tan(\theta_2) & \cos(\theta_1)\tan(\theta_2) \\ 0 & \cos(\theta_1) & -\sin(\theta_1) \\ 0 & \sin(\theta_1)\sec(\theta_2) & \cos(\theta_1)\sec(\theta_2) \end{bmatrix} \begin{bmatrix} \omega_1 \\ \omega_2 \\ \omega_3 \end{bmatrix}. \tag{8.18}$$

If we assume $\theta = \overline{\theta} + \delta\theta$, $\omega = \overline{\omega} + \delta\omega$, and $\tau = \overline{\tau} + \delta\tau$ and linearize this equation about the equilibrium $[\overline{\theta}, \overline{\omega}, \overline{\tau}] = 0$, then we will get $\delta\dot{\theta} \approx \delta\omega$. This result follows from the sine and cosine addition formulas applied to $\cos(\theta_{0_i} + \delta\theta_i)$ and $\sin(\theta_{0_i} + \delta\theta_i)$ while assuming $\cos(\delta\theta_i) \approx 1$ and $\sin(\delta\theta_i) \approx \delta\theta_i$ and disregarding any products of quadratic or higher in $\delta\theta_i$.

Now, let's consider and $\omega = \overline{\omega} + \delta\omega$ as the perturbed solutions from equilibrium angular rate. Furthermore, plugging $\omega = \overline{\omega} + \delta\omega$ into Euler's rotational equations of motion and disregarding terms quadratic in $\delta\omega$, we have

$$\begin{aligned} \dot{\omega} &= 0 + \delta\dot{\omega} = \delta\dot{\omega} \\ &= -J^{-1}\left(\overline{\omega}^{\times} J\overline{\omega} + \overline{\omega}^{\times} J\delta\omega + \delta\omega J\overline{\omega} - \delta\tau\right) \\ &= -J^{-1}\left(0 + \overline{\omega}^{\times} J\delta\omega + \delta\omega^{\times} J\overline{\omega}\right)\delta\dot{\omega} + J^{-1}\delta\tau \\ &= J^{-1}\left[(J\overline{\omega})^{\times} - \overline{\omega}^{\times} J\right]\delta\omega + J^{-1}\delta\tau \end{aligned} \tag{8.19}$$

where $\overline{\omega}^{\times} J\overline{\omega} = 0$ since we assumed $\overline{\omega}$ was an unforced equilibrium of the rotational equations of motion.

For a rigid body with Eqs. (8.18) and (8.19) at the equilibrium $[\overline{\theta}, \overline{\omega}, \overline{\tau}] = 0$, we get the closed-loop linearized equation of motion for attitude and angular rate control

$$\begin{bmatrix} \delta\dot{\theta} \\ \delta\dot{\omega} \end{bmatrix} = \begin{bmatrix} 0 & 1 \\ 0 & 0 \end{bmatrix} \begin{bmatrix} \delta\theta \\ \delta\omega \end{bmatrix} + \begin{bmatrix} 0 \\ J^{-1} \end{bmatrix} \delta u \tag{8.20}$$

where $\delta u = \delta\tau$. Note, $\overline{\omega} \neq 0$ cannot be an equilibrium of the combined attitude and angular rate equations because it is not an equilibrium of Eq. (8.18) for any $\overline{\theta}$. Let us consider the linearized attitude control problem in the following example.

Example 8.1 Here we will provide an example of the possible controls for the closed-loop linear attitude control system in Eq. (8.20). Consider the state and input matrices of Eq. (8.20),

$$A = \begin{bmatrix} 0 & 1 \\ 0 & 0 \end{bmatrix}, \quad B = \begin{bmatrix} 0 \\ J^{-1} \end{bmatrix} \tag{8.21}$$

Also, the full state $\delta x = [\delta\theta, \ \delta\omega]^T$. Furthermore, assume that we have full-state feedback, and our controller has the form $\delta u = \delta\tau = -K\delta x = -k_1 J\delta\theta - k_2 J\delta\omega$. Then our closed-loop system becomes

$$\delta\dot{x} = (A - BK)\delta x = \begin{bmatrix} 0 & 1 \\ -k_1 1 & -k_2 1 \end{bmatrix} \delta x \tag{8.22}$$

(continued)

The eigenvalues of the matrix $(A - BK)$ determine the stability of the closed-loop system. The eigenvalues for $(A - BK)$ here are

$$\begin{aligned}\lambda_1 &= \frac{1}{2}\left(-k_2 - \sqrt{(k_2^2 - 4k_1)}\right)\\ \lambda_2 &= \frac{1}{2}\left(\sqrt{(k_2^2 - 4k_1)} - k_2\right)\end{aligned} \tag{8.23}$$

both with λ_1 and λ_2 with multiplicity 3. Since we assumed $k_1, k_2 > 0$, all six of these eigenvalues have negative real parts and therefore, the linear system here is exponentially stable with decay rates limited by the minimum magnitude between $|\text{Re}(\lambda_1)|$ and $|\text{Re}(\lambda_2)|$.

Now, if we are considering attitude control with a spacecraft momentum control system, then $\tau = -\dot{h} - \omega^{\times} h$. We would like to use $\delta u = -\delta\dot{h}$ as the control variable here. However, we need to somehow incorporate its integral, h into the closed-loop equations of motion. This is possible via Eq. (8.12). Here in lies a major difference between attitude control via internal torques versus attitude control through external torques. In the absence of external torques, $H = J\omega + h$ has constant magnitude. Under this assumption, the total magnitude of angular momentum is constant and if $\|H\| \neq 0$ we cannot even reach the equilibrium $\overline{\theta} = \overline{\omega} = \overline{h} = 0$ (i.e., the system is not controllable to this point). Therefore in this case, you cannot linearize about $\overline{\theta} = \overline{\omega} = \overline{h} = 0$ since $\|H\| \neq 0$ would require that the array angular momentum, h has a nonzero magnitude terminal (i.e., for a maneuver ending at rest, $\omega = 0$) value. We would have to know this angular momentum before linearization or assume it is exceptionally small in magnitude. When $\overline{\omega} = 0$ and $\overline{h} \neq 0$

$$\begin{bmatrix}\delta\dot{\theta}\\ \delta\dot{\omega}\\ \delta\dot{h}\end{bmatrix} = \begin{bmatrix}0 & 1 & 0\\ 0 & J^{-1}(\overline{h})^{\times} & 0\\ 0 & 0 & 0\end{bmatrix}\begin{bmatrix}\delta\theta\\ \delta\omega\\ \delta h\end{bmatrix} + \begin{bmatrix}0\\ -J^{-1}\\ -1\end{bmatrix}\delta\dot{h} \tag{8.24}$$

or

$$\begin{bmatrix}\delta\dot{\theta}\\ \delta\dot{\omega}\end{bmatrix} = \begin{bmatrix}0 & 1\\ 0 & J^{-1}(\overline{h})^{\times}\end{bmatrix}\begin{bmatrix}\delta\theta\\ \delta\omega\end{bmatrix} + \begin{bmatrix}0\\ -J^{-1}\end{bmatrix}\dot{h} \tag{8.25}$$

since $\overline{h}$ is assumed to just be a constant and we are only interested in attitude control and not the evolution of δh.

If by chance, $\|H\| = 0$, then we can linearize about $\overline{\theta} = \overline{\omega} = \overline{h} = 0$ since this point is reachable, but the Eq. (8.13) has almost the same Jacobian matrix A as Eq. (8.21) except for an additional column and row matrix of zeros for the new

Fig. 8.6 NASA Kepler telescope (Courtesy of NASA)

state h in Eq. (8.12). Therefore, when $\|H\| = 0$, we have exactly the same result for Example 8.1 with $\delta u = -\delta \dot{h}$.

Let us consider an example problem here of small angle set-point underactuated attitude control utilizing solar radiation pressure as an additional source of torque, although external. In this example, we are allowing $\|H\| \neq 0$ to not be constant so we are assuming the constant $\overline{h} \neq 0$. This example is motivated by the NASA Kepler spacecraft after two RWA failures and the recent successes of its K2 mission. A illustration of the NASA Kepler Telescope is shown in Fig. 8.6. With two RWA the pairs of attitude and angular velocity that can be matched is reduced heavily since the array angular momentum is now restricted to a plane. As a consequence of having only two RWA, Kepler was no longer able to point at the initial set of stars long enough to meet integration time for acceptable images (i.e., light integration time). However, with two noncolinear RWA and a weak external torque, a different fixed orientation with respect to the sun that exploits the material properties of the spacecraft and considering a new set of stars for pointing, provides the degrees of freedom to offset the components of angular momentum perpendicular to the plane spanned by the two remaining RWA. The goal is to find a good inertial attitude and solar panel configuration where the solar radiation pressure along with the two RWA help you remain long enough (i.e., for light integration time) in a small error box to image the surrounding areas of specific stars. This is in part, what Ball Aerospace actually did to successfully extend Kepler's life to the K2 mission [85]. In fact, as shown in reference [86], the linear controllability of a spacecraft is increased when the moments of inertia are less symmetric and in the case of space telescopes in general, they typically have one small momentum of inertia around the telescope boresight.

Here we will provide a simplified example taken from reference [86], where the spacecraft is assumed to be cuboid shape with unequal moments of inertia (see Fig. 8.7).

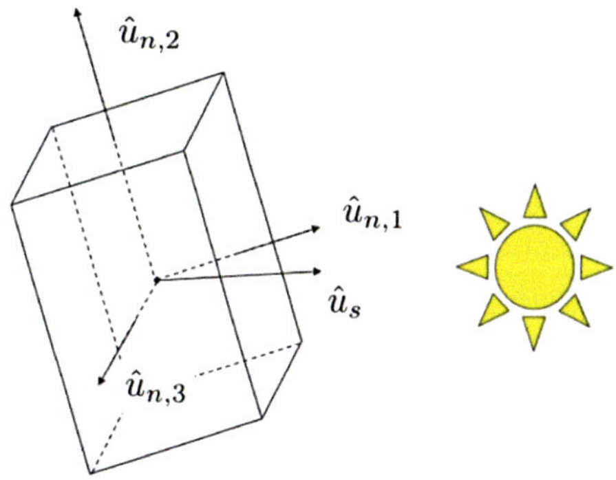

Fig. 8.7 Cuboid spacecraft affected by SRP torques from direction $\hat{u}_s$

Example 8.2 For our simplified example, if we assume that we are making small angle adjustments about an inertially fixed orientation, then we can linearize the entire system including the external attitude dependent solar radiation pressure torques about $\overline{\theta} = \overline{\omega} = 0$, $\overline{h} \neq 0$. For these assumptions to provide a linear set of ordinary differential equations, we must assume specific properties on the external disturbance torques, ε. We will consider the example here from the problem discussed in reference [86]. If we assume that ε associated with solar radiation pressure is very small magnitude and a function of attitude (i.e., due to orbit location), we arrive now at the linearized closed-loop equations of motion

$$\begin{bmatrix} \delta\dot{\theta} \\ \delta\dot{\omega} \end{bmatrix} = \begin{bmatrix} 0 & 1 \\ J^{-1}T & J^{-1}(\overline{h})^{\times} \end{bmatrix} \begin{bmatrix} \delta\theta \\ \delta\omega \end{bmatrix} + \begin{bmatrix} 0 \\ -J^{-1} \end{bmatrix} \dot{h} \tag{8.26}$$

or more specifically for control using RWA

$$\begin{bmatrix} \delta\dot{\theta} \\ \delta\dot{\omega} \end{bmatrix} = \begin{bmatrix} 0 & 1 \\ J^{-1}T & J^{-1}(I_w W\overline{\Omega})^{\times} \end{bmatrix} \begin{bmatrix} \delta\theta \\ \delta\omega \end{bmatrix} + \begin{bmatrix} 0 \\ -J^{-1}I_w W \end{bmatrix} \delta\dot{\Omega}, \tag{8.27}$$

where we are linearizing about $\overline{\theta} = \overline{\omega} = 0$ and $\overline{h}$ constant is the equilibrium spacecraft momentum control system angular momentum at rest with $T = 0$, and T is the linearized matrix for the solar radiation pressure torques about small angles. For Eq. (8.27), the (A, B) pair of matrices is

$$A = \begin{bmatrix} 0 & 1 \\ J^{-1}T & I_w J^{-1}(W\overline{\Omega})^{\times} \end{bmatrix}, \quad B = \begin{bmatrix} 0 \\ -I_w J^{-1} W \end{bmatrix}, \tag{8.28}$$

(continued)

where $\overline{h} = I_w W \overline{\Omega}$, I_w is the rotor inertia component about the rotor axis for the RWA assumed uniform along the array, $\overline{\Omega} = [\overline{\Omega}_1 \ \overline{\Omega}_2]^T$ is a 2×1 column matrix for the equilibrium rotor angular velocity, $W = [w_1 \ w_2]$ is a 3×2 matrix of rotor axis directions in the body frame. Considering this (A, B) pair, controllability can be checked by the controllability Gramian or matrix (see Appendix B). Under specific assumptions, not requiring $H = 0$ or constant provides the (A, B) pair in Eq. (8.28) is linear controllable, and thus can be sent to the desired equilibrium $\overline{\theta}, \overline{\omega}$ from any initial $(\theta(t_0), \omega(t_0))$ within the region of attraction, in a finite time $\Delta t = (t_f - t_0)$. The term $J^{-1}T$ added to A from solar radiation pressure torque assumed attitude dependent to first order by the linearization combine with the internal torques of the spacecraft momentum control systems to allow for local linear controllability. Other external torques may also be considered for control or as perturbations. Here it is assumed that the external disturbance torque model is accurate enough to use it directly. In practice, and estimate will need to be used.

Recall, point set attitude control allows for some wiggle room in attitude and angular state matching that allows for control design at the level of actuator states. Arrays of RWA should avoid the states of $\Omega = 0$ (i.e., zero crossings) due to friction issues with the bearings at low speeds. When designing controllers for this system, constraints on the actuator limits must be taken into account, because unlike external torque actuators, the momentum capacity of spacecraft momentum control systems can be used up very quickly. This would definitely be the case for even a small external disturbance torque that has a DC offset.

This example illustrates that it is possible to exploit the weak external torques to supplement internal torques in space to the benefit holding an attitude (i.e., set-point attitude control) or desaturating momentum control actuators.

8.7 (Small) Angular Rate Control

There are cases when it is desirable to track closely a desired angular velocity equilibrium $\overline{\omega} \neq 0$. In this case, we cannot linearize about attitude coordinates as well since $\overline{\omega} \neq 0$ can never be an equilibrium for the differential equations associated with attitude kinematics. Therefore we would ignore attitude as a state here, and our new rigid body linear system is

$$\begin{aligned}\delta\dot{\omega} &= J^{-1}\left[(J\overline{\omega})^{\times} - \overline{\omega}^{\times}J\right]\delta\omega + J^{-1}\tau \\ &= A\delta\omega + B\delta u\end{aligned} \tag{8.29}$$

Since we are now ignoring the attitude here, we are only concerned with controlling to the desired angular rate equilibrium, $\overline{\omega}$. Clearly, if we choose a proportional control law such as $\tau = -K\delta\omega = -JK_0\delta\omega - [(J\overline{\omega})^{\times} - \overline{\omega}^{\times}J]\delta\omega$ for a positive definite K_0, we would get

$$\begin{aligned}\delta\dot{\omega} &= J^{-1}\left[(J\overline{\omega})^{\times} - \overline{\omega}^{\times}J\right]\delta\omega - J^{-1}K\delta\omega \\ &= (A - BK)\delta\omega \\ &= -K_0\delta\omega\end{aligned} \tag{8.30}$$

where all of the eigenvalues for the closed-loop system are in the open left half plane and therefore, the system here is actually exponentially stable with the exponential decay rate bounded below by the smallest magnitude eigenvalues of K_0. To make sure that K_0 also provides a robust solution in terms of gain and phase margin, you can either use frequency based methods (e.g., Bode plots) to determine the right K_0 to give a sufficient gain or phase margin or design this proportional controller as an infinite time horizon linear quadratic regulator (LQR) which automatically provides a controller with acceptable gain and phase margin.

If you are using a spacecraft momentum control system to match the internal torque τ required for angular rate control, you must make sure that the angular rate states are reachable. If the angular momentum of the system is constant and $\|H\| = 0 =$ constant, then the A matrix is trivial from the nonlinear equations of motion, Eq. (8.13) since $H = 0$ when $\|H\| = 0$. To achieve $\omega = \overline{\omega}$ (i.e., $\delta\omega = 0$), we simply need to make sure that $\delta u = \delta\tau = -\delta\dot{h}$ controls the angular momentum to the desired $\overline{h}$ value since we pulled the $\overline{\omega}^{x}\overline{h}$ term back into the dynamics. Therefore, we desire to have $\overline{h} = -J\overline{\omega} = 0$ and $\delta\dot{h}$ control to $\overline{h}$. In order for us to do that, we can choose $\delta\dot{h} = J\delta\omega$ provides the exponential convergence of $\omega \to \overline{\omega}$. Indeed, if we use Lyapunov's direct method with the candidate Lyapunov function

$$\begin{aligned}V &= \frac{1}{2}\delta\omega^T\delta\omega \\ &= \frac{1}{2}(\omega - \overline{\omega})^T(\omega - \overline{\omega})\end{aligned} \tag{8.31}$$

we get

$$\begin{aligned}\dot{V} &= \delta\omega^T \delta\dot{\omega} \\ &= \delta\omega^T \dot{\omega} \\ &= \delta\omega^T (-J^{-1}\delta\dot{h}) \\ &= -\delta\omega^T \delta\omega\end{aligned} \tag{8.32}$$

so $V(t) = V(0)\exp(-t)$. When $\|H\| \neq 0$ and is constant magnitude, the A matrix is not trivial and the total angular momentum's direction is now dependent on attitude of the spacecraft body with respect to the inertial frame. For this case, we derive the linearized equations with $\overline{\omega}, \overline{h} \neq 0$. Let us denote $\overline{H} = J\overline{\omega}+\overline{h}$ to be the equilibrium total angular momentum in body coordinates. We need $\overline{\omega}^\times \overline{H} = 0$ (i.e., $\overline{\omega} \parallel \overline{H}$) for $\overline{\omega}, \overline{h}$ to be an actual equilibrium for $\dot{\omega}$. Therefore, we are restricted in what $\overline{\omega}, \overline{h}$ we can choose for linearization. Assuming we choose an $\overline{\omega}, \overline{h}$ such that $\overline{\omega}^\times \overline{H} = 0$, the linearized equations of motion are

$$\begin{bmatrix}\delta\dot{\omega} \\ \delta\dot{h}\end{bmatrix} = \begin{bmatrix}-J^{-1}\left(\overline{\omega}^\times J - (J\overline{\omega})^\times - \overline{h}^\times\right) & -J^{-1}\overline{\omega}^\times \\ 0 & 0\end{bmatrix}\begin{bmatrix}\delta\omega \\ \delta h\end{bmatrix} + \begin{bmatrix}-J^{-1} \\ 1\end{bmatrix}\delta\dot{h} \tag{8.33}$$

There exists a very interesting angular rate control problem known as nutation control where there is no needed to linearize about an $\overline{\omega}$. Nutation is the periodic variation of the inclination angle (precession angle) of a rotating body. Nutation control is to dampen out this periodic variation and leave the vehicle rotating about an axis, in our case assume this axis is $\overline{\omega} = [0\ 0\ \eta]^T$.

Example 8.3 Here we will provide an example of the possible controls for closed-loop nutation rate control. We will assume that the closed-loop linear system is a symmetric spinner represented by Eq. (8.17) with a symmetric inertia matrix

$$J = \begin{bmatrix} I_t & 0 & 0 \\ 0 & I_t & 0 \\ 0 & 0 & I_s \end{bmatrix} \tag{8.34}$$

where I_t, I_s are the transverse and spin-axis principal moments of inertia. For a rigid body in the absence of external torques, we have an analytic solution for angular velocity profile along the principal inertia axes as

$$\omega(t) = \begin{bmatrix} \hat{\omega}\cos(\lambda t) \\ \hat{\omega}\sin(\lambda t) \\ \eta \end{bmatrix} \tag{8.35}$$

(continued)

with λ, $\hat{\omega}$ constants. This angular velocity solution provides us a linear system for the angular velocity kinematics

$$\begin{bmatrix} \dot{\omega}_1 \\ \dot{\omega}_2 \\ \dot{\eta} \end{bmatrix} = \begin{bmatrix} 0 & -\lambda & 0 \\ \lambda & 0 & 0 \\ 0 & 0 & 0 \end{bmatrix} \begin{bmatrix} \omega_1 \\ \omega_2 \\ \eta \end{bmatrix} \tag{8.36}$$

Now to converge to our desired equilibrium, $\overline{\omega} = [0\ 0\ \eta]^T$, we choose the PD controller

$$\begin{aligned} \tau_1 &= -k_1\omega_1 - k_2\dot{\omega}_1 \\ \tau_2 &= -k_1\omega_2 - k_2\dot{\omega}_2 \\ \tau_3 &= 0 \end{aligned} \tag{8.37}$$

which by Eq. (8.36) gives

$$\tau = \begin{bmatrix} k_1 & -\lambda k_2 & 0 \\ \lambda k_1 & k_2 & 0 \\ 0 & 0 & 0 \end{bmatrix} \begin{bmatrix} \omega_1 \\ \omega_2 \\ \eta \end{bmatrix} = -K\omega \tag{8.38}$$

The system we have for the symmetric spinner is actually already linear and has the dynamics

$$\begin{aligned} \dot{\omega} &= \begin{bmatrix} 0 & (I_t - I_s)\eta & 0 \\ (I_s - I_t)\eta & 0 & 0 \\ 0 & 0 & 0 \end{bmatrix} \omega + \begin{bmatrix} \frac{1}{I_t} & 0 & 0 \\ 0 & \frac{1}{I_t} & 0 \\ 0 & 0 & \frac{1}{I_s} \end{bmatrix} \tau \\ &= A\omega + B\tau \\ &= \left(\begin{bmatrix} 0 & (I_t - I_s)\eta & 0 \\ (I_s - I_t)\eta & 0 & 0 \\ 0 & 0 & 0 \end{bmatrix} + \begin{bmatrix} \frac{1}{I_t} & 0 & 0 \\ 0 & \frac{1}{I_t} & 0 \\ 0 & 0 & \frac{1}{I_s} \end{bmatrix} \begin{bmatrix} -k_1 & \lambda k_2 & 0 \\ -\lambda k_1 & -k_2 & 0 \\ 0 & 0 & 0 \end{bmatrix} \right) \omega \\ &= \frac{1}{I_t} \begin{bmatrix} -k_1 & -\lambda(I_t - k_2) & 0 \\ \lambda(I_t - k_1) & -k_2 & 0 \\ 0 & 0 & 0 \end{bmatrix} \omega \\ &= (A - BK)\omega \end{aligned} \tag{8.39}$$

(continued)

where $I_t\lambda = (\sigma - 1)\eta$ with $\sigma = I_s/I_t$ by equating Eq. (8.36) and (8.39) when $\tau = 0$. Clearly, $(A - BK)$ for Eq. (8.39) will always have a zero eigenvalue. Therefore, we might as well chose $k_1 = k_2 = I_t$ which will provide us with $\text{eig}(A - BK) = \{-1, -1, 0\}$ with $\omega_i = \omega_i(0)e^{-t}$, $i = 1, 2$ exponentially converging to the equilibrium, $\overline{\omega} = [0\ 0\ \eta]^T$.

We would like to know what the result from the previous example would be for control restricted to internal torques from a spacecraft momentum control system. In order for us to do this, we need to consider different equations of motion.

Example 8.4 Here we will provide an example of the possible controls for closed-loop nutation rate control using spacecraft momentum control systems. We will assume a symmetric spinner with a symmetric inertia matrix

If there are no external torques present and $\|H\| = 0$, then we can control to $h = -J[0, 0, \eta]^T$ in principle coordinates as the control and it would guarantee by the conservation of angular momentum that the equilibrium is our desired value $\overline{\omega} = [0, 0, \eta]^T$. Therefore, our control becomes an angular momentum regulator which implicitly leads to our desired angular rate control. Consider an angular momentum error $h_e = h - J\overline{\omega}$. Then with a control $u = \dot{h}_e = \dot{h} = -Kh_e$ for a positive definite symmetric matrix K, we would get exponential convergence of the system to the desired angular velocity $\overline{\omega} = [0, 0, \eta]^T$. Indeed,

$$\dot{h}_e = -Kh_e \tag{8.40}$$

gives an exponential convergence of $h_e \to 0$.

Note, $\dot{h}$ is typically represented in body coordinates and here we are considering it in principal axis coordinates so the values must be rotated back to body coordinates for actual implementation.

What if $\|H\| \neq 0$ but is still constant. Then we can denote $\overline{H} = {}^{\mathrm{N}}Q^{\mathrm{P}}(I\omega + h)$ as the total angular momentum in inertial coordinates that is constant, and ${}^{\mathrm{N}}Q^{\mathrm{P}}$ is the DCM that takes you from principal coordinates to inertial coordinates. The matrix $\overline{H}$ is constant in these coordinates. Therefore, it is only possible to reach the point $\overline{\omega} = [0, 0, \eta]^T$ if we can achieve

$$\begin{aligned} h &= {}^{\mathrm{P}}Q^{\mathrm{N}}\overline{H} - J\omega \\ &= {}^{\mathrm{P}}Q^{\mathrm{N}}\overline{H} - \begin{bmatrix} 0 \\ 0 \\ I_s\eta \end{bmatrix} \end{aligned} \tag{8.41}$$

(continued)

Since this is an angular momentum regulator problem that is now a function of attitude, a nonlinear control scheme should be incorporated since the system is no longer naturally linear with the attitude dependence of h and we cannot linearize simultaneously about attitude, angular rate, and angular momentum with $\overline{\omega} \neq 0$.

8.8 Linear Prescribed Trajectory Tracking

Let us consider a prescribed trajectory-control pair $(x^*(t), u^*(t))$ which we would like to track. This pair is usually provided by the attitude guidance system and is a function of time. Since the guidance system may only provide an open-loop control result and not take into account uncertainty and disturbances, it is desired to track close to this trajectory-control pair. If we assume that we are only tracking very small displacements to the trajectory and control, we can linearize about $(x^*(t), u^*(t))$. Recall, we use Eq. (8.16) to derive the linear explicitly time-varying system for our problem.

For this linearization, it makes the most sense to use an attitude parameterization that has bilinear kinematics. For that reason we will use unique quaternions to describe our attitude. Recall from Sect. 4.2 that Euler parameters (unit quaternions) have the following kinematics

$$\dot{q} = S(q)\omega = F(\omega)q \tag{8.42}$$

where $S(q)$ is a 4×3 matrix linear in elements of q and $F(\omega)$ is a 4×4 matrix linear in elements of ω. However, the linearized equation is actually simpler than Eq. (8.42) since the scalar terms fall out. Considering only the eigenaxis elements of the quaternion $\hat{q}$, the linearized quaternion kinematics are

$$\delta\dot{\hat{q}} = -\omega^*(t)^\times \delta\hat{q} + \frac{1}{2}\delta\omega \tag{8.43}$$

See reference [81] for details on linearization of quaternion kinematics for propagation of the quaternion kinematics for the Extended Kalman Filter. In reference [81], the estimate of the angular velocity is analogous to the angular velocity trajectory that we are linearizing about for attitude control and the true angular velocity is the same as our true angular velocity, assumed to be known exactly. For the state vector

$x = [\hat{q}, \omega, h]$ we have its linearization about a trajectory control pair, $(x^*(t), u^*(t))$ as

$$\begin{aligned}\delta\dot{x} &= \begin{bmatrix}\delta\dot{\hat{q}}\\ \delta\dot{\omega}\\ \delta\dot{h}\end{bmatrix} \approx \begin{bmatrix}-\omega^*(t)^\times & \frac{1}{2}1 & 0\\ 0 & \Xi(\omega^*(t), h^*(t)) & -J^{-1}\omega^*(t)^\times\\ 0 & 0 & 0\end{bmatrix}\begin{bmatrix}\delta\hat{q}\\ \delta\omega\\ \delta h\end{bmatrix} + \begin{bmatrix}0\\ J^{-1}\\ -1\end{bmatrix}\delta u\\ &= A(t)\delta x + B(t)\delta u\end{aligned} \tag{8.44}$$

where $\Xi(\omega^*(t), h^*(t)) = J^{-1}\left[(J\omega^*(t))^\times - \omega^*(t)^\times J + h^*(t)^\times\right]$, and $\delta u = -\delta\dot{h}$. Note, unit quaternions are not closed under addition or subtraction. Therefore, Jacobian matrix linearization for Eq. (8.44) was actually replaced directly with the result from reference [81]. Recall that in the absence of external torques, $h^*(t)$ is constrained by $\omega^*(t)$ and $\hat{q}^*(t)$ by the conservation of angular momentum. Therefore, with spacecraft momentum control systems, you cannot simply pick $h^*(t)$ and $\omega^*(t)$ independently with which to linearize. Controllability of a linear time varying system, and stability, and robustness of an associated controller can be checked by a multitude of methods. Some of these analysis methods appear in Appendix B.

In satellite operations, very often a simple maneuver may be chosen as the prescribed trajectory. For the purposes of satisfying constraints on angular velocity, acceleration, and jerk, that maneuver may as well be a versine slew (see Sect. 4.4.3). Here we will assume that we have a prescribed trajectory in terms of a versine slew with the same closed-loop equations of motion from Eq. (8.44) except we now have an analytic expression for $q^*(t), \omega^*(t)$. The prescribed trajectories are

$$\begin{aligned}q^*(t) &= \begin{bmatrix}\sin\left(\frac{\phi(t)}{2}\right)c\\ \cos\left(\frac{\phi(t)}{2}\right)\end{bmatrix}\\ \omega^*(t) &= \dot{\phi}(t)c\end{aligned} \tag{8.45}$$

where ϕ is the eigenangle about an eigenaxis c. Recall from Sect. 4.2, that

$$\phi(t) = \frac{\phi_2}{2}(1 - \cos at) \tag{8.46}$$

where ϕ_2 is the desired terminal angle and $0 \leq t \leq \frac{\pi}{a}$ is a time-based parameter. The eigenaxis elements of $q^*(t)$ in Eq. (8.44) can now be replaced with those of Eq. (8.45) with eigenangle from the versine slew in Eq. (8.46).

8.9 Angular Momentum Regulation

Consider the goal regulating to an angular momentum point for the spacecraft momentum control system by using external torques whether from on-board actuators such as thrusters or magnet torquers, or from the environment such as in Example 8.2. Consider that with the desired point, $\overline{h}$ we also would like $\overline{\omega} = 0$. For $\overline{\omega} = \overline{h} = \overline{\varepsilon} = 0$, we would need to use the external torques to zero out the total system angular momentum. The closed-loop equations of motion for this problem are then

$$\begin{bmatrix} \dot{\omega} \\ \dot{h} \end{bmatrix} = \begin{bmatrix} J^{-1}\left(-\omega^{\times} J + h^{\times}\right) & -J^{-1}(\omega)^{\times} \\ 0 & 0 \end{bmatrix} \begin{bmatrix} \omega \\ h \end{bmatrix} + \begin{bmatrix} J^{-1} & J^{-1} \\ 0 & -1 \end{bmatrix} \begin{bmatrix} (\tau + \varepsilon) \\ -\dot{h} \end{bmatrix}, \tag{8.47}$$

where τ is the external torques provided by thrusters and ε are the external torques from disturbances. Clearly in Eq. (8.47), you must design both τ and $\dot{h}$ simultaneously to dump angular momentum and converge to the desired array angular momentum state. Now if we assume that both h and ω are very close to their desired equilibrium values $\overline{\omega} = \overline{h} = \overline{\tau} = \dot{\overline{h}} = 0$, $\varepsilon \approx 0$, we then can linearize Eq. (8.47) giving

$$\begin{aligned} \begin{bmatrix} \delta\dot{\omega} \\ \delta\dot{h} \end{bmatrix} &= \begin{bmatrix} J^{-1}\left[(J\overline{\omega})^{\times} - \overline{\omega}^{\times} J + \overline{h}^{\times}\right] & -J^{-1}(\overline{\omega})^{\times} \\ 0 & 0 \end{bmatrix} \begin{bmatrix} \delta\omega \\ \delta h \end{bmatrix} + \begin{bmatrix} J^{-1} & J^{-1} \\ 0 & -1 \end{bmatrix} \begin{bmatrix} \delta\tau \\ -\delta\dot{h} \end{bmatrix} \\ &= A\delta x + B\delta u \end{aligned} \tag{8.48}$$

Now, for the moment, let us chose $\delta\tau = -K_1\delta\omega$ and $\delta\dot{h} = -K_2\delta h$. Then we have

$$\begin{aligned} \begin{bmatrix} \delta\dot{\omega} \\ \delta\dot{h} \end{bmatrix} &= \begin{bmatrix} J^{-1}\left[(J\overline{\omega})^{\times} - \overline{\omega}^{\times} J + \overline{h}^{\times} - K_1\right] & -J^{-1}\left[(\overline{\omega})^{\times} + K_2\right] \\ 0 & -K_2 \end{bmatrix} \begin{bmatrix} \delta\omega \\ \delta h \end{bmatrix} \\ &= (A - BK)\delta x \end{aligned} \tag{8.49}$$

If we choose $K_1 = (J\overline{\omega})^{\times} - \overline{\omega}^{\times} J + \overline{h}^{\times} + J\overline{K}_1$ for some positive definite symmetric matrix $\overline{K}_1$, we get

$$\begin{bmatrix} \delta\dot{\omega} \\ \delta\dot{h} \end{bmatrix} = \begin{bmatrix} -\overline{K}_1 & -J^{-1}\left[(\overline{\omega})^{\times} - K_2\right] \\ 0 & -K_2 \end{bmatrix} \begin{bmatrix} \delta\omega \\ \delta h \end{bmatrix} \tag{8.50}$$

The matrix $(A - BK)$ in Eq. (8.50) is upper triangle so its eigenvalues are the eigenvalues of $-\overline{K}_1$ and $-K_2$ which can be designed separately to achieve exponential convergence to the desired state equilibrium $\overline{\omega} = \overline{h} = 0$. Note, with $\varepsilon \neq 0$, if we have directly knowledge of this disturbance, we could compensate it with more torque from τ. However, this would not allow us to approach the total desired equilibrium of $\overline{\omega} = \overline{h} = \overline{\tau} = \dot{\overline{h}} = 0$ since we could never approach $\overline{\tau} = 0$. If we assume that the external disturbances are very small and take a long time to build up angular momentum, then we can ignore them when dumping angular momentum.

8.10 Chapter Summary

Linear attitude, angular rate, and angular momentum control was discussed in this chapter. Since momentum control systems preserve angular momentum, for attitude trajectory tracking, all linearized sets of dynamics and control systems developed assumed that the prescribed trajectories did not ask for control that would violate this conservation. It was shown that there are a large swath of methods based on linearized attitude and angular rate dynamics, all of which have decades of heritage in use and validation. This chapter only provided an exposure to some of the valid methods of control for these systems, and so further reading should be carried out for the designer to truly understand control engineering. Books on multi-input-multi-output control and digital control should be considered as critical further reading for designers.

8.11 Exercises

1. About which axis is the following spacecraft stable while spinning without any form of active control? Provide your answer in terms of the B basis vectors (ignore numerical issues—this matrix is meant to be physically realizable).

$$
{}^{\mathrm{B}}I = \begin{bmatrix} 425 & 25 & 61.23724 \\ 25 & 425 & 61.23724 \\ 61.23724 & 61.23724 & 350 \end{bmatrix} \text{kg-m}^2
$$

2. Consider the linear system in Eq. (8.24). For this system,

    1. Design an LQR controller, plot the response of this controller for a rest-to-rest maneuver from $\theta_0 = [0,\ 0,\ 0]^{\mathrm{T}}$ to $\theta_f = [\frac{\pi}{2},\ 0,\ 0]^{\mathrm{T}}$, $\theta_0 = [0,\ 0,\ 0]^{\mathrm{T}}$ to $\theta_f = [0,\ \frac{\pi}{2},\ 0]^{\mathrm{T}}$, and $\theta_0 = [0,\ 0,\ 0]^{\mathrm{T}}$ to $\theta_f = [0,\ 0,\ \frac{\pi}{2}]^{\mathrm{T}}$.
    2. Plot the response $\delta\theta(t), \delta\omega(t)$ and with the same controller, compare to the nonlinear responses $\theta(t), \omega(t)$.

3. Consider the linear system in Eq. (8.25). For this system,

1. Design an PID controller, plot the response of this controller for a rest-to-rest maneuver from $\theta_0 = [0,\ 0,\ 0]^{\mathrm{T}}$ to $\theta_f = [\frac{\pi}{2},\ 0,\ 0]^{\mathrm{T}}$, $\theta_0 = [0,\ 0,\ 0]^{\mathrm{T}}$ to $\theta_f = [0,\ \frac{\pi}{2},\ 0]^{\mathrm{T}}$, and $\theta_0 = [0\ ,0,\ 0]^{\mathrm{T}}$ to $\theta_f = [0,\ 0,\ \frac{\pi}{2}]^{\mathrm{T}}$.
2. Plot the response $\delta\theta(t), \delta\omega(t)$ and with the same controller, compare to the nonlinear responses $\theta(t), \omega(t)$.
3. What are the gain and phase margin?

4. (Attitude Tracking) Consider the linear system in Eq. (8.25). For this system,

    1. Design a full-state feedback controller to asymptotically track the angles $\theta^*(t)(t) = [s(t),\ c(t),\ 0]^{\mathrm{T}}$.
    2. Plot the error response in $\delta\theta_e(t) = \delta\theta(t) - \theta^*(t)(t)$.

    Feel free to use methods from Appendices B and C.
5. (Angular Momentum Regulation) Consider the linear system in Eq. (8.50). For this system,

    1. Design a full-state feedback controller to asymptotically to regulate the angular momentum to $\overline{h} = 0$.
    2. Plot the response in $\delta h(t), \delta\omega(t)$.

6. (Angular Momentum Regulation) Consider the nonlinear system in Eq. (8.47). Use the same controller designed for the system in Eq. (8.50) above and plot the response in $\delta h(t), \delta\omega(t)$.
    Feel free to use methods from Appendices B and C.

Chapter 9
Steering Algorithms

This chapter discusses the methods of CMG torque allocation, i.e. algorithms for calculating CMG gimbal rates to provide a commanded torque to the vehicle. Solutions to this problem are analogous to the problem of inverse kinematics in robotics. The literature often refers to these allocation techniques as steering algorithms or steering laws, of which pseudoinverse solutions are the most commonly discussed. The chapter surveys many classes of steering algorithms, starting with pseudoinverse methods, continuing with the more conservative methods based on limited gimbal angles and angular momentum, and ending with the less conventional and state-of-the-art optimal methods.

9.1 CMG Steering Algorithms

CMG steering algorithms allocate inputs to the actuators (gimbal speed and/or rotor acceleration) to provide a commanded internal torque to the spacecraft. Thruster allocation and magnetic torque allocation are examples of similar control allocation problems on spacecraft, in which the actuators create an external torque on the spacecraft (force and torque in the case of thrusters). In contrast, steering algorithms for momentum devices work with limited angular momentum and abide by the constraint of conservation of angular momentum. RWA steering is rarely more sophisticated than a simple pseudoinverse of a constant matrix, the Jacobian matrix discussed in Chap. 4. CMG steering is vastly more involved because the Jacobian matrix changes in ways that depend on the past usage of the CMGs. It is one of the most challenging aspects of momentum systems and consequently the topic of primary interest in the literature.

A CMG steering algorithm serves as an interface between the attitude compensation and the individual actuators. The attitude-control system calculates a torque needed to accelerate the spacecraft toward a commanded attitude and

© The Author(s), under exclusive license to Springer Nature Switzerland AG 2025

F. A. Leve et al., *Spacecraft Momentum Control Systems*, Space Technology Library 46, https://doi.org/10.1007/978-3-031-96165-6_9

angular velocity. The steering algorithm transforms this torque into individual CMG gimbal-rate commands. These commands are sent the gimbals, which track the rate commands through closed-loop control as discussed in Chap. 10. However, it is very rarely the case that the command sent to the gimbals produces exactly the commanded gimbal rate or acceleration. Some reasons include uncertainty in gimbal-bearing friction and losses in the gear train, as well as the inherent limitations of motors, which cannot respond with arbitrarily high or fast speed changes. Therefore, inner-loop control may be accompanied by feedforward: a known, time-dependent open-loop input. The gimbal's rate loop then rejects any remaining (unknown) disturbances. Such a control system can settle faster (be higher in bandwidth) without risking saturation due to overshoot. This architecture is common in CMG-driven spacecraft.

CMG steering algorithms may be considered successful when they accurately map spacecraft-level torque into actuator behaviors while avoiding singularities and thus high gimbal rates and accelerations. Successful algorithms comprise three categories, as shown in Fig. 9.1: singularity escape algorithms, singularity avoidance algorithms, and singularity avoidance and escape algorithms. Other approaches also appear in the literature, but this chapter is limited to these three types.

Singularity avoidance and inner-loop control are separate phenomena, but both impact feedback-control performance of a spacecraft with momentum control devices. For arrays of RWAs or CMGs, saturation can occur at the array level as limits on array angular momentum (see Chap. 5) or at the actuator level as limits on the rotor and gimbal rates and accelerations. Chapter 3 discusses these issues in detail.

For the likely case in which rotor angular momentum is identical for all CMGs in an array, the location and form of singularities depends solely on the CMGs'

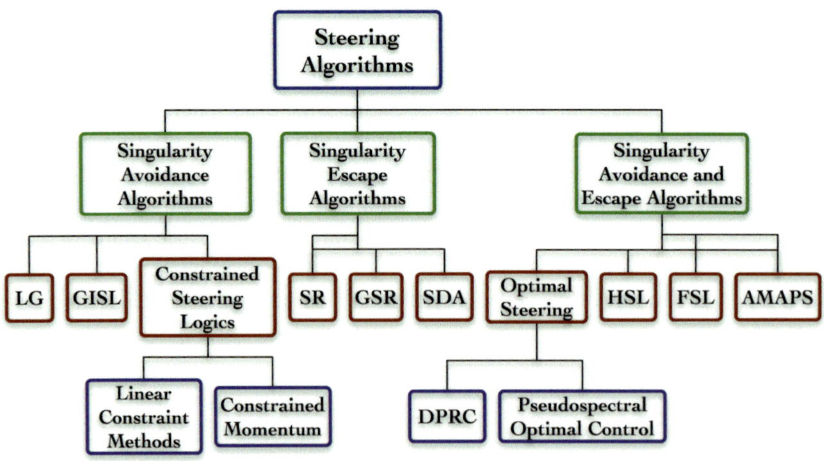

Fig. 9.1 Steering algorithms for SGCMGs

orientations or alignments relative to one another. As discussed in Chap. 6, the behavior of momentum systems is independent of the devices' location within the spacecraft.

Steering laws enable the array to impart a commanded output torque, $\dot{h}$ by "inversion" of the Jacobian matrix A to compute $\dot{\Delta}$:

$$\dot{h} = A\dot{\Delta} \tag{9.1}$$

Decomposition of A into its range and null spaces is central to an understanding of steering algorithms. The gimbal rate solutions can be decomposed as

$$\dot{\Delta} = \dot{\Delta}_f + \dot{\Delta}_n \tag{9.2}$$

where $\dot{\Delta}_f$ is a column of gimbal rates in the range space of A, so that $\dot{h} = A\dot{\Delta}_f$. With four or more CMGs, an array can exhibit so-called null motions. This motion consists of non-zero gimbal rate combinations that neither produce torque on the spacecraft nor (equivalently) change the array's angular momentum. These combinations correspond to $\dot{\Delta} = \dot{\Delta}_n$. The quantity $\dot{\Delta}_n$ lies entirely in the nullspace of A, so that $A\dot{\Delta}_n$=0.

With this nomenclature in place, we can state three objectives for the majority of steering algorithms in terms of forced and null motion: (1) impart a commanded torque, (2) escape singularity through the addition of torque error τ_e and/or avoid singularity through null motion $\dot{\Delta}_n$, and/or (3) follow a commanded gimbal rate trajectory $\dot{\Delta}_d$ while simultaneously achieving (1) and/or (2). A cost function that is often used to assess a steering law's success in meeting these objectives is

$$J = \frac{1}{2}\tau_e^{\mathrm{T}} R \tau_e + \frac{1}{2}\dot{\Delta}_e^{\mathrm{T}} Q \dot{\Delta}_e \tag{9.3}$$

where Q and R are positive definite, symmetric matrices,

$$\tau_e = \dot{h} - A\dot{\Delta} \tag{9.4}$$

and

$$\dot{\Delta}_e = \dot{\Delta}_d - \dot{\Delta} \tag{9.5}$$

Most published steering algorithms are known as pseudoinverse methods. They are found from taking the first variation of Eq. (9.3) and setting it equal to zero, at which point the solution for gimbal rates is a minimum-norm column matrix of values. Most steering algorithms do not consider constraints on the gimbal rates and accelerations. These constraints are typically considered in the inner-loop controllers after the fact, where input power and accuracy may be sacrificed at the boundaries of the constraint, or they are accommodated in the trajectory-planning step, where the only permissible maneuvers are those that never cause the momentum devices to reach these limits.

9.2 Origin of Pseudoinverse Methods

Pseudoinverse methods are generally analytic solutions for a set of gimbal rates or accelerations that produce a commanded torque. They are all based on pseudoinverses of a nonsquare matrix, giving them their name. It should be noted however, that in practice, these analytic solutions are almost never implemented. Only solutions that are never ill-conditioned are utilized. See Appendix C.

9.2.1 *Blended Inverse*

Taking the first variation of Eq. (9.3) and setting it equal to zero leads to

$$\begin{aligned}\dot{\Delta} &= (Q + A^{\mathrm{T}}RA)^{-1}(A^{\mathrm{T}}R\dot{h} + Q\dot{\Delta}_d)\\ &= (Q + A^{\mathrm{T}}RA)^{-1}A^{\mathrm{T}}R\dot{h} + (Q + A^{\mathrm{T}}RA)^{-1}Q\dot{\Delta}_d \end{aligned} \tag{9.6}$$

which has been given the name Blended Inverse (Tekinalp and Yavuzoglu [88]). Equation (9.6) is a steering law that achieves singularity escape through the addition of torque error, singularity avoidance through the addition of null motion, and tracking of commanded gimbal rates. While mathematically convenient, the torque error and indeterminacy near singularity may make such an approach unappealing in practice. Typically A has at most full row rank. The most common case is $n \geq 3$ for CMGs to control 3 axes of torque, where the column size is n and the row size is 3. Therefore, in the following, we will choose $Q \in \mathbb{R}^{3\times 3}$ over $Q \in \mathbb{R}^{n\times n}$ and $R \in \mathbb{R}^{n\times n}$ over $R \in \mathbb{R}^{3\times 3}$ separately for both parts of the RHS of Eq. (9.6). Therefore, in an effort to use the minimum-norm solution, i.e., use $R^{\mathrm{T}}A^{\mathrm{T}}(Q + AR^{T}A^{\mathrm{T}})^{-1}\dot{h}$ rather than $(Q + A^{\mathrm{T}}RA)^{-1}A^{\mathrm{T}}R\dot{h}$, the inverted quantities of the first term on the RHS of Eq. (9.6) will always be 3×3.

9.2.2 *Moore-Penrose Inverse*

When A is full rank, the simplest and possibly the most prevalent pseudoinverse in the literature is known as the Moore-Penrose inverse and it is found through the first variation of Eq. (9.6) with parameters ($Q = 0_{3\times 3}$, $R = I_{n\times n}$, and $\dot{\Delta}_d = 0$):

$$\dot{\Delta} = A^{\mathrm{T}}(AA^{\mathrm{T}})^{-1}\dot{h} \tag{9.7}$$

The Moore-Penrose pseudoinverse in Eq. (9.6) has but one objective, which is obvious from the choice of ($Q = 0_{3\times 3}$, $R = I_{n\times n}$, and $\dot{\Delta}_d = 0$). This objective is to follow the torque command, sometimes known as perfect "torque tracking." Another consequence of the Moore-Penrose solution is that the norm of $\dot{\Delta}$ is minimized, eliminating null-space gimbal motion, if a solution exists. But its simplicity also comes with limitations: the Moore-Penrose pseudoinverse is incapable of avoiding or escaping singularities and does not accommodate constraints on gimbal angles, rates, or accelerations. Furthermore, this is only the representation for Moore-Penrose solution when A is full rank. If A is rank deficient, the Moore-Penrose pseudo-inverse exists, but it is not calculated this way. Appendix C provides more detail on this calculation.

9.3 Singularity-Escape Algorithms (Torque-Error Algorithms)

Singularity-escape algorithms have the goal of passing through singularities and recovering with minimal (but not necessarily zero) deviation from the commanded torque. The addition of a scalar or matrix whose only purpose is to maintain numerical conditioning of a matrix inverse gives these methods the name "dirty inverses:" they are inexact mappings from commanded torque to gimbal rates. This approach to numerical conditioning constitutes the "dirt" that can cause inaccuracy in the calculation near singularity.

9.3.1 Singularity Robust (SR) Inverse

One of the earliest singularity-escape methods is the Singularity-Robust (SR) inverse (Bedrossian et al. [89]). This pseudoinverse is found through substitution of the parameters ($Q = \alpha I_{3\times 3}$, $R = I_{n\times n}$, and $\dot{\Delta}_d = 0$) in Eq. (9.6) as

$$\dot{\Delta} = A^{\mathrm{T}}(AA^{\mathrm{T}} + \alpha I_{3\times 3})^{-1}\dot{h} \tag{9.8}$$

with a possible choice of a singularity parameter

$$\alpha = \alpha_0 e^{-\mu m} \tag{9.9}$$

where α_0 and μ are positive constants. The term $m = \sqrt{\det(AA^{\mathrm{T}})}$ is known as the singularity measure or manipulability in the field of robotics (Yoshikawa [38]). The singularity measure is the product of the singular values of the Jacobian matrix. Therefore, it provides a measure of the distance of the CMG array from singularity.

The SR-inverse is effective in escaping most singularities eventually, and the potentially long escape time can be traded against the torque error through the choice of tuning parameters, α_0 and μ. Specifically, a system that can accept high torque error can escape the singularities quickly. Nevertheless, this escape time is not deterministic and must be bounded through Monte Carlo analysis. This non-determinism may be unacceptable for spacecraft architectures where delay in following a specified trajectory adversely impacts data collection.

The objective of the singularity parameter is to regularize the singular values of the matrix AA^{T} at singularity so that the matrix to be inverted remains positive definite, rather than losing rank. The solution, of course, is not exact when $\alpha \neq 0$. It therefore achieves a torque that deviates from the command when the array is nearly singular. In fact, the singularity parameter α adds torque error in all directions, which is clear from an SVD of Eq. (9.8):

$$\dot{\Delta} = A^{\mathrm{T}}(AA^{\mathrm{T}} + \alpha I_{3\times 3})^{-1}\dot{h} = V \begin{bmatrix} \frac{\sigma_1}{\sigma_1^2+\alpha} & 0 & 0 \\ 0 & \frac{\sigma_2}{\sigma_2^2+\alpha} & 0 \\ 0 & 0 & \frac{\sigma_3}{\sigma_3^2+\alpha} \\ 0 & 0 & 0 \end{bmatrix} U^{\mathrm{T}}\dot{h} = A^{\mathrm{SR}}\dot{h} \tag{9.10}$$

The torque error resulting from singularity escape for many pseudoinverse methods is $\tau_e = \dot{h} - A\dot{\Delta}$. For the SR-inverse this error is

$$\tau_e = \dot{h} - A\dot{\Delta} = \dot{h} - AA^{\mathrm{T}}(AA^{\mathrm{T}} + \alpha I_{3\times 3})^{-1}\dot{h} = U \begin{bmatrix} \frac{\alpha}{\sigma_1^2+\alpha} & 0 & 0 \\ 0 & \frac{\alpha}{\sigma_2^2+\alpha} & 0 \\ 0 & 0 & \frac{\alpha}{\sigma_3^2+\alpha} \\ 0 & 0 & 0 \end{bmatrix} U^{\mathrm{T}}\dot{h} = \epsilon\dot{h} \tag{9.11}$$

where ϵ is an torque error amplification parameter that approaches zero as the singularity parameter α goes to zero.

Equation (9.11) clearly shows that the torque error is not only scaled by the size of the singularity parameter but also by the size of the torque command. Therefore, the attitude-control subsystem cannot assume a fixed amount of torque error at every singularity.

9.3.2 Singularity Direction Avoidance (SDA) Pseudoinverse

To reduce the torque error introduced for singularity escape, the Singularity Direction Avoidance (SDA) pseudoinverse localizes torque error to the singular direction only. The SDA pseudoinverse does so by regularizing only the smallest singular value. The parameters defining SDA from Eq. (9.6) are ($Q = X$, $R = I_{n\times n}$, and $\dot{\Delta}_d = 0$), giving the expression

$$\dot{\Delta} = A^{\mathrm{T}}(AA^{\mathrm{T}} + X)^{-1}\dot{h} \tag{9.12}$$

where $X = U\text{diag}(0, 0, \alpha)U^{\mathrm{T}}$ and α is the same singularity parameter from Eq. (9.9). Substitution of X into Eq. (9.12), taking the SVD, and sorting the singular values leads to

$$\dot{\Delta} = A^{\mathrm{T}}(AA^{\mathrm{T}} + X)^{-1}\dot{h} = V \begin{bmatrix} \frac{1}{\sigma_1} & 0 & 0 \\ 0 & \frac{1}{\sigma_2} & 0 \\ 0 & 0 & \frac{\sigma_3}{\sigma_3^2+\alpha} \\ 0 & 0 & 0 \end{bmatrix} U^{\mathrm{T}}\dot{h} = A^{SDA}\dot{h} \tag{9.13}$$

with torque error

$$\tau_e = U \begin{bmatrix} 0 & 0 & 0 \\ 0 & 0 & 0 \\ 0 & 0 & \frac{\alpha}{\sigma_3^2+\alpha} \\ 0 & 0 & 0 \end{bmatrix} U^{\mathrm{T}}\dot{h} = \epsilon\dot{h} \tag{9.14}$$

Compared to Eq. (9.11), Eq. (9.14) clearly reduces torque error in an overall sense. Numerically stable and computationally efficient Methods exist to compute the SVD. Therefore, realtime implementation of the SDA pseudoinverse in deterministic flight code is feasible. Also, a byproduct of the SVD is that there is no need to invert AA^{T}: the inverse can be reconstructed from the product of the unitary matrices U and V and the singular values once the SVD has been performed as shown in Eq. (9.13).

Like the Moore-Penrose inverse and the SR inverse, the SDA pseudoinverse is susceptible to gimbal lock (see Chap. 5). This weakness becomes clear when one takes the SVD of MP, SR, and SDA pseudoinverses. At singularity, all of these pseudoinverse solutions for gimbal rates take the form

$$\dot{\Delta} = V \begin{bmatrix} \bullet & 0 & 0 \\ 0 & \bullet & 0 \\ 0 & 0 & 0 \\ 0 & 0 & 0 \end{bmatrix} U^{\mathrm{T}}\dot{h} \tag{9.15}$$

where $\bullet$ indicates a nonzero number. Now if the array is singular, $\dot{h} \propto u = u_3$, the third column of the unitary matrix U for a rank 2 singularity. At singularity, Eq. (9.15) gives a zero result for the MP, SR, and SDA pseudoinverse methods, even when $\dot{h} \neq 0$ (see Chap. 5). As discussed in Chap. 6, for parallel-type (roof) arrangements, there is a possibility of a rank 1 singularity where there is a plane of singular directions and u is in that plane.

9.3.3 Generalized Singularity Robust (GSR) Inverse

In an effort to mitigate the risk of gimbal lock in a pseudoinverse-based steering algorithm, the Generalized Singularity Robust (GSR) inverse has been proposed (Wie et al. [90]). The parameters for Eq. (9.6) that define the GSR inverse are ($Q = E$, $R = I_{n\times n}$, and $\dot{\Delta}_d = 0$). They yield

$$\dot{\Delta} = A^{\mathrm{T}}(AA^{\mathrm{T}} + \alpha E)^{-1}\dot{h} = A^{GSR}\dot{h} \tag{9.16}$$

where α can be chosen as in Eq. (9.9), and

$$E = \begin{bmatrix} 1 & \epsilon_0 \sin(\omega_3 t + \phi_3) & \epsilon_0 \sin(\omega_2 t + \phi_2) \\ \epsilon_0 \sin(\omega_3 t + \phi_3) & 1 & \epsilon_0 \sin(\omega_1 t + \phi_1) \\ \epsilon_0 \sin(\omega_2 t + \phi_2) & \epsilon_0 \sin(\omega_1 t + \phi_1) & 1 \end{bmatrix} > 0 \tag{9.17}$$

where ϵ_0 is a positive amplitude, and ω_i and ϕ_i are the frequencies and phase shifts of the dither used to nudge or wiggle the array out of the local minimum of gimbal lock. Because this pseudoinverse is an explicit function of time, dithering may be successful in driving the array out of a singularity at gimbal lock. However, there is no guaranteed upper bound on the time required for escape. The time is impacted by uncertainty in the Jacobian matrix due to unavoidable measurement noise, delay, and/or quantization in gimbal angles. Also, the effect of gimbal-lock escape is embedded in the pseudoinverse non-linearly. Therefore, although this technique may be assessed with statistical modeling of slews and found to be successful in a large fraction of cases, using GSR or any torque error/singularity escape methods for CMG array steering is likely impractical when guaranteed accuracy in tracking a specific torque trajectory and/or predictability in maneuver execution are of primary importance.

If some torque error is acceptable, a more deterministic way to escape gimbal lock is to perturb the commanded torque along directions orthogonal to the singular direction. Consider the commanded torque as a projection along the basis created by the unitary matrix U from the SVD of A as

$$\dot{h} = \beta_1 u_1 + \beta_2 u_2 + \beta_1 u_3 \tag{9.18}$$

where

$$\beta_i = \dot{h}^{\mathrm{T}} u_i \tag{9.19}$$

In a gimbal-lock singularity, $\dot{h} = u_3$, and the majority of pseudoinverse solutions provide an improper result as shown in Eq. (9.15). Therefore such a method perturbs the commanded torque in Eq. (9.18) as

$$\dot{h} = \beta_1^* u_1 + \beta_2^* u_2 + \beta_1 u_3 \tag{9.20}$$

where the parameters β_1^* and β_2^* are nonzero away from singularities. The appropriate choice of these parameters ensures that gimbal-lock is impossible analytically, because at singularity, the commanded torque always has components orthogonal to the singular direction (Leve [91] and Leve et al. [92]). Away from singularity, the singularity parameters in Eq. (9.20) approach their true values β_1 and β_2. However, the method in Eq. (9.20) is still a torque-error method and does not track accurately the commanded torque trajectory near a singularity.

9.4 Singularity Avoidance Algorithms

Some singularities, i.e., non-degenerate hyperbolic singularities, offer the possibility of avoidance without the use of torque error (see Chap. 5). Such methods can be posed as pseudoinverse solutions with the addition of null motion used solely for singularity avoidance. These solutions are considered exact solutions and impart precisely zero torque error except at saturation singularities. Developing this steering-law software goes hand-in-hand with defining the array architecture so that only non-degenerate and, ideally, only globally passable hyperbolic singularities are encountered internally. Some of the most popular, published singularity-avoidance methods are the local-gradient methods, as discussed in this section.

9.4.1 Generalized Inverse Steering Law (GISL)

The Generalized Inverse Steering Law (GISL) is found through substitution in Eq. (9.6) of the parameters ($Q = B^{\mathrm{T}}A$, $R = I_{3\times3}$, and $\dot{\Delta}_d = Q^{-1}B^{\mathrm{T}}\dot{h}$) and made to be the minimum-norm solution as

$$\dot{\Delta} = [(A+B)^{\mathrm{T}}A]^{-1}(A+B)^{\mathrm{T}}\dot{h} \Longrightarrow (A+B)^{\mathrm{T}}[A(A+B)^{\mathrm{T}}]^{-1}\dot{h} = A^{\mathrm{GISL}}\dot{h} \tag{9.21}$$

where B is a $3 \times n$ matrix of torque directions (sine and cosine functions of gimbal angle), the ith column of which is orthogonal to the corresponding column of A, i.e., $B_i^{\mathrm{T}}A_i = 0 \;\forall\; i = 1, \ldots, n$ and not $B_i^{\mathrm{T}}A_j = 0 \; j \neq i$ and $\forall\; i, j = 1, \ldots, n$. Pre-multiplying both sides of Eq. (9.21) by A yields

$$A\dot{\Delta} = \dot{h} \tag{9.22}$$

Therefore, the GISL provides an exact solution for gimbal rates to deliver the commanded torque. The matrix B is responsible for this exactness because it adds null motion within the pseudoinverse. However, even for a GISL there is no possibility of escape from elliptic singularities because null motion does not exist for those cases. Therefore, GISL is classified as a singularity- avoidance algorithm as shown in Fig. 9.1.

9.4.2 Local Gradient (LG) Methods

Equation (9.6) provides the general form of pseudoinverse steering algorithms. Decomposing it with the Matrix Inversion Lemma (see reference [93]), separates range-space from null-space terms:

$$\dot{\Delta} = (Q + A^{\mathrm{T}}RA)^{-1}A^{\mathrm{T}}R\dot{h} + (Q + A^{\mathrm{T}}RA)^{-1}Q\dot{\Delta}_d \tag{9.23}$$

Applying the Matrix Inversion Lemma yields

$$(Q + A^{\mathrm{T}}RA)^{-1} = [Q^{-1} - Q^{-1}A^{\mathrm{T}}(R^{-1} + AQ^{-1}A^{\mathrm{T}})^{-1}AQ^{-1}]$$

Now if the desired gimbal rate, $\dot{\Delta}_d$ is within the null space, so as to provide singularity avoidance, the gimbal rates are $\dot{\Delta}_d = \beta[I_{n\times n} - A^{+}A]d$, where β is a singularity parameter of the same form as Eq. (9.8), and $[I_{n\times n} - A^{+}A]$ is a matrix that projects any vector d onto the null space of A. Substituting into Eq. (9.23) the quantity $\dot{\Delta}_d = \beta[I_{n\times n} - A^{+}A]d$ and the second part of Eq. (9.23) we get

$$\dot{\Delta} = (Q + A^{\mathrm{T}}RA)^{-1}A^{\mathrm{T}}R\dot{h} + \beta[I_{n\times n} - A^{+}A]d \tag{9.24}$$

Note that the second term on the RHS of Eq. (9.24) is invariant to choice of Q and R with the constraint that $\dot{\Delta}_d$ lies in the nullspace of A. The local gradient is derived from substituting the parameters ($Q = 0_{3\times 3}$, $R = I_{n\times n}$, and $\dot{\Delta}_d = \beta[I_{n\times n} - A^{+}A]d$) into Eq. (9.23):

$$\dot{\Delta} = A^{\mathrm{T}}(AA^{\mathrm{T}})^{-1}\dot{h} + \beta[I_{n\times n} - A^{+}A]d \tag{9.25}$$

The gradient-search direction d in Eq. (9.25) can be chosen to minimize a function that maximizes the array's distance from singularity. A convenient function to use in this optimization is $f = -m^2$, where

$$d = \frac{\partial f}{\partial \Delta} \tag{9.26}$$

The gradient-search direction in Eq. (9.26) has been shown to be effective in avoiding non-degenerate and globally passable hyberbolic singularities for SGCMGs and singularities for DGCMGs in the literature (Kurokawa [52]). The LG method in Eq. (9.25) is an exact solution that can avoid non-degenerate globally passable singularities but is ineffective in avoiding elliptic and locally passable singularities.

9.4.3 Constrained Gimbal Angle or Angular Momentum Methods

A particular class of single-gimbal CMG steering laws uses constraints (whether in hardware or software) to avoid singularities while instantaneously specifying gimbal rates without future knowledge and without intentionally inducing torque error. Several variations of this steering law exist in the literature (Kurokawa [52] and Havill and Ratcliff [94]). Here, we offer a generalized form of such a steering law that frees the algorithm development from specific array geometries (Jones *et al.* [95]). The following generalized mathematical description of steering laws with linear constraints uses two examples to explain the formulation: the familiar architecture of three scissored pairs and six CMGs arranged in two triplets.

The justification for this approach is articulated in Kurokawa's steering law survey [52]. He and others agree on the capabilities of an ideal steering law, which should be

1. Capable of handling singularities
2. Exact (error-free)
3. Instantaneous (requiring no future knowledge)
4. Technologically feasible in terms of computational power and hardware implementation
5. General with regard to maneuvers and CMG array

Null motion and constraint-based methods are the only categories that clearly meet the first three criteria. Equation (9.1) shows the general form of the equation for the torque in terms of an array of gimbal rates $\dot{\Delta}$ and a Jacobian matrix A,

CMG steering laws often solve Eq. (9.1) using a version of the Moore-Penrose pseudoinverse solution:

$$\dot{\Delta} = A^{\mathrm{T}}(AA^{\mathrm{T}})^{-1}\dot{h} \tag{9.27}$$

Linearly constrained steering laws are no exception; however, in addition to describing the dynamics of the problem via the Jacobian matrix, these steering laws also include linear constraint equations:

$$D = C\dot{\Delta} \tag{9.28}$$

Equation (9.27) can be augmented with Eq. (9.28) to include the constraints:

$$\begin{bmatrix} \dot{h} \\ D \end{bmatrix} = \begin{bmatrix} A \\ C \end{bmatrix} \begin{bmatrix} \dot{\Delta} \\ 0 \end{bmatrix} = \overline{A} \begin{bmatrix} \dot{\Delta} \\ 0 \end{bmatrix} \tag{9.29}$$

The matrix $\overline{A}$ combines both of these partitions, with the pseudoinverse solution

$$\dot{\Delta} = (\overline{A}\overline{A}^{\mathrm{T}})^{-1}\overline{A}^{\mathrm{T}}\begin{bmatrix}\dot{h}\\0\end{bmatrix} \tag{9.30}$$

This description generalizes all variations of this class of steering laws where A is not necessarily square. However, if it is, the solution simplifies further and becomes

$$\dot{\Delta} = \overline{A}^{-1}\begin{bmatrix}\dot{h}\\0\end{bmatrix} \tag{9.31}$$

In order for the $\overline{A}$ matrix to be invertible, the constraint matrix C must contain rows that are linearly independent of the range space of A. When these constraints are guaranteed to maintain linear independence for the range of possible motions, the array becomes singularity-free. To achieve this goal, C must consist of constraint equations with some component orthogonal to the directions described by the rows of the Jacobian matrix.

These additional equations provide a basis for the singularity-avoidance properties of the steering laws involved. Holonomic constraints specify the gimbal angles directly. They can be incorporated into this formulation by simply taking the derivative of the constraint involved and specifying a carefully chosen, non-singular initial gimbal-angle condition. The scissored-pair array constraint is an example and is summarized below. Non-holonomic constraints specify the path or some function of the gimbal rates without specifying angles *per se*. Further research may identify non-holonomic constraints of interest, but to date only holonomic constraints have been identified.

For simplicity, the examples here consider only exactly constrained systems, where the number of constraints m is equivalent to the number of CMGs beyond the minimum of three. Using exactly three CMGs leaves no null space for constraints to be added. For example, controlling a two-dimensional momentum space with three SGCMGs offers the opportunity for one constraint equation. Such a situation may be preferred: inverting the now square A matrix represents an opportunity for minimal computation, compared to the more computationally burdensome calculation of a numerically robust pseudoinverse.

An important practical consideration is that in order for a constraint-based steering law to operate properly, the constraint must be exactly enforced. The finite precision of typical software environments demand that some low-bandwidth feedback be used to avoid an accumulation of numerical error that would cause the CMGs to drift off the path consistent with the initial conditions and the constraint.

9.4.3.1 Linear-Constraint Methods

In a scissored-pair configuration, two SGCMGs are constrained such that their gimbal displacements are equal in magnitude but opposite in direction (Cunningham and Driskill [49]). The motion of the two momentum vectors resembles that of a pair of scissors, hence the name. The angles, as defined in Fig. 9.2, are constrained to be the same ($\delta_1 = \delta_2$). The scissoring motion caused by this constraint produces an output torque in a direction fixed along the reflection line between the two CMGs, as shown in Fig. 9.2. Thus, complete three degree-of-freedom control of a spacecraft requires three scissored pairs of SGCMGs. One potential arrangement is shown in Fig. 9.3.

Scissored-pair arrangements of SGCMGs are singularity-free except at the saturation limits, where the pair produces the maximum momentum of which it is capable in that direction. Such a configuration of CMGs is generally not considered mass efficient despite its singularity-avoidance properties and its relative simplicity because the angular-momentum envelope is considerably smaller than other six CMG arrays of similar mass. Nevertheless, the modularity of such a configuration and the prospect of commercially available, off-the-shelf scissored-pair units suggests that the configuration has some value at the system level.

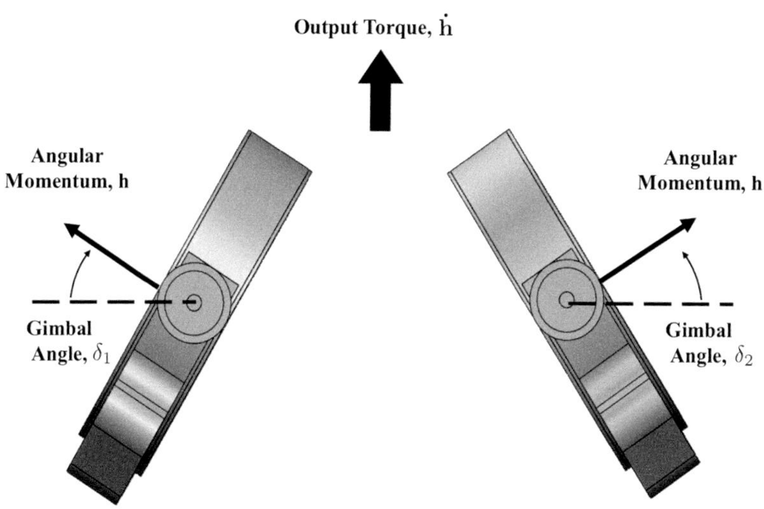

Fig. 9.2 Scissored pair constraint, *The scissored pair of CMGs is constrained such that the gimbal angles are equal in magnitude and opposite in direction, which results in a net torque along a single axis*

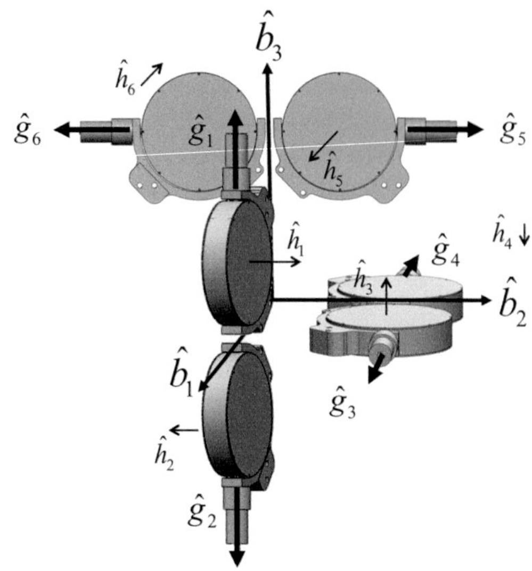

Fig. 9.3 Arrangement of 3 scissored pairs. *This figure illustrates a possible arrangement of the three scissored pairs of CMG that are required for 6-DOF control of a spacecraft*

The Jacobian matrix for the array of three scissored pairs is

$$A = \begin{bmatrix} -c\delta_1 & -c\delta_2 & 0 & 0 & -s\delta_5 & s\delta_6 \\ -s\delta_1 & s\delta_2 & -c\delta_3 & -c\delta_4 & 0 & 0 \\ 0 & 0 & -s\delta_3 & s\delta_4 & -c\delta_5 & -c\delta_6 \end{bmatrix} \tag{9.32}$$

Each pair is constrained as follows: $\delta_1 - \delta_2 = 0$; $\delta_3 - \delta_4 = 0$; $\delta_5 - \delta_6 = 0$. Taking the derivative of these holonomic constraints produces a similar constraint on the gimbal rates:

$$\begin{bmatrix} 0 \\ 0 \\ 0 \end{bmatrix} = \begin{bmatrix} 1 & -1 & 0 & 0 & 0 & 0 \\ 0 & 0 & 1 & -1 & 0 & 0 \\ 0 & 0 & 0 & 0 & 1 & -1 \end{bmatrix} \tag{9.33}$$

This constraint equation provides the C and D matrices. Augmenting the Jacobian matrix with the constraint matrices provides $\overline{A}$:

$$\begin{bmatrix} \dot{h} \\ D \end{bmatrix} = \begin{bmatrix} -c\delta_1 & -c\delta_2 & 0 & 0 & -s\delta_5 & s\delta_6 \\ -s\delta_1 & s\delta_2 & -c\delta_3 & -c\delta_4 & 0 & 0 \\ 0 & 0 & -s\delta_3 & s\delta_4 & -c\delta_5 & -c\delta_6 \\ 1 & -1 & 0 & 0 & 0 & 0 \\ 0 & 0 & 1 & -1 & 0 & 0 \\ 0 & 0 & 0 & 0 & 1 & -1 \end{bmatrix} \begin{bmatrix} \dot{\Delta} \\ 0 \end{bmatrix} \tag{9.34}$$

The well-known singularity avoidance properties of this constraint-based steering law can be confirmed by showing that the rows of the constraint matrix in Eq. (9.33) are linearly independent of A. Using the fact that the column rank is equivalent to the row rank,

$$\begin{bmatrix} \dot{h} \\ D \end{bmatrix} = \begin{bmatrix} \lambda_1 \\ \lambda_2 \\ \lambda_3 \\ \lambda_4 \\ \lambda_5 \\ \lambda_5 \end{bmatrix} = 0_{6\times 1} \tag{9.35}$$

$$\begin{aligned} -c\delta_1\lambda_1 - c\delta_2\lambda_2 - s\delta_5\lambda_5 + s\delta_6\lambda_6 = 0 \\ -s\delta_1\lambda_1 + s\delta_2\lambda_2 - c\delta_3\lambda_3 - c\delta_4\lambda_4 = 0 \\ -s\delta_3\lambda_3 + s\delta_4\lambda_4 - c\delta_5\lambda_5 - c\delta_6\lambda_6 = 0 \\ \lambda_1 = \lambda_2, \lambda_3 = \lambda_4, \lambda_5 = \lambda_6 \end{aligned} \tag{9.36}$$

The solution is trivial $(0=0)$ for angles below $\pi/2$, meaning that the system has full rank for gimbal angles below saturation, as seen in Eq. (9.37).

$$\begin{aligned} -c\delta_1\lambda_1 - c\delta_1\lambda_1 - s\delta_5\lambda_5 + s\delta_5\lambda_5 = -2c\delta_1\lambda_1 = 0 \\ -s\delta_1\lambda_1 + s\delta_1\lambda_1 - c\delta_3\lambda_3 - c\delta_2\lambda_3 = -2c\delta_2\lambda_3 = 0 \\ -s\delta_3\lambda_3 + s\delta_3\lambda_3 - c\delta_5\lambda_5 - c\delta_5\lambda_5 = -2c\delta_5\lambda_5 = 0 \\ \lambda_1 = \lambda_2, \lambda_3 = \lambda_4, \lambda_5 = \lambda_6 \end{aligned} \tag{9.37}$$

9.4.3.2 Triplet Steering Law

If three CMGs are placed in a plane (a "triplet"), the additional CMG adds a degree of freedom, creating null space So, for example, an infinite set of gimbal angles can correspond to the zero momentum state, as shown in Fig. 9.4.

For the case of three CMGs, one constraint equation is sufficient to constrain the system fully in a two-dimensional workspace. This constraint can be designed to avoid the internal singularity as in Fig. 9.5 while also providing the required torque by exploiting the null space of the Jacobian matrix. Again, the constraint equation that meets this need must accomplish three objectives: (1) avoid singular configurations; (2) capture the maximum possible range of momentum values and (3) introduce no unwanted torque.

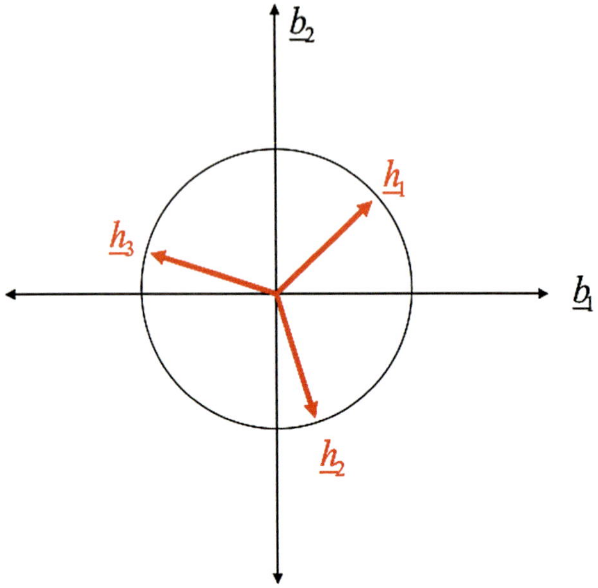

Fig. 9.4 Planar representation of CMG momentum. *The angular momentum vectors of three planar CMGs span the two-dimensional momentum manifold in a non-deterministic way. This configuration shows one of the many zero-momentum configurations for the array*

Fig. 9.5 Internal singularity for a triplet. *The angular-momentum vector one of the CMGs cancels that of a second CMG. The third CMG's momentum is also parallel to one of those two, creating a situation where torque cannot be imparted in the direction parallel to the CMGs*

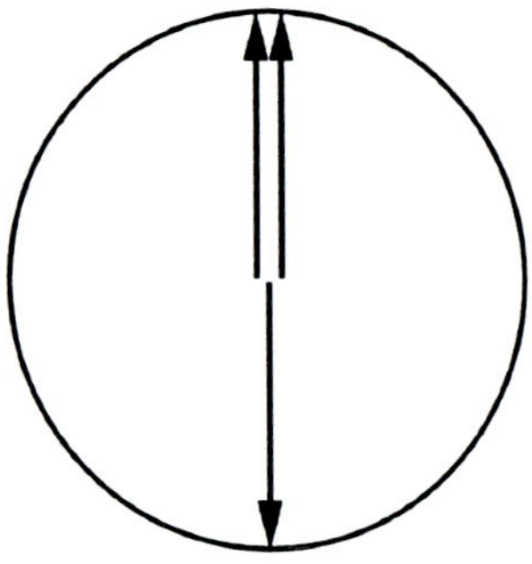

The triplet configuration has a 2×3 Jacobian matrix, as shown in Eq. (9.38), where A_1 and A_2 are the top and bottom rows of the planar Jacobian matrix, respectively.

$$A_{Planar} = \begin{bmatrix} -s\phi_1 & s\phi_2 & s\phi_3 \\ c\phi_1 & c\phi_2 & c\phi_3 \end{bmatrix} = \begin{bmatrix} A_1 \\ A_2 \end{bmatrix} \tag{9.38}$$

Singularities occur when the matrix being inverted in the Moore-Penrose pseudoinverse (Eq. (9.30)), in this case the augmented $\overline{A}$ matrix, is not full rank. In order to ensure that the constraint equation does not introduce more singularities, it is sufficient to write a constraint equation that is always linearly independent of the two existing rows of the Jacobian matrix: for example, the cross-product of two linearly independent vectors. Thus, a singularity-free constraint equation must include a scaled component in the direction of the cross-product of the two rows of the Jacobian matrix:

$$D = d = [\beta_1 A_1 + \beta_2 A_2 + \beta_c (A_1 \times A_2)]\dot{\Delta} = C\dot{\Delta} \tag{9.39}$$

where d is a scalar solution to the constraint equation, β_1 and β_2 are scaling values for the component of the vector in the direction of the Jacobian matrix rows, and β_c is a non-zero scaling value for the constraint equation. This example uses a value of zero for β_1 and β_2 and a value of one for β_c, such that the constraint equation is as shown in Eq. (9.40).

$$D = [A_1 \times A_2]\dot{\Delta} = [\mathrm{s}(\delta_3 - \delta_2)\ \mathrm{s}(\delta_1 - \delta_3)\ \mathrm{s}(\delta_2 - \delta_1)]\dot{\Delta} = C\dot{\Delta} \tag{9.40}$$

D determines the null motion needed to condition the array and steer it away from the internal singularity at $1h$. Because C is orthogonal to the two rows of A, the singularities of the augmented matrix $\overline{A}$ (Eq. (9.29)) are simply the singularities of A. In this example, null-space commands keep the CMGs' momentum vectors in a trapezoid configuration. As shown in Fig. 9.6, with the momentum vectors head-to-tail, the trapezoid configuration is one in which one of the CMG momentum vectors is kept parallel to the total momentum vector (the vector sum of all three CMG vectors), while the other two momentum vectors form the sides of a trapezoid. All three vectors are parallel only at the edge of the momentum envelope. At the internal singularity located at $1h$, the two vectors forming the sides are orthogonal to the

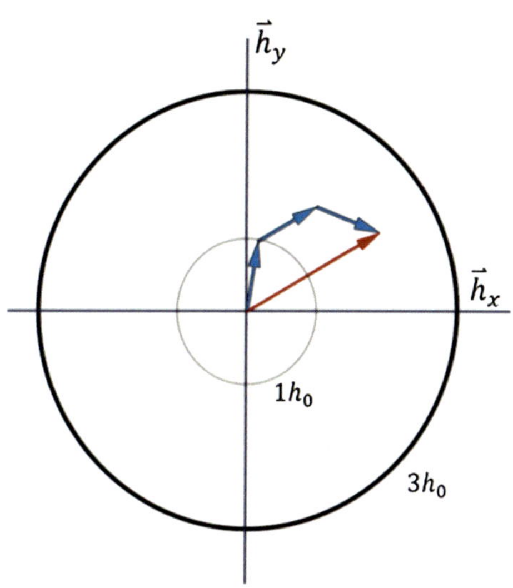

Fig. 9.6 Triplet CMG momentum visualized as a trapezoid. *A trapezoid configuration keeps the array non-singular by maintaining one of the three CMG momentum vectors parallel to the total momentum vector at all times. The three CMGs are shown in blue and the total momentum vector in red*

vector parallel to the total momentum vector, an ideal configuration for traversing the singularity.

Given any total momentum vector within the array's envelope, there are six ways to arrange the CMGs such that they are in a trapezoidal configuration and their momentum vectors sum to the same total. The first step is to determine which of the six trapezoidal configurations is closest to the current (non-trapezoidal) configuration. That is, determine which trapezoid configuration would require the least gimbal motion to reach starting from the current configuration. The next step is to obtain a value for D as shown in Eq. (9.41).

$$D = [A_1 \times A_2]\dot{\Delta} = [\mathrm{s}(\delta_3 - \delta_2)\ \mathrm{s}(\delta_1 - \delta_3)\ \mathrm{s}(\delta_2 - \delta_1)]K \begin{bmatrix} \delta_{1t} - \delta_1 \\ \delta_{2t} - \delta_2 \\ \delta_{3t} - \delta_3 \end{bmatrix} \tag{9.41}$$

δ_{nt} is the gimbal angle the nth CMG would have if the array were in the trapezoid configuration closest to the current configuration, and K is a 3×3 diagonal matrix of gains. The values of the entries of K determine how much effort is used to drive the CMGs to their closest trapezoid configuration and are chosen such that as much null-space effort as possible is used to keep the array close to a trapezoid without exceeding CMG gimbal limits. The selection of entries in K can be done with a deterministic number of numerical iterations or by optimizing via a cost function. The values that produce the highest gimbal rates without exceeding the limits offer the fastest conditioning. They are then used to compute D as shown in Eq. (9.41). Once the D term is obtained, the gimbal rates can be found through a direct inversion of A as shown in Eq. (9.31).

When implemented, this steering law is not used when the array is near its zero-momentum state because the ambiguity of keeping a CMG momentum vector parallel to the zero vector can cause large gimbal rates. The constraint is therefore modified such that within a small radius of the zero momentum state, a Moore-Penrose pseudoinverse rule is used instead. Since the radius in which the pseudoinverse rule is used is much smaller than $1h$, no problems with singularities are encountered, and the issue of very large gimbal rates near zero momentum is avoided. For example, the pseudoinverse rule may be used only within $0.1h$, which may allow for enough null motion between $0.1h$ and $1h$ to properly condition the array. In the general case, the radius is chosen to be as small as possible but still large enough so that it is possible to traverse the zero-momentum state without exceeding the maximum gimbal rate. The choice of this radius depends on the application-specific requirement on simultaneously available torque and momentum.

9.4.3.3 Expansion to a 6 CMG Roof

This triplet analysis can be expanded to six SGCMGs by exploiting the fact that two sets of triplets can exert torque in three directions. Such a configuration results

Fig. 9.7 Two triplets of CMG at right angles

in internal singularity-free control over two planes at an angle to one another, and thus spans $\mathbb{R}^3$. In the example considered here, the gimbal axes are orthogonal to one another, as shown in Fig. 9.7. A six CMG triplet steering law would require two constraint equations:

$$\begin{bmatrix} \dot{h}_x \\ \dot{h}_y \\ \dot{h}_z \\ D_1 \\ D_2 \end{bmatrix} = \begin{bmatrix} 1\,0\,0\,1\,0\,0 \\ 0\,1\,0\,0\,0\,0 \\ 0\,0\,0\,0\,1\,0 \\ 0\,0\,1\,0\,0\,0 \\ 0\,0\,0\,0\,1\,0 \end{bmatrix} \begin{bmatrix} A_1 & 0 \\ C_1 & 0 \\ 0 & A_2 \\ 0 & C_2 \end{bmatrix} \begin{bmatrix} \dot{\delta}_1 \\ \dot{\delta}_2 \\ \dot{\delta}_3 \\ \dot{\delta}_4 \\ \dot{\delta}_5 \\ \dot{\delta}_6 \end{bmatrix} \tag{9.42}$$

A_1 represents the Jacobian matrix of the first CMG triplet and A_2 the Jacobian matrix of the second triplet. C_1 and C_2 are the constraint equations, as described in Eq. (9.40), and D_1 and D_2 are solved for as described in Eq. (9.41). The transformation matrix in Eq. (9.37) is determined by the orientation of the two triplets. In this case, the gimbal axes are aligned with the y and z directions, such that the momentum envelope is greater in the x direction.

9.4.3.4 Gimbal-Angle Constraint Methods (Collinear Steering)

A CMG array is said to be "collinear" when its geometry includes sets of CMGs whose gimbal axes are parallel. The previous section used a collinear array as an example of linear constraints. This section discusses collinear arrays in general. The common case in which there are two such sets is referred to as a "roof" array (see Chap. 6). For example, consider a 4 CMG roof array. Each "plane" of the roof contains two collinear CMGs whose gimbal axes are normal to the plane. Each CMG sweeps a circle of momentum in the plane. Adding the two circles associated with the two CMGs produces a momentum state anywhere within a disc of radius $2h$. To produce a net zero momentum, i.e., the "origin" of the disc, the CMGs must be directed oppositely. When the gimbals are in that orientation, both produce torque about the same axis. Therefore that plane, and possibly the whole array, is singular.

The solution is to recognize that there are four CMGs in this array, and therefore the array has one degree of freedom of null space. As a result, an infinite number of different combinations of gimbal angles can produce the same array momentum vector. The steering law designer has the freedom to choose which set of gimbal angles to use, or put another way, to apply a gimbal-angle constraint. Imposing a constraint on the null space degree of freedom like this is referred to as "fixed map" steering. As a result of such a constraint, each momentum state maps to a unique set of gimbal angles.

In applying this approach to the 4 CMG roof array, we refer to the axis where the two planes of the roof intersect as a "shared" axis. Both planes can produce momentum in this direction. Therefore one possible constraint is to "bias" the two discs against each other along the shared axis, as illustrated in Fig. 9.8. When the roof is producing zero net momentum, the discs have equal and opposite nonzero momentum along the shared axis; thus, both are well separated from the singularity at the disc origin, and the system is well conditioned.

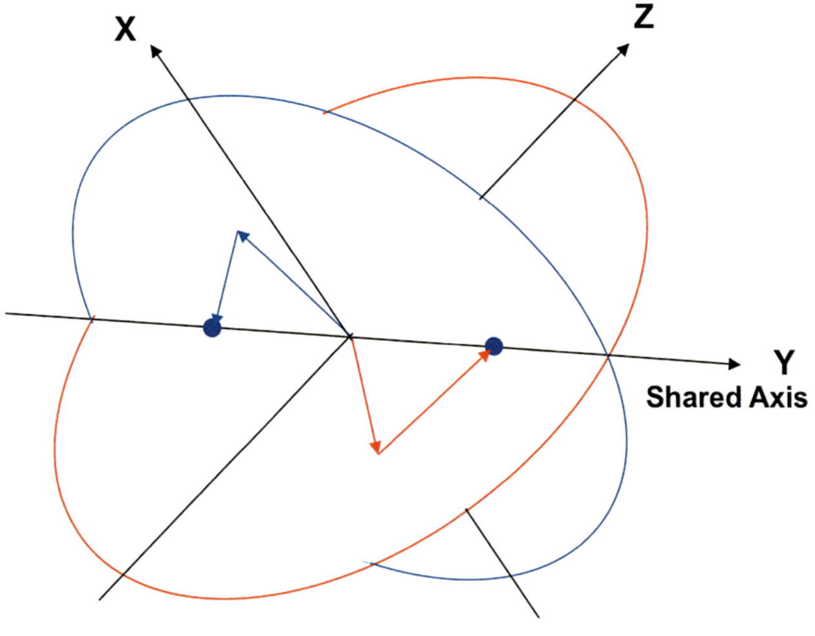

Fig. 9.8 Biasing along the shared axis

9.4.3.5 Polynomial Constraints

The methods of Sects. 9.4.3.1–9.4.3.4 impose linear constraints on the gimbal angles. This section considers polynomial expressions that can provide larger momentum-envelope performance with specific torque guarantees at the boundary of these envelopes. It follows primarily the work of Elliott in 2019 [96, 97], which centers on instantaneous steering that requires no path planning. Such a technique confers two practical benefits: it eliminates the impact of uncertainty in predicting inevitably imperfect attitude paths and sidesteps the operational cost of creating, validating, and uploading such paths to the space vehicle.

Elliott considers arrays with three features:

1. All CMGs in the array have identical angular momentum and mechanical capabilities.
2. The CMG rotor speeds are constant and equal to one another.
3. Base-rate effects, compliance, misalignments, and imbalance dynamics are absent.

This method begins, as in the case of linear constraints, with the output torque vector:

$$\tau_o = \sum_{i=1}^{n} h_i \dot{\delta}_i \hat{\mathbf{o}}_i \tag{9.43}$$

As a matrix equation, where $\hat{o}_i$ is the 3x1 matrix representation of $\hat{\mathbf{o}}_i$ in a spacecraft-fixed coordinate system, the torque for an array of uniform CMGs (i.e., $h_i = h_0$) is

$$\tau_o = h_0 \begin{bmatrix} \hat{o}_1 \ \hat{o}_1 \cdots \hat{o}_n \end{bmatrix} \begin{bmatrix} \dot{\delta}_1 \\ \dot{\delta}_2 \\ \vdots \\ \dot{\delta}_n \end{bmatrix} = h_0 A \dot{\Delta} \tag{9.44}$$

The general optimization problem to maximize torque capability (TC) for any gimbal motion U that moves the gimbals toward optimality is shown in Eq. (9.45), where $\|\tau_o\|_2$ is the magnitude of the output torque, a, b, and c specify the direction of the torque, $h(\dot{\delta}_1, \dot{\delta}_2, \cdots, \dot{\delta}_n)$ is a function that maps the gimbal angles to the array's angular momentum, $\dot{\delta}_{max}$ is the maximum gimbal-rate magnitude a CMG can achieve, and h_d is the desired array angular momentum.

$$\max_{\delta_1,\delta_2,\ldots,\delta_n} \|\tau_o\|_2$$

$$\text{subject to}$$

$$\left\|\left(A^+\frac{T_o}{h_0}+\text{Null}(A)U\right)\right\|_\infty \le \dot{\delta}_{max} \;\; \forall\{a,b,c\}| \left\|\begin{bmatrix} a \\ b \\ c \end{bmatrix}\right\|_2 = 1$$

$$h(\delta_1,\delta_2,\cdots,\delta_n) = h_d$$

$$T_o = \begin{bmatrix} a \\ b \\ c \end{bmatrix} \|\tau_o\|_2 \tag{9.45}$$

Maximizing $\|\tau_o\|_2$ maximizes the TC of the array. With U assumed to be zero, the optimization shown in Eq. (9.45), can be simplified to Eq. (9.46).

$$\max_{\delta_1,\delta_2,\ldots,\delta_n} \|\tau_o\|_2$$

$$\text{subject to}$$

$$\left\|A^+\frac{T_o}{h_0}\right\|_\infty \le \dot{\delta}_{max} \;\; \forall\{a,b,c\}| \left\|\begin{bmatrix} a \\ b \\ c \end{bmatrix}\right\|_2 = 1$$

$$h(\delta_1,\delta_2,\cdots,\delta_n) = h_d$$

$$T_o = \begin{bmatrix} a \\ b \\ c \end{bmatrix} \|\tau_o\|_2 \tag{9.46}$$

For general array architectures this metric can be simplified to Eq. (9.47), with a detailed derivation as provided by Elliott [96]. The result is

$$\max_{\delta_1,\delta_2,\ldots,\delta_n} \|W\|_\infty$$

$$\text{subject to}$$

$$W = \frac{1}{\left\|AA^T\right\|_*}\begin{bmatrix} \pm\sqrt{A^*(1,1)^2+A^*(1,2)^2+A^*(1,3)^2} \\ \pm\sqrt{A^*(2,1)^2+A^*(2,2)^2+A^*(2,3)^2} \\ \vdots \\ \pm\sqrt{A^*(n,1)^2+A^*(n,2)^2+A^*(n,3)^2} \end{bmatrix} \qquad A^* = A^T\text{adj}(AAT)$$

$$h(\delta_1,\delta_2,\cdots,\delta_n) = h_d \tag{9.47}$$

Here, the matrix W represents the maximum possible gimbal rate magnitude for each CMG that could result from a unity torque produced in any direction.

A set of gimbal-angle constraint functions that globally optimizes Eq. (9.47) for all angular momentum values globally maximizes TC. The optimization shown in Eq. (9.47) is difficult to perform analytically for general array architectures. However, for planar array architectures, the expression for TC can be simplified and analytically maximized, as shown in Eq. (9.48).

$$\begin{aligned}
&\max_{\delta_1,\delta_2,\ldots,\delta_n} \|W\|_\infty \\
&\text{subject to} \\
&W = \frac{1}{\left\|AA^T\right\|_*}\begin{bmatrix} \pm\sqrt{A^*(1,1)^2 + A^*(1,2)^2} \\ \pm\sqrt{A^*(2,1)^2 + A^*(2,2)^2} \\ \vdots \\ \pm\sqrt{A^*(n,1)^2 + A^*(n,2)^2} \end{bmatrix} \\
&A^* = A^T \mathrm{adj}(AAT) \\
&h(\delta_1, \delta_2, \cdots, \delta_n) = h_d
\end{aligned} \tag{9.48}$$

The Jacobian matrix of a planar array is shown in Eq. (9.49). This form can be used for any planar array.

$$A = \begin{bmatrix} -\sin(\delta_1) & -\sin(\delta_2) & \cdots & -\sin(\delta_n) \\ \cos(\delta_1) & \cos(\delta_2) & \cdots & \cos(\delta_n) \end{bmatrix} \tag{9.49}$$

When combined with W, this Jacobian matrix yields the simplified expressions of Eqs. (9.50) and (9.51).

$$\begin{aligned}
&\sqrt{A^*(i,1)^2 + A^*(i,2)^2} \\
&= \sqrt{CC^T(CC^T - 2nc_i{}^2) + n^2 c_i{}^2 + SC^T(SC^T - 2ns_i c_i)}
\end{aligned} \tag{9.50}$$

$$\begin{aligned}
&\left\|AA^T\right\|_* = CC^T(n - CC^T) - SC^T SC^T \\
&C = \begin{bmatrix} c_1 \; c_2 \; \cdots \; c_n \end{bmatrix} \\
&S = \begin{bmatrix} s_1 \; s_2 \; \cdots \; s_n \end{bmatrix} \\
&c_i = \cos(\delta_i) \\
&s_i = \sin(\delta_i)
\end{aligned} \tag{9.51}$$

For planar arrays, Eq. (9.49). can be maximized analytically to determine the gimbal-angle constraint functions that globally maximize TC. The gimbal-angle constraint functions provided here globally optimize Eq. (9.48). for any planar array of CMGs with four or more CMGs. The provided gimbal-angle constraint functions are not unique; the functions represent one set of constraint functions of an infinite number. For the constraint functions, the array's angular momentum is assumed to lie on one axis and to be positive. So, in what follows, h_d is taken to be scalar, not a matrix. This simplification does not compromise the generality of the solutions because the vector direction of h_d can be chosen arbitrarily.

Two sets of gimbal-angle constraint functions are provided: gimbal-angle constraint functions for arrays with an even number of CMGs of four or more, and gimbal-angle constraint functions for array with an odd number of CMGs of three or more. Both constraint functions have two solutions, corresponding to two different angular-momentum regions. The region corresponding to angular-momentum below the critical angular momentum, h_c, is referred to as the static region. In the static region, the rows of W are equal in magnitude and are constant, and $\left\|AA^T\right\|_*$ is equal to $\frac{n}{2}$. The region where angular momentum exceed h_c is referred to as the declining region. The declining region is a region in which the value of $\left\|AA^T\right\|_*$ progresses from $\frac{n}{2}$ to zero, as the array approaches saturation, resulting in the rows of W tending towards infinity in the limit. For each set of gimbal-angle constraint functions, there is continuity of the gimbal angles between the static and declining angular-momentum regions.

The globally optimal gimbal-angle constraint functions for any array with an even number of CMGs are shown in Eqs. (9.52)–(9.56). In the solutions, $\lceil x \rceil$ denotes the integer that results from rounding x up, and $\lfloor x \rfloor$ denotes the integer that results from rounding x down.

$$\delta_{1,2,\ldots,k_1} = \left\{ \begin{matrix} \cos^{-1}\left(\frac{h_d}{n} - \frac{\sqrt{k_1 n - 2k_1(-2{h_d}^2 + n^2)}}{2nk_1}\right), \; 0 \le h_d \le h_c \\ \cos^{-1}\left(\frac{h_d}{n}\right), \; h_c < h_d \le n \end{matrix} \right\} \tag{9.52}$$

$$\delta_{k_1+1,k_1+2,\ldots,\frac{n}{2}} = \left\{ \begin{matrix} \cos^{-1}\left(\frac{h_d}{n} - \frac{\sqrt{k_1 n - 2k_1(-2{h_d}^2 + n^2)}}{n(n-2k_1)}\right), \; 0 \le h_d \le h_c \\ \cos^{-1}\left(\frac{h_d}{n}\right), \; h_c < h_d \le n \end{matrix} \right\} \tag{9.53}$$

$$\delta_{\frac{n}{2}+1,\frac{n}{2}+2,\ldots,k_2} = -\delta_{k_1+1,k_1+2,\ldots,\frac{n}{2}} \tag{9.54}$$

$$\delta_{k_2+1,k_2+2,\ldots,n} = -\delta_{1,2,\ldots,k_1} \tag{9.55}$$

$$\begin{aligned} h_c &= \frac{n}{\sqrt{2}} \\ k_1 &= \lceil \frac{n}{4} \rceil \\ k_2 &= \frac{n}{2} + \lfloor \frac{n}{4} \rfloor \end{aligned} \tag{9.56}$$

The globally optimal gimbal-angle constraint functions for any array with an odd number of CMGs are shown in Eqs. (9.57)–(9.62).

$$\delta_{1,2,\ldots,k_1} = \left\{ \begin{matrix} \cos^{-1}\left(\frac{h_d-1}{n-1} - \frac{\sqrt{k_1 n-2k_1-1(-2h_d{}^2+4h_d+n(n-3))}}{2k_1(n-1)}\right), \; 0 < h_d \le h_c \\ \cos^{-1}\left(\frac{h_d-1}{n-1}\right), \; h_c < h_d \le n \end{matrix} \right\} \tag{9.57}$$

$$\delta_{k_1+1,k_1+2,\ldots,k_2} = \left\{ \begin{matrix} \cos^{-1}\left(\frac{h_d-1-2\cos(\delta_{1,2,\ldots,k_1})}{n-1-2k_1}\right), \; 0 < h_d \le h_c \\ \cos^{-1}\left(\frac{h_d-1}{n-1}\right), \; h_c < h_d \le n \end{matrix} \right\} \tag{9.58}$$

$$\delta_{k_2+1} = 0 \tag{9.59}$$

$$\delta_{k_3,k_3+1,\ldots,k_4} = -\delta_{1,2,\ldots,k_1} = -\delta_{k_1+1,k_1+2,\ldots,k_2} \tag{9.60}$$

$$\delta_{k_4+1,k_4+2,\ldots,n} = -\delta_{1,2,\ldots,k_1} \tag{9.61}$$

$$\begin{aligned} h_c &= 1 + \sqrt{\frac{(n-1)(n-2)}{2}} \\ k_1 &= \lceil \frac{n-1}{4} \rceil \\ k_2 &= \frac{n-1}{2} \\ k_3 &= k_2 + 2 \\ k_4 &= k_3 + \lfloor \frac{n-1}{4} \rfloor \end{aligned} \tag{9.62}$$

The resulting array configurations for the static and declining regions are shown in Fig. 9.9. In the static region, gimbal angles are constrained to four different values, two of which are the negative of the other two. In the declining angular-momentum region, the gimbal angles are constrained to two different angles that are the negative of one another.

The gimbal-angle constraint functions that globally optimize TC for five CMGs are shown in Eqs. (9.63)–(9.68).

$$\delta_1 = \left\{ \begin{matrix} \cos^{-1}\left(\frac{\tilde{h}_d-1}{4} - \frac{5\sqrt{-\tilde{h}_d{}^2+2\tilde{h}_d+5}}{4}\right), \; 0 \le \tilde{h}_d < \tilde{h}_c \\ \cos^{-1}\left(\frac{\tilde{h}_d-1}{4}\right), \; \tilde{h}_c \le \tilde{h}_d \le 5 \end{matrix} \right\} \tag{9.63}$$

$$\delta_2 = \left\{ \begin{matrix} \cos^{-1}\left(\frac{\tilde{h}_d-1}{4} + \frac{5\sqrt{-\tilde{h}_d{}^2+2\tilde{h}_d+5}}{4}\right), \; 0 \le \tilde{h}_d < \tilde{h}_c \\ \cos^{-1}\left(\frac{\tilde{h}_d-1}{4}\right), \; \tilde{h}_c \le \tilde{h}_d \le 5 \end{matrix} \right\} \tag{9.64}$$

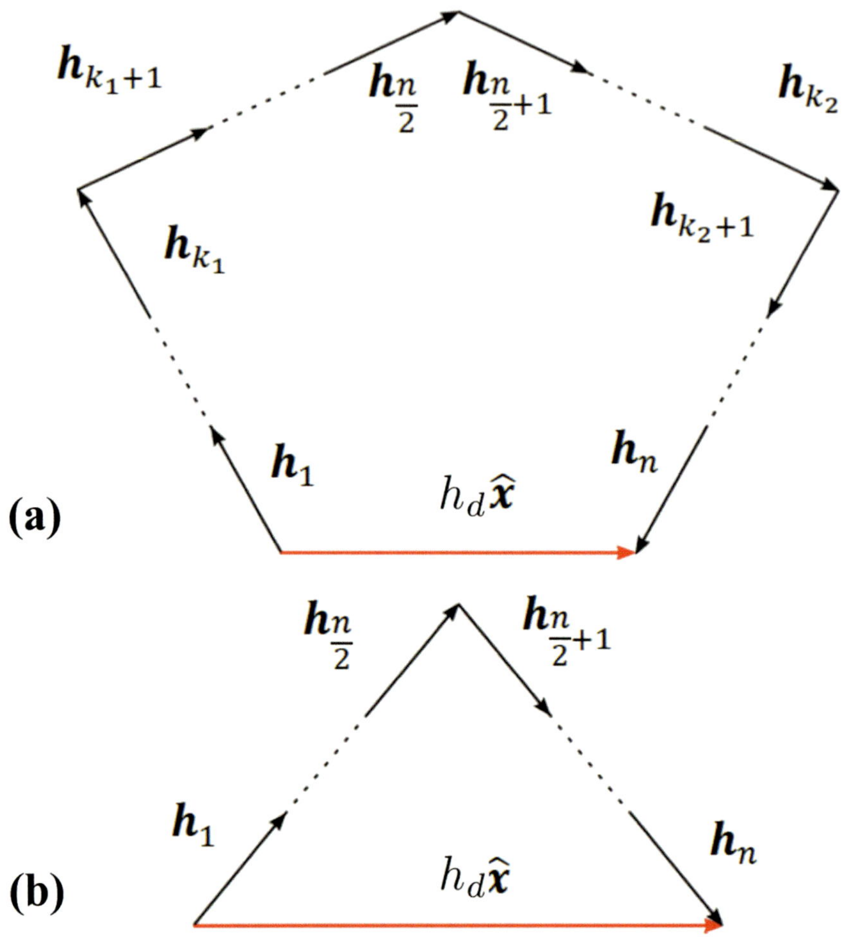

Fig. 9.9 (**a**) represents the array constraints for the static angular-momentum region. (**b**) represents the array constraints for the declining angular-momentum region

$$\delta_3 = 0 \tag{9.65}$$

$$\delta_4 = -\delta_2 \tag{9.66}$$

$$\delta_5 = -\delta_1 \tag{9.67}$$

$$\tilde{h}_c = 1 + \sqrt{6} \tag{9.68}$$

The resulting array constraints for the static and declining angular-momentum regions are shown in Fig. 9.10. In the declining angular-momentum region, the five

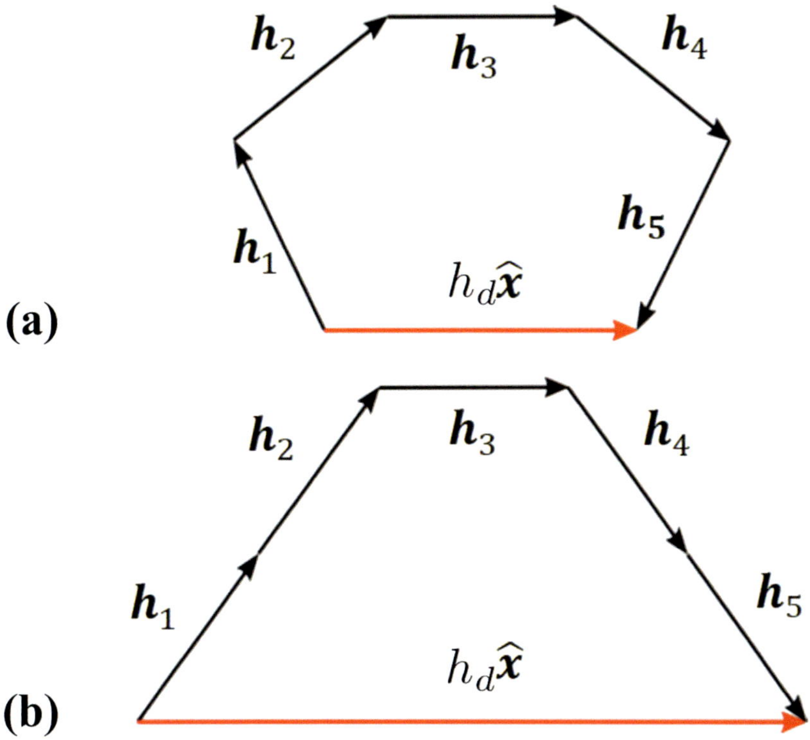

Fig. 9.10 (**a**) represents the array constraints for the static angular-momentum region. (**b**) represents the array constraints for the declining angular-momentum region

gimbal angles are constrained to three different angles, resulting in the trapezoidal shape shown.

The gimbal-angle constraint functions that globally optimize TC for seven CMGs are shown in Eqs. (9.69)–(9.76).

$$\delta_1 = \begin{Bmatrix} \cos^{-1}\left(\frac{\tilde{h}_d-1}{6} - \frac{5\sqrt{-\frac{1}{2}\tilde{h}_d{}^2+7+\tilde{h}_d}}{6\sqrt{7}}\right), \; 0 \le \tilde{h}_d < \tilde{h}_c \\ \cos^{-1}\left(\frac{\tilde{h}_d-1}{6}\right), \; \tilde{h}_c \le \tilde{h}_d \le 7 \end{Bmatrix} \tag{9.69}$$

$$\delta_2 = \begin{Bmatrix} \cos^{-1}\left(\frac{\tilde{h}_d-1}{6} + \frac{\sqrt{-\frac{1}{2}\tilde{h}_d{}^2+7+\tilde{h}_d}}{6\sqrt{7}}\right), \; 0 \le \tilde{h}_d < \tilde{h}_c \\ \cos^{-1}\left(\frac{\tilde{h}_d-1}{6}\right), \; \tilde{h}_c \le \tilde{h}_d \le 7 \end{Bmatrix} \tag{9.70}$$

$$\delta_3 = \left\{ \begin{matrix} \cos^{-1}\left(\frac{\tilde{h}_d - 1}{6} + \frac{2\sqrt{-\frac{1}{2}\tilde{h}_d{}^2 + 7 + \tilde{h}_d}}{3\sqrt{7}}\right),\ 0 \le \tilde{h}_d < \tilde{h}_c \\ \cos^{-1}\left(\frac{\tilde{h}_d - 1}{6}\right),\ \tilde{h}_c \le \tilde{h}_d \le 7 \end{matrix} \right\} \tag{9.71}$$

$$\delta_4 = 0 \tag{9.72}$$

$$\delta_5 = -\delta_3 \tag{9.73}$$

$$\delta_6 = -\delta_2 \tag{9.74}$$

$$\delta_7 = -\delta_1 \tag{9.75}$$

$$\tilde{h}_c = 1 + \sqrt{15} \tag{9.76}$$

The resulting constrained array configurations for the static and declining angular-momentum regions are shown in Fig. 9.11. Similar to the constraints for arrays with five CMGs, in the declining region, the seven gimbal angles are constrained to three angles, resulting in the trapezoidal shape shown in figure.

Elliott [96] proves the global optimality of the solutions for an even number of CMGs greater than three for any angular-momentum state. For arrays with

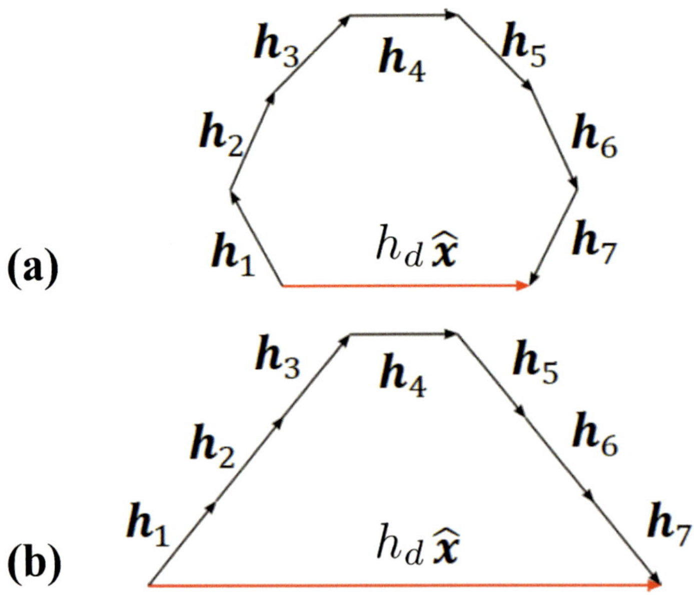

Fig. 9.11 (**a**) represents the array constraints for the static angular-momentum region. (**b**) represents the array constraints for the declining angular-momentum region

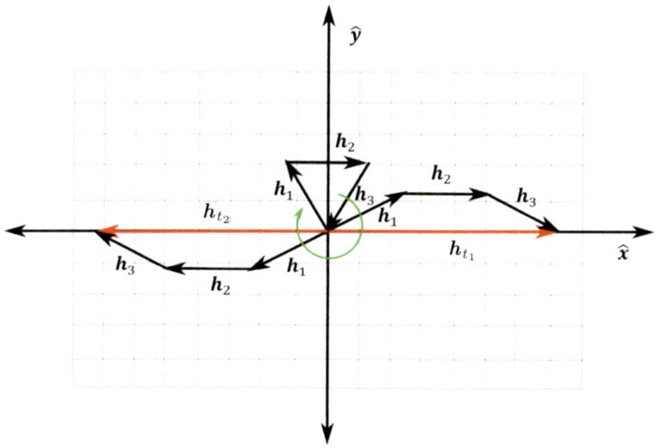

Fig. 9.12 Three CMG algorithmic singularity. Arrows represent the angular momentum vectors for the CMGs in the array

an odd number of CMGs, a proof of global optimality is provided for the static angular-momentum region. For the declining angular-momentum region, numerical evidence of optimality is provided for planar arrays with five, seven, nine, and eleven CMGs, but optimality is not proven.

The constraint functions described here have algorithmic singularities. One such algorithmic singularity occurs at the zero-momentum state. Infinite gimbal rate is required to continue to track the constraint through zero, as shown in Fig. 9.12 for three planar CMGs. In Fig. 9.12, at time t_1, the array produces the angular momentum $h(t_1)$. The array then progresses along $\hat{\mathbf{x}}$ to $h(t_2)$ at time t_2. At zero array momentum, the array must rotate along the green arrow instantaneously to track the constraint. The same phenomenon occurs for all of the constraints provided in this chapter.

To eliminate these algorithmic singularities, linear feedback can be used implement the constraints. Implementing the constraints using linear feedback complicates the development of performance guarantees, because it necessarily allows the array to progress away from the constrained orientation.

To illustrate the performance of the provided gimbal-angle constraint functions, these constraint functions are compared to the so-called circular-backbone constraint [98, 99]. The circular-backbone constraint restricts the links of a planar serial manipulator to approximate a circular arc, avoiding all internal singularities. This type of constraint is not closed-form, complicating a general comparison. Instead, performances for arrays with four, five, six, seven, eight, and nine CMGs are compared here. Both constraints have algorithmic singularities. For this comparison, no algorithmic singularities are considered. Instead, the arrays are assumed to achieve a gimbal-angle set that continuously satisfies the constraints.

Unlike Elliott's constraint functions, the circular-backbone constraint does not necessarily optimize TC. However, the circular-backbone constraint still provides a

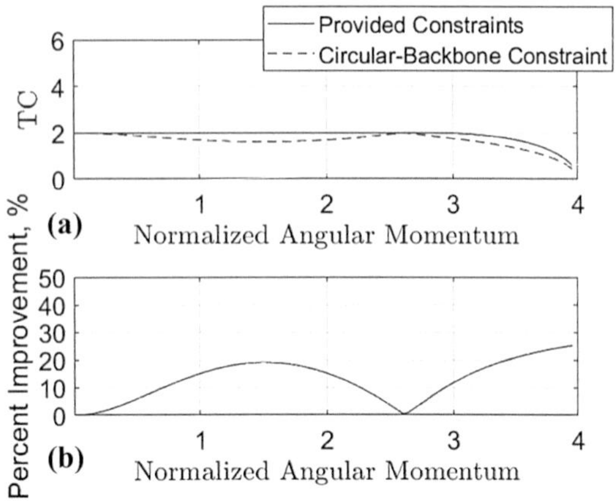

Fig. 9.13 (**a**) represents the values for TC calculated using the provided constraint functions and the circular-backbone constraint for arrays with four CMGs. (**b**) represents the percent difference between the value of TC for both constraints

good reference for comparison because the constraint functions represent the only other set of general constraints for planar arrays. The metric used for comparison is TC for the case of the Moore-Penrose pseudoinverse.

Figure 9.13 shows the value of TC versus the normalized angular momentum for the provided gimbal- angle constraint functions as well as the circular-backbone constraint for an example of four planar CMGs. The momentum range shown is from 0 to $0.99n$. The momentum of $0.99n$ represents 99% of the saturation limit. Larger momentum values are not evaluated because at greater momentum states, the arrays are too near a singularity to yield credible results.

These gimbal-angle constraint functions achieve a value for TC that equals or exceeds that of the circular-backbone constraint. Additionally, because the provided constraint functions are closed-form, they do not require a numerical solver when implemented, eliminating the risk of numerical instability and reducing the computation required in flight software. This analysis also quantifies the percent improvement, which is computed by dividing the difference in TC by the constraint's TC. The value of TC computed using the circular-backbone constraint is less than that computed using the provided constraint functions, by as much as approximately 25% in the static angular-momentum region and approximately 35% in the declining angular-momentum region. Caution should be exercised when interpreting the percent difference near the saturation of the array because as the array approaches saturation, the numerical error that occurs due to the numerical solver error when implementing the circular-backbone constraint has a large effect on the value of TC. The numerical error in the circular-backbone constraint is less relevant in this comparison when the array is further from saturation.

These constraint functions maximize TC for any planar array of CMGs with four or more CMGs. Proofs or numerical evidence of global optimality confirm this fact for each constraint function. They also further generalize and improve the performance of constraint-based steering laws. The solutions can be applied to planar and non-planar arrays consisting of planar segments of CMGs, such as roof

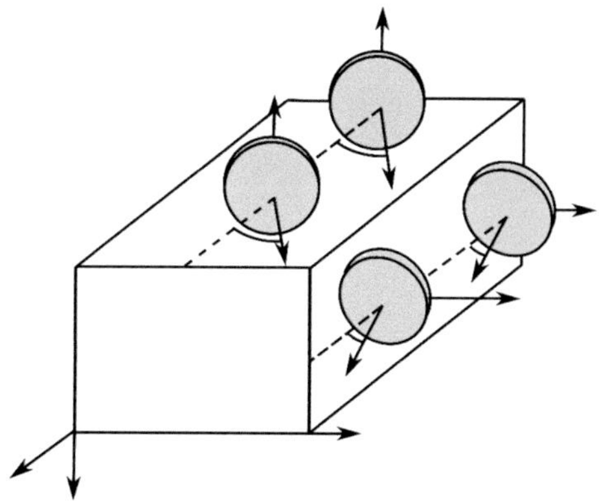

Fig. 9.14 Four-CMG box-90 array. Shows the configuration of the CMGs within a four-CMG box-90 array

arrays or box arrays. The resulting steering law progresses the array to a solution that globally optimizes each plane's TC. The gimbal-angle constraint functions also enable fault-tolerant steering laws, because the constraints apply to any number of CMGs greater than four. Thus, if a CMG fails within the array, the steering law can still perform optimally for the remaining CMGs, so long as four or more CMGs remain in the plane.

The next simplest array architectures are the roof arrays, which consist of two orthogonal planes of CMGs. Such an array also represents one of the most common array architectures used in practice. Specifically, the four-CMG box-90 array, shown in Fig. 9.14, is practical because it represents a high-performance array with a minimally redundant suite of actuators and the option of a dormant spare. This section provides a constraint-based steering law and performance guarantees for the four-CMG box- 90 array that avoids all internal singularities. This steering law also guarantees a specified magnitude of array-level TC for any direction (within the limitations of the hardware).

The four-CMG box-90 array has both avoidable and elliptical internal singularities. One plane of CMGs, consisting of CMGs 1 and 2 as shown in Fig. 9.14, can produce angular momentum in the $\hat{\mathbf{x}}-\hat{\mathbf{y}}$ plane, but not in the $\hat{\mathbf{z}}$ direction. The second plane, consisting of CMGs 3 and 4, can produce angular momentum in the $\hat{\mathbf{x}} - \hat{\mathbf{z}}$ plane, but not in the $\hat{\mathbf{y}}$ direction. $\hat{\mathbf{x}}$ is the only common axis of both planes. Therefore, if the subset of CMGs in the $\hat{\mathbf{x}} - \hat{\mathbf{y}}$ plane are singular for the $\hat{\mathbf{y}}$ direction and/or the CMGs in the $\hat{\mathbf{x}}-\hat{\mathbf{z}}$ plane are singular for the $\hat{\mathbf{z}}$ direction, the overall array is singular.

Elliptical array-level singularities result from elliptical singularities for each plane of CMGs. For each plane, angular momentum cannot progress from a positive angular momentum along $\hat{\mathbf{x}}$ to a negative angular momentum along $\hat{\mathbf{x}}$, or vice versa, without infinite null motion or becoming singular in either the $\hat{\mathbf{y}}$ direction for the CMGs in the $\hat{\mathbf{x}} - \hat{\mathbf{y}}$ plane or the $\hat{\mathbf{z}}$ direction for the CMGs in the $\hat{\mathbf{x}} - \hat{\mathbf{z}}$ plane. For the CMGs in the $\hat{\mathbf{x}} - \hat{\mathbf{y}}$ plane, the elliptical singularity is shown in Fig. 9.15, where $\hat{\mathbf{s}}_i(t_j)$ ith CMG at the $\hat{\mathbf{s}}$ time point and the vector $\mathbf{h}(t_j)$ represents the total angular momentum of the plane at the jth time point. In Fig. 9.15, the plane of CMGs starts

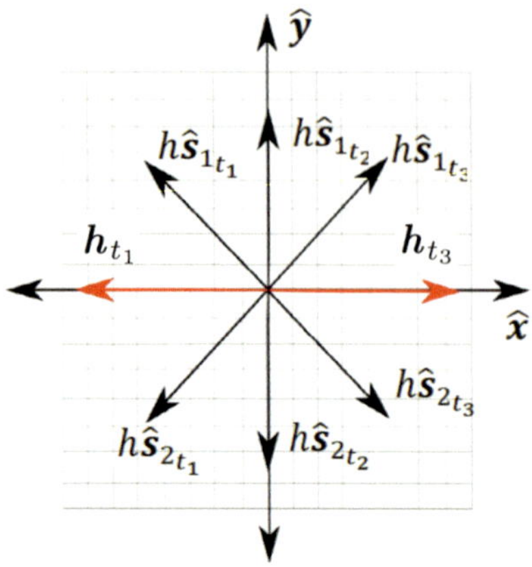

Fig. 9.15 Elliptical singularities for plane 1

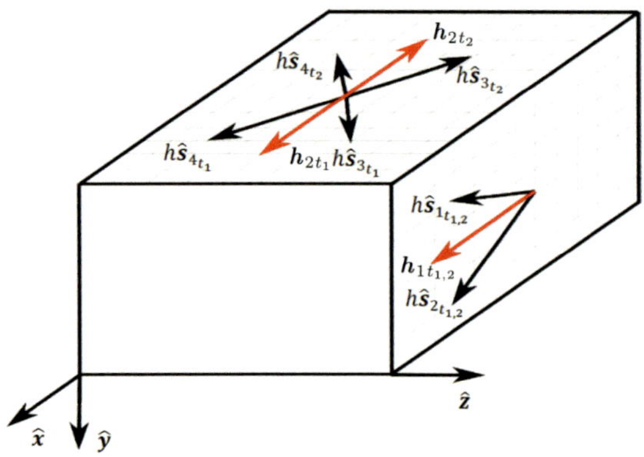

Fig. 9.16 Elliptical array singularities for the Box 90

at t_1 producing a negative angular momentum along $\hat{\mathbf{x}}$, which is represented by $\mathbf{h}(t_1)$. The array then progresses along $\hat{\mathbf{x}}$ to the state at t_3. To progress from the state at t_1 to the state at t_3 the array must pass through the state at t_2, assuming finite null motion, resulting in the array becoming singular along $\hat{\mathbf{y}}$. These elliptical singularities for each plane of CMGs, result in an elliptical singularity for the entire array anytime a plane of the array must pass through the zero angular-momentum state along $\hat{\mathbf{x}}$.

These elliptical singularities can be encountered in many different scenarios, such as the one shown in Fig. 9.16. Here, $\mathbf{h}_i(t_j)$ represents the total angular momentum of the ith at the jth time point. In Fig. 9.17, at a time t_1, the array's angular momentum along $\hat{\mathbf{x}}$ exceeds $2h$, and the momentum along $\hat{\mathbf{y}}$ and $\hat{\mathbf{z}}$ is zero. At this angular-momentum state, each plane of CMGs must produce a positive angular momentum along $\hat{\mathbf{x}}$ as shown in Fig. 9.17. If at time t_1, a continuous torque is commanded in the negative $\hat{\mathbf{x}}$ direction until the array's angular momentum along $\hat{\mathbf{x}}$ is equal to zero, which occurs at time t_2, then one of the planes must have a negative angular momentum along $\hat{\mathbf{x}}$ (scenario shown in Fig. 9.17), or both planes must produce zero angular momentum. In either scenario, one or both of the planes

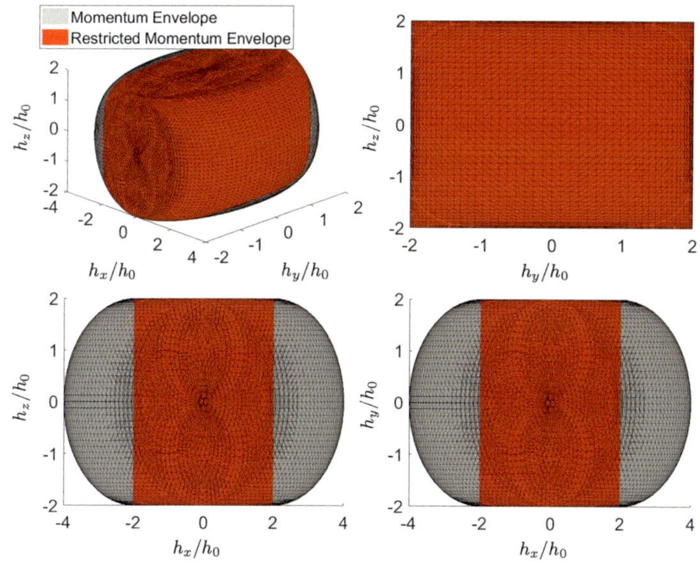

Fig. 9.17 Angular-momentum envelopes for the Box 90

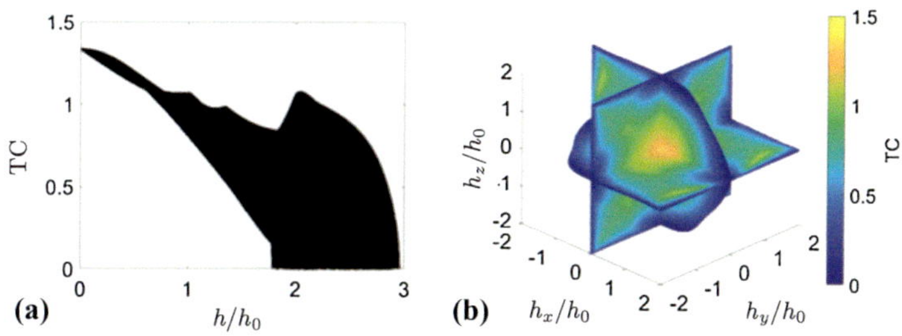

Fig. 9.18 Torque capability of the array within the constrained angular-momentum envelope

become singular in the $\hat{\mathbf{y}}$ and/or $\hat{\mathbf{z}}$, resulting in the array becoming singular in those directions. No finite trajectory for U can avoid this singularity.

By restricting the operational envelope to all angular-momentum states that lie within $\pm 2h$ along $\hat{\mathbf{x}}$, shown in Fig. 9.18, all singularities internal to the restricted envelope are avoidable. The values of TC for the augmented Jacobian method and the four-CMG box-90 array are shown in Eq. (9.77), where $A^{-1}(i, j)$ represents the value in the ith row and the jth column of A^{-1}. A detailed derivation of Eq. (9.78) is provided in Elliott [97].

$$TC = \frac{1}{\|W_{CB}\|_\infty} \tag{9.77}$$

$$W_{CB} = \begin{bmatrix} \sqrt{A^{-1}(1,1)^2 + A^{-1}(1,2)^2 + A^{-1}(1,3)^2} \\ \sqrt{A^{-1}(2,1)^2 + A^{-1}(2,2)^2 + A^{-1}(2,3)^2} \\ \sqrt{A^{-1}(4,1)^2 + A^{-1}(4,2)^2 + A^{-1}(4,3)^2} \end{bmatrix} \tag{9.78}$$

The constraint function for the four-CMG box-90 array is shown in Eqs. (9.79) and (9.80)

$$\tilde{h}_{x_1} - \frac{\tilde{h}_x - \Psi}{2} = 0 \tag{9.79}$$

$$\begin{aligned} \Psi &= b\left(1 - \left(\frac{\tilde{h}_x}{2}\right)^2\right) + \left(1 - \left(\frac{\tilde{h}_y}{2}\right)^2\right) + \left(1 - \left(\frac{\tilde{h}_z}{2}\right)^2\right) + P(\tilde{h}_x) \\ P(\tilde{h}_x) &= -\frac{67}{3840}\tilde{h}_x^8 + \frac{1}{6}\tilde{h}_x^6 + \frac{1}{24}\tilde{h}_x^4 + \frac{47}{60}\tilde{h}_x^2 \\ \tilde{h}_x &= \cos(\delta_1) + \cos(\delta_2) + \cos(\delta_3) + \cos(\delta_4) \\ \tilde{h}_{x_1} &= \cos(\delta_1) + \cos(\delta_2) \\ \tilde{h}_y &= \sin(\delta_1) + \sin(\delta_2) \\ \tilde{h}_z &= \sin(\delta_3) + \sin(\delta_4) \\ b &= \frac{4\sqrt{6}}{3} \end{aligned} \tag{9.80}$$

The formulation of the constraint function is based on the principle that $\hat{\mathbf{x}}$ is the shared axis for both planes of CMGs. Thus, the angular momentum along $\hat{\mathbf{x}}$ for each plane of CMGs can be biased against each other, enabling the gimbal-angle set to be changed without changing the angular momentum of the array. The amount each plane is biased against one another is dictated by Ψ.

The formulations of Ψ are restricted to only those that enable the array to reach the extents of the restricted angular-momentum envelope along $\hat{\mathbf{x}}$, $\hat{\mathbf{y}}$ and $\hat{\mathbf{z}}$, which motivates constraint functions whose constrained angular-momentum envelope is maximized. For the array to be capable of reaching the extents of the angular-momentum envelope along $\hat{\mathbf{x}}, \hat{\mathbf{y}}$ and $\hat{\mathbf{z}}$, Ψ must achieve the values shown in Eqs. (9.81)–(9.83).

$$\tilde{h}_x = 0,\ \tilde{h}_y = \pm 2,\ \tilde{h}_z = 0,\ \Psi = 0 \tag{9.81}$$

$$\tilde{h}_x = 0,\ \tilde{h}_y = 0,\ \tilde{h}_z = \pm 2,\ \Psi = 0 \tag{9.82}$$

$$\tilde{h}_x = 0,\ \tilde{h}_y = 0,\ \tilde{h}_z = 0,\ \Psi = \pm 2 \tag{9.83}$$

The formulation of Ψ in Eq. (9.80) includes a cubic constraint function and a univariate polynomial in $\tilde{h}_x$. The cubic constraint function satisfies Eqs. (9.81) and (9.82) and results in the greatest performance. Equations (9.84), (9.87) show four other constraint function formulations that perform less well, but all of which satisfy Eqs. (9.81) and (9.82). Here, s and c are tuning constants.

$$\left(1-\left(\frac{\tilde{h}_x}{c}\right)^2-\left(\frac{\tilde{h}_y}{c}\right)^2-\left(\frac{\tilde{h}_z}{c}\right)^2\right)^s \tag{9.84}$$

$$\left(1-\left(\frac{\tilde{h}_y}{2}\right)^2-\left(\frac{\tilde{h}_z}{2}\right)^2\right)^s \tag{9.85}$$

$$\left(\left(1-\left(\frac{\tilde{h}_y}{2}\right)^2\right)\left(1-\left(\frac{\tilde{h}_z}{2}\right)^2\right)\right)^s \tag{9.86}$$

$$\left(\left(1-\left(\frac{\tilde{h}_x}{c}\right)^2\right)\left(1-\left(\frac{\tilde{h}_y}{2}\right)^2\right)\left(1-\left(\frac{\tilde{h}_z}{2}\right)^2\right)\right)^s \tag{9.87}$$

The spherical constraint function, shown in Eq. (9.84), offers marginally greater minimum TC at most angular-momentum magnitudes than does the cubic constraint function. However, the spherical constraint function's constrained angular-momentum envelope is considerably smaller than that of the cubic constraint function.

The polynomial added to the spherical constraint is formulated such that Ψ satisfies Eq. (9.83). Infinite polynomial formulations also satisfy Eq. (9.83). The polynomial's coefficients and order are tuned to provide the greatest minimum TC at all angular-momentum states.

The following example shows the performance of the constraint implemented with the augmented Jacobian method. TC is evaluated at evenly distributed angular-momentum grid points within the Nms restricted angular-momentum envelope. At each grid point, the gimbal-angle sets that satisfy each angular-momentum state are computed analytically. There are multiple gimbal-angle set solutions for each angular-momentum state, because the gimbal-angle solution for a CMG within a plane can be applied to the other CMG within that plane. However, the multiple solutions at each angular-momentum state result in the same value for TC.

Figure 9.18a shows TC vs. the normalized angular-momentum magnitude of the array, where h represents the angular-momentum magnitude of the array. For a single value of the angular-momentum magnitude, the values of TC represent all possible TC values that would occur at that angular-momentum magnitude. The lower edge of the black band represents the minimum TC that is guaranteed at

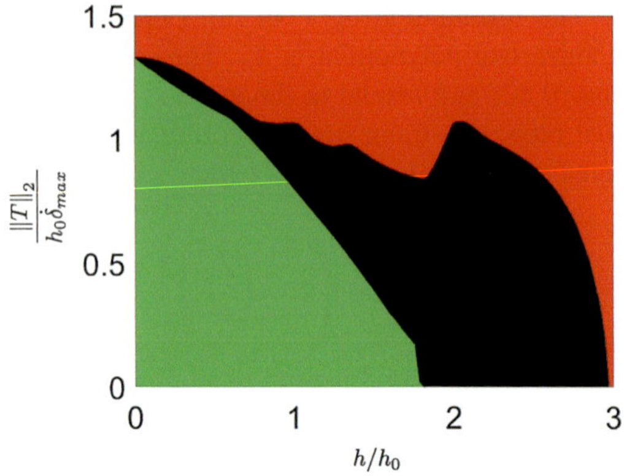

Fig. 9.19 Producible torque

a given array angular-momentum magnitude. Figure 9.18b shows TC at slices of angular momentum where one of $\tilde{h}_x$, $\tilde{h}_y$ or $\tilde{h}_z$ are equal to zero. Figure 9.19 shows the torque zones for the constraint-based steering law, where $\dot{\delta}_{max}$ represents the maximum gimbal rate of the CMGs in the array. For a specified torque magnitude, $\|T\|_2$, along the $\hat{\mathbf{y}}$ axis, the green zone represents angular-momentum magnitudes that enable the specified torque magnitude to be producible in all torque directions for any angular-momentum direction. Thus, for a specified torque magnitude, the right edge of the green zone represents the radius of the largest angular-momentum sphere, within which the array is capable of producing the specified torque magnitude in all torque directions at all angular-momentum states. The black zone represents the angular-momentum magnitudes, where a specified torque magnitude can be produced in all torque directions, but not in all angular-momentum directions. For certain angular-momentum directions, the array is capable of producing the specified torque magnitude in all torque directions at greater angular-momentum magnitudes. The greatest angular-momentum magnitude at which the array can still produce the specified torque magnitude in all torque directions, is represented by the right most edge of the black zone. The red zone represents angular-momentum magnitudes at which the specified torque is not guaranteed in all torque directions. However, the torque magnitude may be producible in some torque directions.

Figure 9.20 shows the overall constrained angular-momentum envelope, where $TC \sim 0$ at the boundary. From Fig. 9.20, the constrained angular-momentum envelope is nearly the same size as the restricted angular-momentum envelope. However, algorithmic singularities prevent the array from accessing the whole restricted angular-momentum envelope. The minimum angular-momentum magnitude that lies on the constrained angular-momentum envelope boundary is 1.79, which represents the radius of the largest sphere, centered on the origin, that can be inscribed within the constrained angular-momentum envelope.

Figure 9.21 shows the angular-momentum boundary for when TC is equal to 0.5 compared to the overall constrained angular-momentum envelope. I.e., all angular-momentum states that are within the volume can produce a torque of a magnitude

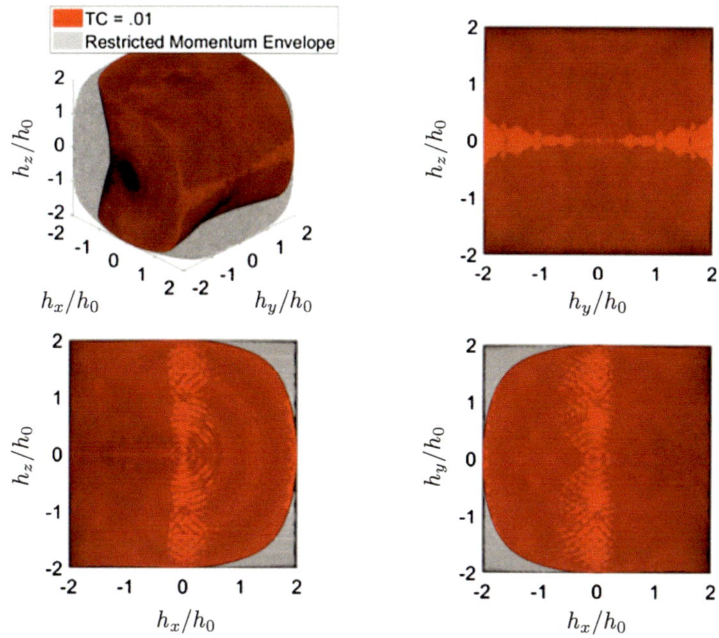

Fig. 9.20 Constrained angular-momentum envelope (TC nearly zero)

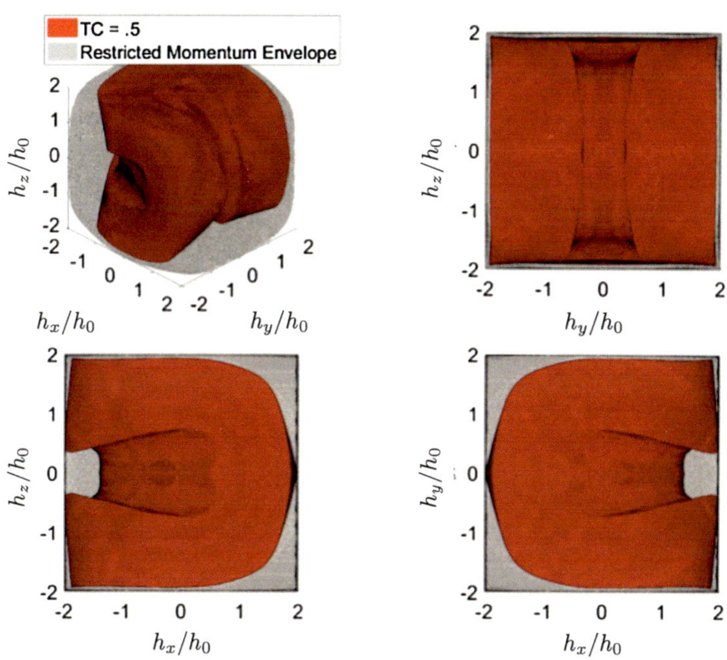

Fig. 9.21 Angular-momentum boundary for $TC = 0.5$

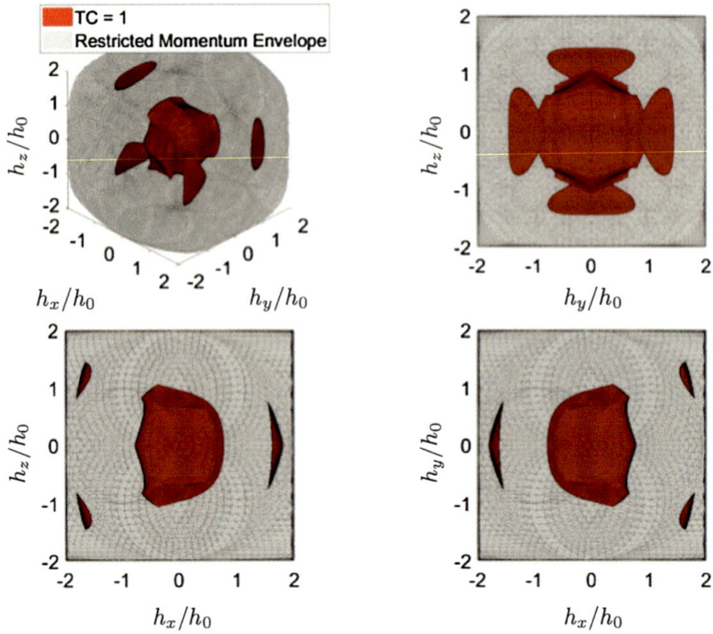

Fig. 9.22 Angular-momentum boundary for $TC = 1$

of at least $0.5h\dot{\delta}_{max}$. The minimum angular-momentum magnitude that lies on the angular-momentum boundary for a TC of 0.5 is $1.39h$.

Figure 9.22 shows the angular-momentum bounds for a TC of 1 compared to the restricted angular-momentum envelope. All angular-momentum states within the volume have $TC > 1$. Thus, all angular-momentum states within the volume can produce a torque of a magnitude of at least $h\dot{\delta}_{max}$. There are small clusters of angular-momentum states that have a TC of at least 1. These small clusters are represented by the small lobe of TC centered at an angular momentum magnitude of approximately 2, shown in Fig. 9.18a. The minimum magnitude that lies on the angular-momentum boundary for a TC of 1 is 0.73. These clusters are unlikely to serve a useful purpose in a practical setting, where convenient, simple volumes like spheres, boxes, and ellipsoids may be best for clarity in requirements development and systems-engineering communications.

The comparisons among these steering laws highlight the main benefit of these constrained steering laws is the guarantee of performance. They not only guarantee singularity avoidance, ensuring that the array can actuate the body in all possible directions, but they also specify the maximum torque magnitude that the array can always produce in all directions. These guarantees result in a robust control architecture. They enable maneuvers to be designed such that the required torque remains less than the guaranteed torque. Additionally, the performance guarantees enable the CMGs in the array to be more easily sized for a particular application by scaling the gimbal rates such that torque commands during operation are equal to or less than the torque magnitude that is guaranteed.

9.4.4 Limited Angular Momentum Methods

In many cases, including the 4 CMG roof array of the previous section, applying a gimbal-angle constraint does not eliminate the singularities; it merely changes the point in momentum space at which they are encountered. This constraint is particularly useful for the 4 CMG roof array, since the singularity is at the origin. But if we start at our biased, net-zero momentum state (Fig. 9.8) and continue to apply torque about the shared axis, eventually one of the discs must encounter its origin singularity.

Strategies for managing this encounter are discussed elsewhere in this chapter. However, depending upon the requirements of the application, it may be acceptable simply to impose a limit on the net array momentum prior to reaching this point. This limit results in a simple and conservative control scheme with singularities avoided because they cannot be encountered. Some applications, such as missions that execute small slews in a short amount of time, do not have time for the gimbals to rotate very far during a slew. As such, they rarely access the extremes of the momentum envelope anyway and suffer little penalty from this practice.

Some care must be taken in how a momentum limit is implemented. Consider a large-angle eigenaxis slew, where the momentum limit is reached in the eigenaxis direction, resulting in maximum vehicle rate for the majority of the slew. One approach might be to arrest all gimbal motion until the commanded torque is in the direction that reduces momentum at the end of the maneuver. Doing so would not only impose the momentum limit but would also essentially stop all command response during the slew, leaving the ACS open. This open-loop behavior is a problem, as there is no cross-axis torque available to make small corrections and keep the slew trimmed about the correct axis.

9.5 Singularity Avoidance and Escape Algorithms

Some steering algorithms simultaneously avoid singularities through null motion or constraints and escape from singularities through torque error. Such algorithms benefit arrays of CMGs that contain both elliptic and hyperbolic internal singularities. Fully exploiting exact solutions at hyperbolic singularities consists of using null motion when it is available and torque error only at elliptic singularities, i.e., singularity escape when no null motion is possible. Such steering laws can be implemented only when the form of the singularity (see Chap. 5) is known.

9.5.1 Hybrid Steering Logic (HSL)

Parameters that identify the type of singularity must be incorporated in the steering algorithm. Chapter 5 explains that the parameters from the eigenvalues of the singularity definition matrix S and projection matrix P determine the form of a singularity: whether it is hyperbolic or elliptic. The steering algorithm must take this information into account for CMG arrays that contain internal elliptic singularities. The Hybrid Steering Logic (HSL) does just that in the form of a pseudoinverse. The pseudoinverse expression for the HSL arises when the parameters ($Q = X = U\text{diag}(0, 0, \alpha)$, $R = I_{n\times n}$, and $\dot{\Delta}_d = \beta[I_{n\times n} - A^{+}A]d$) are substituted into Eq. (9.6), taking into account the separation of the null- and forced-spaces in Eq. (9.24):

$$\dot{\Delta} = A^{\mathrm{T}}(AA^{\mathrm{T}} + X)^{-1}\dot{h} + \beta[I_{n\times n} - A^{+}A]d \tag{9.88}$$

Equation (9.88) may appear to be simply a hybrid of the SDA algorithm in Eq. (9.12) and the LG method in Eq. (9.25), but there are some subtle differences. First, the null-space projection matrix $[I_{n\times n} - A^{+}A]$ is singular when A is singular. For Eq. (9.25), the equivalent matrix $V\text{diag}(0, 0, 0, 1)V^{\mathrm{T}}$ is now non-singular. Second, the singularity parameters—α for A^{SDA} and β for the LG—are different. They are defined by the form of singularity, i.e., hyperbolic or elliptic. For a 4 CMG pyramid array, the expressions α and β for this method are

$$\alpha = \alpha_0 e^{-a\overline{\alpha}} e^{\mu_1 m} \tag{9.89}$$

$$\beta = \beta_0 e^{-a\overline{\beta}} e^{\mu_2 m} \tag{9.90}$$

where α_0, a, μ_1, β_0, b, and μ_2 are positive scalar constants and m is the singularity measure. Now, the form of singularity is found through the states $\overline{\alpha}$ and $\overline{\beta}$ which depend directly on the singularity definition matrix S as

$$\overline{\alpha} = |S_0 - \det(S)| \tag{9.91}$$

$$\overline{\beta} = \frac{1}{\overline{\alpha}} \tag{9.92}$$

where S_0 is the maximum value that $\det(S)$ can achieve. S_0 is found heuristically or through optimization. The absolute value in Eq. (9.48) compensates for numerical error, producing a negative term. It should be clear from Eqs. (9.88) through (9.92) that at a hyperbolic singularity, $\det(S) \leq 0$. Therefore, Eq. (9.89) would reduce the effect of torque error in Eq. (9.88), thus reducing torque error and providing a more precise LG method. At an elliptic singularity, $\det(S) > 0$. Therefore, at an elliptic singularity, when null motion is ineffective, higher torque error is required to efficiently escape singularity.

The HSL is unique among pseudoinverse-based steering algorithms because it uses the information on the form of singularity to maximize steering precision and escape internal singularities, thus trading off the properties of LG and SDA depending on the type of singularity encounter. The singularity parameter metrics identified in Eqs. (9.89) and (9.90) must to be chosen with care to provide the desired performance (Leve and Fitz-Coy [100]).

9.5.2 *Angular Momentum Artificial Potential Steering (AMAPS)*

Singularities can be thought of as obstacles to trajectories in momentum space. Such obstacles can be avoided through algorithms that utilize barrier functions known as repulsive artificial potential functions (APFs). An APF is a mathematical framework that represents vector-field sources and sinks as repulsive and attractive potentials. The gradient of the sum of these functions provides a trajectory that discourages the trajectory from approaching sources and encourages motion toward commanded trajectories in the form of sinks.

A set of APFs can be defined that combines objectives of an attractive potential for tracking a torque, angular momentum, or attitude and repulsive potentials for avoiding angular momentum saturation and singularities. Angular Momentum Artificial Potential Steering (AMAPS) is a method that uses such a framework for singularity avoidance. It trades torque error for singularity avoidance through soft constraints posed by APF Munoz and Leve [101].

The use of APF may introduce local minima where neither objective is satisfied (e.g., high torque error and/or close proximity to singularities), and yet the steering may converge here. Furthermore, repulsive APFs used to avoid singularities are inherently destabilizing because they introduce torque error to push the array configuration away from singularity. However, on a positive note, these functions are analytic, involve only minimal computation, and with the right tuning may provide local minima.

9.5.3 *Feedback Steering Law (FSL)*

All steering algorithms presented up to this point require matrix inversions. Matrix inversions are expensive and require stable numerical algorithms regardless of singularities. Some steering algorithms avoid inverses altogether, taking an approach that differs from the pseudoinverse solutions. The possible benefit in computation by comes at the expense of nonzero torque error.

An example of an inverse-free solution, different from that in reference [102], is known as the Feedback Steering Law (FSL), in which the solution for gimbal rates needed to provide a commanded torque is found through construction of a linear system. This method is unique in that it does not correspond to a solution found from Eq. (9.6). Yet it is familiar, since it is based on linear-system analysis. Without going into detail, the linear system constructed for FSL has the dynamics

$$\dot{x} = A_K x + B_K u \tag{9.93}$$

$$\dot{\Delta} = C_K x \tag{9.94}$$

where the coefficient matrices in Eqs. (9.93) and (9.94) A_K, B_K, C_K are found by minimizing the cost function

$$J = \frac{1}{2}\left[\tau_e \; \dot{\Delta}\right]^{\mathrm{T}} \underset{\dot{\Delta}\in\mathbb{R}^4}{\begin{bmatrix} K_1 & 0 \\ 0 & K_2 \end{bmatrix}} \left[\tau_e \; \dot{\Delta}\right] \tag{9.95}$$

which is exactly the SR inverse when $K_1 = 1$ and $K_2 = \lambda 1$. However, for FSL, $K_1 = 1$ and the compensator $K_2 = K(s)$ and the coefficient matrices in Eqs. (9.92) and (9.93) are found through an H_∞ minimization of

$$\min_{K(s)\in\mathbb{R}^{3\times 3}} \left\| \begin{bmatrix} w_1(s)[1 + AK(s)]^{-1} \\ w_2(s)K(s)[1 + AK(s)]^{-1} \end{bmatrix} \right\|_\infty \tag{9.96}$$

where the sensitivity weights

$$w_1(s) = \begin{bmatrix} A_K & B_K \\ C_K & 0 \end{bmatrix} \tag{9.97}$$

comprise a matrix function that bounds the bandwidth of the steering algorithm, and

$$w_2(s) = \frac{1}{\omega} I_{4\times 4} \tag{9.98}$$

provides upper bounds on the gimbal rates. The final result is in the controller-observer form

$$\begin{aligned} \dot{h} &= A_K h + B_K \tau_e \\ \dot{\Delta} &= C_K h \end{aligned} \tag{9.99}$$

where $C_K = A^{\mathrm{T}} b\omega \overline{P}$ (Pechev [103]). The constant b is the required bandwidth of the steering algorithm, and $\overline{P}$ is the solution to the steady-state, i.e., infinite frequency Riccati Equation. The elegant steering algorithm in Eq. (9.99), unlike pseudoinverse solutions, is based entirely on linear systems theory and optimization and provides a way to limit the bandwidth and gimbal rate of the SGCMGs used for control while avoiding singularities. Also, it accomplishes these goals without matrix inversions.

However, this result derives from a steady-state Jacobian matrix and therefore may not always produce an exact solution. Also, like the other methods based on the cost in Eq. (9.3), its solution involves mixed weighting of objectives and therefore may not produce an exact torque while attempting to limit gimbal rates.

9.5.4 Optimal CMG Attitude Control

Allocation of torque among CMGs tends to optimize over some aspect of their control effect, such as the output-torque error from the CMG array, the magnitude of the gimbal rates resulting from their torque tracking, or some combination thereof, in terms of the objective function of Eq. (9.3).

However, most pseudoinverse methods that are derived from the cost function in Eq. (9.3) are locally optimal at best, and their performance is dominated by the choice of the weighting matrices R and Q (Leve [91]). One may want to optimize the CMG gimbal rates to minimize both the torque error and the electromechanical power, which is not exactly equivalent to the 2-norm of the gimbal rates, particularly when decelerating a gimbal consists of regenerative braking that stores energy in the spacecraft's batteries. At the same time, such a steering approach may have to obey constraints on the gimbal kinematics. Alternatively, optimal steering may simultaneously include attitude trajectory and CMG gimbal rates, obviating the need to minimize array-level torque. For such cases, the operations concept may call for solving the minimum-time attitude trajectory along with minimizing the gimbal rate magnitudes. This section addresses optimal methods of control allocation that simultaneously optimize the attitude trajectory, error in attitude, rate, and torque, and electromechanical power/energy of the CMG, and/or some combination of these objectives.

Normally, the commanded torque is sent from the attitude control system and then mapped by a steering algorithm to the gimbals rates. From there, the inner-loop gimbal controller offsets friction while enforcing hard constraints for the maximum gimbal rate and acceleration. This methodology is neither optimal with respect to power and accuracy nor efficient for utilizing the entire performance envelope. With the exception of the FSL algorithm of Eq. (9.99), analytic solutions for steering algorithms are unable to handle such constraints. Constrained optimization techniques can permit a steering algorithm to be better than locally optimal, although not necessarily globally optimal.

9.5.4.1 Convex Optimization: Minimum Error-Dissipative Power Reduction Control

Steering a redundant array of RWAs and/or CMGs may prioritize accuracy in torque tracking more highly than other metrics such as the minimum input power to the actuators as secondary objectives. Such priority-based optimization processes are analogous to how we think as humans but are not how optimization techniques are typically structured. Consider, for example, a redundant RWA or CMG array control-allocation problem, in which there are infinitely many ways to impart a given torque, i.e., due to the null-space degree(s) of freedom in Eq. (9.2). These solutions constitute a searchable set, as shown in Fig. 9.23.

As an algorithm optimizes over torque-tracking accuracy, it reduces the set to only those solutions that have the optimal accuracy. Searching within this set can yield a subset of solutions that provide the minimal power for control allocation. The second priority may result in a multiple-solution set or a singleton. Now to provide a real-time, polynomial convergent solution of this priority-based optimization, the constraints and the cost may be structured as a convex function (not necessarily linear). This method of priority-based optimization is known as "Lexicographic Optimization" and relies on interior-point methods of convex optimization to find real-time, polynomial convergent solutions to the optimal control allocation of redundant arrays of RWA and CMG (Dueri et al. [104]). An example of the priority

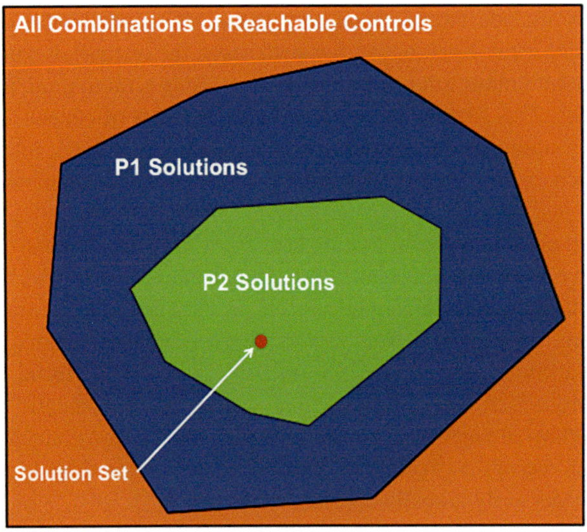

Fig. 9.23 Searchable sets used for lexicographic optimization

optimization for Minimum Error-Dissipative Power Reduction Control or ME-DPRC from Dueri et al. [104] for n RWA is

$$\begin{aligned} \min_{\Omega_{r,k+1},\dot{\Omega}_{r,k}} \quad & J_1 = \left\| \dot{h} - J_r A_s \dot{\Omega}_{r,k} \right\|_1 \\ & \text{subject to} \\ \Omega_{r,k+1} &= \Omega_{r,k} + \dot{\Omega}_{r,k} \Delta t, \\ -\Omega_{max} &\leq \Omega_{k+1} \leq \Omega_{max}, \\ -\dot{\Omega}_{max} &\leq \dot{\Omega}_{r,k} \leq \dot{\Omega}_{max}, \end{aligned} \tag{9.100}$$

where Δt is the discretization time step, $\dot{\Omega}_{r,k}$ is the rotor accelerations at the kth timestep, Ω_{max} is a vector of symmetric upper bounds on RWA flywheel rates, $\dot{\Omega}_{max}$ is a column matrix of symmetric upper bounds on RWA flywheel accelerations, and $\leq$ represents an element-wise inequality. Using the 1-norm of the torque error makes the above problem an LP [105]. Let J_1^* be the minimum torque error. Then the second optimization (minimizing power) is posed as:

$$\begin{aligned} \min_{\Omega_{r,k+1},\dot{\Omega}_{r,k}} \quad & J_2 = \left\| \left(J_r \dot{\Omega}_{r,k} - \tau_{Wf,k} \right) \odot \Omega_{r,k} \right\|_1 \\ & \text{subject to} \\ \Omega_{r,k+1} &= \Omega_{r,k} + \dot{\Omega}_{r,k} \Delta t, \\ & \left\| \dot{h} - J_r A_s \dot{\Omega}_{r,k} \right\|_1 \leq J_1^*, \\ -\Omega_{max} &\leq \Omega_{r,k+1} \leq \Omega_{max}, \\ -\dot{\Omega}_{max} &\leq \dot{\Omega}_{r,k} \leq \dot{\Omega}_{max}. \end{aligned} \tag{9.101}$$

where $\tau_{Wf,k}$ is the wheel friction torque at the kth time step from Appendix A, and the first inequality imposes minimum-torque error. Clearly, the first inequality constrains the second-priority optimization to search within the solutions left over from the first-priority optimization. That first inequality constraint is $\left\| \dot{h} - J_r A_s \dot{\Omega}_{r,k} \right\|_1 \leq J_1^*$. This method does exactly what is shown in Fig. 9.9. To keep the second cost convex, the shaft power is multiplied with current rotor rates. This assumption is reasonable for RWAs because rotor accelerations are typically not large enough that the rotor rates will be in a different regime at the next time step.

Unlike many methods of optimal control, lexicographic programming can be used as a real-time method of control allocation. However, it does rely on the ability to design a practical convex cost function and set of constraints. It also does not guarantee anything better than a locally optimal, i.e., a greedy-algorithm solution for the electromechanical power unless a receding-horizon approach is used such as model predictive control (Fisher et al. [106]).

9.5.4.2 Gauss Pseudo-Spectral Collocation Methods

Nonlinear optimization techniques may be appropriate when a spacecraft requires a global, or at least better-than-locally optimal, solution for attitude control, particularly when the optimization is inherently nonlinear and nonconvex. The most common of these techniques include direct and indirect methods. One approach is to solve an optimal control problem as a two-point boundary value problem, which is in turn numerically solved by shooting methods. Another is to construct a discrete nonlinear program to be solved as a set of algebraic equations at discrete points in place of differential equations.

Shooting methods have their challenges: they require accurate initial guesses for convergence to the optimal solution; they may use integration methods that are not amenable to control and trajectories with discontinuities; and because they consider roughly the entire trajectory and control of the optimal solution, they are not in general, computationally efficient. On the other hand, collocation methods discretize the controls and the trajectories at a set of points and use interpolation, typically by splines between the points, to provide a complete solution to the optimal-control problem. Such methods are more computationally efficient, while also able to handle discontinuities.

In simplest terms, collocation methods replace differential equations with algebraic equations solved at discrete nodes, called collocation points (Divya et al. [107] and Fahroo and Ross [108]). At the nodes, optimality is usually proved through costate mapping theorems. The proof shows that each discrete point where the optimal control solution is solved meets the optimality conditions.

For CMG attitude control, the most popular objective for optimization in the literature is minimum-time slewing. A frequently discussed collocation method for this objective comes from Gauss Pseudo-Spectral methods (Ross and Karpenko [54]). These methods use Gauss Quadrature to approximate the integrals and the derivatives of the differential equations at the nodes. These methods have been shown in simulation and test-bed hardware to provide gimbal rates and rotor accelerations of an array of CMGs and RWAs for minimum time maneuvers (Ross and Karpenko [54]). Also, such methods have been used in flight, specifically on the International Space Station, for optimal offloading of angular momentum while minimizing thruster use. These "zero-propellant maneuvers" have saved millions of dollars' worth of propellant (Bedrossian et al. [84]).

A drawback to solving the entire nonlinear optimization problem in its original form is that the solution is rarely guaranteed to be globally optimal. It is also not clear how fast such an algorithm converges to the optimal solution. So, its real-time implementation must be evaluated on a case-by-case basis. Furthermore, the interpolation between the discrete points where the solution is solved may not in fact be optimal or even satisfy constraints. In addition, it is not practical to add additional points to the nonlinear program to avoid errors between points from interpolation because the number of discrete points in general exponentially increases the size of the problem and computational cost.

For these many reasons, such methods are typically most effective if used as an open-loop feedforward for a feedback controller to follow. When the parameters of the spacecraft—mass properties, sensor biases and scale factors, flexible dynamics, and actuator alignments and biases—are very well known, this approach can be effective. For precise attitude control, or attitude control for systems where knowledge of the full trajectory is not possible, the smallest amount of model uncertainty can cause unacceptable inaccuracies in spacecraft attitude tracking. In such a situation, care must be taken in choosing a nonlinear program approach such as Gauss Pseudo-spectral methods.

9.6 Variable Speed Control Moment Gyroscopes

VSCMGs and the steering algorithms that go along with them have been suggested as a solution to the singularity-avoidance problem. Some effective VSCMG steering algorithms have been developed by Shaub and Junkins [45] and Yoon and Tsiotras [41]. However, there are significant drawbacks from a hardware perspective, as discussed in Chaps. 4 and 5. These drawbacks explain why VSCMGs are not common in contemporary spacecraft.

9.7 Chapter Summary

The purpose, derivation, and differences in performance of steering algorithms for CMGs are the subject of this chapter. Singularities can be avoided/escaped through the choice of algorithm or through the choice of a CMG array architecture with non-degenerate hyperbolic singularities. In such arrays, null motion will always be effective in avoiding internal singularities. Singularities can also be avoided through constraining the gimbal angles or usable angular momentum, i.e., workspace for steering algorithm solutions. This limitation leads to a method that provides an exact solution within the constraints without addressing the singularities directly on-the-fly and provides some solutions that do not require a redundant set of CMGs.

9.8 Exercises

1. Write out the four properties for the following types of generalized inverses in Appendix C:

   (a) Generalized inverse
   (b) Generalized reflexive inverse
   (c) Moore-Penrose pseudoinverse

Why aren't the weaker generalized inverses used for steering algorithms?

2. Does the Moore-Penrose always exist? If so, is it unique? If it exists and is unique, how can it be constructed?
3. What is more beneficial in terms of maximizing internal output torque to input power (See Chap. 4 and Appendix A)?

    (a) A pseudoinverse as a steering algorithm mapping desired internal output torque to gimbal accelerations and considering internal torque from gimbal rates as a disturbance.
    (b) A pseudoinverse as a steering algorithm mapping desired internal output torque to gimbal rates and considering internal torque from gimbal accelerations as a disturbance.

4. Assuming finite gimbal rates and accelerations, does a steering algorithm exist that avoids all internal and external singularities while providing an exact mapping of an internal torque?

Chapter 10
Inner Loop Control of Momentum Devices

Implicit in a discussion of how to command the devices in momentum control system is the assumption that the devices faithfully track those commands. And yet, momentum devices exhibit several significant nonlinearities and can experience significant disturbances and errors. For this reason, momentum devices almost always include "inner loops" applying feedback control to the quantities of interest to ensure that those commands are tracked.

This chapter examines some of the key considerations in the inner loops of momentum devices. The goal is not to teach the art of inner-loop design but rather to explore the ways in which the inner loops can impact the spacecraft designer. Because these devices contain active feedback systems, they may interact with the spacecraft structure or controls. These interactions are usually of great interest in spacecraft design.

10.1 Spin Speed Control

10.1.1 CMG Spin Speed Loops

Both RWAs and CMGs provide means for adjusting the spin speed of the rotor. In the case of the CMG, the goal is generally to hold the spin speed constant, at a speed that is traditionally about 6000 RPM. Some smaller devices and next-generation CMGs may feature even faster spin speeds, perhaps tens of thousands of RPM. In most cases, CMGs include an integral feedback loop to perform this function. So, the user need only command the desired spin speed. Spin loops' compensation usually includes a free integrator, allowing them to offset the effects of spin bearing drag, both Coulomb and viscous.

During normal operation of a CMG, various influences cause the measured spin speed to vary. For example, the spin bearings exhibit an increase in drag torque

© The Author(s), under exclusive license to Springer Nature Switzerland AG 2025
F. A. Leve et al., *Spacecraft Momentum Control Systems*, Space Technology Library 46, https://doi.org/10.1007/978-3-031-96165-6_10

when exposed to radial loads, such as the forces that result when the CMG is producing output torque. This additional drag reduces the rotor speed. Another more subtle example results because the rotor spin speed is measured not relative to an inertial reference, but rather with respect to the gimbal housing, which moves with the spacecraft. Imagine rotating the gimbal 180°, thus imparting momentum to the spacecraft in the direction opposing spin. The rotor speed is now measured relative to a body moving in the opposite direction, and thus speed appears to increase.

While these changes in speed are small, and in some cases, cyclic, the spin-speed control loop responds to them nevertheless. It is therefore advisable for the spin loop to be very low in bandwidth, for example, 0.1 Hz or less. A high bandwidth spin loop would expend considerable power attempting to follow a specific speed precisely, when this is not really necessary.

CMG users often impose a tight specification on how well the spin loop should hold speed, and this is often because they have assumed in their steering law design that the rotor speeds are constant. The user of a CMG array should recognize that the spin system draws additional power to hold speed during periods when the array is maneuvering the vehicle. If power is a concern, the user should relax the spin-speed control specification, permitting variation in spin speed at these times. By allowing the rotor's speed to vary by even a few percent, the spin system power can be limited to a specified value, with the result that the CMGs recover more slowly after a period of active maneuvering. A good steering-law design accommodates such variations in speed by taking into account the instantaneous rotor speed in computing CMG commands, resulting in no appreciable degradation in performance. It's clearly an advantageous trade: lower power for a little more algebra.

10.1.2 RWA Spin Speed

Most reaction wheels accept a torque command, which is mapped by the motor K_T to a spin-motor current command. The RWA electronics produce this current using a high-bandwidth current loop, on the order of kHz. The current loop uses integral control to overcome significant motor back-emf and variations in resistance with temperature, leaving only variations in the motor K_T as the primary error in motor torque. The rotor speed changes as a result of the accumulation of momentum while these torques are applied, but most RWA applications treat momentum as a byproduct, where the primary action of the device is to produce torque.

The biggest drawback in operating an RWA in this manner is the error introduced by bearing friction. Friction torques act on the rotor and react on the spacecraft in exactly the same way as motor torques. Therefore, any friction torque represents an error between the commanded motor torque and the actual torque delivered to the spacecraft. The Coulomb drag of the bearings can be estimated by the flight software and added to the command, but near zero speed the friction changes rapidly and is difficult to compensate. This difficulty motivates spacecraft RWA-array steering-law architectures that keep the speeds away from zero as much as possible.

Some RWAs include an integral speed-control loop. This is a useful way to mitigate the effects of friction. Thanks to conservation of angular momentum, the change in rotor momentum is equal in magnitude but opposite in direction to the change in momentum imparted to the vehicle. Since torque $\tau = dH/dt$ and the rotor inertia is well known, it follows that the change in rotor speed provides a precise measurement of the integrated torque applied to the spacecraft, regardless of whether it arose from the motor or friction torque.

Therefore, regardless of whether feedback control is integral to the RWA hardware or realized on the spacecraft side, the preferred approach is to integrate the desired torque, scale (by inertia) to the expected rotor speed, and close a speed loop to produce this calculated speed. A torque feedforward can also be employed to improve performance. Recognizing that the speed command to this loop is often characterized by ramping up and down at the maximum torque, the loop compensation should be chosen such that minimal error results from a ramp command.

We recognize that an RWA speed loop is a hierarchical "inner loop" in the forward path of the ACS. As such, the bandwidth of the speed loop can impose a limit on the available ACS bandwidth. However, in general, this is not an issue with typical ACS bandwidths at tenths of a Hz, and speed loops at a few Hz. These bandwidths are far enough apart that the control loops are uncoupled.

More sophisticated methods of RWA control and friction mitigation exist. In general, the performance of these control approaches is limited by the quality of rotor-motion feedback available. These limitations must be taken into consideration in specifying an RWA if loops are to be closed around it via spacecraft bus electronics.

10.2 Gimbal Rate Loops

In the case of CMGs, the motion of the gimbal(s) is also regulated with feedback control. As the output torque of the CMG is gyroscopic ($\boldsymbol{\omega} \times \mathbf{h}_r$), the precision with which the scalar gimbal rate, ω_g, is controlled directly impacts torque performance. This impact on precision motivates virtually all CMGs to feature integral gimbal rate loops and to respond to gimbal-rate commands. The design of these loops can influence the design and performance of the ACS, and of the spacecraft itself.

10.2.1 Errors Mitigated by Gimbal Rate Loops

One of the key motivations for closing gimbal rate loops is the need to manage gyroscopic precession-induced motor torques. When the vehicle is in motion, the gimbal motor must apply sufficient torque to prevent the gimbal from precessing, and this torque can be large (see Sect. 3.2.5). Indeed, for direct-drive CMGs (no gears between the motor and the gimbal axis), it is common for the precession due to vehicle rate to drive the design of the gimbal motor, requiring many times the torque needed for gimbal acceleration alone. This large "disturbance" must be attenuated by an assertive loop, usually including a free integrator to increase gain at low frequencies and minimize rate error. Gimbal rate loops usually have bandwidth in excess of 10 Hz. In contrast, ACS loops for spacecraft with such CMGs are unlikely to include feedback control with a bandwidth above a few Hz.

In addition to rejecting the precession disturbance, gimbal rate loops also mitigate forward-path errors such as motor ripple, cogging and accuracy. The rate performance of the CMG thus approaches the performance of the gimbal rate sensor. Historically, gimbal rate has been measured by a precisely wound tachometer. This electromagnetic device is essentially a motor/generator that produces a back-emf in response to rotor rate. The device must be commutated, and traditionally the low voltage back-emf must be amplified with high-quality circuitry to avoid apparent rate offset due to offsets in the amplifier. With the advent of digital electronics, some CMGs have employed digital encoder technology to measure both gimbal angle and rate with high precision.

10.2.2 Impact on Vehicle Structure

In the design of positioning servos in general, a common mistake is to assume that the positioner itself is rigidly fixed and that only the body being positioned is moving. We understand that when the torque motor acts on the gimbal, it also reacts into the structure that attaches the motor stator to "ground." In our case, "ground" is the interface that mounts the momentum device to the spacecraft bus structure.

Of concern is the case illustrated in Fig. 10.1, where the spacecraft interface appears as "local" structure, connected through compliance to the rest of the vehicle. An example of such a condition might be a CMG mounted to a flat plate, or deck, with local compliance. When the CMG is mounted to such a plate, its inertia lowers the frequency of bending modes in which this plate participates.

When the motor acts/reacts into the gimbal/local inertia, both bodies move. As a result, any structural resonance appears in the open-loop response of the gimbal

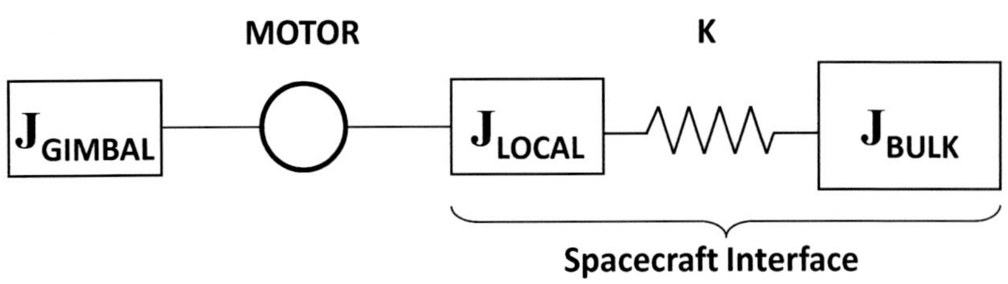

Fig. 10.1 A compliant spacecraft interface

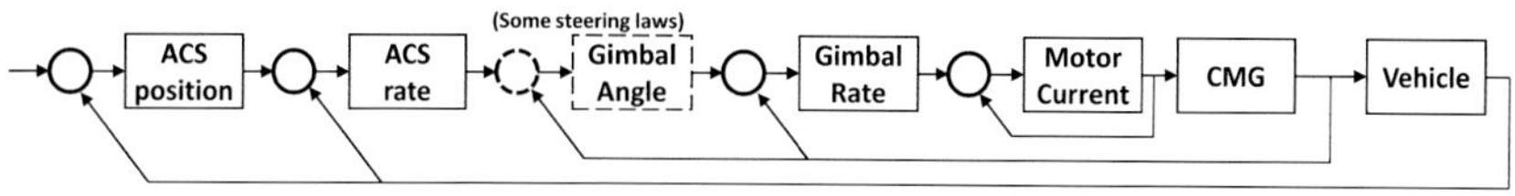

Fig. 10.2 Hierarchy of control

rate loop. If the structural frequency is near the rate-loop crossover, the closed-loop system may be unstable. Spacecraft designers should recognize the importance of ensuring that CMG mounting interfaces are stiff and well coupled to the bulk of the spacecraft. This is particularly important when the CMGs are direct-drive, where the effect is more pronounced than in the case of a geared transmission.

10.2.3 Bandwidth Considerations

The CMG gimbal rate loop typically lies within a hierarchy of nested loops, as illustrated in Fig. 10.2.

As in all such hierarchical systems, loop stability demands frequency separation (a factor of 3 or 4 is typical) between the bandwidths at each level. This separation is necessary because the closed-loop phase rolloff of each level is evident in the open-loop response of the next outer loop, compromising its phase margin.

For example, to support an ACS with a position bandwidth of a few tenths of a Hz, the ACS rate bandwidth might be around 1 Hz. If the steering technology in use closes any loops (for example, on gimbal angle or momentum state), it would require a bandwidth of a few Hz, and the gimbal rate loop would therefore need bandwidth in the vicinity of 10–15 Hz. This bandwidth is typical for many CMGs. There is little margin in this strategy, and thus it is important for the ACS designer to be aware of the bandwidth of the CMG hardware, as well as any closed-loop dynamics in the steering technology chosen. These matters can directly influence the ACS design, or, alternatively, the ACS design might drive CMG performance requirements.

These design interdependencies might lead one to expect that the CMG rate-loop bandwidth would simply be over-specified, to reduce risk. The problem here, of course, is the one discussed earlier. If the CMG bandwidth is increased, its stability is threatened by higher-frequency structural modes. Stabilizing it would flow to a need for stiffer and therefore heavier spacecraft structure.

10.2.4 Friction and ACS Limit Cycling

The ball bearings used to support the CMG gimbal exhibit friction characteristics that are generally categorized based on their physical mechanism. Viscous friction opposes gimbal rate, and the torque increases nearly linearly with rate. Coulomb friction is “running drag,” which has a fairly constant torque over a wide range of gimbal rates, but always opposes the gimbal rotation. It therefore reverses polarity abruptly as gimbal rate passes through zero. Static friction, or “stiction,” causes a torque greater than the running drag to be required to get the gimbal moving again once it has stopped during a zero-rate crossover. Finally, when the gimbal is “stuck” at zero rate, a “pre-breakaway stiffness” is observed. So, the axis is never completely “stuck.”

All of this nonlinear behavior is generally negligible when the CMGs are used for agile slews requiring large torque. The large gimbal-rate commands quickly overcome the behavior near zero rate, and the viscous term is very small for the low rates typically associated with gimbals.

Of greater concern is the behavior when the vehicle is under precise control at a constant rate. When spacecraft accelerations are low, the CMG gimbal rates are near zero, and friction nonlinearities become apparent. When a rate reversal is required, the free integrator in the gimbal rate loop compensation must have time to “wind up” enough torque to overcome static and Coulomb friction before the gimbal can rotate. And since the rate command is very small, the rate error is small, and the integrator ramps slowly. In a worst-case scenario, this apparent deadband behavior can result in a limit cycle in the ACS at a low frequency and amplitude.

In general, high-precision, preloaded ball bearings do not exhibit excessive static friction, but the reversal of the Coulomb friction is always present. The CMG manufacturer should be able to provide information on this behavior or include it in a dynamic model of their rate loop (for example, as a Dahl friction model). Armed with this information, the ACS designer should check for nonlinear limit cycles early in the design process. A describing-function analysis of the system is the best tool for determining both margin from limit cycle and strategies for modifying the design if necessary. Time-domain computer modeling can be used to validate the results of the analysis.

10.3 Chapter Summary

The inner loops of momentum devices can influence the spacecraft design. Each of the spin and gimbal loops commonly found is discussed. Choices surrounding this discussion can influence system power, performance and stability. Early decisions ranging from how the momentum devices are specified to the design of the spacecraft structure near the momentum devices are advised.

10.4 Exercises

1. You plan to use an array of CMGs and your steering law assumes all run at the same speed. You specify to the CMG manufacturer that wheel speed must be controlled to ± 1 rpm. Why is this a bad idea?
2. An RWA manufacturer offers rotor speed telemetry obtained by measuring the time between pulses from hall devices. These pulses occur every 60° of rotor rotation. To mitigate the effects of friction on output torque accuracy, you plan to close a speed loop using this telemetry. What will go wrong?
3. A precision CMG uses a gimbal rate sensor which has 3% ripple at 128 cycles per revolution. The gimbal axis closes a very aggressive rate loop with 20 Hz bandwidth. Will this be sufficient to provide less than 0.1% output torque ripple?

Chapter 11
Motors in Space

All momentum devices require a spin motor of some type to actuate the rotor spin axis. CMGs also require a motor on the gimbal axis. The torque and speed requirements for the gimbal motor design are typically much different from those for the rotor. These issues, among others, motivate the following discussion of motors commonly employed in spaceborne momentum devices. The objective of this chapter is to familiarize the reader with some of the key considerations with the goal of informing the analysis, implementation, and operation of momentum-control systems and the spacecraft that use them. Motor design is a broad specialty field that demands much more depth than offered here.

11.1 Motor Technology

Several types of motors find application in space products. AC induction motors can provide a simple solution for constant-speed applications where AC power is available, for example mechanical rate-sensing gyroscopes found in many aircraft. They find no application in momentum systems, however. Even for CMGs, in which the spin speed is held constant, that constant speed needs to be adjustable on orbit to provide the option of mitigating excitation of structure vibrations.

DC motors, both brushed and brushless, are found in space applications. However, the brushed motor, while simpler, is far less preferable for long-life applications. Brushes are a source of debris, EMI-producing arcing, and other problems. The brushless DC motor has a permanent-magnet rotor and two- or three-phase coils on the stator, as illustrated in Fig. 11.1.

When the coils are energized in sequence, a rotating field is produced that interacts with the field of the rotor to produce torque. Commutating the stator windings requires a position measurement and more complex drive electronics than do brushed motors. High-speed motors are often commutated discretely using Hall-

© The Author(s), under exclusive license to Springer Nature Switzerland AG 2025

F. A. Leve et al., *Spacecraft Momentum Control Systems*, Space Technology Library 46, https://doi.org/10.1007/978-3-031-96165-6_11

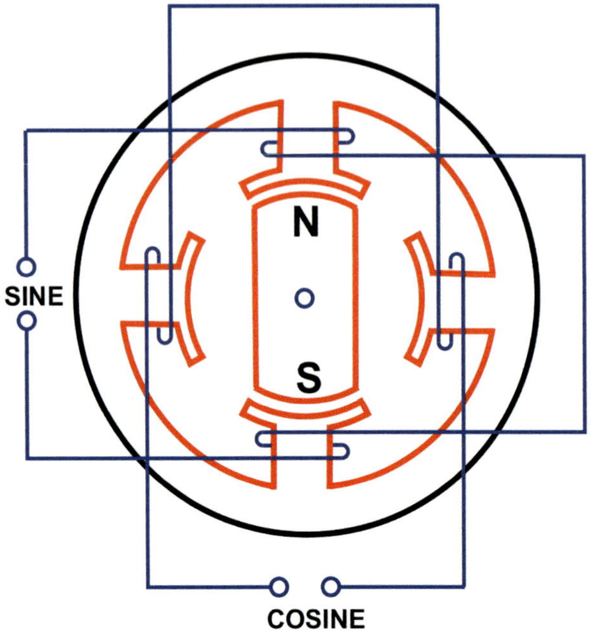

Fig. 11.1 One-pole, two-phase brushless DC motor

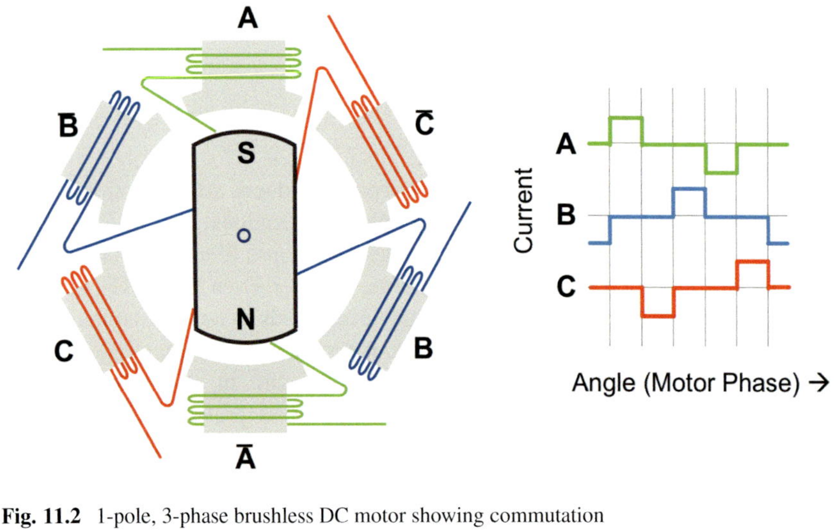

Fig. 11.2 1-pole, 3-phase brushless DC motor showing commutation

effect devices to switch the appropriate coils on or off at the appropriate times, as shown in Fig. 11.2.

Commutation methods like this are most common for spin motors. Lower-speed, higher-precision applications use an actual rotor-angle measurement, such as from a resolver or encoder, and the coil currents are commutated sinusoidally to produce a nearly constant torque. Most CMG gimbal motors use this technology. Brushless DC motors tend to yield the most efficient and lightest design for typical space applications. A design based on brush motors would be suitable only for low-cost, short-duration technology-demonstration missions and even then would likely not be fully traceable to an operational space system.

Stepper motors are a special case of the brushless DC motor. In this case, rather than commutate the stator field to continually "lead" the rotor field, a stator coil is energized, and the rotor is allowed to pull (opposite magnetic poles) into alignment, come to equilibrium, and stop. This action constitutes a "step." Then the next coil in sequence is activated, resulting in another step. Designed to allow this to happen quickly, a stepper motor can then be operated without commutation as such; no position measurement is required. The distance travelled can be precisely controlled by counting the number of "steps," and the motor is designed with a large number of magnetic poles for precise positioning. This simple scheme has found application in many space devices, including some CMG gimbal drives. However, this type of operation produces a broad vibration spectrum and is fairly power inefficient. The disturbance from such a spectrum is problematic because it is not readily attenuated with simple vibration-isolation techniques such as passive spring/damper devices.

One additional technology that is found in these devices is the "voice coil," or "ironless armature" motor. A voice coil derives its name from its similarity to an amplified, electromagnetic audio speaker. This type of motor is used for high-speed applications, in which the separate windings must be commutated very quickly. As the time-varying stator magnetic fields of a conventional motor increase in frequency, problems arise with eddy currents that cannot be addressed using conventional laminations. Rather than exploiting the interaction of two magnetic fields, a voice-coil design directly uses the Lorentz force between a field and a current. As shown in Fig. 11.3, windings on an insulated bobbin are placed into the permanent magnet field of the rotor and commutated to obtain the necessary

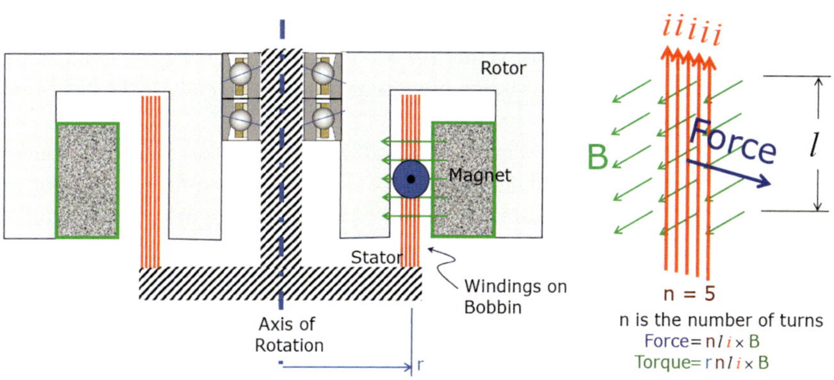

Fig. 11.3 Ironless armature spin motor

torque. The rotor design produces a uniform radial magnetic field, and coil current flows vertically through the field. The result is force directed out of the page and thus constitutes torque about the centerline of the motor.

The rotor field is constant in the rotor frame of reference; however, it is still rotating rapidly with respect to the stator. Care must therefore be taken to minimize leakage of the rotor field, and any material that can conduct eddy currents must be moved well away from the rotor field. For very high-speed motors, the care with which eddy-current losses are managed by design is critical to minimizing motor drag, which can easily result in excessive heat generation and power expenditure. Another key consideration in this type of motor is the thermal design. Unlike other motors, the heat dissipated in the motor coils is being conducted through a bobbin made of a non-conductive material. In general, materials that do not conduct electrical current also do not conduct heat well. Thus this design must take particular care to provide adequate heat-conduction capability.

Regardless of the technology chosen, motors for space must be designed with the same concerns as any device for space. These concerns include avoiding materials that outgas in a vacuum, emphasizing conduction in thermal management, and robustness in the presence of launch vibration, particularly where small-gauge wiring is implemented. For these reasons, choosing an "off-the-shelf" motor must be carefully considered for these environments. It is essential that any "off-the-shelf" motor design be verified for appropriate materials and construction, and tested in vacuum to prove that the winding temperatures do not exceed the manufacturer's ratings as well as the program deratings. Most motor manufacturers do not have an appreciation for the thermal demands operating in vacuum, so a dedicated test is the only way to confirm the suitability of any design for operation in space.

11.2 Sizing a Brushless DC Motor

This section offers general insights into the nature of motor sizing. This information is useful in understanding the general trends and design dependencies, and rapidly obtaining a rough idea of motor sizing for use in CMG design trades. For specific space-systems design, the best source of assistance is likely vendor data sheets and application engineers.

Motor sizing consists of two largely separable parts. One part is the mechanical design of the motor, i.e. the diameters and axial lengths of the rotor and stator, as shown in Fig. 11.4. These parameters size the permanent magnets, the amount of iron, and a weight allocation for copper based on the envelope available. Instead of going into detail on the trades available between length and diameter, this discussion simply assumes a typical ratio. The sizing of the motor determines the amount of torque it can produce.

Second, the copper allocation is detailed into a coil design by the choice of the wire gauge and number of turns. The motor torque constant K_T (ft-lb per amp) is affected by this choice.

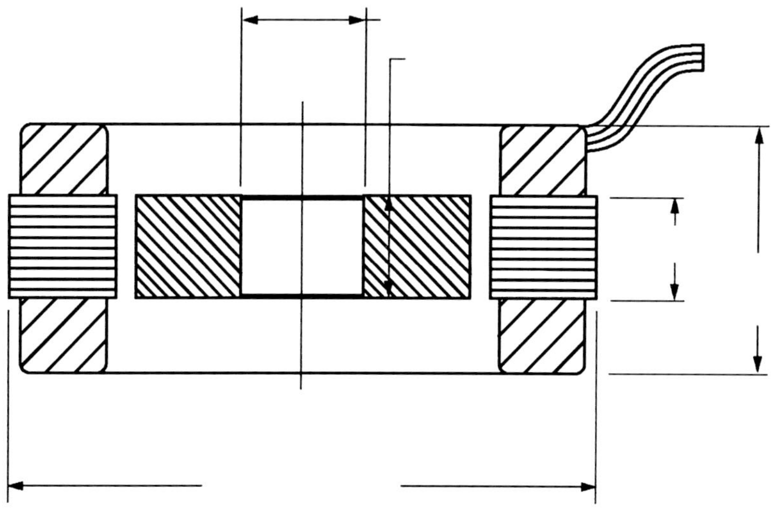

Fig. 11.4 Brushless DC motor mechanical design

There are, perhaps, as many ways to design a motor as there are designers. Each vendor has its own unique approach to the details. However, the laws of physics allow us to get the big picture without going into much detail. For example, we can estimate the power dissipation and weight of a motor knowing only the required torque. This is because the torque required directly influences the magnetic circuit design. Motors are often scored by their "motor constant," known as K_M, which is expressed in ft-lb/$\sqrt{\text{W}}$. This constant expresses the amount of torque the motor can produce for a given amount of power dissipated as heat. K_M depends upon the magnetic flux density, pole cross-sections, etc. It can be increased (thus dissipating less power) by increasing the weight of the motor (bigger magnets, larger poles), and, in general, this increase in K_M scales nearly linearly with weight. Therefore, we can expect that the "K_M density," K_{MD}, in ft-lb/$\sqrt{\text{W}}$ per lb would be a constant, over at least a fairly wide range of designs. For the class of motors typically found in momentum devices, a rule of thumb would be $K_{MD} = 0.07$ ft-lb/$\sqrt{\text{W}}$ per lb. This rule assumes high quality magnets and optimized, unpackaged designs.

Another consideration in space motor sizing is the ability to conduct dissipated heat out of the device. Conventional wisdom has it that a certain number of Watts can be effectively transferred out of a motor (without excessive temperature rise) per pound of motor. In a small motor, less heat dissipation can be managed than a larger motor. This "power density," K_{PD}, is expressed in Watts per pound. For the class of motors typically found in momentum devices, a rule of thumb would be $K_{PD} = 8$ W/lb.

11.2.1 Sizing a Brushless DC Motor

Given these two constants K_{MD} and K_{PD} as guidelines, we can approximate the weight and power of a motor necessary to apply a particular torque. Using nothing more than dimensional analysis, we find that the weight of the motor must be:

$$W = \left(\frac{\tau}{K_{MD}\sqrt{K_{PD}}}\right)^{2/3} = \left(\frac{\text{ft-lb}}{\frac{\text{ft-lb}}{\sqrt{\text{W}}\ \text{lb}}\sqrt{\frac{\text{W}}{\text{lb}}}}\right)^{2/3} = \left(\text{lb}^{3/2}\right)^{2/3} = \text{lb} \tag{11.1}$$

and the power must be:

$$P_{heat} = \left(\frac{\tau K_{PD}}{K_{MD}}\right)^{2/3} = \left(\frac{\text{ft-lb}\ \frac{\text{W}}{\text{lb}}}{\frac{\text{ft-lb}}{\sqrt{\text{W}}\ \text{lb}}}\right)^{2/3} = \left(\sqrt{\text{W}}^{3/2}\right)^{2/3} = \text{W} \tag{11.2}$$

In the above, the power being calculated is the I^2R power dissipated in the motor coils as heat. The total electrical power drawn at the terminals includes mechanical power delivered to the load.

Equations (11.1) and (11.2) show that both weight and power increase as the 2/3 power of the required torque. Furthermore, neither of these quantities varies with how the coil is wound. These principles allow us to quickly approximate the cost of producing torque. If this cost is unacceptable, in some cases gearing can be employed to reduce the torque required.

11.2.2 Electrical

In the mechanical design equations above, an allocation is included in the weight for the copper wire. Next, we must take this copper and partition it into wire size and number of turns. This allocation dictates the resistance of the coil and the torque constant K_T (ft-lb/A). The fact that the same motor can be wound with different coils is evident in motor catalogs, where each motor size is available with different "dash numbers," characterized by different K_T values.

If expressed in metric units of Nm/A, K_T has the same numerical value as the back-emf constant K_B, in Volts per rad/s. The reason for this identity derives from fundamental motor physics and is beyond the scope of this discussion. But as a result, increasing K_T has two direct effects. An increase means that it takes less current to produce a given torque, but it takes more voltage to overcome back-emf at a given speed. Generally, a given application has limited bus voltage. Therefore, the limit drives the decision. Starting from the spacecraft power bus, we must subtract the voltage drops related to power conditioning, drive transistors, wiring harness, etc. to arrive at the terminal voltage seen by the motor. In order to apply rated torque at rated speed, this available voltage must comply with

$$V > K_B\omega + iR = K_B\omega + \frac{\tau}{K_T}R = K_B\omega + \frac{\tau K_T}{{K_M}^2} \tag{11.3}$$

The previous equation uses the relationship below, which can be verified by dimensional analysis.

$$R = \left(\frac{K_T}{K_M}\right)^2 \tag{11.4}$$

In Eqs. (11.3) and (11.4), K_M is the motor constant itself, not the K_M density K_{MD} discussed earlier. This can be evaluated after the mechanical design phase because the weight is known.

Using Eq. (11.3) above, we choose a K_T that allows operation with the given bus voltage. This is the step where the required motor speed comes into play, as the back-emf is usually the dominant driver in required voltage. As a larger K_T implies less current to get the same torque, we might think increasing it would reduce power; but this is not the case. The resistance also changes, such that the power dissipated in the motor remains constant at the value derived in the previous section. Nevertheless, there are distinct advantages to a motor design with less current. Lower current simplifies the drive electronics, reduces power dissipation throughout the electronics and harnessing, reduces the weight of cabling, reduces EMI, etc. Therefore, the largest possible K_T should be chosen. If the motor requires high torque and high speed, the current increases, which drives the electronics design, and a point is eventually reached where the motor-design engineer must push back and require a higher system bus voltage. Machines with high torque and speed, i.e., mechanical power, cannot be expected to work efficiently, if at all, at low voltage.

Once K_T has been chosen, it is a simple matter to evaluate the coil resistance (from Eq. (11.4)), the current (from the definition of K_T), and any remaining parameters of interest. From this simple approach, we know enough about the motor to allocate a preliminary weight budget and flow requirements to the electronics design. We can even estimate the motor volume by assuming a reasonable ratio of iron and copper, the dominant weight drivers. To proceed beyond this point, it is advisable to work with a motor vendor applications engineer.

11.2.3 Numerical Example

Example 11.1 Suppose we are given the following requirements: design a motor that must produce 0.75 ft-lb at 30 rad/s. Assume that 20 V is available at the terminals of the motor. Assume $K_{MD} = 0.07$ ft-lb/$\sqrt{\text{W}}$/lb, and $K_{PD} = 8$ W/lb.

$$W = \left(\frac{\tau}{K_{MD}\sqrt{K_{PD}}}\right)^{2/3} = \left(\frac{0.75}{0.07\sqrt{8}}\right)^{2/3} = 2.4\text{lb} \tag{11.5}$$

$$P_{heat} = \left(\frac{\tau K_{PD}}{K_{MD}}\right)^{2/3} = \left(\frac{0.75(8)}{0.07}\right)^{2/3} = 19.4\text{W} \tag{11.6}$$

$$K_M = K_{MD}\text{W} = 0.07(2.4) = 0.17\frac{\text{ft-lb}}{\sqrt{\text{W}}} \tag{11.7}$$

Next, we solve Eq. (11.3) for K_T, getting

$$K_T = \frac{V}{\frac{4.448(12)}{39.37}\omega + \frac{\tau}{K_M^2}} = 0.3\frac{\text{ft-lb}}{\text{A}} \tag{11.8}$$

$$i = \frac{\tau}{K_T} = \frac{0.75}{3} = 2.5A \tag{11.9}$$

$$R = \left(\frac{K_T}{K_M}\right)^2 = \left(\frac{0.3}{0.17}\right)^2 = 3.1\Omega \tag{11.10}$$

Note that the heat power is $i^2R = 2.52^2 \times 3.1 = 19.4$ W, as expected. The total power drawn by the motor when delivering the spec torque at the spec speed includes both the power dissipated as heat, and the mechanical power delivered to the load (torque times speed, in appropriate units). So,

$$P_{Total} = 19.4 + 0.75\left(\frac{4.448(12)}{39.37}\right)30 = 50\,\text{W} \tag{11.11}$$

If we consider this result from the motor terminal perspective, the total power should be voltage times current $= 20 \times 2.5 = 50$ W.

If we now consult some vendor data sheets for high quality brushless DC motors, we should be able to find designs that are similar to this. Bear in mind that this approach provides only a rough approximation. In the final design, one must include margins for manufacturing tolerances, variations with temperature and life, etc.

11.3 Chapter Summary

This chapter discussed the various motor technologies found in spaceborne momentum systems. Simple equations were developed for rough sizing studies and rules of thumb for key constants were provided along with a numerical example. Detailed motor design should always be done in consultation with a motor vendor.

Chapter 12
Bearings in Space

Any angular momentum control device requires a precise and reliable rotational degree of freedom to spin a rotor, or to gimbal a spinning rotor. This is most commonly achieved through the use of ball bearings. Due to the extremely high numbers of failures and unexplained anomalies of ball bearing in spaceflight mechanisms, they have attained a mystique of being very difficult to implement, and therefore subjected to highly sensitive and specific design rules, delicate handling practices, stringent cleanliness requirements, and extremely conservative lifetime factors. While such best practices are always advised for the objectives of launching high reliability equipment into space, numerous examples of unexpected bearing problems in space have been encountered even after diligent adherence to these requirements and guidelines (Shapiro [109]). So, in this chapter we discuss the basic design variables and tradeoffs for bearings used in momentum control devices. The goal is to impart a broad understanding of the most important considerations of ball bearings when selecting or reviewing the design of a component which is used in a momentum control system, or when one is involved in unexpected anomalies or failures of ball bearings.

12.1 Bearing Types and Tradeoffs

There are many types of bearings available which are used to support a spinning rotor. Terrestrially, ball bearings are also the most common, but hydrodynamic or hydrostatic fluid film bearings, journal bearings, roller bearings and magnetic bearings are also used for some applications. Most alternatives to ball bearings are used in applications requiring much higher loading than that required to control the attitude of a spacecraft in microgravity, so more robust bearing types are rarely, if ever, considered for spaceflight mechanisms. Magnetic bearings have been used for spaceflight momentum control devices, since there is no contact between the

© The Author(s), under exclusive license to Springer Nature Switzerland AG 2025
F. A. Leve et al., *Spacecraft Momentum Control Systems*, Space Technology Library 46, https://doi.org/10.1007/978-3-031-96165-6_12

rotating and non-rotating surfaces, eliminating wear. However, the general result compared to ball bearings has been higher power consumption, more mass, higher disturbances and much higher cost (Howard et al. [110]). Intuitively, it seems to make sense that in microgravity a magnetic bearing would be an ideal solution, but in practice the need for complex control electronics and active magnetic actuators are much larger penalties than the potential benefits of a magnetic suspension. While this may not always be the case, presently it is almost impossible to achieve the stiffness of a ball bearing with magnetics, without severe cost, weight and power penalties compared to a simple ball bearing. Although the magnetic bearing is non-contacting, even the predicted reliability is often lower for the magnetic bearing, since the additional electronics have a higher theoretical failure rate than a simple ball bearing. We therefore will pragmatically focus solely on ball bearings.

Spin bearings are the most crucial component of a reaction wheel, momentum wheel, or control moment gyro. The bearing must provide reliable, quiet and low-friction support for a rotating high-inertia rotor. The rotor may be used at low speeds or very high speeds, or both, and the design of the bearing can be different based on the operating speed range. For simplicity, a universal bearing design which can be used reliably at both high and low speeds, and cross through zero speed for bi-directional reaction wheels, is typically employed. The tradeoff is usually slightly higher than necessary friction at high speeds. For most reaction wheel applications, little time is spent at the highest speeds, so this is usually only a minor consequence.

Gimbal bearings are used in control moment gyros to allow an actuator to force a cross-axis rate on a spinning rotor assembly. In this application, they are used only at low rotation rates, and the requirements are driven by need for smooth operation and longevity. Since a gimbal bearing spends most of its time not moving, its design is largely driven by capacity to survive launch loads, and proper lubrication to keep stiction low when the balls and races have no hydrodynamic lubricant film at zero speed.

12.2 Ball Bearing Selection

Selection of ball bearings for a particular application is best done by working directly with an application engineer from an established bearing manufacturer. There are a few special considerations for operation in space, but it should be understood that there is no black magic that makes one bearing manufacturer more suitable for spaceflight than another, provided the quality control standards are adequate and the required precision can be attained and verified. This section gives an overview of the types of design options that are available, and the fundamental tradeoffs involved. Most of this type of information is provided in detail in literature available from bearing manufacturers. Ball bearings require extremely precise components with a host of sensitive variables. Modeling of specific bearing designs is complex, and best left to bearing modeling experts or technical resources at bearing manufacturers.

12.2.1 Anatomy of a Ball Bearing

A ball bearing has four components; an inner ring, an outer ring, a complement of balls, and a retainer, as shown in Fig. 12.1. The rings incorporate a precise groove, or raceway which has a radius that is slightly larger than the radius of the balls. The amount of this difference is quantified as the conformity ratio, or the raceway groove radius, r, divided by the ball diameter, d, as shown in Fig. 12.2. The result is a conformity ratio that is always slightly greater than 50%. The conformity is an important variable in designing a bearing, as it has a significant impact on the load capacity, the contact angle, operating stresses, and friction torque. If the conformity is open, the contact stresses are high. If the conformity is too tight, the friction torque is higher because there is more contact with the races. So, finding a happy medium is part of the bearing design process, and the starting point is typically around 52% for the inner race and 54% for the outer race.

The ball makes point contact with the race when no load is applied. The deformation of the ball and race under load results in a Hertzian contact stress, and which takes the shape of an ellipse, as shown in Fig. 12.3. It is a narrow ellipse for the inner race and a wider ellipse for the outer race since the inner race has a convex relationship with the ball in the rotation axis and the outer race has a concave relationship with the ball in the rotation axis. As a result of this, the outer race conformity ratio is intentionally higher by design than the inner race conformity to result in similar level of stress on both the inner and outer race for any given applied load.

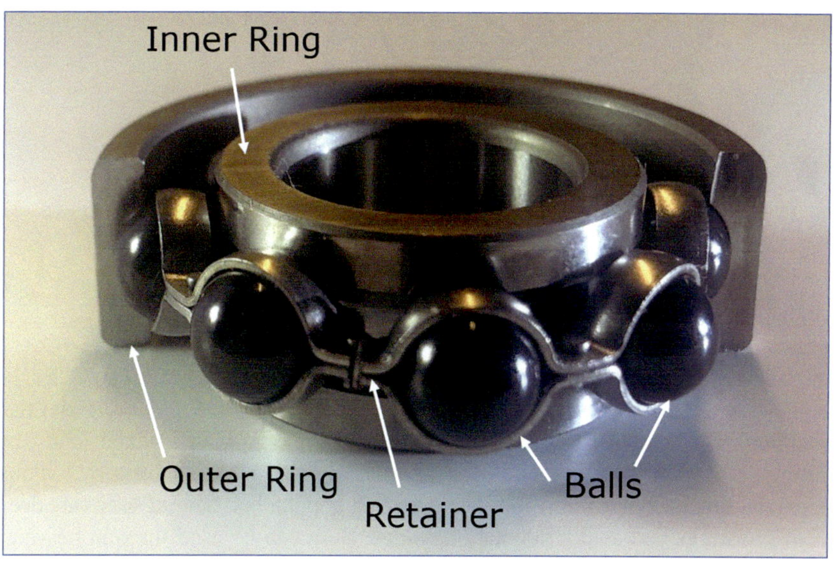

Fig. 12.1 Main components of a ball bearing

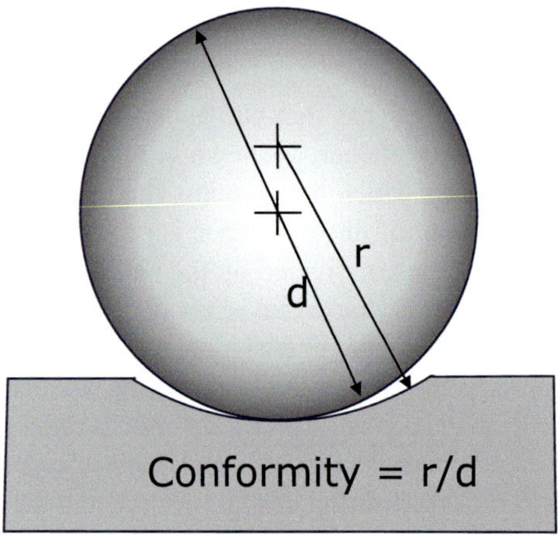

Fig. 12.2 Ball and raceway groove contact showing conformity ratio

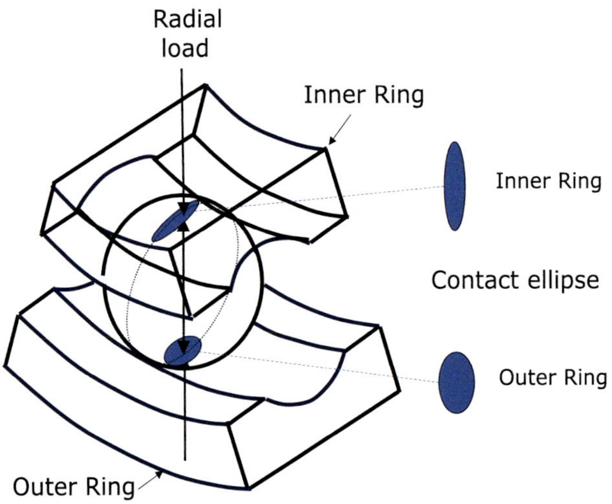

Fig. 12.3 Contact ellipses in a ball bearing. (Hamrock and Anderson [111])

12.2.2 Ball Bearing Types

The ball bearing was invented by Leonardo Da Vinci around the year 1500 CE. His original thrust bearing for his helicopter concept is shown in Fig. 12.4. From that inception, ball bearings evolved to handle radial loads as well as axial loads, by putting grooves in the races, and even greater advances were achieved by making them out of metal instead of wood. The two most popular ball bearing types are deep groove bearings and angular contact bearings. Both can be used for spin bearings and gimbal bearings, and we will discuss the primary tradeoffs between the two.

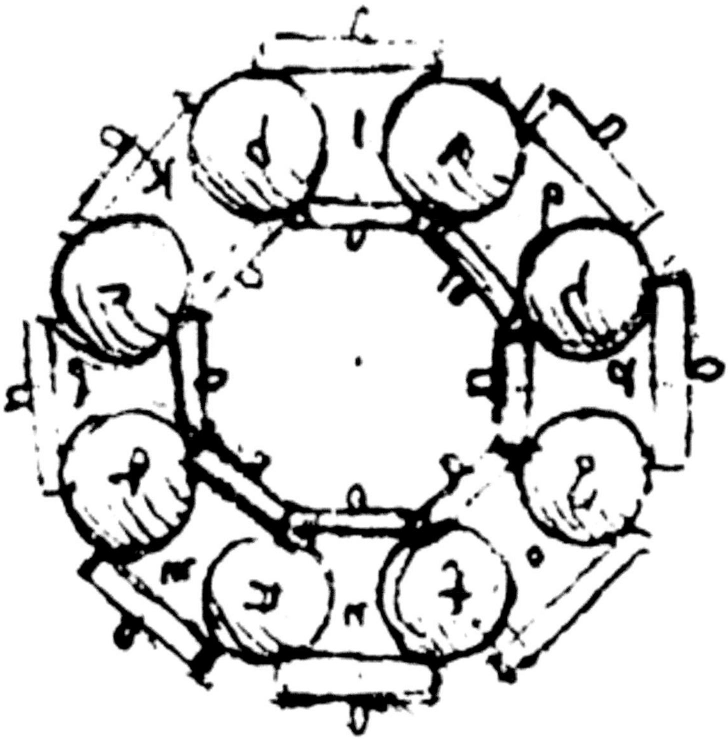

Fig. 12.4 Ball bearing concept by Leonardo Da Vinci

Deep groove bearings are the classic ball bearing type, patented by Robert Conrad in Britain in 1903, and therefore commonly referred to as a Conrad bearing. It has grooves in both the inner and outer races, and a retainer that is installed after the balls have been introduced. The assembly process of bearings is important to understand when assessing the tradeoffs and features. In a deep groove bearing, the balls can be biased to one side during the initial assembly, as shown in Fig. 12.5a, and then the ball complement can be evenly distributed around the circumference as shown in Fig. 12.5b. Finally a retainer is installed as the final step of the assembly process, as shown in Fig. 12.5c. The primary benefit of this arrangement is that the bearing can accept high axial loads in either direction.

The Angular Contact Ball Bearing is derived from the deep groove bearing, but with one land relieved so that a retainer can be installed prior to the final assembly, as shown in Fig. 12.6. This opens up options for the construction of a single-piece retainer, since the ball complement can be distributed before the final assembly process, which is done with a simple press and snap of the balls over a small ball-retention dam. It should be noted that the drawback of the angular contact ball bearing is that it can only accept axial loads in one direction. Loading in the

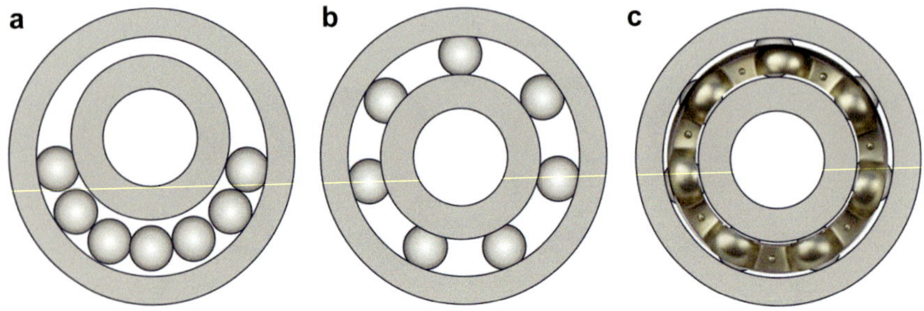

Fig. 12.5 Assembly process of a deep groove bearing, (**a**) biased balls, (**b**) distributed balls and (**c**) two-piece retainer installation

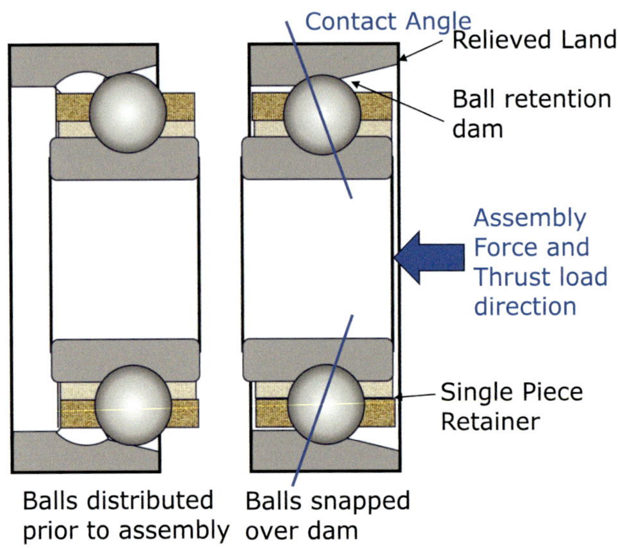

Fig. 12.6 Angular Contact Bearing assembly process

opposite direction may overstress the ball on the small assembly dam, which would cause denting or damage of the balls.

Angular contact ball bearings are frequently used in duplex pairs as shown in Fig. 12.7, for a precise preload and to allow bi-directional loading. It is also important to note that deep groove bearings in most applications are used as an angular contact configuration when preload is applied.

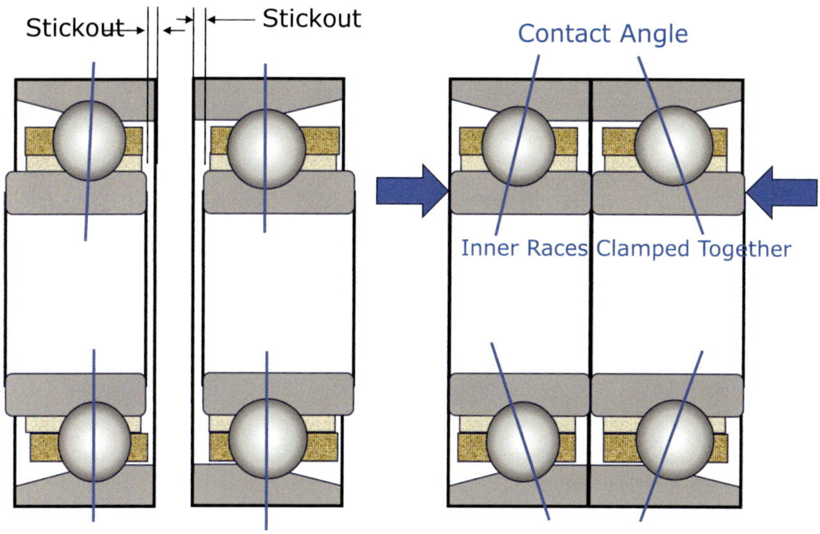

Fig. 12.7 Angular Contact Bearings preload in a back-to- back duplex arrangement, before and after clamping together

12.2.3 Ball Bearing Retainers

The purpose of a retainer in a ball bearing is to keep the balls from running into each other. Sometimes called a separator, or a cage, in a nominally operating bearing the retainer only provides an occasional nudge to advance a ball that is slightly slower than the average, or retreat a ball that is slightly faster than the average. One of the most common types of retainers in terrestrial applications is a pressed steel ribbon retainer, as shown in Fig. 12.8a. This is used with a deep groove bearing, where the two-piece ribbon is installed from both sides and riveted in place in a very simple operation. The result is a very robust retainer that is guided by the balls. Another common type of retainer is a composite cotton phenolic retainer shown in Fig. 12.8b (that can be snapped in place from one direction in deep groove bearings) and Fig. 12.8c (which is assembled with the balls in the case of an angular contact ball bearing). In these cases, either the inside diameter or outside diameter land is used to guide the retainer. The retainer can also be made of a variety of materials, and some can theoretically provide a small reservoir of lubricant right inside the bearing.

Clearances exist inside the contacting surfaces of a bearing due to the critical design feature that the ball radius is slightly smaller than the raceway radius. These clearances are precisely designed in order to achieve the desired contact angle when the bearing is loaded, and these internal clearances are referred to the axial and radial free play. This free play must be taken out during operation, and this is achieved by applying a constant load, or preload, to the bearing. The preload can be a soft

Fig. 12.8 (**a**) Pressed steel ribbon retainer (left), (**b**) Crown retainer for deep groove (center), (**c**) Halo for angular contact (right)

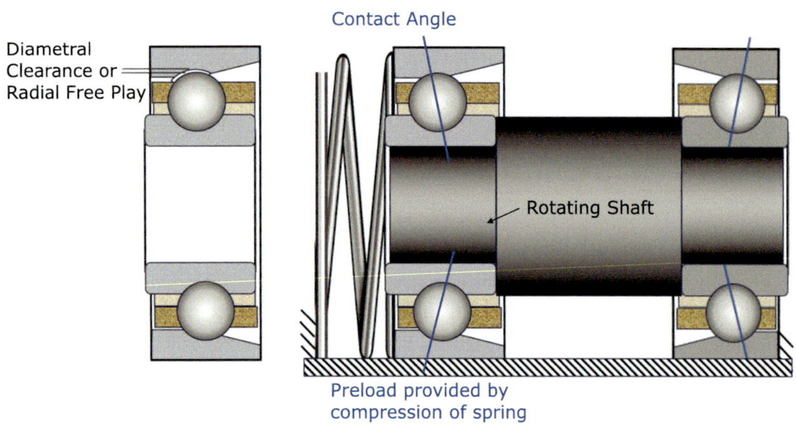

Fig. 12.9 Bearing showing radial free play and preload by a compliant spring

preload, applied by a compliant spring as shown in Fig. 12.9, or a hard preload, which is attained by mounting two bearings against each other with a precise amount of stick-out that is compressed. This is known a duplex pair, as shown in Fig. 12.7, and which is tuned to a specified preload at the bearing manufacturer. The advantage of a soft preload is a relatively constant preload over a wide temperature range, even with a large bearing span and differential thermal expansion. It is important to note that in a rotating mechanism, the preload spring should be applied to the non-rotating race.

A typical momentum control device such as a reaction wheel or momentum wheel will have a relatively large rotating mass, and typically just enough preload is applied to keep the bearings under preload in any orientation with respect to gravity.

The sizing of bearings for a momentum control device is driven by the loads during the random vibration exposure during launch. The balls can dent the races (called Brinelling) when loaded, and the static load capacity is derived from the maximum load that can be applied without exceeding 400 Ksi in general, and without exceeding 335 Ksi mean stress for quiet running applications [111]. The general capacity was derived from the amount of loading that has been demonstrated to not affect fatigue life, and the quiet running capacity was derived empirically for a group of typical instrument bearings (Leveille and Murphy [112]). Using the values calculated by the bearing manufacturer based on these criteria are a starting point, but what is most important for any given application is that the loads should not exceed the amount that will adversely affect the payload, and testing of the device to the required vibration environment and measuring noise before and after is the only real way to confirm sufficient load capacity for the bearings. Brinelling is detectable by the emitted vibration or high-precision Talyround measurements of the raceway, and rarely by visual inspection of the bearing, since the stresses are on conforming surfaces making the visual deformations subtle.

Because of the relatively high mass of inertia rotors for momentum control devices, the response during vibration exposure is usually very close to a single degree of freedom (SDOF) system. Therefore, the bearing loads for random vibration can usually be approximated from a basic structural model if you know the rotor mass and fundamental resonant frequency by Miles' equation, (Miles [113]) which estimates a quasi- static statistical response acceleration to a random vibration power spectral density typically specified for space hardware.

$$A = 3\sqrt{\frac{\pi}{2}(PSD)(fn)(Q)}$$

where:

$A = 3$ sigma response acceleration in gs
$PSD =$ power spectral density input at the resonant frequency in g2/Hz fn = natural frequency of SDOF system in Hz
$Q =$ the transmissibility at the resonant frequency – unitless

One must make an assumption for the transmissibility, which is a very significant variable, and which can really only be known by performing a modal survey of the actual device and measuring the response vs. input acceleration at the resonant frequency (Bialke [114]). The transmissibility will vary significantly depending on the preloading mechanism, and is strongly dependent on the vibration environment input levels. A hard preload will tend to have a high transmissibility, and a soft preload will tend to have more damping and thus lower transmissibility.

12.3 Ball Bearing Lifetime

Ball bearing lifetime is a very mysterious topic for spaceflight mechanisms. Conventional longevity models are occasionally proven wrong, but in fact there are many experiences of spaceflight ball bearings greatly exceeding their expected lifetimes. Despite this, premature failures of ball bearings attract all the attention as the consequences can be dramatic, and sometimes mission ending. We will start by discussing the conventional modes of bearing failure, and then delve into the lesser-known effects of electrical discharge in bearings.

12.3.1 Fatigue Life

In terrestrial applications, fatigue life is the driving factor for bearing design, since most here on Earth have demanding loads at the limits of bearing capacity, since machines are often designed to accomplish demanding work. In space, the loads in zero gravity are mostly very gentle and the bearings are sized by the launch load capacity far more frequently than the operating loads. The fatigue life is calculated with a bearing parameter known as the L10 Dynamic Load Rating, which is an extrapolation of the expected number of revolutions of a bearing under a specified load, for which 90% of bearings will theoretically survive. This calculation includes the bearing load, number of revolutions, and some quality factors, and estimates a statistical lifetime. The L_{10} rated fatigue life is be calculated by:

$$L_{10} = (C/P_r)^3 \times 10^6 \text{ revolutions}$$

where:

C = dynamic radial load rating (Force units)
P_r = dynamic equivalent radial load (Force units)

It must be noted that bearing Dynamic Load Rating as determined by the bearing manufacturer has been frequently misinterpreted as a static load rating during dynamic loading such as random vibration. To be clear, the dynamic load rating is only a statistical prediction of the number of revolutions that can be expected for a given radial load without fatigue failure. There are modifiers for higher than 90% reliability, better than standard materials, and lubrication conditions, but the Dynamic Load Rating is only an input to the fatigue life equation, not an allowable dynamic load during vibration exposure.

12.3.2 Wear

There are many accounts of bearing wear being deemed the culprit for premature bearing failures, but few have credibility when you stand back and look at the whole picture. Wear is fairly well-understood phenomenon which occurs when two surfaces are rubbed together (the term Tribology is Greek for "to rub") and the harder of the two surfaces wears material from the softer of the two surfaces. That is why hardness of bearing materials is so important. The wear rate of two surfaces is determined by Archard's equation: (Archard [115])

$$V = kFs$$

where F is the normal load, s the sliding distance, V the wear volume and k the specific wear rate coefficient.

This is a handy equation in general, but it is rarely applicable in ball bearings, since most of the relative motion is rolling, and the only sliding component is due to ball control on one race. Nearly all ball bearings in space applications will have a contact angle. As such, either the inner race or outer race will be pure rolling, and the other race will have a sliding component. A simple way to think about this is that a pure radial bearing could have perfect rolling on the inner and outer race, but once you introduce a contact angle, the difference in the height of the radius of contact between the inner and outer race results in a rotational sliding component in either the inner race or outer race surface, whichever has the least amount of resisting torque on the ball. In most cases, the sliding will occur on the inner race, since it has the longer contact ellipse due to the tighter conformance, as shown in Fig 12.10. But, this small sliding distance in well lubricated conditions typically results in insignificant wear, which leads us to the topic of lubrication.

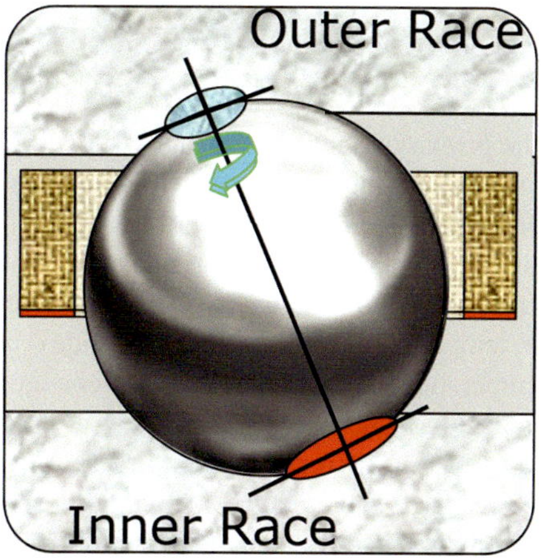

Fig. 12.10 Typical Inner Race Control with rotational sliding component on outer race due to shorter contact ellipse

12.3.3 Lubrication

Proper and reliable lubrication is the cornerstone of most mechanisms, including spaceflight mechanisms. It is hard to overstate the importance of reliable lubrication in ball bearings. We will discuss the key performance variables associated with a lubricant to be used in space.

12.3.3.1 Vapor Pressure

Vapor pressure is extremely important for any lubricant used in vacuum. It is essentially a measure of the evaporation rate in vacuum, which is important to avoid lubricant depletion, but also to avoid contamination of other surfaces on the spacecraft, especially sensitive optics. A lubricant must meet the required standards for outgassing in space, and if it meets these, it will by default not be susceptible to depletion by evaporation. Early in the space age, there were many lubricants used which were refined to specifically reduce the vapor pressure. These were natural hydrocarbons (oils literally pumped from below the ground) that were refined by introduction to vacuum and high temperature to essentially boil off the low vapor pressure components until only the heavy molecules were left, making them vacuum-hard, to a point. Later, some synthetic fluorocarbons were introduced such as Bray 815Z ©which were large molecules with an extremely high vapor pressure. In the early 1990s, synthetic hydrocarbons were formulated with very high vapor pressures, such as Pennzane X2000 ©. The extremely high vapor pressure of these synthetic lubricants successfully prevented lubricant loss by evaporation, but the chemistry of long-chained pure molecules results in a significant increase in electrical resistance, which can make the bearings more susceptible to electrical discharge.

12.3.3.2 Viscosity

The primary effect of lubricant viscosity is easy to imagine, since thicker lubricant will make a bearing harder to spin. But thinner lubricant will not lubricate as well under load and will more easily migrate away from the bearing. This is a fundamental tradeoff of lubricant selection. An additional parameter is the viscosity index, which is a measure of how much the viscosity changes as a function of temperature. A low viscosity index means the viscosity changes a lot with temperature, a high viscosity index changes less with temperature. So, low viscosity index lubricants may be too thick at cold temperature resulting in high friction, and too thin at high temperature to provide adequate lubricant films. Synthetic lubricants typically have a high viscosity index making them desirable for the extreme temperature ranges typically specified for spaceflight.

12.3.3.3 Grease vs. Oil

A grease lubricant is merely a base oil with a thickener. That thickener makes grease less likely to migrate out of the bearing. A disadvantage is the thickener will also result in higher viscous friction losses when it is being sheared by the relative moving surfaces in a bearing. There are some greases that will channel, which means that the bulk grease gets pushed out of the way of any relative moving surfaces, and therefore does interfere with the rotation of the bearing and friction drops to a low level. In this case, the grease filler merely acts like a matrix of material that keeps a supply of the oil lubricant in close proximity to the ball and race contacts. If the grease does not channel, it just increases the viscosity of the lubricant and adds viscous friction losses, which works well for many highly loaded terrestrial applications, but is just a big drag in spaceflight mechanisms. Testing the friction torque vs. speed of oil and grease formulations in an application in a variety of quantities and after adequate run-in is the only real way to quantify this tradeoff.

12.3.3.4 Lubricant EHD Films

EHD is an acronym for Elasto-Hydro-Dynamic film, a name coined when Hamrock and Dawson [116] combined the elastic properties of bearing balls and races with the hydrodynamic behavior of a separating lubricant to predict the thickness of a separating film between moving parts. Once the speed of the bearing reaches a point where the EHD film thickness exceeds the composite surface finish of the ball and race, there is no longer any direct contact between the moving surfaces. The thought is that when this speed is reached, there is no longer any potential for wear, and therefore it becomes a desirable minimum operating speed for bearing longevity.

There is very little published evidence to support that low operating speed is a significant contributor to bearing failures. In fact, the opposite has been reported (Dube [117], Bialke [118]). While low speeds may theoretically have more tribological influence due to sliding contact, they also have proportionally fewer revolutions and lower generation of heat.

12.3.3.5 Electrical Discharge

In terrestrial applications, electrical discharge is a well-known cause of bearing failures. A voltage potential across a bearing, or a current running through a bearing, can result in electrical discharge erosion, or physical scarring, to the point that damage occurs (Busse, Erdman [119]). This is typically rectified by grounding the shaft of the machine or using ceramic balls to prevent damage. This must be considered in space as well, since the space charging environment is more volatile than on grounded Earth.

There is strong evidence that electrical discharge may be the cause of many unexplained ball bearing anomalies and failures in space (Bialke, Hansell [120]).

Mechanisms using ball bearings have failed prematurely or had anomalies in space after successful life tests on Earth which did not include a space charging environment (Bruegman [121], Bendix [122], McMahon [123], McMahon [124], Pantaleoni [125]). It is clear that electrical discharge must be considered when considering ball bearing lifetime in spaceflight mechanisms.

If a potential is generated across the bearings by either induced voltages from the motor, or by volatile space charging events causing the voltage of the spacecraft to change rapidly, bearing damage can result. The instantaneous charge potential of an electrically isolated flywheel on a dielectric EHD lubricant film can lag the instantaneous potential of the spacecraft to the point that a significant voltage is generated across the bearing, and an electrical discharge or breakdown occurs. Eventually this can affect bearing race surface finish to the point that friction increases due to ball speed variation and cage windup [120].

One proven countermeasure for electrical discharge is the use of a hybrid ceramic bearing, with Silicon Nitride balls and steel races. This is called a hybrid bearing, and since it is non-conducting, it eliminates the possibility of any electrical discharge by increasing the breakdown voltage into the KV range. Since the electrical resistance is so high, breakdown can still occur under some charging conditions, so the location of the electrical discharge (usually at the smallest gap) must be understood to ensure mechanical or electrical damage does not occur if or when a discharge occurs at the high voltage [126].

12.3.4 Life Testing

Verification of lifetime of ball bearings is typically done by a controlled life test in a suitable unit in the application. Ball bearing manufacturers usually have a disclaimer that life of a bearing can only be verified by life testing or field testing in the application. Real time life testing of a sample of units usually satisfies this requirement. For space applications, this can be difficult when a newly designed component for a specific mission needs to be validated for a 15-year life, but launch is only a few years down the road. As a result, accelerated life testing can be a necessity. Life tests of bearings in an application can be accelerated by speed, preload and temperature, all of which have some significant caveats. Speed affects lubricant film thickness, preload only really affects fatigue life, and temperature effects are difficult to quantify. As a result, any accelerated test will have significant uncertainty. This does not mean it should not be run to weed out obvious design problems, but it is usually no guarantee of success in space.

In addition, it must be recognized that terrestrial life tests of aerospace mechanisms to date rarely include the variable of the space charging environment.

12.4 Ball Bearing Friction Anomalies and Failure Modes

Understanding the behavior of ball bearings in space has sometimes been described as difficult as 'herding kittens', since there seems to be no way to predict the behavior of each one at any given time. As previously mentioned, there have been many friction anomalies and failures on spacecraft momentum control devices which have gone unexplained. This section provides some background of a few of these events and a plausible explanation of why they may have occurred, in order to inform the reader of the potential effects of electrical discharge as a cause.

12.4.1 *The Ball Bearing Paradigms*

Here are some common paradigms regarding ball bearings used in aerospace mechanisms.

- It is generally concluded that most bearing friction anomalies and failures in space are the result of violation of mechanical tolerances, lubricant starvation, or over-lubrication. This has even been concluded by some to be the case for anomalous friction in bearings exhibiting behavior of Coulomb friction increases (which are assumed to be due to lubricant starvation), and viscous friction increases and decreases, (which are assumed to be due to over-lubrication) on the same bearings at the same time (Tungseth [127]).
- Ball bearings are assumed to be fragile and extremely sensitive to particulate or chemical contamination, and therefore must be handled with special care and in ultra-clean environments. The lubrication process is also viewed as an easily-compromised step in bearing assembly, like a secret recipe handed down from generation to generation. The truth is that bearings used in space are very lightly loaded, and usually oversized to handle launch loads, rather than operational duties. As long as some lubricant is present in the bearings at beginning of life and the bearings are not abused, most bearings in space should last for many years of continuous operation. Many tests have shown that gross physical overloading and even intentional lubricant starvation does not have a significant impact on the reliability of bearings for many years [120]. This is not to say that bearings should not be treated with respect and kept clean—rather that poor performance in orbit should not be simply blamed on mishandling or contamination.
- Because of the many unexplained failures in space mechanisms which occurred after a successful life test campaign, gravity is often concluded to be the culprit since microgravity cannot be simulated in terrestrial life testing. When all other known causes are eliminated, it is the only thing left on the failure tree.
- One last paradigm is that electrical discharges across bearings cannot possibly have any significant impact on bulk materials such as bearing surfaces, dry lubricants, etc. The rationale is that a tiny spark cannot possibly have an impact on bearing performance.

All of these paradigms were systematically debunked during the course of a Relentless Root Cause Analysis of reaction wheel failures and anomalies [120]. Over 150 bearings were life tested over the course of 16 years, which included extreme variables of abrasive contamination, under-lubrication, over-lubrication, severe misalignment, high preload, low preload, zero preload, ball defects, high temperature with high preload and misalignment, and operating sub-EHD and high speeds. The only variable to demonstrate any friction increase with a signature seen in space anomalies was applying as low as 6 V across the bearings. This resulted in immediate friction increases, increased wear, and prompted cage instability, which could not be duplicated with any other set of variables.

12.4.2 Ball Bearing Failures in Space

It was noted in the preface to this chapter that ball bearings in space have a bad reputation, so we will discuss some of the background, and some very plausible explanations to consider with respect to electrical discharge during space weather events.

The earliest infamous bearing failures in the space age were the Control Moment Gyros on the Skylab space station. Skylab was launched in May of 1973. The first bearing anomaly of CMG-1 occurred on November 3 after only several months of operation; it subsequently failed on November 23. CMG-2 started exhibiting similar behavior shortly thereafter on December 5, and was shut down in January of 1974 after frequent and increasing friction increases indicated it was also doomed for failure [128]. CMG-3 performed flawlessly for the entire mission. The fact that two out of three CMGs failed after a relatively short period of time was surprising, since for the five years prior to launch, a total of 480,000 hours of testing were accumulated on 22 units (Bendix [122]). After an exhaustive failure analysis, the final conclusion was that the testing in Earth's gravity was not representative since most testing was performed spin axis horizontal, and the lubricant replenishment experienced in gravity would not exist in space, resulting in lubricant starvation.

A second occurrence of systemic ball bearing failures early in the space age was on the GPS Block 1 and Block IIA constellations. A number of reaction wheels failed in the Spring of 1983, and in one case 2 reaction wheels failed simultaneously on the same spacecraft (Losik [129]). This occurred only three years after launch. The cause was concluded to be lubricant starvation, due to the orientation of the reaction wheel in the shipping/storage container, with spin axis horizontal preferable to spin axis vertical to minimize loss of bearing lubricant due to gravitational migration. However, two reaction wheels systematically failing simultaneously on the same spacecraft does not fit the random deterioration which would be expected with gradual lubricant starvation, and the same design had many reaction wheels that lasted 10–20 years, which begs the question whether the failures were due to lubricant starvation [130].

In June 1998 GOES 9 developed a problem with one of its two momentum wheels, which were used to control its orientation. The momentum wheel was drawing more power than expected, and generating noise, visible in some of the images the spacecraft returned, so it was turned off. The spare wheel started developing the same problem shortly there after, when a decision was made to replace GOES 9 with the backup satellite, GOES 10. This is another case where lubricant starvation was also concluded to be the cause of two nearly concurrent failures.

A series of reaction wheels manufactured by ITHACO Space Systems had systemic failures and anomalies starting in 2001 on the NASA Far Ultraviolet Spectroscopic Explorer (FUSE) spacecraft, and later including failures on the NASA Kepler Space Telescope, Japan's Hayabusa asteroid rendezvous, and NASA'S DAWN interplanetary spacecraft. In 2011, a commercial constellation had multiple reaction wheel failures and anomalies very soon after launch (Hacker [131]). The failure investigation previously discussed in 12.4.1 positively eliminated lubricant starvation, contamination, preload, misalignment, zero crossings, handling, workmanship and bearing geometry variables. The relentless root cause analysis eventually determined that the root cause was associated with space charging [120]. The turning point of the failure analysis was when it was discovered that the friction anomalies and failures of reaction wheel bearings had a very strong correlation with geomagnetic storms. An example of the timing of the FUSE RWA spacecraft failures is shown in Table 12.1. Not only are there two RWAs on the same spacecraft with

Table 12.1 FUSE RWA friction increase dates, with concurrent events and high ranking geomagnetic storms highlighted in red

Friction increase	Friction initiation date	Geomagnetic storm ranking since 1994 and category
RW-X		
5 mN-m	13 February 2001	Not ranked G1
700 mN-m	25 November 2001	20th G4
RW-Y		
3 mN-m	15 July 2000	5th G5
2 mN-m	27 July 2000	Not ranked G1
5 mN-m	3 July 2000	Not ranked
4 mN-m	5 October 2000	16th G4
1.2 mN-m	25 November 2001	20th G4
1400 mN-m	10 December 2001	Failed after operating at low speed in elevated friction condition for 15 days
RW-Z		
5 mN-m	12 September 2002	Not ranked G1
>40 mN-m	December 2004	N/A
RW-Skew		
7 mN-m	30 April 2007	Not ranked G1
>40 mN-m	12 July 2007	ot ranked G1

friction events during a G4 geomagnetic storm on November 25, 2001, but the odds of four of the events randomly being during G4 or G5 events in that time period is only 1 in 18,000 (Bialke et al. [126]). During geomagnetic storms, the charged plasma environment can be very volatile, and cause rapidly changing ground potential of the spacecraft. This fact then pointed towards electrical discharge as a potential cause, and eventually testing confirmed that bearing friction and bearing health could indeed be affected by voltages as low as 6 V across the bearings. In addition to reaction wheels, there have been several other strong correlations of geomagnetic storms with space mechanism failures and anomalies [126].

The Skylab CMG failures previously noted were concluded to be most likely caused by lubricant starvation, but this could not be duplicated in the lab. While testing of intentionally starved bearings did result in friction failures, they did not manage to manifest the same friction signatures as observed with the failed CMGs, and the only strong conclusion from the exhaustive failure analysis was that the bearing in CMG-1 failed due to ball speed variation, with lubricant starvation suggested as a likely cause of the ball speed variation (Bendix [122]). It should be noted that ball speed variation has been identified in the sequence of events for reaction wheel failures concluded to most likely be due to electrical discharge [120].

The International Space Station (ISS) incorporated four CMGs of the same basic design used on Skylab, with strategic design modifications to eliminate the potential for lubricant starvation. CMG-1 failed on ISS on June 8, 2002, after only 18 months of operation. The CMG-1 failure on Skylab and the CMG-1 failure on ISS both occurred within hours of significant attitude maneuvers of the large structures. CMG-3 on ISS exhibited anomalous friction also within hours of a significant attitude maneuver on September 29, 2006. Since this experience in 2006, the severe attitude changes have no longer been required on the ISS (Gurrisi et al. [132]). The remaining CMGs have operated continuously at 6600 rpm for at least the last 18–23 years without incident. There is a significant correlation of the CMG friction events on Skylab and ISS with major maneuvers, which was recognized by the failure review teams, but the assumption was that the radial loading from the CMG gimballing during the maneuvers affected the spin bearings. However, what was never considered is that the maneuver itself can cause differential charging of the large structures. NASA-HDBK-4002B [133]) resulting in chassis currents and potential electrical discharges which can affect the bearings. This is not mentioned here as a conclusive cause of these cases, but just to point out that the possibility of electrical discharge has been neglected in the past, and therefore needs to be considered in all spaceflight bearing friction anomalies and failures.

Ball bearings in space can have various friction anomalies, and some of them can be blamed on just the general random nature of the bearing wear evolution and lubricant condition. However, if a ball bearing in space is observed to have a gradual increase in Coulomb friction over the period of a few days, or a step change in viscous friction, electrical discharge should be considered as a possible cause. The details of the mechanism for these behaviors are described in detail in Reference (Bialke [134]). This is especially true if the behavior occurs during or shortly after a geomagnetic storm.

In terrestrial applications, electrical discharge can also be caused by induced motor voltages, and a simple measurement of the rotating shaft voltage relative to the housing can diagnose such an issue. Also, one must be aware that motor state changes, such as direction reversals, can induce a voltage spike across the bearing and result in torque bumps on the raceway.

12.5 Chapter Summary

We have summarized the key features of ball bearings which are intended to operate in the space environment. Design of this type of equipment requires much more in-depth awareness of ball bearing design, but when using equipment designed for space, it is important to have a general understanding of some important details of any device with ball bearings to ensure that the manufacturer has sufficient awareness of all variables to provide reliable and well-performing hardware.

Chapter 13
Modeling Simulation and Test Beds

Ground-based validation of spaceborne momentum control systems and the attitude control systems that depend upon them offers a unique set of challenges. Computer simulations must include a variety of nonlinear phenomena found in momentum devices, which fundamentally limits the way these simulations can be architected. And because momentum systems rely on the conservation of angular momentum for proper operation, hardware-in-the-loop test facilities must reproduce the dynamics of a free body with great precision.

13.1 Math Modeling and Computer Simulation

Chapter 4 and Appendix A discuss the equations of motion for a spacecraft with an onboard momentum system. These equations can be of great value in understanding the physics of the problem, evaluating candidate control strategies, and other steps in the spacecraft-design process. As always in the development of such equations, various simplifications keep the scope of the development tractable. These simplifications include, for example, that all bodies are rigid, that all constraints are infinitely stiff, and that all rigid-body degrees of freedom are free and frictionless.

When we reach the point in a development project where a time-domain computer simulation is justified, we must go beyond such simplifications. The intent of the effort is generally to create a tool that can evaluate the performance and/or stability of an intended design in the presence of real-world imperfections not included in these simple equations of motion. There are many aspects of momentum systems that directly impact overall vehicle performance that a simulation should address, for example:

- Friction, gear backlash, driver saturation and other nonlinear effects
- Compliance in bearings and other joints between bodies

© The Author(s), under exclusive license to Springer Nature Switzerland AG 2025

F. A. Leve et al., *Spacecraft Momentum Control Systems*, Space Technology Library 46, https://doi.org/10.1007/978-3-031-96165-6_13

- Flexibility of the elements of the MCS, and the spacecraft itself
- Ripple and cogging effects in motors and other electromagnetics
- Induced vibrations from rotor balance and bearing eccentricities
- Gyroscopic coupling between flexible modes, that can change both their frequency and mode shape

The following sections offer a proven approach to the architecture of such a non-linear simulation. The examples shown depict MATLAB/Simulink models. However, the basic principles are suitable for any general-purpose computer-modeling environment. These Simulink diagrams closely resemble general feedback-control block diagrams; so, a reader familiar with feedback controls ought to be able to interpret them regardless of experience with MATLAB.

13.1.1 First Principles

When introduced into the model, many of the effects discussed above appear at the interfaces between rigid bodies and violate some of the simplifying assumptions of Chap. 4. As a result, those equations of motion cannot be used as the cornerstone of a precise time-domain model. Fortunately, it is not necessary to go back and attempt to include these effects in the derivation. Today's computers are fast enough to solve all of the complex interactions between bodies "on the fly" if we simply organize the model to allow it. Taking advantage of this capability requires a paradigm shift in which we abandon the instinct to perform complex multi-body derivations and code the results. Instead, we should model each body separately, in a way that presumes nothing about what it will be connected to. Then we model the connections among the bodies, presuming no knowledge about what they are connecting. This is the method of "first principles" modeling. In fact, assembling such a model is generally easier and faster because it tends to be modular. And because each of the building blocks is a fairly simple model, the incidence of human error is reduced, along with debug time (or the risk of incorrect results). The method of first principles can be employed for models of any fidelity. For example, a component can be modeled as a rigid body with one degree of freedom, or three, or six, depending the on rigid-body constraints known to be present. Or it could be modeled as a flexible body with arbitrarily many of degrees of freedom. With some care in bookkeeping, models with differing fidelity can be simply dropped into a simulation depending upon the needs of the analysis.

13.1.2 What Is the Question?

One of the most common mistakes in creating a computer model of a system is doing so without a clear understanding of its purpose. A model is created to answer certain questions, and these questions must be allowed to dictate the scope and fidelity of the model. Today's computers are fast, but there are still practical limits. Here are some examples of how the question drives the model:

- A momentum system model is to be created as part of a spacecraft model to study slew times through day-in-the-life representations of targeting scenarios. This model needs to run hours of simulation time. So, in the case of current desktop computers, it is not practical for the integration step time to be a microsecond, or for bodies to have thousands of degrees of freedom. The runs would simply take too long. Phenomena producing poles or action at frequencies well beyond the bandwidth of the ACS under study should be neglected (e.g. above 20 Hz). Things that might have a direct effect on slew times, like limits on rate, acceleration, jerk, actuator current, etc. should be included. Things that bear more on inner-loop stability and fine performance can be neglected.
- A momentum system model is to be assembled to verify the stability of the gimbal rate loop design in a CMG system. In this case, a run of only a few seconds is more than enough, but higher frequencies are involved. If, for example, a discrete-time loop is closed in a gate array, the sample time of the loop might be 10 kHz and has to be modeled. In general, subtleties like motor current loops and commutation can be neglected. Interactions with flexible modes of the surrounding structure are possible, and therefore body flexibility should be included. Effects of nonlinearities like friction and backlash can impact stability. Certain saturation limits might be important, but many can be neglected since stability analyses are usually performed in the domain of "small signals." Sources of vibration are forcing functions but do not directly affect loop stability. Therefore, they can be neglected.
- A model is to be assembled to verify that a high-speed three-phase wye spin-motor drive system will function properly at the rated speed. In this case, the model must incorporate the motor-coil inductance, pulse-width modulator, commutation and current loop. We are studying sub-microsecond phenomena, but simulating only a handful of commutation cycles is enough to determine if the driver is working. At these frequencies the loops closed around the motor, and even the body flexible modes are so slow as to be negligible. So, we can focus only on the motor.

These examples illustrate that two key aspects of a simulation should be determined by the question being asked. First, the fidelity of the model: what degrees of freedom and phenomena must be included? The second aspect is the integration time of the model: what frequencies are important? A day may come when computers are so fast we can just model everything. But for now, we must use engineering judgment to consider these questions carefully.

13.1.3 Modularity

From the discussion in the previous section, one might be quick to conclude that the wide range of possible questions will result in the creation of many different models. Indeed, this is often the case. However, in today's reality of limited resources, it behooves us to be efficient and try to minimize such replication. One of the most compelling reasons for doing so is configuration control. At the same time these simulations are being assembled and used, the hardware design itself is often maturing. When changes happen, it becomes a logistical nightmare to maintain a large number of separate models in lockstep. But the penalty for not doing so is different teams on a spacecraft program all analyzing different versions of the design, likely leading to errors and delaying the completion of the work.

The solution to this dilemma is to construct modular building blocks with varying levels of fidelity, so that all of the different models on a program can be assembled from the same blocks. When a change comes, the affected blocks can be updated in a common library and will flow to the various analyses as they are performed. Modeling by first principles facilitates modularity because each "block" is modeled without knowledge of the rest of the model. All that is necessary is some initial agreement on interfaces and architecture.

13.1.4 Choices of States

Consider a simple example of two single-degree-of-freedom masses connected by parallel spring and damper elements. Intuitively, we understand that this system has four states: the rate and position of each body. The block diagram of Fig. 13.1 illustrates how such a model might be connected.

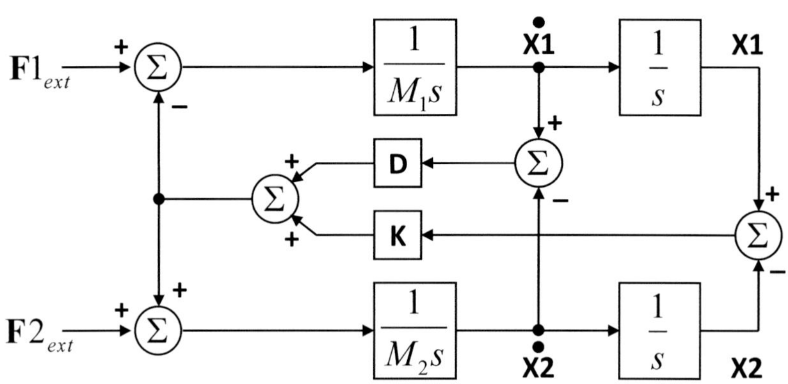

Fig. 13.1 Two-mass system

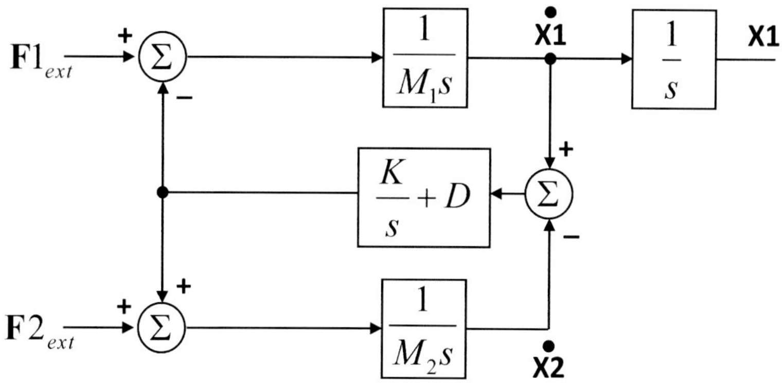

Fig. 13.2 Two-mass system feeding back rates only

This topology can be substantially simplified if, as is often the case, we do not really need to know the positions of some bodies. Consider the example of Fig. 13.2.

In this case, the topology of the interconnecting bodies is much cleaner. The state in the spring model "K/s" integrates the displacement between two bodies, rather than the position of the second body. In all other respects, the models are identical.

In multibody models that involve transformations among various coordinate systems, calculations of moment arms from points of force application, etc., the need to perform all of these calculations on both positions and rates as in Fig. 13.1 becomes a burden on computational resources. We therefore find the approach of Fig. 13.2 to be a useful means of constructing multibody models. Bodies take as inputs forces and/or torques, and they output rates. Connections, such as springs and dampers, take rates as inputs and output the resulting forces and/or torques that the connections will apply between the two bodies. Interconnecting these building blocks becomes trivial. Body positions are integrated only when required as outputs.

13.1.5 Polarities

One of the most common errors in modeling is incorrect signs on summations. By adhering to a simple set of rules, these errors can be all but eliminated.

- Positive forces or torques cause positive rate on a body.
- Positive differential rate into a passive attachment causes negative forces/torques, such as in Hooke's Law, $F = -KX$.
- All forces/torques and rates add positively.
- Minus signs are placed between an attachment model and one of the two bodies, suggesting action/reaction. The choice is arbitrary in the sense that either can represent physically correct behavior. The choice merely defines whether positive displacement of the attachment is extension or compression.

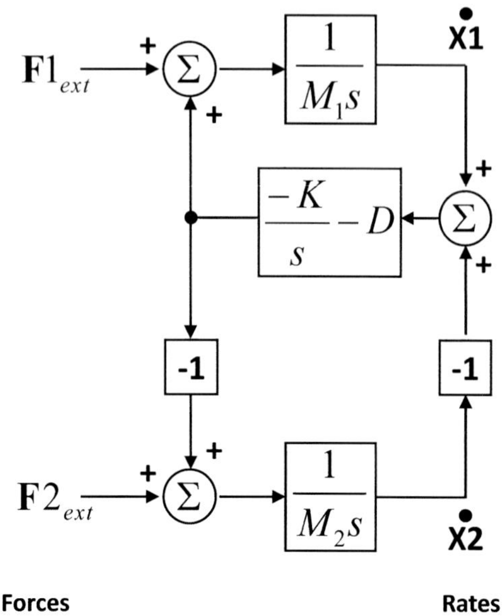

Fig. 13.3 Managing polarities

An example of this approach is shown in Fig. 13.3. By definition, all feedback loops closed in such an architecture have either one or three minus signs, indicating that they are stable.

13.1.6 Body Models

As above, the model of a single body takes as inputs forces and/or torques, and it outputs rates. The rigid body can be modeled with anywhere from one to six degrees of freedom, depending upon the needs of the model. Furthermore, the rotational degrees of freedom can be constrained to small angles, or allowed to rotate freely. The small-angle approximation allows independent state integrations on each degree of freedom according to $\tau = J\alpha$. Without this approximation, the model must utilize the complete equations that capture the rotating-frame effects, i.e. Euler's Equation and the Transport Theorem, for all six degrees of freedom. Furthermore, if position information is required, it must be done using quaternion integration or some other non-singular attitude representation.

While it is customary to model the body as located at its center of mass, in its own body-fixed coordinate system, it is convenient to express the inputs (F, τ) and outputs (rates) of the body model in the coordinate system of the attachments to other bodies or forces, at the points where they attach. This is illustrated in Fig. 13.4.

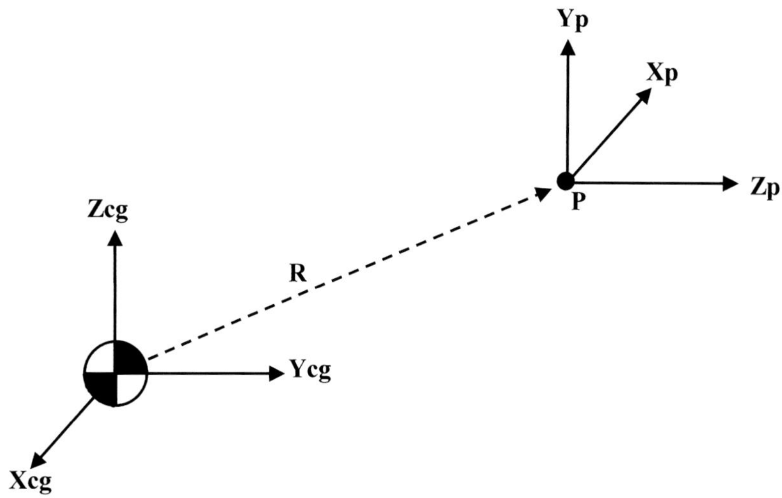

Fig. 13.4 Reconciling forces and torques applied at an attachment point "P"

Transformations that include both the change in coordinates and the moment arm from CG to attachment point can be expressed as a 6×6 matrix, as shown in Eq. (13.1).

$$\left[\frac{F}{\tau}\right]_{CG} = M \left[\frac{F}{\tau}\right]_{P}$$

where: (13.1)

$$M = \begin{bmatrix} I_{3\times3} & 0_{3\times3} \\ R^{\times} & I_{3\times3} \end{bmatrix} \begin{bmatrix} A_P^{CG} & 0_{3\times3} \\ 0_{3\times3} & A_P^{CG} \end{bmatrix}$$

where: A_P^{CG} is the 3×3 transformation from the P frame to CG frame, and $R^{\times}$ is the 3×3 skew symmetric "tilde" matrix for the column vector R.

Figure 13.5 shows an example that illustrates the suggested approach to a modular body model. The block at the center of the model represents the vacuum housing of the four conventional CMGs in an array. The representation is a state-space linear system stored in the variable "hous.s." Each housing is modeled at its center of mass as a small-angle rigid body with six degrees of freedom. Therefore, the state-space system has 24 inputs, outputs and states. This state-space model is simply the concatenation of many $1/Ms$ and $1/Js$ integrations. This body attaches to the rest of the model at four points: two spin bearings and two gimbal bearings. Each attachment is modeled with six degrees of freedom, thus a total of 96 inputs. The 24×48 input transformation matrices are assembled by block concatenation of the M matrices from Eq. (13.1) for the spin bearing and gimbal

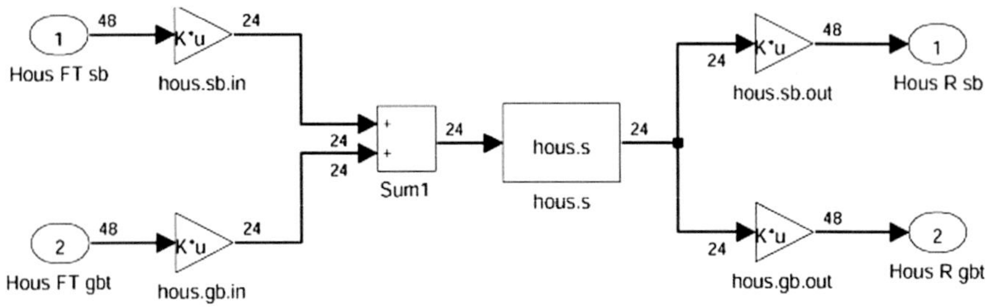

Fig. 13.5 Example modular body model

bearing attachment points, respectively. They transform the forces and torques applied at each attachment point to the body frame at the body CG, where all such inputs are summed. Each of the 48×24 output matrices is the transpose of the analogous input matrix, and they transform the body CG rates to the rates at each attachment point, in the attachment coordinate system. This simple, modular construction is easy to assemble with minimal errors.

If the body is to be modeled as flexible, the recommended approach is to split the body into rigid states and flexible states using modal superposition. The rigid states can be modeled using either small or large-angle attitude representations. The flexible model is obtained by a Craig-Bampton [135] extraction from a NASTRAN model or other flexible representation, using the boundary nodes required for the inputs and outputs. For the flexible model, the transformations are buried in the NASTRAN model. The inputs are not summed but applied directly at the individual model nodes. The outputs of the flexible model should be at the same points, in the same coordinate system as the rigid model, allowing the modal displacement rates to be added to the outputs of the rigid model.

13.1.7 Attachment Models

The model of a single attachment should take as inputs the rates of the bodies being attached, and it should output forces and/or torques. The inputs and outputs should already be in the chosen coordinates of the attachment model after conversion by the body models, allowing quantities to be directly combined without further coordinate transformation. The attachment can be modeled with anywhere from one to six degrees of freedom, depending upon the needs of the model. For example, a simple hexapod isolator model might use only a single degree of freedom for each isolator: force and displacement along the isolator axis. The key elements of this module are the representations of $K/s + D$, and the differential rates using balanced negations (reaction) as discussed earlier in Sects. 13.1.4 and 13.1.5. To that basic model, we then add any relevant nonlinear elements and other complications.

Figure 13.6 models the eight spin bearings (two each) in the four CMGs of an array. They are modeled in six degrees of freedom, producing the 48 signals shown. Note that the output forces and torques from the state-space representation of $K/s + D$ (sb.s) act on the rotor and react on the vacuum housing. In the state-space model sb.s, the values for K and D are zero for the bearing spin axes. Instead, the four spin torques are calculated at the bottom of the model and added through the "$toRz$" matrix. The spin axis torques include the spin motor itself, models of Coulomb and viscous drag, and a model of how the drag varies when the bearings are subjected to load (such as during CMG torquing). For the spin bearings of a CMG, the Coulomb drag is essentially a constant. However for RWA spin bearings (or CMG gimbal bearings) a better model is required. The Dahl friction model is recommended in these cases (Dahl [136]). Finally, we see a model of spin-bearing disturbances due to bearing eccentricities. This model takes spin speed as input and produces a comb of sinusoids at various frequencies representing the bearing tones. Note that bearing eccentricities are properly modeled as forced displacements across the interface, because the bearings are so much stiffer than the surrounding structure that they might as well be rigid constraints. Modeled in this way, the actual forces and torques that arise from the eccentricity depend on the surrounding stiffness, rather than just the bearings themselves.

13.1.8 Integration

Using building blocks such as those illustrated above, very high-fidelity representations of MCS-based attitude control systems can be produced using the architecture described herein. Such a model can be easily reduced or enhanced when necessary by changing the fidelity of the blocks. In keeping with the modular philosophy, the only place where any gyroscopic terms are found is in the model of the rotor body. All resulting gyroscopic effects, such as CMG output torque amplification, gimbal back torque due to precession when the vehicle is in motion, augmented gimbal inertia and coupling of flex modes simply arise as the simulation solves the interactions among the modular bodies. The rotor can be modeled simply as a full, large-angle body that spins, but integrating the resulting motion would use substantial computer resources without a meaningful increase in fidelity. In general, it is advisable to use the simplifying assumption that the rotor spin speed is much greater than the angular velocity in other axes, allowing a simple gyroscopic coupling matrix to be derived and combined with a simpler body model. In this matrix, the rotor's angular momentum is taken to be the spin axis inertia times the rotor's spin speed. Numerical integration of the rotor's spin axis angular state need not be performed.

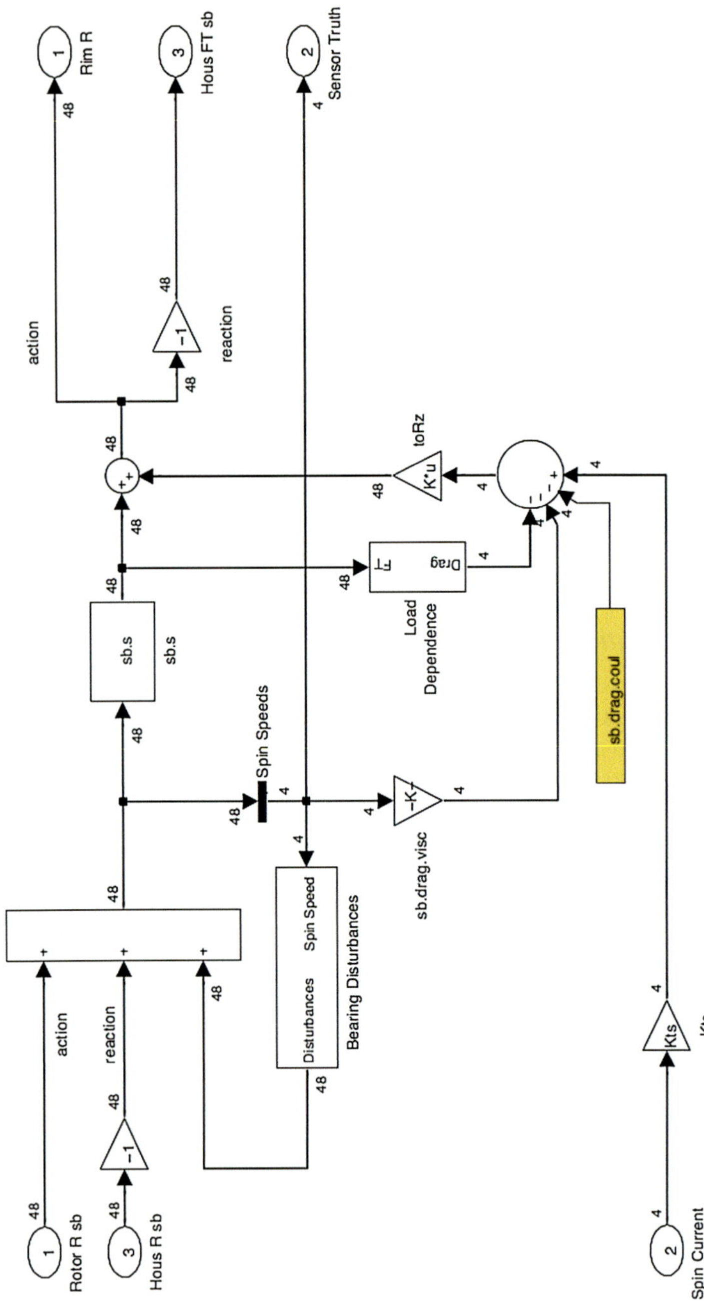

Fig. 13.6 Example modular attachment model

13.2 Hardware-in-the-Loop Test Beds

While modern computer simulations offer more risk reduction than ever before, there is no substitute for working with hardware. However, performing ground testing of MCS-based attitude control systems is particularly challenging. Momentum systems rely on the conservation of angular momentum within a closed system to exchange momentum with the spacecraft and affect control torques. As a result, hardware-in-the-loop test facilities must provide a surrogate spacecraft that approaches a truly closed system, i.e. no external torques.

The quality with which this objective is achieved is the single overriding characteristic that defines the success of a test facility. Various technologies exist to simulate zero gravity, including such options as buoyant support or complex offloading mechanisms. But the most popular is the air bearing. This technology, familiar to anyone who has played air hockey, is simple, inexpensive, and can provide a nearly perfect zero-friction suspension. Sputnik launched in 1957. Explorer-1 launched in 1958. The earliest air-bearing spacecraft simulator is documented in 1960. Truly these systems have played an integral role in verifying space technology since the beginnings of space exploration (Schwartz et al. [137]).

Planar air bearings provide an ideal test bed for simulating two-vehicle dynamics, with two translational degrees of freedom and one rotational. Control techniques for relative orbital maneuvers and formation flying, rendezvous and docking can be fully developed and tested prior to launch. An example of a planar test bed is the Two-Link Manipulator Arm in Stanford University's Aerospace Robotics Laboratory, shown in Fig. 13.7.

The Air Force Research Laboratory (AFRL) has been using large spherical air-bearing spacecraft simulators since the late 1980s. One of the first of these simulators was known as Advanced Structural Technology Research Experiment (ASTREX). ASTREX, shown in Fig. 13.8, housed four large Honeywell CMGs and had a spherical air bearing capable of supporting approximately 15,000 lbs. The AFRL used it for studies in structural control and damping for spacecraft beam-director pointing.

Fig. 13.7 Stanford's two-link manipulator arm (image courtesy of NASA)

Fig. 13.8 AFRL ASTREX spacecraft simulator (Image courtesy of AFRL)

Since its use on ASTREX, this air bearing has been used on multiple project platforms such as the Flywheel Attitude Control and Energy Transmission System (FACETS) and mini-Agile Multi-Purpose Satellite Simulator (mini-AMPSS) (See Chap. 6). The air bearing is currently being used to test state-of-the-art spacecraft fault detection and isolation algorithms as well as fault-tolerant control techniques at the AFRL Space Vehicles Directorate at Kirtland AFB in New Mexico. Currently, the AFRL Space Vehicle Directorate no longer houses ASTREX, but in its place, the new and improved ReBEL testbed featured in Fig. 13.9. Honeywell has also operated a precision spherical air bearing testbed at its Glendale, AZ facility (Fig. 13.12) for two decades.

Some test beds use rotational air bearings atop a planar carrier to achieve six degrees of freedom. An example of such a facility is Marshall Space Flight Center's "Flight Robotics Laboratory," shown in Fig. 13.10. The lab has a 44 ft × 86 ft precision flat floor. The Air Bearing Spacecraft Simulator used on the planar floor provides a 400 lb payload six degree-of-freedom motion via a floating spherical air bearing coupled with a cylindrical lift. While such a sophisticated facility can enable

Fig. 13.9 AFRL ReBEL spacecraft simulator (Image courtesy of AFRL)

Fig. 13.10 MSFC's flight robotics laboratory (Image courtesy of NASA)

certain specific experiments, six degrees of freedom are not necessary for complete studies of an MCS-based attitude control system.

13.2.1 Precision Rotational Air Bearing Systems

Rigid-body studies of MCS-based attitude control systems require only the three rotational degrees of freedom. Flexible behaviors can be incorporated in such test beds if the moving components conform to the rules of gyrostats discussed in Chap. 4. In any case, external torques must be limited to the greatest extent possible. The scope of this problem is well understood. An important paper on the subject was delivered by G. Allen Smith at the Role of Simulation in Space Technology conference held at Virginia Tech in 1964. Smith presented a description of several systems, along with an overview of the torques which act on the rotor of an air bearing. Smith defined four classes of disturbance torques and listed particular sources for each group, as follows:

I. Torques Arising from Platform
   - Static Unbalance
   - Dynamic Unbalance
   - Anisoelasticity
   - Material Instability (stress, temperature, humidity, evaporation)
   - Gravity Gradient
   - Equipment Motion (solenoids, relays)

II. Torques from Bearing
   - Aerodynamic Turbine Effect
   - Exhaust Air Impingement

III. Torques from Environment
   - Air Damping
   - Air Currents
   - Magnetic Fields
   - Vibration
   - Radiation Pressure

IV. Torques from Test System
   - Electrical Wire to Base
   - Mass Shift in Bearings and Loose Fits
   - Battery Discharge
   - Reaction Jet Supply Discharge
   - Replacement of Components

Most torques from groups I and IV can be mitigated through test bed design: well-designed structures outfitted with well-chosen components. Group II effects received more attention in the early development of air-bearing systems than they do now; although internal bearing effects may be important in the design and operation of industrial gas bearings, they impart a negligible effect upon most systems we are considering. Several facilities have developed large-scale means to mitigate environmental torques (group III). Thermal and air currents often cause a significant effect and are simplest to eliminate: several NASA facilities are installed within vacuum chambers. The facility designed for the Boeing Company's Lunar Orbiter Attitude Control Simulator could not make use of this solution, as it was piloted. Instead, the room design included full air circulation and thermal control. Further, the system was mounted on a 90,000 lb concrete slab supported by seven air springs; thus the system was effectively isolated from seismic effects. Marshall Space Flight Center installed one of their systems within a set of Helmholtz coils in order to cancel the effect of the terrestrial magnetic field on the payload.

13.2.2 Anisoelasticity

Perhaps the most difficult of the sources listed above is anisoelasticity. This effect, while relatively unknown, directly influences balance stability as the test article rotates. To understand it, ones needs only recognize that the test article cannot be infinitely stiff. As a result, there is some sag between the support point and the test article's extremities due to gravity. How much sag exists depends on the distribution of mass and stiffness in the direction of the gravity load. If the stiffness of the structure varies with the direction of loading, as most do, it is said to be "anisoelastic." The center of mass of such structures varies with attitude, and the changing mass center introduces a secular pendulum-like disturbance torque.

Consider the simple structure of Fig. 13.11. At left, the large dumbbell is sagging. So, an additional mass is added above the pivot (red) to raise the CG and balance the system. At right, after rotating clockwise 90°, the large dumbbell, which is much stiffer in this direction, is no longer bending. The balance mass that compensated for the sag in the initial orientation now makes the system asymmetric (and sags itself), resulting in both horizontal and vertical CG shifts.

13.2.3 Active Mass Balancing

The effects of anisoelasticity prevent a physically realizable structure in gravity from remaining passively, perfectly balanced at all attitudes. The solution is an active system that can move mass to make CG adjustments under computer control. One facility which makes use of this technique is Honeywell's Momentum Control System test bed, shown in Fig. 13.12 (Hamilton [138]).

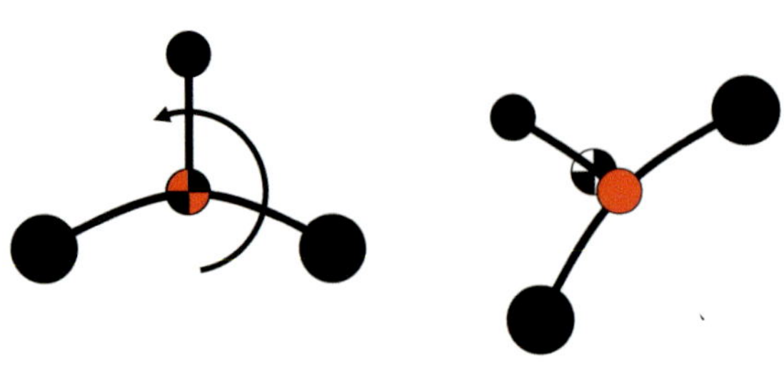

Fig. 13.11 Simple anisoelastic structure

Fig. 13.12 Honeywell's Momentum Control System test bed

This facility features three servo-driven ballscrew linear stages that move small blocks of aluminum along rails. This arrangement allows the test bed's mass center to be adjusted in three degrees of freedom. The paths of motion of the masses are directed at the center of rotation of the spherical air bearing so that the acceleration of the masses produces no torque. These masses eliminate torque due to mass-center offset. They actively place the mass center of this 1400 kg test bed to within a few microns of the air bearing's center of rotation.

13.2.4 Controlling the Mass-Balancing System

When contemplating the design of a control system using these actuators, the first and foremost requirement must be to prevent the mass-balancing system (MBS) from interacting with the ACS under test. The actions of the MBS must be transparent so as not to interfere with or corrupt the experiment being performed on the test bed. One key step towards this goal is the choice of feedback for the controls.

We recall that a key goal of the test bed is to behave as a closed system for the conservation of angular momentum. This is necessary for proper operation and testing of MCS-based systems. It follows, therefore, that the purpose of the MBS is to make the net angular momentum of the test bed system remain constant in the inertial frame (implying a lack of external torque). The net angular momentum of the test bed vehicle consists of the sum of the momentum vectors of the vehicle itself, and all of the CMGs mounted thereon. We can calculate this vector from the following information, all of which is readily available:

- Vehicle rate (from the rate gyros onboard)
- Vehicle inertia
- CMG wheel speeds (from the CMG tachometer telemetry)
- CMG wheel inertias
- CMG gimbal angles (from the CMG resolver telemetry)

It is important to note the significance of this principle. It means we can develop a quantity to be used as feedback for the MBS system that is entirely based on the state of the hardware. No information from the ACS is required. The ACS need not even be closed and operating. The feedback quantity is always valid, whether the system is quiescent or in the middle of a slew. Under ideal conditions, actions of the ACS do not change the quantity, because the change of the CMG array momentum will be equal and opposite the change of the vehicle momentum. This again underscores the independence of the MBS and ACS controls.

Figure 13.13 illustrates the architecture of the MBS controls. It shows the development of the rate of change of net vehicle momentum **H**, which the central path tries to drive to zero. Having calculated a desired rate of change for the external torque, the algorithm uses a cross-product with the gravity vector to determine the desired rate of change of the body CG. This command then goes to the MBS actuators. We wish to keep the bandwidth of this loop closure low (tenths of Hz). As such, it does a fairly poor job at rejecting rapidly changing external torques, such as may arise during an agile slew. Turning to a common controls solution, we have added a "feedforward" path at the top of Fig. 13.13. This path anticipates the needed correction as soon as the body begins to move and can apply the correction with the bandwidth of the MBS actuators.

Finally, an additional feature is evident at the bottom of Fig. 13.13. When working with momentum systems in the lab, just as in space, it is useful to have a way to "dump" any momentum accumulated over time. The MBS controls described

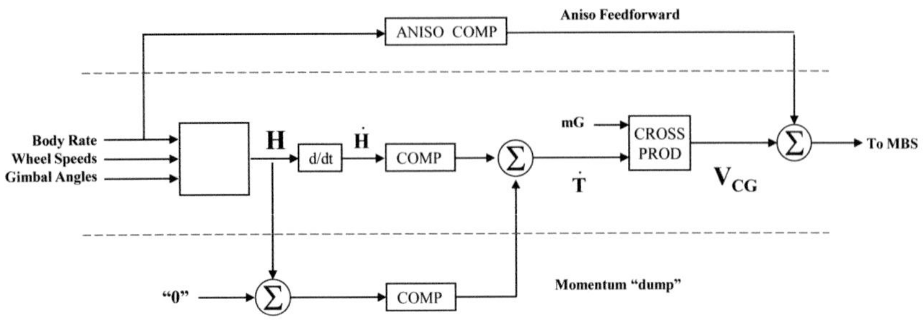

Fig. 13.13 Control of a mass-balance system

thus far have only the objective of preventing the net momentum from changing. This additional loop closure acts to drive the net momentum to zero, thus providing the dumping function. This functionality should be provided at a bandwidth much lower than that of the main balance loop for stability, and it can easily be removed by opening the loop.

For clarity and simplicity in this description, only the fundamental philosophy of an MBS control system has been provided. Someone working in the field will note, for example, that we make no distinction between the moving body frame (on which the MBS rides) and the lab frame (where the vector mG is constant). Furthermore, effects due to the rotation of the Earth are neglected here, an omission that is justified because of the significant difference in frequency content involved. As such, we take momentum to be conserved in the lab frame.

13.3 Chapter Summary

This chapter has discussed methods used in testing and validation of MCS-based attitude control systems, both in computer simulations and hardware-in-the-loop tests.

A proven architecture for computer modeling of these nonlinear devices has been offered. The architecture employs first-principles methods, and a disciplined approach to the choice of states, placement of negation, and organization of the model. The approach lends itself to the inclusion of nonlinear elements where necessary throughout the model, and avoids the violation of any assumptions that might have been made had pre-computed multibody equations of motion been used.

The preferred method for hardware-in-the-loop testing of MCS-based systems was shown to be the three degree-of-freedom spherical air-bearing suspension. The most difficult challenge in creating such a test facility is the effect of structural anisoelasticity on mass balance in gravity. A method for overcoming this problem using an active system was described.

Chapter 14
Flight Experiment: Honeywell's MMCS on the NG-11 Cygnus Spacecraft

14.1 Background

Control Moment Gyroscopes have been the actuator technology of choice for agile satellites for over 40 years. It would be difficult to imagine a question at this point for which flight data does not exist to support an answer. Unfortunately, as discussed in Chap. 1, all of the early work in this arena supported government programs from which access to flight data is restricted. Industry and academia have only recently fielded flight programs from which on-orbit data could be drawn. The authors, in our desire to provide sample data from an actual flight program, have chosen a specific experimental mission on which some of us were directly involved.

14.1.1 Hardware Development

The development of the mini-Momentum Control System (MMCS) was undertaken in 2008 with support from both Honeywell and the US Air Force Research Lab (AFRL). The objective was to explore next-generation CMG technology in the context of application to small spacecraft. For a small momentum system to be feasible, it was necessary to find ways of dramatically reducing size, weight and power. The resulting design was a departure from Honeywell's heritage CMG products in many regards, some of which are listed below:

- Increased rotor spin speed, allowing reduction in the rotor size
- Housed an array of CMG assemblies in a single vacuum enclosure
- Housed drive electronics for the entire array in a single enclosure
- Increased gear ratio to reduce gimbal motor size and improve performance
- Selected a harmonic drive over conventional gears to reduce size
- Eliminated the electromagnetic resolver and tach in favor of an optical encoder

© The Author(s), under exclusive license to Springer Nature Switzerland AG 2025

F. A. Leve et al., *Spacecraft Momentum Control Systems*, Space Technology Library 46, https://doi.org/10.1007/978-3-031-96165-6_14

- Integral isolation of the CMG array (reduces induced vibration, bearing launch loads)
- Integral steering laws allow turnkey array-level functionality
- Single FPGA hosts low-level hardware control in fabric, and higher-level software in a soft-core processor on the same chip.

A prototype unit was manufactured to demonstrate the approach. In the design and manufacture of the prototype, materials and components selected were all available in flight-qualified and rad hard versions. While this approach demonstrated a clear path to flight hardware, many of the actual components in the experimental prototype were of commercial quality to save cost. No redundancy was designed into the prototype. The possibility of flying the unit was always considered, but it was assumed it would be a short experimental mission.

14.1.2 The Guest Investigator Program

The guest investigator (GI) program was work performed by Honeywell for the Air Force Research Lab (AFRL) under an extension to the Mini-Momentum Control System (MMCS) Program. In this effort, Honeywell acted as the AFRLs agent to perform a series of evaluations of various steering law technologies provided by independent guest investigators from government, industry and academia. These evaluations were conducted using the MMCS experimental protoflight unit mounted to Honeywell's MCS Testbed facility in Glendale, AZ (see Sect. 13.2), as shown in Fig. 14.1.

Since both Honeywell and the Air Force had submitted steering technologies as "guests" for the study, great care was taken to standardize a set of profiles and scoring metrics used in the tests to make sure that the assessment of all guest technologies was performed on a level playing field.

A total of four "scenarios" were selected for use in the investigation. The scenarios explored several key facets of CMG array steering technology. The first, called the "simple slews", examined the strategies employed by the steering laws for limiting torque and momentum. The second, called the "mission-relevant" scenario, scored how well the array supported acquisition and tracking of a series of ground targets from low earth orbit. The final facet was how well the controls dealt with the presence of singularities. "Academic" scenarios forced the array to approach and pass through singularities of different type and class to evaluate resulting stability, error torques, etc.

AFRL's plan from the beginning was to repeat these same evaluations on orbit. Unfortunately, with shifting priorities in the Air Force mission, finding a "ride" for the MMCS experiment proved difficult. Years passed with the hardware sitting on a shelf.

Fig. 14.1 MMCS on Honeywell's MCS testbed facility

14.1.3 The Cygnus Mission

In 2017, Honeywell was contracted by (then) Orbital ATK to retrofit and test the unit in preparation for a flight experiment on the Cygnus spacecraft. Orbital had reached agreement with the Air Force, who retained ownership of the prototype, to allow them to use it in exchange for a report on the resulting orbital experiments.

For reasons related to funding, none of the original "guest" steering laws (including AFRL's) were included in the flight experiment. Only Honeywell's steering was demonstrated on-orbit.

After successful retrofit and environmental testing, the unit was delivered to Northrup Grumman (NG) and installed on the NG-11 vehicle. Launched in May, 2019, the NG-11 docked with the ISS (Fig. 14.2) and performed its primary resupply mission. On August 6, it departed the ISS, and began a series of tests using the MMCS.

14.2 Differences Between Lab and Flight Experiments

AFRLs original plan was to repeat all the same experiments on-orbit that were done on the ground and show similar results and scores. This lofty goal proved impossible for a variety of reasons. In lieu of a direct comparison, a high-fidelity MATLAB

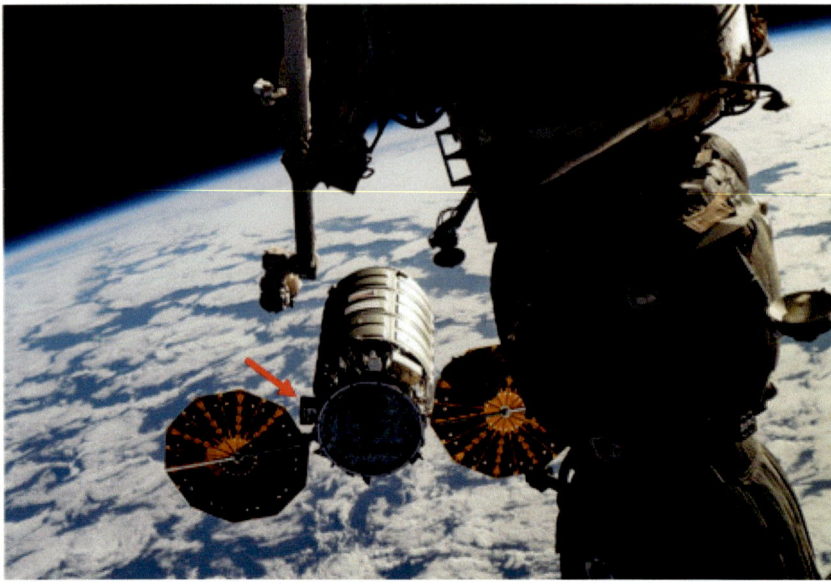

Fig. 14.2 NG-11 approaching ISS with MMCS highlighted (Image courtesy of NASA)

nonlinear time-domain simulation was shown to compare favorably to both the ground and flight results, forming a bridge between the two that compensated for the differences. The nonlinear simulation was developed using the first-principles methodology described in Sect. 13.1.

This section discusses the differences that prevented direct comparison of flight and lab data. Some of these differences were simply due to the use of the Cygnus vehicle to perform the experiment. Others resulted from some bad luck on orbit.

14.2.1 *Attributable to the Cygnus Vehicle*

14.2.1.1 Scaling

The most immediate and profound difference is the matter of scaling. The Cygnus vehicle is big. Its largest principle inertia is 23,294 kg-m, compared with 2436 kg-m for the testbed facility in the lab—almost 10 times the size. The TacSat-2 satellite that was used as the sizing basis for MMCS had an inertia of 190 kg-m, which is another factor of 10 smaller than the testbed. The fact that the Cygnus vehicle was more than 100 times larger than MMCS was designed for eliminated any possibility of performing "agile" maneuvers.

The approach taken in performing the GI scenarios on Cygnus was to scale the commanded vehicle motions such that the same momentum and torque profiles seen in the lab would result. In this way, the MMCS hardware was exercised in the same way it was in the lab. Typical maximum vehicle rates on Cygnus during the scenarios varied from about 0.03 to 0.2°/s.

14.2.1.2 Feedback

With such low maximum rates, an attempt was made to locate a high-quality rate sensor that could be attached to MMCS and used by its internal ACS. When no solution was found within the program's budget, plans were made to have NG send us quaternion and rate information from the existing Cygnus sensors, which they could provide at a maximum update rate of 2 Hz. Initial concerns regarding sensor noise were unfounded—flight data shows less than 0.001°/s rms noise in the Cygnus rate data, which was sufficient. The update rate of 2 Hz, however, did require that we significantly reduce the bandwidth of our ACS. This is a fundamental difference from the experiments run in the lab where our sample rate was 50 Hz, and resulted in slower settling and larger errors during maneuvers. In order to compare flight data to lab data, we used the MATLAB time-domain simulation to predict expected flight performance. Since this same simulation was also validated against lab performance, this provided an acceptable bridge.

14.2.1.3 Momentum Accumulation

The MCS Testbed facility utilized an active mass balance system to minimize the accumulation of momentum due to balance errors in gravity. It successfully limited the accumulation of momentum to no more than 3 Nms throughout the ground testing. The Cygnus vehicle normally experiences much larger accumulations, due in part to the effect of gravity gradient. Engineers at NG determined an attitude that minimized this accumulation, and the experiments were all run at this attitude. During the longest scenario performed (about 7.5 minutes), the accumulation was only about 6 Nms. This corresponds to an external torque of about 0.013 Nm. Since the MMCS was designed to guarantee 15 Nms in any direction, 6 Nms represents a significant fraction of the available momentum, and thus this had a measurable effect on the unit's response. Nevertheless, the scenarios were able to be performed, and the simulation was used to predict the effect that the momentum accumulation had on the results.

14.2.2 Attributable to In-Flight Difficulties

14.2.2.1 Hardware Failure

One of the four CMGs in the MMCS experienced a spin system failure before the planned GI scenarios could be executed. The problem was eventually traced to a failure of one of the hall devices used for spin motor commutation. This was the unfortunate result of the use of commercial-grade parts and lack of redundancy in the prototype unit. The MMCS still functioned properly with 3 CMGs, however it was no longer feasible to compare the flight data to the ground data directly, even if the ACS bandwidth had not been reduced.

Unfortunately, this dealt a serious blow to the science value of the experiments. With only 3 CMGs, it was not possible to produce the same momentum and torque profiles as the GI experiments. Furthermore, the goal of the experiments was to assess steering law performance, some notably in regions where singularities were present. Running with 3 CMGs eliminates the null space degree of freedom, the management of which is one of the key differences between steering laws in these regions. Once again, the MATLAB simulation provided the bridge. Flight data was compared to simulation predictions of how the unit should behave with 3 CMGs operating. While less compelling than a 4-CMG demonstration, showing good agreement between the flight data and simulation tied the results to the lab data, which matched the same simulation when demonstrating the scenarios in the correct 4-CMG configuration.

14.2.2.2 Latency

It was determined that the rate feedback from the Cygnus rate sensors had a single 500 ms frame of delay in reaching the MMCS ACS. This could have been happening in the Cygnus software that sends the data to MMCS, or it could have been in the MMCS software that was added for Cygnus. No attempt was made to locate the source of the delay, since nothing could be done about it anyway. The presence of this delay served to compound the phase loss already present from the 2 Hz sampling, and further impact ACS performance. When the latency was added to the MATLAB simulation, the loop response matched flight data almost exactly.

14.3 CMG Array Geometry

Before we present numerical examples with actual flight data, we must first describe the orientation of the CMGs on the Cygnus spacecraft. We shall work in the Cygnus body-fixed frame of reference that was chosen for communications between Cygnus and the MMCS. The illustration in Fig. 14.3 shows the CMG array along with the chosen body frame (blue).

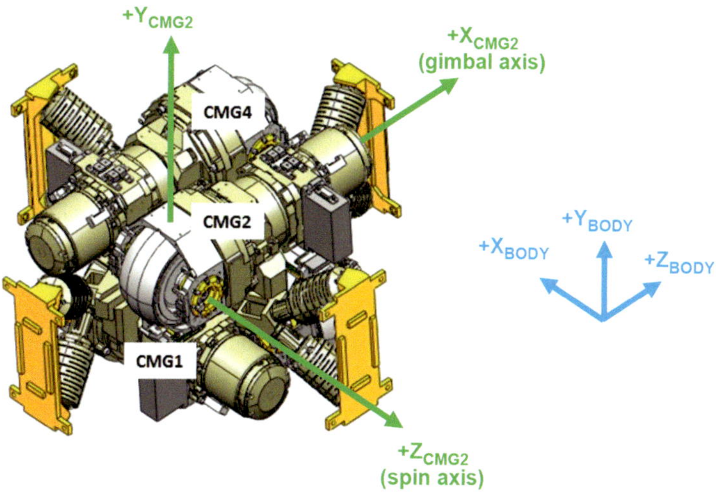

Fig. 14.3 MMCS CMG array and body frame

14.3.1 CMG Orientations and Momentum

As an example, the local frame of CMG2 is shown in green. The CMG X axis is the gimbal axis, and the CMG Z axis is the spin axis, shown at a gimbal angle of zero degrees. We can see, therefore, that for CMG2, its momentum vector at gimbal angle zero is in the −X body direction. If the gimbal is rotated 90° (positively about the CMG X axis), the CMG momentum will now be directed along the body −Y axis. Below, we construct the unit vectors in the spin direction for each of the 4 CMGs at 0 and 90° gimbal angles:

$$GA0 = \begin{bmatrix} 0 & -1 & 0 & 1 \\ 0 & 0 & 0 & 0 \\ -1 & 0 & 1 & 0 \end{bmatrix}$$

$$GA90 = \begin{bmatrix} 0 & 0 & 0 & 0 \\ -1 & -1 & -1 & -1 \\ 0 & 0 & 0 & 0 \end{bmatrix}$$

With all 4 CMGs running, we can say that for a column vector of four gimbal angles Δ, the total momentum vector of the CMG array in the body frame would be:

$$10.86(GA0\cos(\Delta) + GA90\sin(\Delta))\,\text{Nms} \tag{14.1}$$

14.3.2 Initial Gimbal Angles

We now consider how to use this array of CMGs. The illustration above is shown with all gimbal angles at 0°. If we evaluate Eq. (14.1) with $\Delta = [0\,0\,0\,0]^T$, we find the array momentum is zero. However, as discussed in Sect. 5.5, this set of gimbal angles is singular and must be avoided. Instead, we choose to bias the CMGs against each other along the shared axis Y as discussed in Sect. 9.4.3.4. This is a non-singular combination of gimbal angles that also produces net zero momentum, as shown below.

These gimbal angles are $\Delta = [-30\ 30\ -30\ 30]^T$ degrees. We note that the choice of 30° is arbitrary, and changes only the amount of bias in the 2 discs (this is a change in null space). In fact, the flight hardware used 31.7° for the 4-CMG configuration.

From these initial gimbal angles, it is easy to see how each pair of CMGs in this roof array can sweep out a disc of radius $2h$ as shown in Fig. 14.4. While all of this momentum is "reachable", each disc has a singularity at the origin to be managed.

In the case of the 3-CMG array (after CMG 4 fails), the locus of momenta in the XY disc changes to a $1h$ circle (blue, as shown in Fig. 14.5) and none of the points inside. With only 3 CMGs there is no null space, so only a handful of initial gimbal angle sets are available that produce net zero momentum. The only possible solutions for CMG 2 are either +90° or −90°, which places its momentum on the shared axis. The only solutions for the other two are combinations of $\pm 30/250°$ that oppose CMG2. The set shown below is $\Delta = [-30\ 90\ -30\ *]^T$.

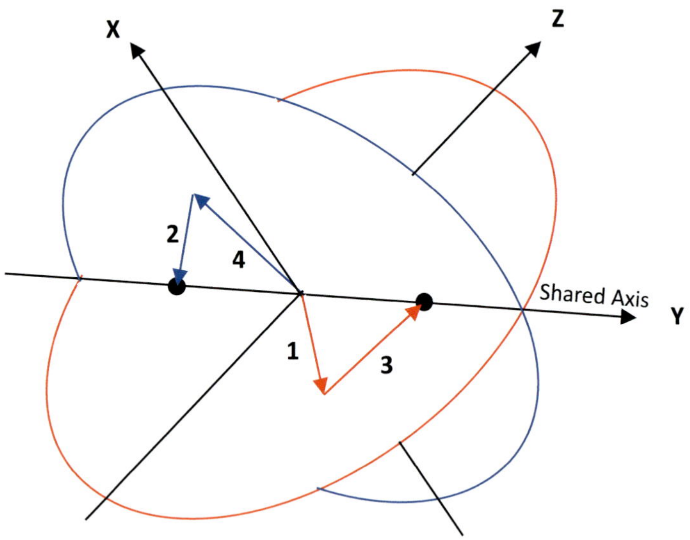

Fig. 14.4 Biased momentum vectors in a 4-CMG array

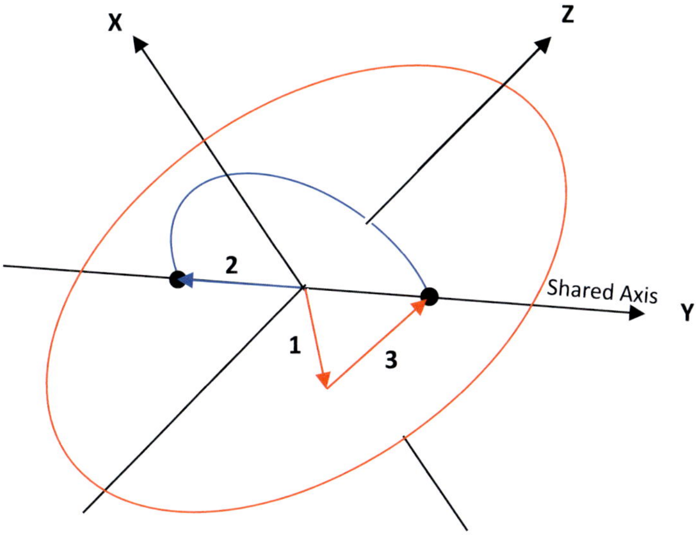

Fig. 14.5 Initial momentum vectors in a 3-CMG array

From this initial set, any required momentum in the X direction must be produced by CMG2, thus setting its gimbal angle. The remaining 2 gimbal angles are solved for the necessary Y and Z momentum. Not all of the possible momentum in this array can be used without error. We note, for example, that if we wanted to reach +3h in the Y direction, CMG2 would have to flip from +90° to −90° at some point, and this would produce a transient error in the X direction. We can see an example of this in Sect. 14.4.3.

14.4 Numerical Examples with Flight Data

This section presents a few of the results from the flight experiment.

14.4.1 Initial ACS Loop Closure

During initial spinup and checkout of the CMGs, the MMCS was commanded to produce a net zero momentum so as not to affect the default ACS. This was an MMCS command mode which adjusted gimbal angles in real time to compensate for the wheels spinning up at different rates. Then, at about 3400 s elapsed in the plots in Fig. 14.6, the Cygnus ACS mode was set so torque commands were sent to the MMCS instead of using the default (thruster-based) system.

Fig. 14.6 MMCS array momentum state and ACS attitude error

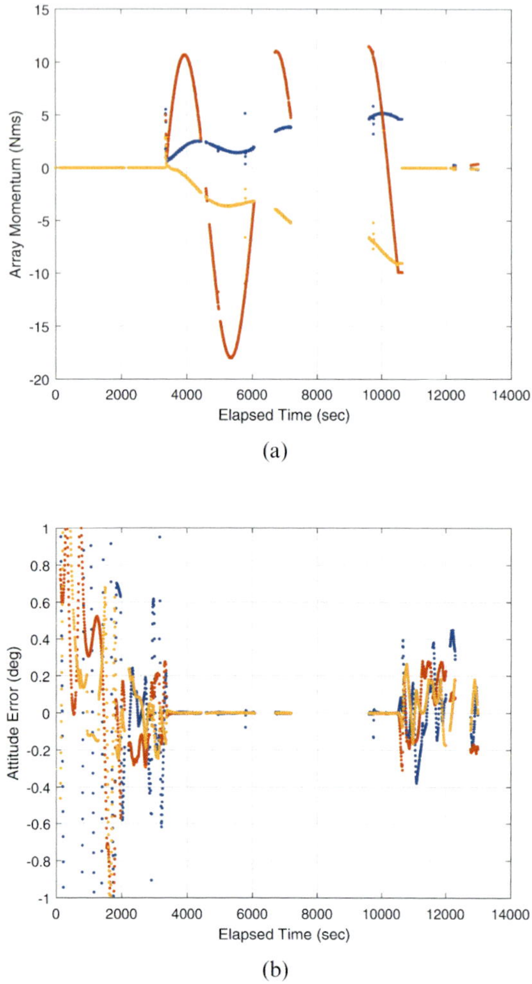

The plots in Fig. 14.6 show the MMCS array momentum state (plot a), and the ACS attitude error (plot b). As we can see, prior to the mode change, the MMCS array momentum was zero and the thrusters were being used to hold attitude. This results in limit- cycling within the allowed attitude error boundaries. When the MMCS mode was engaged instead of the thrusters, the ACS error drops to nearly zero. We then see the external vehicle torques accumulating as momentum in the MMCS array. This is predominantly a sinusoid about the Y axis at two cycles per orbit. This is what the NG engineers expected at this attitude. Of particular interest is the lack of thruster fuel usage during this time, with MMCS absorbing the cyclic component of the stored momentum. The MMCS mode was used for about 2 hours, then control was returned to the thrusters and a momentum command of $[0\ 0\ 0]^T$

was given to the MMCS. The gaps in the data Fig. 14.6 are periods of loss of signal (LOS) from the spacecraft.

14.4.2 Unplanned Steering Law Test

As luck would have it, the hardware failure mentioned in Sect. 14.2.2.1 happened during the test described above. About 5 minutes after the MMCS mode of the ACS was engaged, CMG4's spin system failed, and the wheel began to spin down. Over the next 2 hours, the wheel spun all the way to zero rpm, which effectively changed the MMCS from a 4-CMG array to a 3-CMG array.

This gave us a unique opportunity to observe the steering execute an orderly transition from 4-CMG control to 3-CMG control. This transparent failure management was one of the underlying goals of Honeywell's steering design, and it was shown to function exactly as intended. Throughout the entire spindown, the MMCS remained responsive to the torque commands from the NG hybrid ACS. Examining the plots in Fig. 14.6, one would not immediately notice there had been a CMG fault.

The plot in Fig. 14.7 shows the CMG gimbal angles during this time. We note that at the start of the run, when $H = 0$ with 4 CMGs, the gimbal angles for this roof array were all $\pm 31.67°$ (biased discs). At the end, when $H = 0$ with 3 CMGs, the gimbal angles were $[-30\ 90\ -30\ 76.9]^T$. These are exactly the gimbal angles discussed in Sect. 14.3.2. The steering law abandoned CMG4 at 76.9°. This is simply where it was located when the threshold speed was crossed, while following NG's commands. Its gimbal angle is, of course, no longer relevant since it has spun down.

Again, the missing data on the plot is due to periods of LOS.

The abrupt changes in gimbal angles just before 11,000 s elapsed is the point at which NG returned to zero momentum commands from non-zero stored momentum. Other abrupt changes were the result of steering law decisions as CMG4 spun down.

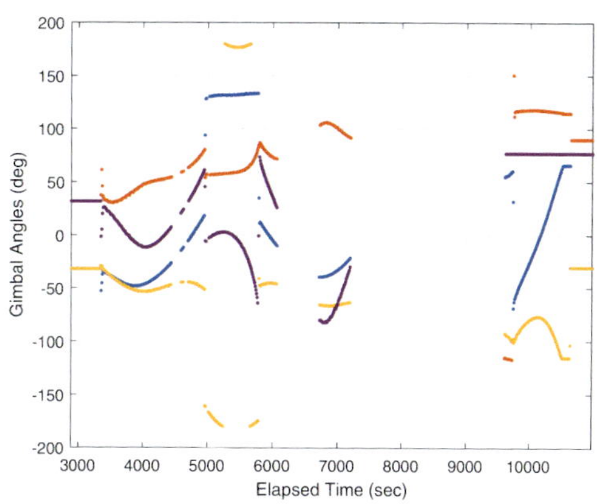

Fig. 14.7 CMG gimbal angles

A closer inspection of the attitude error plot in Fig. 14.6b reveals some brief transient error at the times of these abrupt gimbal angle changes, mostly about the X axis. We understand from the discussion in Sect. 14.3.2 that these were tied to "flips" of CMG2 as the steering law worked to produce the necessary momentum while CMG4 capability was being lost.

14.4.3 GI Scenarios: Academic Slew

As discussed previously, the GI scenarios were designed to test steering laws in a 4-CMG array. Because the failure occurred before any of the GI scenarios were run, this goal was not met and many of the results are not presented herein. One case is provided which illustrates the CMG2 flip discussed above very clearly.

AFRL had designed two "Academic" slews, the purpose of which was to explore the behavior of guest steering laws in the vicinity of singularities. The first, academic slew 1, explored the internal singularities at $2h$. The goal was to start at the origin of momentum space, and increase momentum along the y-axis until reaching approximately 3h. Note that the y axis is the "shared" axis of the roof, in which momentum of $\pm 4h$ would be available with 4 CMGs. This forces passage through the $2h$ singularity. We then decrease momentum back through $2h$, through 0, through $-2h$, arriving at about $-3h$ (passing through another singularity). Finally, we return to zero.

Unfortunately, the objective of this scenario cannot be met with only 3 CMGs operating. The 4th CMG provides the null space degree of freedom that allows steering laws to manage these singularities. With only 3 CMGs, this maneuver is impossible without incurring torque errors. The best the 3-CMG steering law can achieve is to remain robustly in control during the passage, tolerating the resulting error transient.

The plot in Fig. 14.8 shows the recorded array momentum during the Academic slew described above. There was no attitude control at this time, but simply a momentum profile commanded to the array as described above. We do see the y-axis momentum profile intended, which moves between +30 and −30 Nms. Unfortunately, with only 3 CMGs, the singleton (CMG2) must "flip" 180° in order to make this possible in the positive direction. Doing so produces unavoidable errors in the singleton plane. In the plot below, the dashed lines are the momentum profile commanded. If there had been 4 CMGs, we would have gotten a very clear picture of the singularity avoidance capability of the steering, which was the intent of this scenario.

We note on the plot that the error transient takes about 3 s. Since CMG2 must rotate 180° (π radians), and the maximum gimbal rate for MMCS was 1 rad/s, we would expect this to take 3.14 s.

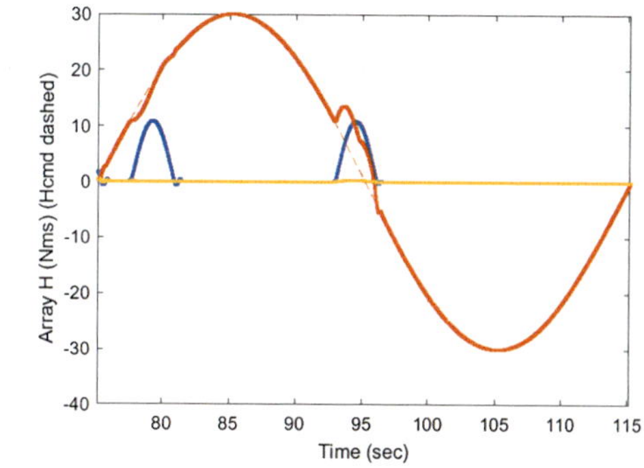

Fig. 14.8 Momentum profile during 3-CMG academic scenario

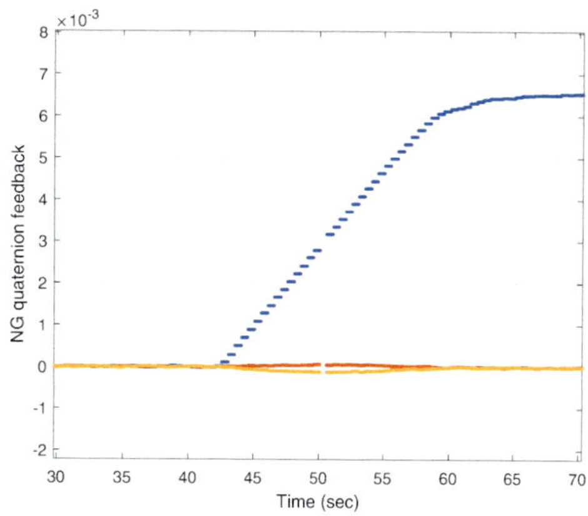

Fig. 14.9 Simple slew (quaternions q1, q2 and q3)

14.4.4 GI Scenarios: Simple Slew

This section presents data from the "simple" guest investigator scenario. This scenario performed a series of three slew maneuvers about various axes, and the first of the three is presented.

A simple repositioning slew is illustrated by rotating the vehicle 0.75° about its X axis. This rotation is represented as a quaternion by $[0.0065\ 0\ 0\ 0.99998]^T$. The plot in Fig. 14.9 shows the vehicle attitude (q1 blue, q2 red and q3 yellow) during the slew as reported by the Cygnus telemetry (2 Hz sample rate).

The gimbal angles of the 3 functioning CMGs during this slew are shown in Fig. 14.10 (CMG1 blue, CMG2 red and CMG3 yellow). Note that this telemetry was recorded at the 100 Hz sample rate of the CMG software.

Neglecting the 4th column of GA0 and GA90 in Eq. (14.1), we can compute the net momentum of the 3-CMG array from the gimbal angle telemetry as shown in Fig. 14.11 (body frame X, Y and Z as blue, red, yellow):

Note that before the slew, there is nonzero momentum in the array even while the vehicle rate is zero. This indicates the presence of "stored" momentum. We saw earlier in Sect. 14.4.1 how stored momentum from external torques accumulates in

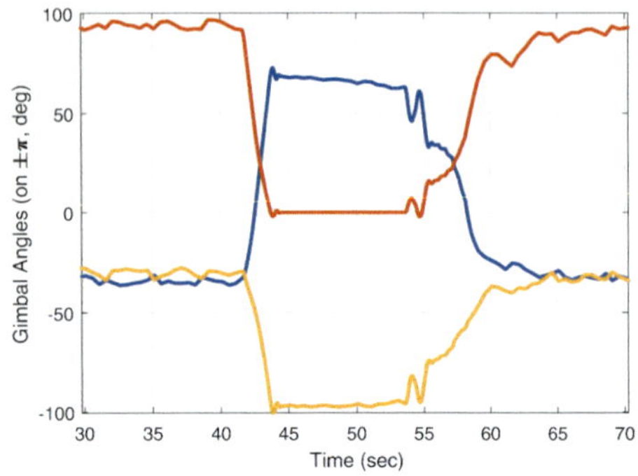

Fig. 14.10 Simple slew gimbal angles

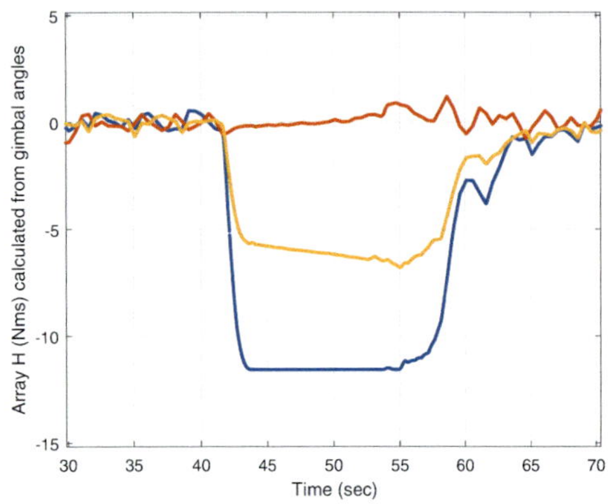

Fig. 14.11 Array momentum computed from gimbal angles

the CMG array to prevent vehicle motion and unnecessary use of thrusters. Since this maneuver lasts less than 30 s, the stored momentum can be taken as constant for our purposes and will be subtracted out of the above data in the next paragraph.

The law of conservation of angular momentum states that in the absence of external torque, the body's momentum will remain constant in the inertial frame. We will simplify this by assuming that the external torque is zero, and the net momentum is also zero (by removing the bias from the plot above). With these assumptions the vehicle momentum is simply equal and opposite to the array momentum, and the vehicle momentum is equal to Jw, the inertia times the vehicle rate. The vehicle inertia in the body frame was determined empirically to be:

$$\begin{bmatrix} 15079 & -568 & 7995 \\ -568 & 22426 & -281 \\ 7995 & -281 & 14474 \end{bmatrix} \text{kg-m}^2$$

With this, we can solve for the vehicle rate that should be produced by the motions observed on the CMG gimbals. The plot in Fig. 14.12 shows that solution as solid lines, along with the telemetry from the Cygnus rate sensor.

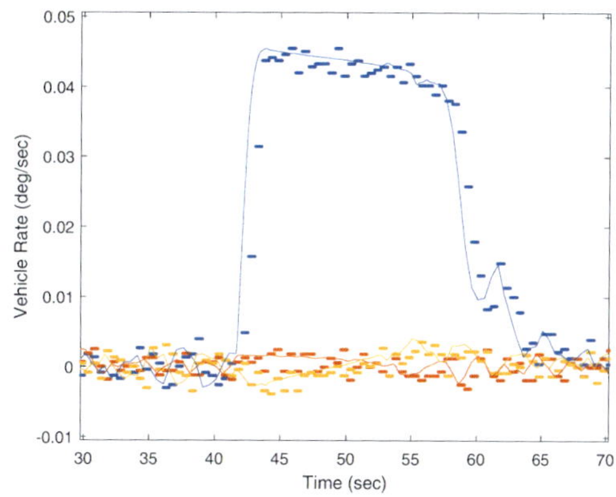

Fig. 14.12 Vehicle rate: expected (solid) vs sensor data

As we can see, these agree quite closely. Note that in addition to the 2 Hz sampling of the Cygnus rate sensor, there is also one frame of transport delay in these data.

This section does not address how the attitude control and CMG steering combined to produce the commands that yielded these gimbal angle trajectories. These topics are discussed elsewhere in the book, but we can make some observations. First, we note that the largest component of the required momentum is in the $-X$ direction, which we know can only be provided by CMG2. Therefore, it is unsurprising that CMG2 has gone to $0°$ gimbal angle, and the X momentum is 10.86 Nms. For the 3-CMG array, this is the largest momentum that can be produced in the correct direction, and the remaining gimbal angles simply point the other two components of momentum correctly. We also note lightly damped oscillation throughout the maneuver. This is the result of the unexpected extra frame of transport delay in the ACS feedback, which degraded the stability margins of the ACS.

14.4.5 Acceleration During the Simple Slew

Because the MMCS is so undersized in this experiment, the maneuver in the previous section was "vehicle rate limited". The CMGs very quickly transferred all available momentum, and then we simply had to wait for the slew to happen at a very slow vehicle rate. If the CMG array had been more appropriately sized, the resulting agile slew would happen much more quickly, and might be "acceleration limited" for much of the maneuver. These concepts are discussed further in Chap. 3. To slew about a certain axis, it is important that both acceleration and rate of the vehicle be in the (same) correct direction. In the plots of the previous section we saw the gimbal angles move to the correct values to achieve vehicle rate in the

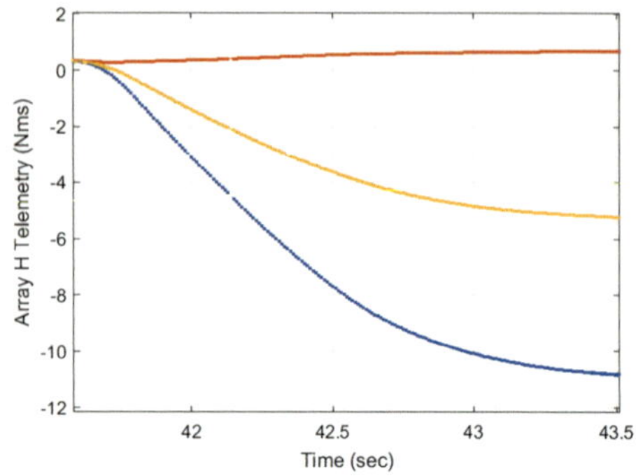

Fig. 14.13 Momentum telemetry during vehicle acceleration

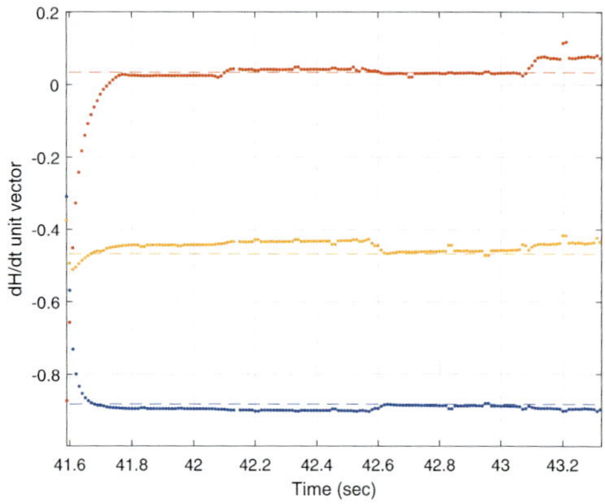

Fig. 14.14 Normalized torque vector during acceleration

correct direction. In fact, positioning the gimbals took approximately 2 s. We will now examine those seconds that it took to accelerate the vehicle to that rate.

The plot in Fig. 14.13 shows array momentum telemetry during those 2 s. As the magnitude of the momentum vector grows, we want the direction of the vector to remain correct.

Let us first consider what the correct direction for momentum and torque is for this maneuver. With our simplifying assumptions, we know that momentum is $J\omega$, and torque (dH/dt) is $J\alpha$. Since we want to slew about the X axis, we want vehicle acceleration and rate to both be about X, so the correct momentum and torque will both have the same unit vector. For rotation about X, this is simply the first column of J, which we normalize to get the unit vector $u = [-0.88\ 0.03\ -0.474]^T$. We recognize this relationship in the plot in Fig. 14.13, but it is difficult to be precise. Further, since there is some stored momentum, the exact direction of the net array momentum would be corrupted. If we want the vehicle to accelerate in the correct direction, we really want the torque (dH/dt) to be correct. In the plot from Fig. 14.14 we have computed the derivative of the momentum data and normalized it to show the unit vector in the direction of torque during those 2 s. The dashed lines are the components of the desired unit vector calculated above.

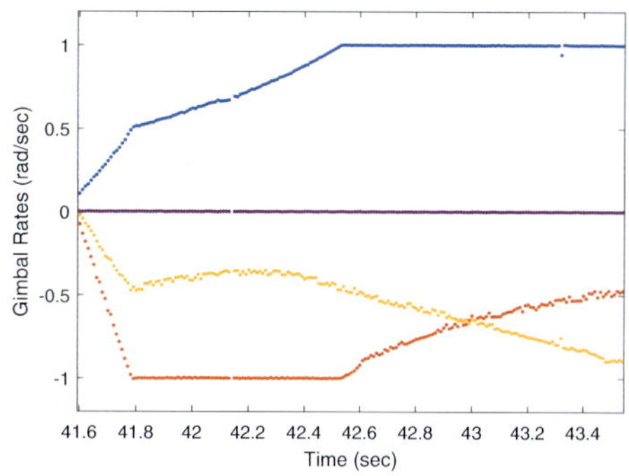

Fig. 14.15 Gimbal rate telemetry during acceleration

As we can see, in only tenths of a second, the torque has converged on the right direction and continues to do so throughout the acceleration phase. Let us see how this was accomplished. One might think that each gimbal was simply given a gimbal angle command and rotated at its maximum gimbal rate of 1 rad/s to get there as quickly as possible, but doing so would have produced transient torque, and thus acceleration, in the wrong direction. In fact, the gimbal rates produced by the MCS during this time are shown in Fig. 14.15 (from flight telemetry).

We recall from Sect. 9.1 that the inverse of the Jacobian matrix $A = (dH/d\Delta)$ can be used to determine the gimbal rates that produce a given dH/dt. The A matrix for the 3-CMG array at its initial gimbal angles is:

$$A = 10.86 * \begin{bmatrix} 0 & 1.0000 & 0 \\ -0.8660 & -0.0000 & -0.8660 \\ -0.5000 & 0 & 0.5000 \end{bmatrix}$$

Since we want dH/dt to be in the direction u, we have

$$d\Delta/dt = A^{-1} * u = [0.041\ -0.081\ -0.045]^T$$

These are the gimbal rates that would produce 1 Nm in the correct direction. Since we have 1 rad/s available from the MMCS hardware, we divide these rates by the largest magnitude and get $d\Delta/dt = [0.5\ -1.0\ -0.55]^T$. These are the gimbal rates that will produce the largest possible torque from the CMG array in the correct direction. from the plot in Fig. 14.15, we can see that at 41.8 s (when the gimbal rates reach their commanded value), these are the gimbal rates the hardware has produced. We also notice that those gimbal rates are constantly changing during the acceleration, which we would expect. Since the gimbal angles are changing, the A matrix changes, and thus the solution for the correct gimbal rates. The steering laws of the MMCS are functioning correctly to determine the gimbal rates that produce the largest possible torque in the correct direction at all times, regardless of which CMG is the limiting factor. Note how this changes from CMG2 to CMG1 at about 42.5 s.

14.5 Chapter Summary

In this chapter we have examined data from the 2019 flight experiment in which the Honeywell MMCS was used to control the attitude of the NG-11 Cygnus spacecraft.

In addition to showing that the vehicle behaved as physics would predict, some insight was also provided into how the CMG array operates and why the results looked the way they do.

The MMCS was grossly undersized for this vehicle (about 100x), therefore the agility for which CMG arrays are most valued was not demonstrated. Because the time frames were long, emphasis herein was placed on gimbal angles, momentum and vehicle rate. In an agile application, precision control is also required on gimbal rates, vehicle torques and acceleration. Section 14.4.5 showed this capability being demonstrated in the flight experiment as well.

Appendix A
Extended Equations of Motion for Spacecraft with CMGs

It is typically assumed when the subject of dynamics is taught that the origin of a system's mass properties, and angular and linear momenta are understood. However, without understanding of the true origin of these characteristics of a system, knowing when terms from the derivation of the equations of motion can be neglected becomes unclear.

The purpose of this appendix is to treat dynamics of a spacecraft with an array of CMGs all the way from the particle, to component, to system angular momenta and from here, rigorously find the generic vector-dyadic equations of motion. The equations of motion will be derived, making note of steps left out in most of the literature. This appendix is not meant to stray away from or repeat material in the Dynamics chapter, but to provide the analytics behind the whole process of deriving the equations of motion of the discussed multi-body system and its corresponding angular momenta and inertias. Also, from this appendix, the reader should be able to follow the steps needed to derive more complex multi-body system equations of motion. Except for some needed additional notation, the notation will be the same notation adopted from Chap. 4.

We would like to acknowledge Prof Norman Fitz-Coy at the University of Florida from whose notes, much of the integral equations of motion in this appendix are directly derived and expanded upon.

A.1 It Begins with a Particle

Imagine that all matter is made up of particles of a zero volume and inertia that contain mass. A set of these particles (at least 2) creates a system. With a system, there is a center of mass and even an effective inertia. Imagine a barbell with a massless bar shown in Fig. A.1.

© The Author(s), under exclusive license to Springer Nature Switzerland AG 2025

F. A. Leve et al., *Spacecraft Momentum Control Systems*, Space Technology Library 46, https://doi.org/10.1007/978-3-031-96165-6

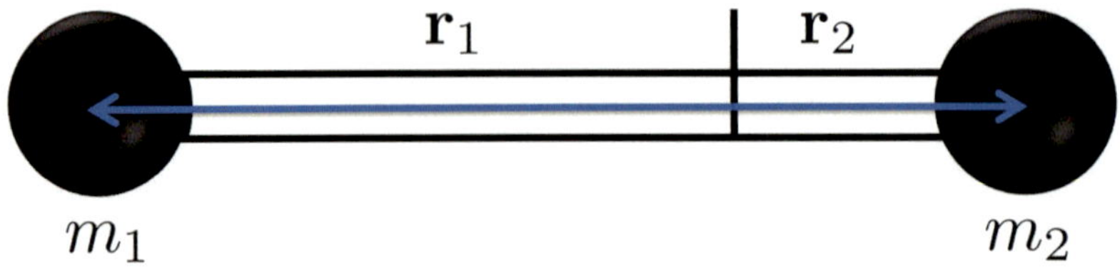

Fig. A.1 Two-particle mass barbell

The center of mass position is equal to the ratio of the masses of the particles multiplied by the vector distance of the bar from a datum position (invariant to where that datum is chosen), over all masses.

$$R_C = \frac{\sum_{i=1}^{n} m_i \mathbf{r}_i}{\sum_{i=1}^{n} m_i} \tag{A.1}$$

where m_i are the masses of the n particles at position vectors, $\mathbf{r}_i$ from the datum position and the products $m_i \mathbf{r}_i$ are known as the first mass moments. Now if the number of particles $n \to \infty$, and a constraint exists so the distances between all of these particles are fixed, a rigid body is formed.

A.2 Components

A system may contain many components and a component, which may or may not be a rigid body (e.g., nonempty propellant tank is not a rigid-body), contains a continuum of particles attached together that are bound together by internal forces. If we take an integral with respect to the differential mass over a continuum of the body, we get its mass

$$m_{\mathrm{B}} = \int_{\mathrm{B}} dm \tag{A.2}$$

Now the equation for the center of mass position is analogous to what was shown in Eq. (A.1), except for a continuum of particles,

$$R_C = \frac{\int_{\mathrm{B}} \boldsymbol{\rho} dm}{\int_{\mathrm{B}} dm} \tag{A.3}$$

The term $\int_{\mathrm{B}} \boldsymbol{\rho} dm$ is known as the first mass moment of a body B and it is equal to zero whenever the datum chosen for measurement of the particles' locations is the center of mass. In general, spacecraft are separated into separate subsystems, and these subsystems are made up of separate components. For example, a typical spacecraft will have the following subsystems or main elements:

- Communications and Ground Station Network
- Command and Data Handling
- Electrical Power System
- Propulsion
- Structures, Thermal, and Health Monitoring
- Payload
- Orbit Control and Trajectory Generation
- Attitude Determination and Control

Within the attitude determination and control system (ADCS), there are sensors such as gyros, accelerometers, RWA and CMG rotor and gimbal encoders and hall sensors, star trackers, horizon sensors, magnetometers, etc. As for actuators, the ADCS may contain magnet coils/torquers, RWAs, CMGs, and/or thrusters.

For a spacecraft hosting an array of CMGs, you can think of three distinct components that make up the system. First you have the spacecraft bus that consists of the spacecraft and its subsystems excluding the CMGs. The CMGs themselves have two separate components or bodies, rotors and gimbals that hold the rotors. Therefore, we will derive the generic integral equations of motion considering the three distinct components of this system, spacecraft, gimbals, and rotors, respectively.

A.3 Reference Frames for Spacecraft n-CMG System

Consider the generic geometric diagram of a spacecraft consisting of a single CMG in Fig. A.2. The system is made up of three distinct components, a spacecraft bus consisting of all components that are not the CMGs (e.g., solar panel, propellant tank, payload, sensors, structure, and electronics) and the CMGs' rotor and gimbals.

The list of symbols at the beginning of the book lists the reference frames, masses, points, position and velocity vectors, masses, and inertias associated with Fig. A.2.

A.4 Spacecraft n-CMG Array Kinematic and Kinetic Equations of Motion

The kinematic equations of motion are those equations which show the relationship of the motion between all components of the systems with respect to the inertial frame without the presence of forces or torques. Therefore, these equations depend solely on the geometry (i.e., angles, positions, and velocities) of all components in

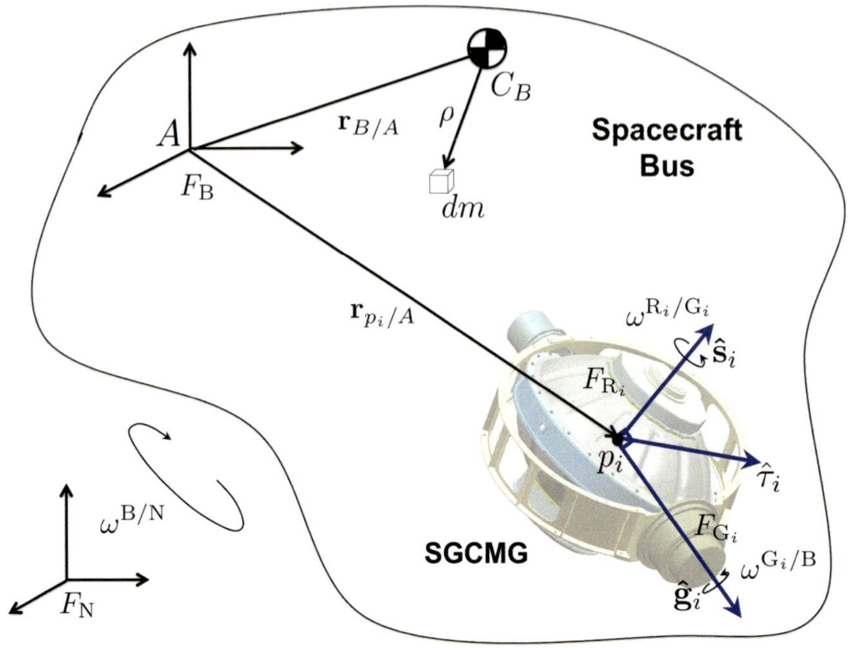

Fig. A.2 Spacecraft-CMG system geometric diagram

the system relative to each other and to the inertial frame. For rigid bodies, these equations are typically written in terms of quaternions, direction cosine matrices, or some other attitude representation and the corresponding angular velocity.

The kinetics equations, or the equations relating the spacecraft's rotational motion with respect to an internal or external torque are found by first deriving all components of angular momentum in the system.

A.4.1 Spacecraft Bus Angular Momentum

The first and largest component of our system is the spacecraft bus. The angular momentum for a body B is of the form $\mathbf{h}_A^{\mathrm{B}} = \int_B \mathbf{r} \times \mathbf{v} dm$ where the mass integral about the body B is taken with total position of the differential mass dm, $\mathbf{r}$ with respect to where the point, A that angular momentum is being derived and the total inertial velocity $\mathbf{v}$ of dm with respect to A.

The mass integral for a spacecraft bus's angular momentum about point A is,

$$\begin{aligned}\mathbf{h}_A^{\mathrm{B}} &= \int_{\mathrm{B}} (\mathbf{r}_{B/A} + \boldsymbol{\rho}) \times \left[\mathbf{v}_A + \boldsymbol{\omega}^{\mathrm{B/N}} \times (\mathbf{r}_{B/A} + \boldsymbol{\rho})\right] dm \\ &= \int_{\mathrm{B}} (\mathbf{r}_{B/A} + \boldsymbol{\rho})\, dm \times \mathbf{v}_A + \int_{\mathrm{B}} (\mathbf{r}_{B/A} + \boldsymbol{\rho}) \times \left[\boldsymbol{\omega}^{\mathrm{B/N}} \times (\mathbf{r}_{B/A} + \boldsymbol{\rho})\right] dm \\ &= \int_{\mathrm{B}} \mathbf{r}_{B/A} dm \times \mathbf{v}_A + \int_{\mathrm{B}} \boldsymbol{\rho} dm \times \mathbf{v}_A + \int_{\mathrm{B}} \boldsymbol{\rho} \times \left(\boldsymbol{\omega}^{\mathrm{B/N}} \times \boldsymbol{\rho}\right) dm \\ &+ \int_{\mathrm{B}} \mathbf{r}_{B/A} \times \left(\boldsymbol{\omega}^{\mathrm{B/N}} \times \mathbf{r}_{B/A}\right) dm + \int_{\mathrm{B}} \boldsymbol{\rho} \times \left(\boldsymbol{\omega}^{\mathrm{B/N}} \times \mathbf{r}_{B/A}\right) dm \\ &+ \int_{\mathrm{B}} \mathbf{r}_{B/A} \times \left(\boldsymbol{\omega}^{\mathrm{B/N}} \times \boldsymbol{\rho}\right) dm \end{aligned} \tag{A.4}$$

The position vector $\mathbf{r}_{B/A}$ can only be pulled out of the mass integral in Eq. (A.4) if it does not depend on the position of the differential masses, $\boldsymbol{\rho}$. For example, fuel slosh or flexible appendages of the spacecraft bus may cause the center of mass position to be variable with respect to the spacecraft body B, for which the mass integral is defined.[1] Therefore, if the center of mass of the spacecraft is assumed to be fixed in body B the spacecraft angular momentum with respect to point A is

$$\begin{aligned}\mathbf{h}_A^{\mathrm{B}} &= \int_{\mathrm{B}} \mathbf{r}_{B/A} dm \times \mathbf{v}_A + \int_{\mathrm{B}} \boldsymbol{\rho} \times \left(\boldsymbol{\omega}^{\mathrm{B/N}} \times \boldsymbol{\rho}\right) dm \\ &+ \int_{\mathrm{B}} \mathbf{r}_{B/A} \times \left(\boldsymbol{\omega}^{\mathrm{B/N}} \times \mathbf{r}_{B/A}\right) dm \\ &= \left(\mathbf{J}_B^{\mathrm{B}} + \mathbf{J}_A^{\mathrm{B}}\right) \cdot \boldsymbol{\omega}^{\mathrm{B/N}} + m_{\mathrm{B}} \mathbf{r}_{B/A} \times \mathbf{v}_A \end{aligned} \tag{A.5}$$

where the terms that are multiplied by $\int_{\mathrm{B}} \boldsymbol{\rho} dm = 0$ by the center of mass definition in Eq. (A.3) and the inertias $\mathbf{J}_B^{\mathrm{B}}$ and $\mathbf{J}_A^{\mathrm{B}}$ are

$$\begin{aligned}\mathbf{J}_B^{\mathrm{B}} &= \int_{\mathrm{B}} (\boldsymbol{\rho} \cdot \boldsymbol{\rho} \mathbb{1} - \boldsymbol{\rho}\boldsymbol{\rho})\, dm \\ \mathbf{J}_A^{\mathrm{B}} &= \int_{\mathrm{B}} \left(\mathbf{r}_{B/A} \cdot \mathbf{r}_{B/A} \mathbb{1} - \mathbf{r}_{B/A}\mathbf{r}_{B/A}\right) dm \end{aligned} \tag{A.6}$$

are found through the vector dyadic identity for any vectors $\mathbf{a}$, $\mathbf{b}$, and $\mathbf{c}$ as

$$\mathbf{a} \times (\mathbf{b} \times \mathbf{a}) = (\mathbf{a} \cdot \mathbf{a}\mathbb{1} - \mathbf{a}\mathbf{a}) \cdot \mathbf{b} \tag{A.7}$$

[1] Note that the coupled integral terms that contain the differential mass element vectors, $\boldsymbol{\rho}$ and the offset vectors are nonzero if the spacecraft bus has flexible appendage. These flexible effects are typically approximated by a summation of a finite number of modes for $\boldsymbol{\rho}$.

where $\mathbb{1}$ is known as the unit dyadic. The term $\mathbf{J}_A^{\mathrm{B}}$ is the parallel-axis inertia associated with the translation of the spacecraft center of mass from point A.

Remark A.1 The way to formulate the angular momentum for a single component of a multi-body system always starts with selecting a reference point for which to take the angular momentum and then all vectors to the center of mass and to the differential masses are taken with respect to that reference point. Therefore, you will need to define a position vector to the center of mass of the component and one from there to a differential mass to the selected reference point. This is the total position vector used in the mass integral. The total velocity is more complicated to formulate. This velocity is the change of the vector lengths in the body reference frame (assumed constant for rigid components), plus the cross product of their inertial angular velocity with the rotating vectors due to the rotation of the frames attached to the component. Note, not all vectors that sum to the total position vector rotate with the same velocity. It helps to bolt reference frames to each rigid component and take its rotation and a vector lying in that frame into account. Here, we choose position vectors that are constant length in each respective frame where time derivatives are taken to form the velocities.

A.4.2 CMG Gimbal Angular Momentum

Consider a single SGCMG as shown in Fig. A.3. We would like to derive the angular momentum of just the CMG gimbal neglecting its rotor for the moment.

This is without any loss to the accuracy of the derivation because in the end, angular momenta about the same base point are vectors and therefore can be added.

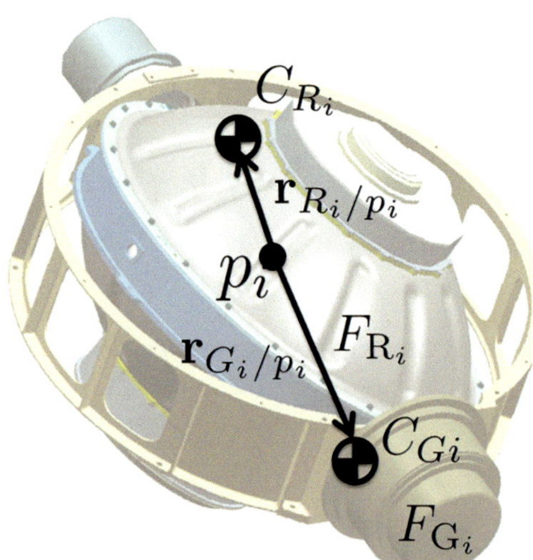

Fig. A.3 A single CMG with center of mass offsets

The angular momentum of a single, SGCMG gimbal with respect to point A, is

$$
\begin{aligned}
\mathbf{h}_A^{\mathrm{G}_i} &= \int_{\mathrm{G}_i} \left(\mathbf{r}_{G_i/A} + \boldsymbol{\rho}\right) \times \left[\mathbf{v}_A + \boldsymbol{\omega}^{\mathrm{B/N}} \times \left(\mathbf{r}_{G_i/A} + \boldsymbol{\rho}\right) + \boldsymbol{\omega}^{\mathrm{G}_i/\mathrm{B}} \times \left(\mathbf{r}_{G_i/p_i} + \boldsymbol{\rho}\right)\right] dm \\
&= \int_{\mathrm{G}_i} \mathbf{r}_{G_i/A} \times \mathbf{v}_A dm + \int_{\mathrm{G}_i} \mathbf{r}_{G_i/A} \times \left(\boldsymbol{\omega}^{\mathrm{B/N}} \times \mathbf{r}_{G_i/A}\right) dm \\
&+ \int_{\mathrm{G}_i} \boldsymbol{\rho} \times \left[\left(\boldsymbol{\omega}^{\mathrm{B/N}} + \boldsymbol{\omega}^{\mathrm{G}_i/\mathrm{B}}\right) \times \boldsymbol{\rho}\right] dm \\
&+ \int_{\mathrm{G}_i} \mathbf{r}_{G_i/A} \times \left(\boldsymbol{\omega}^{\mathrm{G}_i/\mathrm{B}} \times \mathbf{r}_{G_i/p_i}\right) dm,
\end{aligned}
\tag{A.8}
$$

where $\mathbf{r}_{G_i/A} = \mathbf{r}_{G_i/p_i} + \mathbf{r}_{p_i/A}$ and we zeroed out all terms of the form $\int_{\mathrm{G}_i} \boldsymbol{\rho} dm \times (\cdot) = (\cdot) \times \int_{\mathrm{G}_i} \boldsymbol{\rho} dm = 0$, $(\cdot)$ not a function of dm.

For our discussed system, there are multiple moving parts. Therefore, you must use caution when deriving the angular momentum. In Eq. (A.8), it is understood that first, the spacecraft bus has fixed within its reference frame the vector $\mathbf{r}_{p_i/A}$ to the nominal position of the CMG center of mass, p_i assumed to be at the intersection of the gimbal and rotor axes.

If we assume in Eq. (A.8) that the center of mass of the gimbal is at point p_i, then the equation of CMG gimbal angular momentum reduces to

$$
\begin{aligned}
\mathbf{h}_A^{\mathrm{G}_i} &= \int_{\mathrm{G}_i} \mathbf{r}_{p_i/A} dm \times \mathbf{v}_A + \int_{\mathrm{G}_i} \mathbf{r}_{p_i/A} \times \left(\boldsymbol{\omega}^{\mathrm{B/N}} \times \mathbf{r}_{p_i/A}\right) dm \\
&+ \int_{\mathrm{G}_i} \boldsymbol{\rho} \times \left(\boldsymbol{\omega}^{\mathrm{G}_i/\mathrm{N}} \times \boldsymbol{\rho}\right) dm \\
&= \mathbf{J}_A^{\mathrm{G}_i} \cdot \boldsymbol{\omega}^{\mathrm{B/N}} + \mathbf{J}_{G_i}^{\mathrm{G}_i} \cdot \boldsymbol{\omega}^{\mathrm{G}_i/\mathrm{N}} + m_{\mathrm{G}_i} \mathbf{r}_{p_i/A} \times \mathbf{v}_A
\end{aligned}
\tag{A.9}
$$

where $\boldsymbol{\omega}^{\mathrm{G}_i/\mathrm{N}} = \boldsymbol{\omega}^{\mathrm{B/N}} + \boldsymbol{\omega}^{\mathrm{G}_i/\mathrm{B}}$, and

$$
\begin{aligned}
\mathbf{J}_A^{\mathrm{G}_i} &= m_{\mathrm{G}_i} \left[\mathbf{r}_{p_i/A} \cdot \mathbf{r}_{p_i/A} \mathbb{1} - \mathbf{r}_{p_i/A} \mathbf{r}_{p_i/A}\right] \\
\mathbf{J}_{G_i}^{\mathrm{G}_i} &= \int_{\mathrm{G}_i} \left(\boldsymbol{\rho} \cdot \boldsymbol{\rho} \mathbb{1} - \boldsymbol{\rho}\boldsymbol{\rho}\right) dm
\end{aligned}
\tag{A.10}
$$

In practice, CMG vendors tend to balance their gimbals in such a way that they are inherently stiff, their gimbal center of mass offset is small, and can be approximated as a particle along the gimbal axis, thus providing no variable inertia. If one wishes to use this particle assumption and assume the center of the gimbal frame is along the gimbal axis and that a particle of mass, m_{Gp_i} is a position, $\mathbf{r}_{G_{p_i}/p_i}$ away from the gimbal frame origin, then they can construct an additional angular momentum perturbation induced to the system by the rotation of a particle imbalance as

$$\begin{aligned}\mathbf{h}_A^{\mathrm{G}_{\mathrm{p}_i}} &= m_{\mathrm{Gp}_i}\mathbf{r}_{G_{p_i}/A} \times \mathbf{v}_A + m_{\mathrm{Gp}_i}\left(\mathbf{r}_{G_{p_i}/A} \cdot \mathbf{r}_{G_{p_i}/A}\mathbb{1} - \mathbf{r}_{G_{p_i}/A}\mathbf{r}_{G_{p_i}/A}\right) \cdot \boldsymbol{\omega}^{\mathrm{B/N}} \\ &+ m_{\mathrm{Gp}_i}\mathbf{r}_{G_{p_i}/A} \times \left(\boldsymbol{\omega}^{\mathrm{G}_i/\mathrm{B}} \times \mathbf{r}_{G_{p_i}/p_i}\right), \end{aligned} \tag{A.11}$$

where m_{Gp_i} is the mass of the gimbal, $\mathbf{r}_{G_{p_i}/A} = \mathbf{r}_{p_i/A} + \mathbf{r}_{G_{p_i}/p_i}$. Equation (A.11) is found from taking the angular momentum over a single particle mass from Eq. (A.9), reducing the integral to an algebraic equation. Note that gimbal imbalances are typically small, terms associated with spacecraft angular velocity are typically much smaller than that for gimbal angular velocity.

A.4.3 CMG Rotor Angular Momentum

The CMG rotor angular momentum is found analogous to the CMG gimbal angular momentum as

$$\begin{aligned}\mathbf{h}_A^{\mathrm{R}_i} &= \int_{\mathrm{R}_i} \left(\mathbf{r}_{R_i/A} + \boldsymbol{\rho}\right) \times \left[\mathbf{v}_A + \boldsymbol{\omega}^{\mathrm{B/N}} \times \left(\mathbf{r}_{R_i/A} + \boldsymbol{\rho}\right) + \boldsymbol{\omega}^{\mathrm{R}_i/\mathrm{B}} \times \left(\mathbf{r}_{R_i/p_i} + \boldsymbol{\rho}\right)\right] dm \\ &= \int_{\mathrm{R}_i} \mathbf{r}_{R_i/A} \times \mathbf{v}_A dm + \int_{\mathrm{R}_i} \mathbf{r}_{R_i/A} \times \left(\boldsymbol{\omega}^{\mathrm{B/N}} \times \mathbf{r}_{R_i/A}\right) dm \\ &+ \int_{\mathrm{R}_i} \boldsymbol{\rho} \times \left(\boldsymbol{\omega}^{\mathrm{R}_i/\mathrm{N}} \times \boldsymbol{\rho}\right) dm \\ &+ \int_{\mathrm{R}_i} \mathbf{r}_{R_i/A} \times \left(\boldsymbol{\omega}^{\mathrm{R}_i/\mathrm{B}} \times \mathbf{r}_{R_i/p_i}\right) dm, \end{aligned} \tag{A.12}$$

where $\boldsymbol{\omega}^{\mathrm{R}_i/\mathrm{N}} = \boldsymbol{\omega}^{\mathrm{G}_i/\mathrm{N}} + \boldsymbol{\omega}^{\mathrm{R}_i/\mathrm{G}_i}$, and we zeroed out all terms of the form $\int_{\mathrm{R}_i} \boldsymbol{\rho} dm \times (\cdot) = (\cdot) \times \int_{\mathrm{R}_i} \boldsymbol{\rho} dm = 0$, $(\cdot)$ not a function of dm. Again, if we assume in Eq. (A.12) that the center of mass is at point p_i, then the equation of CMG gimbal angular momentum reduces to

$$\begin{aligned}\mathbf{h}_A^{\mathrm{R}_i} &= \int_{\mathrm{R}_i} \mathbf{r}_{p_i/A} dm \times \mathbf{v}_A + \int_{\mathrm{R}_i} \mathbf{r}_{p_i/A} \times \left(\boldsymbol{\omega}^{\mathrm{B/N}} \times \mathbf{r}_{p_i/A}\right) dm \\ &+ \int_{\mathrm{R}_i} \boldsymbol{\rho} \times \left(\boldsymbol{\omega}^{\mathrm{R}_i/\mathrm{N}} \times \boldsymbol{\rho}\right) dm \\ &= \mathbf{J}_A^{\mathrm{R}_i} \cdot \boldsymbol{\omega}^{\mathrm{B/N}} + \mathbf{J}_{R_i}^{\mathrm{R}_i} \cdot \boldsymbol{\omega}^{\mathrm{R}_i/\mathrm{N}} + m_{\mathrm{R}_i}\mathbf{r}_{p_i/A} \times \mathbf{v}_A \end{aligned} \tag{A.13}$$

where m_{R_i} is the mass of the rotor and

$$\mathbf{J}_A^{\mathrm{R}_i} = m_{\mathrm{R}_i}\left[\mathbf{r}_{p_i/A}\cdot\mathbf{r}_{p_i/A}\mathbb{1} - \mathbf{r}_{p_i/A}\mathbf{r}_{p_i/A}\right]$$

$$\mathbf{J}_{R_i}^{\mathrm{R}_i} = \int_{\mathrm{R}_i}(\boldsymbol{\rho}\cdot\boldsymbol{\rho}\mathbb{1} - \boldsymbol{\rho}\boldsymbol{\rho})\,dm \tag{A.14}$$

In practice, CMG vendors tend to statically and dynamically balance their rotors in such a way that they are inherently stiff and precision aligned, their rotor center of mass offset (i.e., eccentricity) is small, and can be approximated as a particle along the rotor axis. However, these imbalances are a main source of jitter on the spacecraft and are more prevalent then gimbal imbalances since the rotor velocities are of a much higher magnitude.

Therefore, if one wishes to use this particle assumption as we did for the gimbal angular momentum, they can assume that the center of mass of the rotor lies along the gimbal axis, and construct an additional angular momentum perturbation induced to the system by the rotation of a particle imbalance as

$$\mathbf{h}_A^{\mathrm{R}_{\mathrm{p}_i}} = m_{\mathrm{R}_{p_i}}\mathbf{r}_{R_{p_i}/A}\times\mathbf{v}_A + m_{\mathrm{R}_{p_i}}\left(\mathbf{r}_{R_{p_i}/A}\cdot\mathbf{r}_{R_{p_i}/A}\mathbb{1} - \mathbf{r}_{R_{p_i}/A}\mathbf{r}_{R_{p_i}/A}\right)\cdot\boldsymbol{\omega}^{\mathrm{B/N}}$$
$$+ m_{\mathrm{Rp}_i}\mathbf{r}_{R_{p_i}/A}\times\left(\boldsymbol{\omega}^{\mathrm{R}_i/\mathrm{B}}\times\mathbf{r}_{R_{p_i}/p_i}\right), \tag{A.15}$$

where $\mathbf{r}_{R_{p_i}/A} = \mathbf{r}_{p_i/A} + \mathbf{r}_{R_{p_i}/p_i}$.

Typically, due to precise alignments and machining, the mass imbalances and the norms of the vectors $\mathbf{r}_{G_{p_i}/p_i}, \mathbf{r}_{R_{p_i}/p_i}$ are exceptionally small. If we assume that the point mass imbalances are negligible, and both the gimbal and rotor center of masses are at the intersection of the gimbal and rotor axes of each CMG, then the total system angular momentum about point A is found through the summation of all component angular momenta from Eqs. (A.5)–(A.10) and (A.12)–(A.14) as

$$\mathbf{h}_A^{\mathrm{S}} = \mathbf{h}_A^{\mathrm{B}} + \sum_{i=1}^{n}\left(\mathbf{h}_A^{\mathrm{G}_i} + \mathbf{h}_A^{\mathrm{R}_i}\right)$$
$$= \left[\mathbf{J}_A^{\mathrm{B}} + \mathbf{J}_B^{\mathrm{B}} + \sum_{i=1}^{n}\left(\mathbf{J}_A^{\mathrm{G}_i} + \mathbf{J}_A^{\mathrm{R}_i} + \mathbf{J}_{G_i}^{\mathrm{G}_i} + \mathbf{J}_{R_i}^{\mathrm{R}_i}\right)\right]\cdot\boldsymbol{\omega}^{\mathrm{B/N}}$$
$$+ \left[m_{\mathrm{B}}\mathbf{r}_{B/A} + \sum_{i=1}^{n}\left(m_{\mathrm{G}_i}\mathbf{r}_{p_i/A} + m_{\mathrm{R}_i}\mathbf{r}_{p_i/A}\right)\right]\times\mathbf{v}_A$$
$$+ \sum_{i=1}^{n}\left(\mathbf{J}_{G_i}^{\mathrm{G}_i} + \mathbf{J}_{R_i}^{\mathrm{R}_i}\right)\cdot\boldsymbol{\omega}^{\mathrm{G}_i/\mathrm{B}} + \sum_{i=1}^{n}\mathbf{J}_{R_i}^{\mathrm{R}_i}\cdot\boldsymbol{\omega}^{\mathrm{R}_i/\mathrm{G}_i} \tag{A.16}$$

With the point A being chosen as the spacecraft n-CMG system center of mass, we have from the definition of the center of mass

$$\left[m_{\mathrm{B}} \mathbf{r}_{B/A} + \sum_{i=1}^{n} \left(m_{\mathrm{G}_i} \mathbf{r}_{p_i/A} + m_{\mathrm{R}_i} \mathbf{r}_{p_i/A} \right) \right] = 0 \tag{A.17}$$

Note, we assumed $\mathbf{r}_{G_i/A} = \mathbf{r}_{R_i/A} = \mathbf{r}_{p_i/A}$, but the result still holds. Therefore, Eq. (A.16) can be represented more compactly when taken about the system center of mass as

$$\mathbf{H} = \left(\mathbf{J}_s + \sum_{i=1}^{n} \mathbf{J}_{gr_i} \right) \cdot \boldsymbol{\omega}^{\mathrm{B/N}} + \left(\sum_{i=1}^{n} \mathbf{J}_{gr_i} \right) \cdot \boldsymbol{\omega}^{\mathrm{G}_i/\mathrm{B}} + \left(\sum_{i=1}^{n} \mathbf{J}_{r_i} \right) \cdot \boldsymbol{\omega}^{\mathrm{R}_i/\mathrm{G}_i} \tag{A.18}$$

where $\mathbf{J}_{gr_i}$ and $\mathbf{J}_{r_i}$ are the gimbal-rotor assembly and rotor inertias about the CMG system center of mass (formally point p_i).

The total angular momentum for a single CMG is

$$\mathbf{h}_i = \mathbf{J}_{gr_i} \cdot \boldsymbol{\omega}^{\mathrm{B/N}} + \mathbf{J}_{gr_i} \cdot \boldsymbol{\omega}^{\mathrm{G}_i/\mathrm{B}} + \mathbf{J}_{r_i} \cdot \boldsymbol{\omega}^{\mathrm{R}_i/\mathrm{G}_i} \tag{A.19}$$

Equation (A.19) can be rewritten with respect to the orthogonal coordinate basis embedded in the ith CMG frame, $\{\hat{\mathbf{s}}_i, \hat{\mathbf{o}}_i, \hat{\mathbf{g}}_i\}$ corresponding to the ith CMG rotor spin vector, gyroscopic torque vector, and gimbal vectors, respectively. Equation (A.19) becomes

$$\mathbf{h}_i = \mathbf{J}_{gr_i} \cdot \boldsymbol{\omega}^{\mathrm{B/N}} + I_g \hat{\mathbf{g}}_i \dot{\delta}_i + I_r \hat{\mathbf{s}}_i \Omega_i \tag{A.20}$$

where $I_g = \mathbf{J}_{gr_i} \cdot \hat{\mathbf{g}}_i$ and $I_r = \mathbf{J}_{r_i} \cdot \hat{\mathbf{s}}_i$ are the just the projections of inertia components along our orthogonal CMG frame assumed to be the same for each CMG, and

$$\begin{aligned} \boldsymbol{\omega}^{\mathrm{G}_i/\mathrm{B}} &= \hat{\mathbf{g}}_i \dot{\delta}_i \\ \boldsymbol{\omega}^{\mathrm{R}_i/\mathrm{G}_i} &= \hat{\mathbf{s}}_i \Omega_i \end{aligned} \tag{A.21}$$

Before we derive the equation of motion we will use the following assumptions

1. The spacecraft body center of mass is fixed in the spacecraft body frame.
2. The gimbal and rotor center of mass are fixed in the spacecraft body frame.
3. The gimbal and rotor centers of mass coincide at the origin of their corresponding CMG body, point p_i, which will also be assumed to be where the rotor and gimbal axes intersect.
4. The rotor spin axis is always orthogonal to the gimbal axis.

The relevance for these assumptions in order are:

1. The perturbations from low frequency disturbances such as fuel slosh or flexible appendages are assumed to be designed for early in the spacecraft and mission design and not a direct function of how the CMG is designed and therefore are not the subject of this appendix.

2. The rotor eccentricity (imbalance) effects are typically designed to be negligible when combined with passive isolation for the majority of high precision pointing spacecraft.
3. Most CMG are designed so that there are no changes in the center of mass of the spacecraft body with rotation of the gimbals, which is a harder constraint to design for than a variation in inertia along the gimbal axis (i.e., counter masses can be added to adjust the gimbal center of mass position but not necessarily the inertia).

To simplify the derivation, it makes sense to redefine the gimbal and rotor angular momenta about the CMG center of mass, $\mathbf{h}^{\mathrm{G}_i} = \mathbf{J}_{gr_i} \cdot \boldsymbol{\omega}^{\mathrm{B/N}} + I_g \hat{\mathbf{g}}_i \dot{\delta}_i$ and $\mathbf{h}^{\mathrm{R}_i} = I_r \hat{\mathbf{s}}_i \Omega_i$ and their associated angular velocities with respect to the inertial frame, $\boldsymbol{\omega}^{\mathrm{G}_i/\mathrm{N}} = \boldsymbol{\omega}^{\mathrm{B/N}} + \dot{\delta}_i \hat{\mathbf{g}}_i$ and $\boldsymbol{\omega}^{\mathrm{R}_i/\mathrm{N}} = \boldsymbol{\omega}^{\mathrm{B/N}} + \dot{\delta}_i \hat{\mathbf{g}}_i + \Omega_i \hat{\mathbf{s}}_i$. Then the derivative of Eq. (A.20) is

$$
\begin{aligned}
\frac{{}^{\mathrm{N}}\mathrm{d}\mathbf{h}_i}{\mathrm{dt}} &= \frac{{}^{\mathrm{N}}\mathrm{d}\mathbf{h}^{\mathrm{G}_i}}{\mathrm{dt}} + \frac{{}^{\mathrm{N}}\mathrm{d}\mathbf{h}^{\mathrm{R}_i}}{\mathrm{dt}} \\
&= \frac{{}^{\mathrm{G}_i}\mathrm{d}\mathbf{h}^{\mathrm{G}_i}}{\mathrm{dt}} + \boldsymbol{\omega}^{\mathrm{G}_i/\mathrm{N}} \times \mathbf{h}^{\mathrm{G}_i} + \frac{{}^{\mathrm{R}_i}\mathrm{d}\mathbf{h}^{\mathrm{R}_i}}{\mathrm{dt}} + \boldsymbol{\omega}^{\mathrm{R}_i/\mathrm{N}} \times \mathbf{h}^{\mathrm{R}_i} \\
&= \mathbf{J}_{gr_i} \cdot \left(\frac{{}^{\mathrm{B}}\mathrm{d}\boldsymbol{\omega}^{\mathrm{B/N}}}{\mathrm{dt}} + \boldsymbol{\omega}^{\mathrm{B}/\mathrm{G}_i} \times \boldsymbol{\omega}^{\mathrm{B/N}} \right) + I_g \ddot{\delta}_i \hat{\boldsymbol{g}}_i + \boldsymbol{\omega}^{\mathrm{G}_i/\mathrm{N}} \\
&\quad \times \mathbf{h}^{\mathrm{G}_i} + \frac{{}^{\mathrm{R}_i}\mathrm{d}\mathbf{h}^{\mathrm{R}_i}}{\mathrm{dt}} + \boldsymbol{\omega}^{\mathrm{R}_i/\mathrm{N}} \times \mathbf{h}^{\mathrm{R}_i} \\
&= \mathbf{J}_{gr_i} \cdot \left(\dot{\boldsymbol{\omega}}^{\mathrm{B/N}} + \dot{\delta}_i \boldsymbol{\omega}^{\mathrm{B/N}} \times \hat{\boldsymbol{g}}_i \right) + I_g \ddot{\delta}_i \hat{\boldsymbol{g}}_i + \left(\boldsymbol{\omega}^{\mathrm{B/N}} + \dot{\delta}_i \hat{\boldsymbol{g}}_i \right) \\
&\quad \times \left(\mathbf{J}_{gr_i} \cdot \boldsymbol{\omega}^{\mathrm{B/N}} + I_g \dot{\delta}_i \hat{\boldsymbol{g}}_i \right) + I_r \dot{\Omega}_i \hat{\boldsymbol{s}}_i + \left(\boldsymbol{\omega}^{\mathrm{B/N}} + \dot{\delta}_i \hat{\boldsymbol{g}}_i + \Omega_i \hat{\boldsymbol{s}}_i \right) \times I_r \Omega_i \hat{\boldsymbol{s}}_i \\
&= \mathbf{J}_{gr_i} \cdot \left(\dot{\boldsymbol{\omega}}^{\mathrm{B/N}} + \dot{\delta}_i \boldsymbol{\omega}^{\mathrm{B/N}} \times \hat{\boldsymbol{g}}_i \right) + I_g \ddot{\delta}_i \hat{\boldsymbol{g}}_i + I_r \dot{\Omega}_i \hat{\boldsymbol{s}}_i \\
&\quad + \boldsymbol{\omega}^{\mathrm{B/N}} \times \left(\mathbf{J}_{gr_i} \cdot \boldsymbol{\omega}^{\mathrm{B/N}} + I_g \dot{\delta}_i \hat{\boldsymbol{g}}_i + I_r \Omega_i \hat{\boldsymbol{s}}_i \right) + \dot{\delta}_i \hat{\boldsymbol{g}}_i \times \left(\mathbf{J}_{gr_i} \cdot \boldsymbol{\omega}^{\mathrm{B/N}} \right) - I_r \dot{\delta} \Omega_i \hat{\boldsymbol{o}}_i .
\end{aligned}
\tag{A.22}
$$

To find the torque of the ith CMG gimbal, we assume that the gimbal motor shaft is infinitely stiff and perfectly aligned and then must only output torque along its gimbal axis. Therefore, the gimbal torque is

$$
\begin{aligned}
\hat{\boldsymbol{g}}_i \cdot \frac{{}^{\mathrm{N}}\mathrm{d}\mathbf{h}_i}{\mathrm{dt}} &= \tau_{g,i} - \tau_{gf,i} \\
&= \hat{\boldsymbol{g}}_i \cdot \Big[\mathbf{J}_{gr_i} \cdot \left(\dot{\boldsymbol{\omega}}^{\mathrm{B/N}} + \dot{\delta}_i \boldsymbol{\omega}^{\mathrm{B/N}} \times \hat{\boldsymbol{g}}_i \right) + I_g \ddot{\delta}_i \hat{\boldsymbol{g}}_i + \boldsymbol{\omega}^{\mathrm{B/N}} \\
&\qquad \times \left(\mathbf{J}_{gr_i} \cdot \boldsymbol{\omega}^{\mathrm{B/N}} + I_r \Omega_i \hat{\boldsymbol{s}}_i \right) \Big]
\end{aligned}
\tag{A.23}
$$

where $\tau_{g,i}$ is the input torque to the gimbal and $\tau_{gf,i}$ is internal friction torque along the gimbal axis.

Similarly, we hold the infinite stiffness and perfect alignment assumptions for the ith CMG rotor to arrive at

$$\begin{aligned}\hat{\boldsymbol{s}}_i \cdot \frac{{}^{\mathrm{N}}\mathrm{d}\mathbf{h}_i}{\mathrm{d}t} &= \tau_{r,i} - \tau_{rf,i} \\ &= \hat{\boldsymbol{s}}_i \cdot \Big[\mathbf{J}_{gr_i}\left(\dot{\boldsymbol{\omega}}^{\mathrm{B/N}} + \dot{\delta}_i \boldsymbol{\omega}^{\mathrm{B/N}} \times \hat{\mathbf{g}}_i\right) + I_r \dot{\Omega}_i \hat{\boldsymbol{s}}_i + \boldsymbol{\omega}^{\mathrm{B/N}} \\ &\qquad \times \left(\mathbf{J}_{gr_i} \cdot \boldsymbol{\omega}^{\mathrm{B/N}} + I_g \dot{\delta}_i \hat{\boldsymbol{g}}_i\right)\Big] + \hat{\boldsymbol{s}}_i \cdot \Big[\dot{\delta}_i \hat{\boldsymbol{g}}_i \times \left(\mathbf{J}_{gr_i} \cdot \boldsymbol{\omega}^{\mathrm{B/N}}\right)\Big]\end{aligned} \tag{A.24}$$

Equations (A.23) and (A.24) can be written to solve for the desired input gimbal and rotor torques,

$$\begin{aligned}\tau_{g,i} &= I_g \ddot{\delta}_i + \tau_{gf,i} \\ &\quad + \hat{\boldsymbol{g}}_i \cdot \Big[\mathbf{J}_{gr_i} \cdot \left(\dot{\boldsymbol{\omega}}^{\mathrm{B/N}} + \dot{\delta}_i \boldsymbol{\omega}^{\mathrm{B/N}} \times \hat{\boldsymbol{g}}_i\right) + \boldsymbol{\omega}^{\mathrm{B/N}} \times \left(\mathbf{J}_{gr_i} \cdot \boldsymbol{\omega}^{\mathrm{B/N}} + I_r \Omega_i \hat{\boldsymbol{s}}_i\right)\Big],\end{aligned} \tag{A.25}$$

and

$$\begin{aligned}\tau_{r,i} &= I_r \dot{\Omega}_i + \tau_{rf,i} \\ &\quad + \hat{\boldsymbol{s}}_i \cdot \Big[\boldsymbol{J}_{gr_i} \cdot \left(\dot{\boldsymbol{\omega}}^{\mathrm{B/N}} + \dot{\delta}_i \boldsymbol{\omega}^{\mathrm{B/N}} \times \hat{\boldsymbol{g}}_i\right) + \boldsymbol{\omega}^{\mathrm{B/N}} \times \left(\mathbf{J}_{gr_i} \cdot \boldsymbol{\omega}^{\mathrm{B/N}} + I_g \dot{\delta}_i \hat{\boldsymbol{g}}_i\right) \\ &\qquad + \dot{\delta}_i \hat{\boldsymbol{g}}_i \times \left(\mathbf{J}_{gr_i} \cdot \boldsymbol{\omega}^{\mathrm{B/N}}\right)\Big]\end{aligned} \tag{A.26}$$

We can also define the actual gimbal and rotor actuator dynamics by the differential equations,

$$\begin{aligned}I_g \ddot{\delta}_i &= \tau_{g,i} - \tau_{gf,i} \\ &\quad - \hat{\boldsymbol{g}}_i \cdot \Big[\mathbf{J}_{gr_i} \cdot \left(\dot{\boldsymbol{\omega}}^{\mathrm{B/N}} + \dot{\delta}_i \boldsymbol{\omega}^{\mathrm{B/N}} \times \hat{\boldsymbol{g}}_i\right) + \boldsymbol{\omega}^{\mathrm{B/N}} \times \left(\mathbf{J}_{gr_i} \cdot \boldsymbol{\omega}^{\mathrm{B/N}} + I_r \Omega_i \hat{\boldsymbol{s}}_i\right)\Big],\end{aligned} \tag{A.27}$$

and

$$\begin{aligned}I_r \dot{\Omega}_i &= \tau_{r,i} - \tau_{rf,i} \\ &\quad - \hat{\boldsymbol{s}}_i \cdot \Big[\mathbf{J}_{gr_i} \cdot \left(\dot{\boldsymbol{\omega}}^{\mathrm{B/N}} + \dot{\delta}_i \boldsymbol{\omega}^{\mathrm{B/N}} \times \hat{\boldsymbol{g}}_i\right) + \boldsymbol{\omega}^{\mathrm{B/N}} \times \left(\mathbf{J}_{gr_i} \cdot \boldsymbol{\omega}^{\mathrm{B/N}} + I_g \dot{\delta}_i \hat{\boldsymbol{g}}_i\right) \\ &\qquad + \dot{\delta}_i \hat{\boldsymbol{g}}_i \times \left(\mathbf{J}_{gr_i} \cdot \boldsymbol{\omega}^{\mathrm{B/N}}\right)\Big]\end{aligned} \tag{A.28}$$

Clearly, the actuator dynamics derived are split into elements of the input torque, friction torque, and the projected dynamics back onto the gimbals and rotors from the spacecraft. The total system equations of motion can now be derived from the total angular momentum and its inertial derivatives. It is well known that for Eulerian equations of motion, the inertial time derivative of the total angular momentum of the system is equal to its external torques. Consider breaking up the total angular momentum in Eq. (A.18) into $\mathbf{H} = \mathbf{h}^{\mathrm{S}} + \sum_{i=1}^{n} \left(\mathbf{h}^{\mathrm{G}_i} + \mathbf{h}^{\mathrm{R}_i}\right)$. We already have the inertial time derivatives for $\mathbf{h}^{\mathrm{G}_i}$ and $\mathbf{h}^{\mathrm{R}_i}$ from Eq. (A.22) and

$$\frac{{}^{\mathrm{N}}\mathrm{d}\mathbf{h}^{S}}{\mathrm{dt}} = \mathbf{J}_s \cdot \dot{\boldsymbol{\omega}}^{\mathrm{B/N}} + \boldsymbol{\omega}^{\mathrm{B/N}} \times \left(\mathbf{J}_s \cdot \boldsymbol{\omega}^{\mathrm{B/N}}\right)$$

Therefore, we have

$$\begin{aligned}
\frac{{}^{\mathrm{N}}\mathrm{d}\mathbf{H}}{\mathrm{dt}} &= \frac{{}^{\mathrm{N}}\mathrm{d}\mathbf{h}^{S}}{\mathrm{dt}} + \frac{{}^{\mathrm{N}}\mathrm{d}\mathbf{h}^{G_i}}{\mathrm{dt}} + \frac{{}^{\mathrm{N}}\mathrm{d}\mathbf{h}^{R_i}}{\mathrm{dt}} = \tau_e \\
&= \left(\mathbf{J}_s + \sum_{i=1}^{n} \mathbf{J}_{gr_i}\right) \cdot \dot{\boldsymbol{\omega}}^{\mathrm{B/N}} \\
&+ \boldsymbol{\omega}^{\mathrm{B/N}} \times \left[\mathbf{J}_s \cdot \boldsymbol{\omega}^{\mathrm{B/N}} + \sum_{i=1}^{n} \left(\mathbf{J}_{gr_i} \boldsymbol{\omega}^{\mathrm{B/N}} + I_g \dot{\delta}_i \hat{\boldsymbol{g}}_i + I_r \Omega_i \hat{\boldsymbol{s}}_i\right)\right] \\
&+ \sum_{i=1}^{n} \left[I_g \ddot{\delta}_i \hat{\boldsymbol{g}}_i + I_r \dot{\Omega}_i \hat{\boldsymbol{s}}_i + \dot{\delta}_i \hat{\boldsymbol{g}}_i \times \left(\mathbf{J}_{gr_i} \cdot \boldsymbol{\omega}^{\mathrm{B/N}}\right)\right. \\
&\qquad \left. + \mathbf{J}_{gr_i} \cdot \left(\dot{\delta}_i \boldsymbol{\omega}^{\mathrm{B/N}} \times \hat{\boldsymbol{g}}_i\right) - I_r \dot{\delta} \Omega_i \hat{\boldsymbol{o}}_i\right].
\end{aligned} \tag{A.29}$$

Remark A.2 Note that the Eqs. (A.25) and (A.26) must be solved simultaneously so that angular momentum is conserved in the presence of internal friction. Specifically, you will integrate Eqs. (A.25) and (A.26) given a set of input torques and knowledge of friction and dynamics at the previous time step, then you will plus the actual and effective gimbal angles, rates, and rotor rates and accelerations into Eq. (A.27).

Internal friction and the additional torques acting against the gimbal and rotor motors of the ith CMG provide a gimbal and rotor acceleration result different from the command. This difference typically shows up as an error in the inner-loop control schemes to track the desired gimbal acceleration, rate, and angle for the gimbal and the desired rotor acceleration and rate for the rotor. However, no matter how well the inner-loop controller is designed, there are physical limits on the how fast or close it can reach the desired gimbal and rotor states and therefore, only the resultant gimbal and rotor states can be fed back into the spacecraft equations of motion in Eq. (A.27). The contributions $\dot{\delta}_i \hat{\boldsymbol{g}}_i \times \left(\mathbf{J}_{gr_i} \cdot \boldsymbol{\omega}^{\mathrm{B/N}}\right) + \mathbf{J}_{gr_i} \cdot \left(\dot{\delta}_i \boldsymbol{\omega}^{\mathrm{B/N}} \times \hat{\boldsymbol{g}}_i\right)$ equal zero when $\mathbf{J}_{gr_i}$ is constant in the body frame by using the transport theorem for the body frame in Eq. (A.22). If this assumption is made, all of these terms should be removed from the equations below Eq. (A.22).

Of particular interest are the most dominant torques associated with the spacecraft equations of motion in Eq. (A.27) and the actuator dynamics in Eqs. (A.25) and (A.26). Furthermore, for CMGs, the rotor speeds are normally held constant. If we reduce Eqs. (A.27) for a spacecraft equipped with n CMGs, keeping only the dominant terms from the conditions $|\Omega_i| >> |\dot{\delta}_i|, |\ddot{\delta}_i| >> \|\boldsymbol{\omega}^{B/N}\|, \|\dot{\boldsymbol{\omega}}^{B/N}\|$, we arrive at

$$\begin{aligned}&\left(\mathbf{J}_s + \sum_{i=1}^{n} \mathbf{J}_{gr_i}\right) \cdot \dot{\boldsymbol{\omega}}^{B/N}\\&+ \boldsymbol{\omega}^{B/N} \times \left[\mathbf{J}_s \cdot \boldsymbol{\omega}^{B/N} + \sum_{i=1}^{n} \left(\mathbf{J}_{gr_i} \boldsymbol{\omega}^{B/N} + I_g \dot{\delta}_i \hat{\mathbf{g}}_i + I_r \Omega_i \hat{\mathbf{s}}_i\right)\right]\\&- \sum_{i=1}^{n} I_r \dot{\delta} \Omega_i \hat{\mathbf{o}}_i = \tau_e.\end{aligned} \tag{A.30}$$

and

$$J_g \ddot{\delta}_i = \tau_{g,i} - \tau_{gf,i} - I_r \Omega_i \hat{\mathbf{g}}_i \cdot \left(\boldsymbol{\omega}^{B/N} \times \hat{\mathbf{s}}_i\right) \tag{A.31}$$

where the actuator dynamics for Ω_i are trivial since we assume it to be held constant speeds for CMGs. Only during spin up or spin down will the rotor dynamics become important. The contribution $I_r \Omega_i \hat{\mathbf{g}}_i \cdot (\boldsymbol{\omega}^{B/N} \times \hat{\mathbf{s}}_i)$ is known as precession back-drive torque (see Chap. 3) and is typically the largest torque besides internal friction of the gimbal bearings, that the gimbal motor must offset.

A.5 Appendix Summary

This appendix complements the dynamics chapter and provides the reader with the generic vector dyadic formulation of the equation of motion as well as their physical significance for a spacecraft hosting an array of n CMGs. It has provided in example, the integral derivation of a generic set of the equations of motion for a spacecraft containing an array of CMGs. With this derivation, the reader should be able to consider the proper assumptions to reduce the equations of motion.

Appendix B
Mathematical Concepts

Attitude dynamics do not evolve on a Euclidian space but on a "nonlinear" compact manifold inside Euclidian space. Since differential equations that represent dynamics are local representations of the vector fields with flows bound to this manifold, the dynamics are always nonlinear. Furthermore, nonlinear control theory is inherently harder than linear control theory as nonlinear dynamical systems contain multiple equilibria, more interesting attractors such as limit cycles, homoclinic, and heteroclinic orbits, and weaker stability results in general. The additional complexity of nonlinear dynamics also makes it harder to synthesize controllers with guaranteed stability, robustness, and safety.

Some of the theoretical development appearing in chapters of this book require some explanation of abstract mathematical concepts. Most of these abstract mathematical concepts deal with the nonlinear nature of the analysis of attitude dynamics and control from spacecraft momentum control systems. This appendix will introduce and explain a minimum of these concepts and serve as a reference complementary to the chapters of the book. It will discuss topological spaces, manifolds and local coordinates, some differential equations theory, as well as the mathematical formalisms behind some attributes of controlled dynamical systems, stability, controllability, robustness, and safety.

B.1 Topological Spaces

For a dynamical system, the state space which consists of points describing positions, orientations, or some other states, lies in what is known as a topological space. These spaces define the universe of available states, their connectivity and paths between them.

© The Author(s), under exclusive license to Springer Nature Switzerland AG 2025

F. A. Leve et al., *Spacecraft Momentum Control Systems*, Space Technology Library 46, https://doi.org/10.1007/978-3-031-96165-6

Definition: (Topological Space and Topology) A topological space is a pair (X, τ) made up of the total space X as a set, and the set of all open sets, τ known as the topology. The topology τ uniquely defines the makeup of the topological space by its open sets and closed sets. Any closed set C is the complement of an open set, $U \in \tau \in X$, i.e., $C = X \setminus U$. Topologies are constructed by the following rules:

1. X and the empty set $\emptyset$ belong to τ and are both open and closed.
2. Any arbitrary (finite or infinite) union of open set of τ is also an open sets of τ.
3. Any intersection of a finite number open sets in τ is also an open set of τ.

Norms which may generate distance metrics are introduced in Appendix C. Open sets can be generated directly by these metrics as open metric balls about a point $x \in X$ with radius $\varepsilon > 0$ as $U_x = \{y \in X | d(x, y) < \varepsilon\}$. When the topology (i.e., the set of all open sets) is defined by a metric, the pair is typically (X, d), where d is understood to be the metric. From now on, we will assume that all topological spaces we discuss are metric spaces.

B.1.1 Properties of Topological Spaces

The maps between topological spaces are known as continuous maps.

Definition: (Continuous maps and homeomorphisms) Continuous maps between topological spaces are defined equivalently in multiple ways. However, we will use the following definition. A map $f : X \to Y$ between topological spaces is considered continuous if, for every open set $V \subseteq Y$, the inverse image $f^{-1}(V)$ is an open set in $U \subseteq X$. Homeomorphisms are maps that have inverses and such that both the map and its inverse are both continuous. Homeomorphisms are also open maps that send open sets of the topology of the domain to open sets to the topology of the codomain and therefore, preserve all topological properties between spaces.

There are many properties of topological spaces and more can be found in references in the field of general (point-set) topology (see reference [139]). Some that we care about for relevant topological spaces associated with spacecraft momentum control are an open cover, compactness, and convergence.

Definition: (Open Cover) An open cover of a subset $K \subseteq X$ of topological space (X, d) is a set of open sets that covers the set K, (i.e., $K \subseteq \bigcup_i U_i$ for the open sets $U_i \in \tau$ of the cover).

Definition: (Compact and Sequential Compact Set) A set is compact if every open cover of this set has a finite subcover. It should be stressed the word "every" here. Another definition that is typically used in control theory is that of sequential compactness. Sequential compactness is equivalent to compactness for metric spaces. Sequential compactness of any subset $K \subseteq X$ is a set such that every sequence of points picked inside K, contains a converging subsequence converging to a point in K. The maximum theorem (i.e., every continuous real-valued function on a compact space attains its maximum) is easily seen to be true for finite spaces. this is a good way in which compactness can be interpreted as a generalization of finiteness.

Definition: (Convergence) A sequence x_n of points is said to converge to a point $x \in X$ if for open metric ball $U = \{x_i | d(x, x_i) < \varepsilon\}$, $\varepsilon > 0$ of x, there is a number $N > 0$ such that

$x_N \in U \; \forall n \geq N$. Clearly no matter how small the radius of the metric ball, U, you can still move into it for a convergent sequence.

Definition: (Hausdorff Property) A topological space is said to have the Hausdorff property if any two points can be separated by open sets. This property is of the utmost importance for convergence of sequences since it provides unique limit points as well as an infinite number of points in any open set about the limit.

Definition: (Second Countable Property and Base) A topological space is said to be second countable if its topology, τ has a countable base. A base is defined as a set of open sets that for which all open sets of τ can be formed by union of elements of the base.

B.1.2 Topological Space Examples

There are some important examples of topological spaces related to spacecraft attitude dynamics and control that need to be introduced. The classical standard topological space is based off of the real numbers $(\mathbb{R}, d)$ where the real line serves as base set for this space and the distance metric, $d(x, y) = +\sqrt{(x-y)^2}$ is the 2-norm distance between any points $x, y \in \mathbb{R}$. The topology generated by d here is just the metric balls $U_x = \{y \in X | d(x, y) < \varepsilon\}$ for all $x \in X$, the open intervals of the real line. This topological space can also be generalized to higher dimensional structures of the real numbers such as vectors and matrices through the cartesian products as $(\mathbb{R} \times \mathbb{R} \cdots \times \mathbb{R} \cong \mathbb{R}^n, d)$, where $d : \mathbb{R}^n \times \mathbb{R}^n \to \mathbb{R}$. For $m \times n$ matrices with real entries, that is, $A \in \mathbb{R}^{n \times n}$, we can use the Frobenius norm $(||A||_F = \sqrt{\text{tr}(A^T A)} = \sqrt{\text{tr}(AA^T)} = \sqrt{\sum_{i,j} a_{i,j}^2}$ to define the metric $d(X, Y) = ||X - Y||_F$. Note, the Frobenius norm is just the 2-norm for matrices. All of these spaces are known as Euclidian spaces.

An important example of topological spaces relevant to attitude kinematics arise as covering spaces shown in Fig. B.1, often acting as surrogates for the base spaces.

Definition: (Covering Map and Space) A covering map $p : \tilde{M} \to M$ of a space M is a continuous map between topological spaces where for every point $x \in M$, there exists an open neighborhood U_x of x and a discrete space discrete space D_x where the inverse image $p^{-1}(U_x) = \sqcup_{d \in D_x} V_d$ and $p|_{V_d} : V_d \to U_x$ is a homeomorphism for all $d \in D_x$. The open sets V_d are known as the sheets of the covering map. The domain $\tilde{M}$ of the covering map p is known as a covering space.

B.2 Manifolds

Manifolds are specialized topological spaces where dynamical systems evolve. They constrain the space of allowable trajectories and for the purpose of classical dynamical systems, they are topological subspaces of $\mathbb{R}^n$ or $\mathbb{R}^{n \times n}$. The most

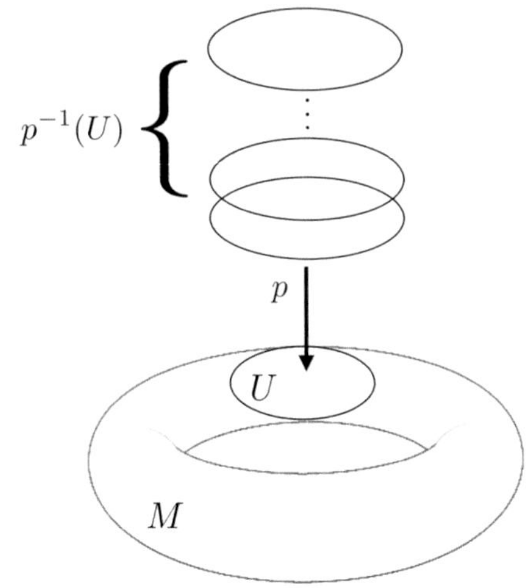

Fig. B.1 Open sets of a covering space above open sets $U \in M$ of topological space M

interesting and complex aspect of attitude control has to do with the fact that attitude trajectories evolve on a compact manifold lying inside $\mathbb{R}^n$.

Definition: (Real Manifold) A real manifold is a topological space that has the following properties:

1. It is a second countable topological space with the Hausdorff property.
2. It is locally equivalent to a real Euclidian space (e.g., $\mathbb{R}^n$).

Like general topological spaces, manifolds may be seen as topological spaces that constrain the trajectories allowable by dynamical systems. They are typically defined as being second countable topological spaces with the Hausdorff property. Locally Euclidian refers to existence of charts with chart maps and chart neighborhoods (see reference [140]).

Remark B.1 Complex manifolds that are locally equivalent to a complex Euclidian space and Banach manifolds that are infinite dimensional manifolds locally equivalent to a complete topological vector space are also possible. We will not discuss those here as they have no direct relevance to the problem of spacecraft momentum control systems.

Definition: (Chart) A chart is a pair (U, ϕ) of homeomorphism combined with an open set satisfying:

1. A chart neighborhood of a manifold, M, is an open set $U \subset M$ that is homeomorphic to an open set V of Euclidian space, in the case of a real manifold, $V \in \mathbb{R}^n$.
2. A chart map, ϕ is the homeomorphism defined on U into an open set of $\mathbb{R}^n$. The direct images of these chart maps provides a set of local coordinate vector $x_i \in \mathbb{R}^n$, where x_i is just an index notation for the ith coordinate of the image of ϕ of an n-dim point in M to a n-dim point in $\mathbb{R}^n$.

Definition: (Atlas and Differentiable Structure) An atlas is an equivalence class of charts that is closed under unions (i.e., if A_1 and A_2 are atlases, then so is $A_1 \cup A_2$). The existence of a maximal atlas is all that is needed and it is defined as the differentiable structure of the differentiable (smooth) manifold.

Calculus is possible via the definitions of derivatives being defined on functions of manifolds by their representations via charts in $\mathbb{R}^n$. Basically, we say a function $f : M \to \mathbb{R}$ is differentiable if its local representation, $\psi \circ f \circ \phi^{-1} : \mathbb{R}^n \to \mathbb{R}$ is differentiable for a chart (U, ϕ) on M. Differentiability can also be formulated for maps between manifolds and we say a map $f : M \to N$ is differentiable for a chart (U, ϕ) on M and (V, ψ) on N if the local representation, $\psi \circ f \circ \phi^{-1} : \mathbb{R}^n \to \mathbb{R}^m$ is differentiable.

Definition: (Differentiable Manifold and Diffeomorphism) A differentiable manifold is a topological manifold where the transition maps are diffeomorphisms. Diffeomorphisms are maps that are invertible and the maps and the inverses are both differentiable up to a certain number of derivatives. If we consider smooth manifolds, then diffeomorphisms are maps that are invertible and differentiable an infinite number of times in both directions. We will require smooth manifolds for our discussion.

Real manifolds lie inside $\mathbb{R}^n$ or $\mathbb{R}^{m \times n}$ as topological subspaces. This provides real manifolds with a locally Euclidian nature and allows us to locally do calculus (i.e., derivatives and integrals) on trajectories that lie on these manifolds. However, for us to use calculus, another requirement is the existence of a maximal atlas.

Remark B.2 It is a well-known theorem that differentiability of functions or maps on manifolds, though a local notion, is actually independent of choice of charts used for its local representation (i.e., local coordinates). Therefore, we can use the tools from regular calculus on the local coordinates defined by charts in the maximal atlas.

Attitude control is with respect to trajectories of a smooth compact manifold, known as the special orthogonal group in three dimensions, SO(3). The points of this manifold are rotation matrices. The reason why SO(3) represents the manifold for attitude is that all orientations between any two frames are uniquely and globally described by rotation matrices. The manifold SO(3) is also a real submanifold SO(3) $\subset \mathbb{R}^{3\times 3}$. In actuality, it is a submanifold of the set of invertible real 3×3 matrices, GL(3) (i.e., SO(3) $\subset$ GL(3) $\subset \mathbb{R}^{3\times 3}$). It is not only a smooth and compact manifold, it is a Lie Group which is a smooth compact manifold with all elements (points) making up a group where the group operations (multiplication or addition) and inverses are smooth in the sense defined by the differentiable structure.

B.2.1 Local Coordinates

Local coordinates for attitude kinematics were discussed in Chap. 4. These coordinates come from charts on SO(3). As noted previously, none of these local coordinates are defined globally on all of SO(3). For example, symmetric and asymmetric Euler angle sequences have coordinate singularities at angles π and

$\pm\frac{\pi}{2}$. These singularities are shown in the chart maps away from their domains. Obviously, these chart map singularities would limit their validity at singular angles and require multiple charts to cover SO(3). Even though SO(3) is a 3-dimensional manifold, no set of local coordinates (i.e., 3 coordinates) can be extended globally without introducing singularities. When you pick rotation matrices as your coordinates, you are already picking the points of SO(3) that are global and unique, but you are no longer just considering real matrices, but real matrices that are rotation matrices.

Some sets of coordinates utilize the notion of a cover space of SO(3) as a surrogate. Euler parameters or unit quaternions are points on the 3-sphere embedded in $\mathbb{R}^4$, $\mathbb{S}^3$ and use this cover space of SO(3) and its global coordinates. However, the $\mathbb{S}^3$ cover space is a double cover of SO(3) meaning each fiber $p^{-1}(u)$ has two solutions in $\mathbb{S}^3$ for each point u of SO(3). Furthermore, these two solutions lead to the quaternion unwinding problem for attitude control in Chap. 8.

B.2.2 Differential Equations

Since real manifolds provide the ability to do calculus locally, we can define differential equations associated with vector fields and their flows as integral paths (trajectories) on manifolds. To do this, we need to provide the notion of a tangent space and tangent bundle for a real manifold.

Definition: (Tangent Space and Tangent Bundle) A tangent space $T_m M$ of a real manifold M at each point $m \in M \subset \mathbb{R}^n$, is an n-dimensional vector space tangent to the point m. Each tangent space has a surjective projection map associated with it, $\pi : T_m M \rightarrow M$ that takes the tangent space to its base point $m \in M$. Tangent spaces of real manifolds are isomorphic to $\mathbb{R}^n$ and that is why we can perform calculus locally and where differential equations appear since tangent spaces also have their own differentiable structure provided by existence of a maximal atlas. Furthermore, by this surjective projection we known that tangent spaces $T_m M$ are also the fibers $\pi^{-1}(m)$. The tangent bundle is union of all of these tangent spaces for every $m \in M$. For smooth manifold M, the tangent bundle is also a smooth manifold. Physically, you can almost think of M as all attitude positions, $T_m M$ as angular velocity directions, and TM the union of all possible angular velocities.

See Fig. B.2 for a visualization of a tangent space of a manifold. The disjoint union of all tangent spaces for each point $m \in M$ with this projection map $\pi : TM \rightarrow M$ is known as the tangent bundle TM. The dimension of the tangent bundle is double that of the tangent spaces since it locally looks like the cartesian product $U_m \times T_m M$. Tangent bundles are more generally known as vector bundles since their fibers are vector spaces themselves.

Definition: (Tangent Vector Fields) A section of the tangent bundle, $X(m)$ is known as a tangent vector field since at every point on the path $m(t) \in M$, there is a corresponding vector $X(m(t)) \in T_{m(t)} M$. The smooth variation of this tangent vector parameterized by t inside TM as a smooth section of TM allows for us to understand how sections in TM project down to paths in M which is analogous to solutions of differential equations in the local sense. This is because, by definition, $\pi \circ X(m(t)) = m(t)$ (i.e., $\pi \circ X = \mathbf{1}_M$) where $\mathbf{1}_M$ is the identity map on M.

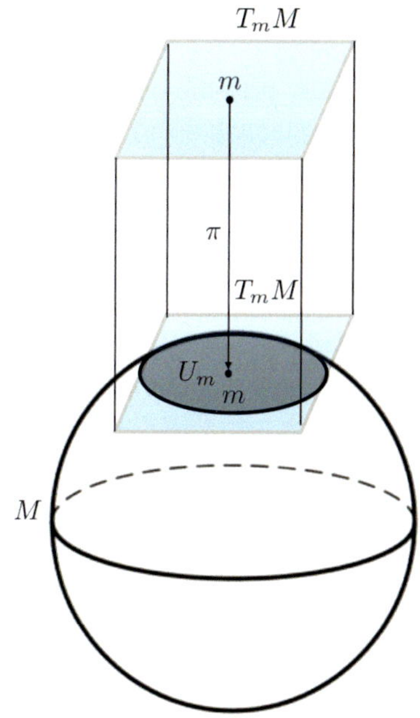

Fig. B.2 Tangent space T_mM above a point $m \in U_m \in M$ of a manifold M

Ordinary differential equations (ODEs) are local coordinate representations in $\mathbb{R}^n$ of vector fields. We are allowed to do calculus to form these differential equations since we have defined the differentiable structure (i.e., maximal atlas of charts) on the manifolds. These differential equations are typically the basis of mechanics and control in general and for Euclidian space as the manifold, the nuances of the differentiable structure, charts, etc. can be ignored. This is not the case for attitude dynamics and control as they evolve in the restricted space SO(3) $\subset \mathbb{R}^{3\times3}$. This is because the differential equations will always be nonlinear. For good elementary references on differential geometry and how it relates to basic mechanics, see references [141] and [140].

The ODEs that describe dynamical systems can be time-invariant or only implicitly time varying and called autonomous or explicitly time-varying and called non-autonomous. The distinction is important for many attributes of the behavior of the dynamical system and both types of ODEs show up in spacecraft momentum control systems use.

B.3 Stability

Stability is one of the most important attributes of a dynamical systems that may or may not be controlled. Basically, stability analyzes the behavior of the trajectories of a dynamical system to remain close to points or sets when they are initially perturbed. Stability is usually measured with respect to the trajectories of a dynamical system and its equilibrium points, attractors and repellers, or more specifically periodic orbits. There are different rules for stability of autonomous and non-autonomous systems of differential equations that represent continuous time dynamics and their counter parts in discrete time represented by diffeomorphic maps in time and state. For this book, we only need to discuss stability of continuous time dynamics with respect to equilibrium points of the dynamical system or of the Lyapunov functions to be discussed later. We will not address stability of attractors and periodic orbits here. See reference [142] for discussion of Poincare maps used for stability of periodic orbits.

We assume that our states are part of a normed vector space (X, d) that induce the distance metric $d(x, y) = ||x - y||$ between any points $x, y \in X$. Furthermore, we will define stability notions for controlled dynamical systems since the subject of this book is spacecraft momentum control. Therefore, we assume our controlled dynamical system is represented by the differential equation,

$$\dot{x} = f(x, u, t) \tag{B.1}$$

for the state x and control u that may or may not be autonomous. For controlled dynamical systems, the notion of stability with respect to perturbations in control and stability with respect to perturbations in state both need to be considered. For the remainder of the discussion of stability, we will refer to reference [143], [144].

Before we begin the discussion of stability, we should conceptually mention the levels of stability that will appear.

Definition: (Levels of Stability for Fixed Points)

- **Basic Stability** is conceptually defined as small changes to trajectories based off of small changes to initial conditions or control input. Depending on which of these you perturb, defines the type of basic stability. Basic stability allows for oscillations and does not guarantee you converge to a specific point, just that you remain close. The perturbation to the initial state and control input will always be measured using the norm induced metric of the space.
- **Attractivity** is the convergence of the trajectories to a point or an orbit. Convergence is measured by the current state in time and the desired state with the norm induced metric. Asymptotic convergence is based off of the limit as time goes to infinity.
- **Asymptotic Stability** is the case when the system is both stable and attractive. Typically, this is the bare minimum result that one wants to achieve.
- **Exponential Stability** is the case when the system is both stable and attractive, but convergence happens as a function of an exponential decreasing with respect to time. This provides a better convergence result than asymptotic stability.
- **Finite-Time Stability** is the case when the system is both stable and attractive, but convergence is in finite time.
- **Prescribed-Time Stability** is the case when the system is both stable and attractive, but convergence happens in a specified time.

In terms of ordering of strongest to weakest stability results, we have: prescribed-time $\subset$ finite-time $\subset$ exponential $\subset$ asymptotic $\subset$ stable.

Note, that the stability definitions can either be:

1. **Locally stable** and defined with respect to an open neighborhood of a point.

2. **Semi-globally stable** if for every compact set of initial conditions that includes the origin, there exists an output feedback controller that stabilizes the origin with the given compact set as a region of attraction.
3. **Almost globally stable** with respect to all but a zero volume set of points in the space (i.e., a set of bad points that contains no open set).
4. **Globally stable** with respect to all points in the space.

Furthermore, finite-time and prescribed time solutions stability will not be discussed further as they are they are not needed for the majority of spacecraft momentum control system maneuvers.

Remark B.3 Stability can also be defined with respect to sets (set stability), with respect to attractors such as periodic orbits or strange attractors, with respect to multiple isolated equilibria (multi-stability), and with respect to a continuum of equilibria (semi-stability). We only restrict the discussion here to fixed points since it is the most relevant to our control objectives utilizing spacecraft momentum control systems.

To analyze stability of a dynamical systems, there are many methods depending on the form of the dynamical system. If the dynamical system is linear, then tools from linear systems theory can be employed.

B.3.1 Linear Stability

For small angle spacecraft attitude control (see Chap. 8) it is usually the case that a linear-time invariant (autonomous) model of the attitude dynamics and output equations is sufficient. Stability of a linear time invariant (LTI) system is typically defined for the closed loop attitude control systems, u, and with output y, are found by the input-output closed-loop transfer function of the following output feedback set of equations:

$$\begin{aligned} \dot{x} &= Ax + Bu \\ y &= Cx + Du. \end{aligned} \tag{B.2}$$

This controlled system's stability is typically analyzed by the entire rational transfer function, in frequency variable s found by a Laplace transform of Eqs. (B.2)

$$T(s) = C(s1 - A)^{-1}B + D. \tag{B.3}$$

If the poles of the rational transfer function in Eq. (B.3) all lie in the open left-half side of the complex plane, then the system is bounded-input-bounded-output (BIBO) stable. BIBO refers to stability with respect to the small perturbations in control, u leading to small perturbations in output, y. Internal stability refers to bounded changes to the trajectories of the dynamical systems given small perturbations of the state. The LTI system in Eq. (B.2) is internally stable if the roots of the polynomial $p(s) = (s1 - A)$ are all in the open left half of the complex plane. Notions of internal stability implies asymptotic stability of the systems equilibrium. Furthermore, internal stability implies BIBO, but not the converse unless the system is linear controllable and observable.

Remark B.4 The dynamics equations of Eq. (B.2) can be changed by a change of coordinates to simplify the output equations to $y = Cx$ and therefore to the transfer function, $T(s) = C(s1 - A)^{-1}B$.

There may be cases where the linear system is time-varying. Such cases may appear when considering disturbance torques that occur along the spacecraft orbit which is periodic in time a part of the linear system dynamics or control. For linear time-varying (LTV) or non-autonomous linear systems of the form

$$\begin{aligned} \dot{x} &= A(t)x + B(t)u \\ y &= C(t)x + D(t)u, \end{aligned} \tag{B.4}$$

stability has to be checked a different way. A uncontrolled linear system, $\dot{x} = A(t)x$ solution flows can usually be analytically described as an matrix operator called the state transition matrix $\Phi(t, t_0)$ on the initial conditions as

$$\dot{x}(t) = \Phi(t, t_0)x_0. \tag{B.5}$$

For LTV systems the state transition matrix is

$$\Phi(t, t_0) = \exp(T)\left[\int_{t_0}^{t} A(\sigma)d\sigma\right], \tag{B.6}$$

where (T) is the time-ordering operator for the Peano-Baker series of $A(t)$ found by limits of product integral of arbitrarily time-interval length on $A(t)$. Given $\Phi(t, t_0)$ in Eq. (B.6), the LTV system in Eq. (B.4) is BIBO stable if it is

1. (Uniformly asymptotically stable): $||\Phi(t, t_0)|| \leq \gamma\exp^{(-\lambda(t-t_0))}, \forall t > t_0$ for constants $\gamma, \lambda > 0$,
2. (Bounded coefficient matrices): $||B(t)|| \leq b$, $||C(t)|| \leq c$, $||D(t)|| \leq d \ \forall t > 0$ with constants $b, c, d > 0$.

B.3.2 Nonlinear Stability

Mathematical formalisms for nonlinear stability generalize to linear systems as well. Nonlinear systems typically have multiple equilibrium points, sometimes in continuum. Therefore, the set of tools to evaluate the stability of nonlinear dynamical systems about equilibrium points requires that we look at them separately. Furthermore, for analysis of nonlinear systems, we will usually require at a minimum, existence and uniqueness of solution trajectories and therefore, we will assume that the dynamical system is represented by ODEs that are at least Lipschitz continuous in time, state, and control. Nonlinear stability of an autonomous and non-autonomous nonlinear dynamical systems are typically defined as

Definition: (Nonlinear Stability) An autonomous nonlinear dynamical system $\dot{x} = f(x)$ is stable at an equilibrium point $\overline{x}$ if for every $\varepsilon > 0$, there is a $\delta(\varepsilon)$ such that if $||x(t_0) - \overline{x}|| < \delta(\varepsilon)$, $||x(t) - \overline{x}|| < \varepsilon$, $\forall t > t_0 > 0$. This definition which uses the norm induced distance metric then of the space basically says that things starting close, remain close for all time. A system that is not stable, is considered unstable.
Similarly, a non-autonomous dynamical system $\dot{x} = f(x, t)$ is stable at equilibrium point $\overline{x}$ if for every $\varepsilon > 0$, and $||x(t_0) - \overline{x}|| < \delta(\varepsilon, t_0)$, $||x(t) - \overline{x}|| < \varepsilon, \forall t > t_0 > 0$. If for the non-autonomous system, there is a $\delta = \delta(\varepsilon)$ independent of t_0, then we say it is uniformly stable.
If in addition to the stability conditions either the autonomous or nonautonomous system case, $||x(t)|| \to 0, t \to \infty$, then we say the equilibrium is asymptotically stable.

B.3.2.1 Lyapunov Direct Method

Let's us first assume we are dealing with an Lipschitz continuous autonomous nonlinear system with full state feedback (i.e., output is the state $y = x$) and that we have already picked a control law $u(x) = g(x)$. Then our dynamical system is simply $\dot{x} = f(x)$ (see reference [144]). For Lyapunov direct methods, we will need to choose the equilibrium to be the origin. Note, that this is always possible since any equilibrium point of a nonlinear system can be shifted to the origin by defining a new dynamical system with the new state defined by this shift. This is usually the case for a nonlinear controlled dynamical system that is based off of the error state (i.e., a state which desired convergence is to zero).

Definition: (Lyapunov Stability for Autonomous Systems)

- **Lyapunov Stable**: An equilibrium point $\overline{x} = 0$ of the dynamical system $\dot{x} = f(x)$ is Lyapunov stable if we have a positive definite Lyapunov function with domain $D \subset \mathbb{R}^n$, $V(x) : D \to \mathbb{R}$ with the following properties:

    1. $V(0) = 0$
    2. $V(x) > 0$ for all $x \in D - \{0\}$
    3. $\dot{V}(x) \leq 0$ for all $x \in D$

- **Lyapunov Asymptotically Stable**: If in addition we have the strict inequality, $\dot{V}(x) < 0$ for all $x \in D$, then we can guarantee that the system is asymptotically stable.
- **Lyapunov Exponentially Stable**: If in addition we have $\alpha \|x\|^p \leq V(x) \leq \beta \|x\|^p$, and the strict inequality, $\dot{V}(x) \leq -kV$ with $\alpha, \beta, k > 0$, $p > 1$, for all $x \in D$, then we can guarantee that the system is exponentially stable.

An important result known as Lasalle's Invariant Principle is used in Appendix D to guarantee global asymptotic stability of an autonomous (i.e., time-invariant) system when we only have $\dot{V}(x) \leq 0$ rather than $\dot{V}(x) < 0$.

Definition: (LaSalle's Invariant Principle): Let $\Omega \subset D \subset \mathbb{R}^n$ be a compact positive invariant set with respect to $\dot{x} = f(x)$ (i.e., all trajectories starting in Ω remain for all

$t \geq t_0$). Let $V(x) : D \to \mathbb{R}$ be a continuously differentiable Lyapunov function candidate with domain D, such that $\dot{V}(x) \leq 0$ in Ω. Let $E = \{x|\dot{V}(x) = 0\}$. Let M be the largest invariant set in E, then every solution starting in Ω approaches M as $t \to \infty$. If M consists of only a point, the equilibrium point $\overline{x} = 0$ of $V(x)$, then that point is locally asymptotically stable.

Remark B.5 Note, Lasalle's invariance principle does not require a positive definite Lyapunov function. Furthermore, for Lasalle's invariance principle it is usually checked with the expression for the control law, $u = g(x)$ to see if it is possible for $\dot{V}(x) = 0$ at $\overline{x} = 0$ and any other points in M. A different result, the Barbashin-Krasovskii theorem, guarantees asymptotic stability to M and globally asymptotic stability when $M = \{0\}$ and $V(x)$ is positive definite and radially unbounded (i.e., $V(x) \to \infty$ when $x \to \infty$).

Lyapunov analysis for non-autonomous systems is more difficult than for autonomous systems. For non-autonomous systems, types of uniform stability are typically analyzed through finding comparison functions to compare the norm of the state for all time and initial time and states (see reference [144]). These comparison functions are defined below.

Definition: (Comparison Functions)

- **Class $\mathcal{K}$ function**: A continuous function $\alpha : [0, a) \to [0, \infty)$, where $a \in [0, \infty)$ belongs to class $\mathcal{K}$ if it is strictly increasing and $\alpha(0) = 0$.
- **Class $\mathcal{K}_\infty$ function**: A continuous function $\alpha : [0, a) \to [0, \infty)$ belongs to class $\mathcal{K}_\infty$ if $a = \infty$, $\alpha(0) = 0$, and $\alpha(r) \to \infty$, $r \to \infty$ (i.e., it is radially unbounded).
- **Class $\mathcal{K}_\mathcal{L}$ function**: A continuous function $\beta : [0, a) \times [0, \infty) \to [0, \infty)$ belongs to class $\mathcal{K}_\mathcal{L}$ if for each fixed s, $\beta(r, s)$ belongs to class $\mathcal{K}$ with respect to r and for each fixed r belongs to class $\mathcal{K}_\infty$ with respect to s.

We can use these comparison functions to define stability in the sense of Lyapunov for systems represented by non-autonomous nonlinear differential equations.

Definition: (Lyapunov Stability for non-autonomous Systems)

- **Lyapunov Stable (uniform in initial time)**: An equilibrium point $\overline{x} = 0$ of the dynamical system $\dot{x} = f(t, x)$ is Lyapunov stable if we have a positive definite Lyapunov function candidate with spatial domain D, so that $V(t, x) : [0, \infty) \times D \to \mathbb{R}$ is continuously differentiable there exists class K functions $\alpha_1(x)$, $\alpha_2(x)$, such that for all $t \geq 0$, $x \in D$
  1. $\alpha_1(x) \leq V(t, x) \leq \alpha_2(x)$
  2. $\frac{\partial V}{\partial t} + \frac{\partial V}{\partial x} f(t, x) \leq 0$
  Then equilibrium point $\overline{x} = 0$ is uniformly stable.
- **Lyapunov Asymptotically Stable (uniform in initial time)**: If in addition we have the strict inequality, $\frac{\partial V}{\partial t} + \frac{\partial V}{\partial x} f(t, x) \leq -\alpha_3(x)$ for a class K function $\alpha_3(x)$ for all $t \geq 0$ and $x \in D$, then we can guarantee that the system is uniformly asymptotically stable.
- **Lyapunov Globally Asymptotically Stable (uniform in initial time)**: If in addition for positive constants r and c we construct the metric ball of radius r, $B_r = \{x| \, ||x|| \leq r\} \subset D$ and $c < \min_{||x||=r} \alpha_1(x)$ then every trajectory in starting in the set $\{x \in B_r | \alpha_2(x) \leq c\}$ satisfies $||x(t)|| \leq \beta(||x(t_0)||, t - t_0)$, $\forall t \geq t_0 \geq 0$ for a class $\mathcal{K}_\mathcal{L}$ function β. If $D = \mathbb{R}^n$ and $\alpha_1(x)$ is radially unbounded, equilibrium point $\overline{x} = 0$ is uniformly globally asymptotically stable.

We did not address how to determine even local asymptotic stability when $\dot{V}(t, x) \leq 0$ (i.e., only a semi-definite Lyapunov function candidate time derivative). One way of checking for asymptotic stability in this situation is to rely on results such as Barbalat's Lemma. We will not discuss this here, but references [144] and [145] provide the necessary content.

B.3.2.2 Lyapunov's Indirect Method

Poles and roots of rational functions and polynomials in the variable s are not the only way of checking if a control LTI system is stable. In fact, Lyapunov's indirect method allows us to check stability for nonlinear autonomous systems in the local sense by creating a linear system by linearizing Eq. (B.1) about the origin as an equilibrium point. The method of linearizing a nonlinear system about an equilibrium point and checking for stability using is known as Lyapunov's Indirect Method. Where general Lyapunov analysis (i.e., Lyapunov's direct method) for nonlinear systems chooses a Lyapunov function candidate $V(x)$ and checks for stability, Lyapunov's indirect method linearizes the system about an equilibrium point and checks if the Jacobian matrix from the linearization, A is Hurwitz. If A is Hurwitz, the equilibrium point is asymptotically stable, else if any eigenvalue of the linearized Jacobian has a positive real part, it is unstable. This is result is equivalent as choosing a specific Lyapunov function and output value and seeing if a solution exists.

Consider the Lyapunov function $V(x) = x^T Px$ for with a symmetric and positive definite matrix, P, and its time derivative $\dot{V}(x) = \dot{x}^T Px + x^T P\dot{x} = x^T(A^T P + PA)x$. We will choose a positive definite symmetric matrix Q for the desired asymptotic stability result $\dot{V}(x) = x^T(A^T P + PA)x = -x^T Qx$. Therefore, we wish to solve for P so that $A^T P + PA = -Q$. If such a solution exists, the equilibrium point $\overline{x} = 0$ of the linear system is locally asymptotically stable to the region of validity for the linearization (see Theorem 4.6 of reference [144]). In fact, the result is actually locally exponentially stable since we can get the time derivative of the Lyapunov function, $\dot{V} \leq -kV$ where $k > 0$ is proportional to the smallest normed eigenvalue of Q.

Lyapunov direct method provides you a different way to establish/determine the stability of a linear or linearized nonlinear system. It is well known that the Jacobian linearized system (see Chap. 8) is stable when the real parts of the eigenvalues of the linearized Jacobian matrix, $Re(\lambda(A)) < 0$ is Hurwitz and asymptotically stable. Furthermore, if there exists a $\lambda \in$ spec(A) with $Re(\lambda(A)) > 0$, then the system is unstable. We have not described what happens when $Re(\lambda(A)) = 0$. In the case of an actual LTI system, we require all $\{\lambda_i | Re(\lambda(A)) = 0\}$ has an algebraic multiplicity, $a_i \geq 2$, and rank$(A - \lambda_i \mathbf{1}) = n - a_i$ where n is the dimension of the state.

Remark B.6 The majority of spacecraft motions lie outside the linear small angle control regime. Therefore, linearization and linear stability analysis is typically not relevant outside of the small-angle maneuvers discussed in Chap. 8.

Remark B.7 (Topological Obstructions and Almost Global Stability) Control in general, requires an error state which is the error "vector" (consists of multiple elements) of coordinates between the desired and current state. It is also important to Lyapunov analysis, but also in general to have a mathematical formalism for zero error (e.g., a norm of the error vector). This is usually in the form of a nonnegative error function.

In the case of attitude control which occurs on the smooth compact Lie group manifold, SO(3) or in the case of control on any smooth compact manifold, there are points of the state associated with any error function that obstruct any smooth or even continuous, time-invariant control laws from being globally asymptotically stabilizing. These "bad" points are additional critical points to error functions on the compact manifold associated to additional undesired equilibrium error states. When at these critical points, you cannot guarantee convergence to the desired critical point on the manifold associated with the desired zero error state since you may get stuck. Furthermore, in the vicinity of these bad points, divergence away from them and convergence to the desired point may be slow.

It is of course possible to construct a time-varying or discontinuous (or hybrid) controller to bump off of these bad points and stay a distance away from them, but this comes at a cost of moving away from continuous-time and time-invariant control which may or may not cause problems for implementation.

B.3.3 Control Lyapunov Functions

The previous discussions of Lyapunov analysis methods assumed a choice of a Lyapunov function candidate. For Lyapunov's direct method and carried out the analysis, or in the case of Lyapunov's indirect method, we solved for a P matrix with a specific quadratic Lyapunov function in mind. Furthermore, we also assumed we already started with a control law, $u = g(x)$. We would like to know if it is possible to even design a feedback control law $u = g(x)$ that asymptotically stabilizes, or even better exponentially stabilizes an equilibrium point of a nonlinear system. Methods to check for the existence and even construct a feedback law for nonlinear autonomous systems that is asymptotically stabilizing rely on the concept of a control Lyapunov function, which generalizes a normal Lyapunov function that was previously discussed.

Definition: (Control Lyapunov Function): **(CLF)**: A control Lyapunov function (CLF) is a continuously differentiable function, $V(x) : D \to \mathbb{R}$, with set of admissible controls U that is

1. (Positive Definite) $V(x) > 0 \; \forall D - \{0\}$ and $V(0) = 0$.
2. (Stabilizing Controller Existence): There exist a control law $u = u(x)$ such that $\dot{V}(x, u) = \inf_{u \in U} \langle \nabla V(x), f(x, u) \rangle < 0, \forall x \in D - \{0\}$.

The general result is that a system is globally asymptotically stable if and only if a CLF exists. See references [146] and [147] for more discussion on these functions.

Definition: (Control Affine Systems): A system that is control affine, linear in control variables and is represented by a differential equation $\dot{x} = f(x) + \sum_{i=1}^{m} g_i(x)u_i$ where the control vector fields are $g_i(x)$.

Remark B.8 Many of strongest results for CLFs rely on the assumption of a control affine system. Some of these results explicitly provide a stabilizing control law.

B.4 Inverse Dynamics

Another method of control synthesis for nonlinear systems constructs a controller with closed-loop feedback dynamics to match a prescribed linear system. Since this system is usually a mechanical system, it is directly constructed using the second order dynamics of the typical Euler-Lagrange equations for a mechanical system. These dynamics are

$$M(x)\ddot{x} + C(x, \dot{x})\dot{x} + G(x) + F(\dot{x}) = u + d \tag{B.7}$$

where the state is x (respectively the error state if state is known perfectly), mass matrix is $M(x)$, Coriolis/centrifugal matrix $C(x, \dot{x})$, gravity matrix $G(x)$, viscous friction matrix $F(\dot{x})$, control input u, and disturbances d. We would like to formulate the closed-loop error system from Eq. (B.7). For a great set of results for inverse dynamics as it applies to spacecraft momentum and general attitude control, see reference [82]. If we consider x as our error state and thus the state we wish to send to zero, we can prescribe our desired control system response in terms of a second-order linear system

$$\ddot{x} + K_d\dot{x} + K_p x + K_i \int_0^T x dt = 0, \tag{B.8}$$

where K_d, K_p, K_i control gain scalars or matrices for the derivative, proportional, and integral terms as in a system constructed from a proportional integral derivative controller. The objective is to design u so that the response of x follows the prescribed dynamical system in Eq. (B.8). To do this, you would simply substitute $\ddot{x}$ from Eq. (B.8) into Eq. (B.7) to find the required control input, u as

$$u = -M(x)\left(K_d\dot{x} + K_p x + K_i \int_0^T x dt\right) + C(x, \dot{x})\dot{x} + G(x) + F(\dot{x}) - d(t) \tag{B.9}$$

Clearly, we are assuming we know the dynamics and disturbance perfectly. This may very well not be the case. However, there are many methods within adaptive control that allow for globally (or almost) asymptotically stable results in the presence of parameter uncertainty, and uniformly ultimately bounded results when an unknown disturbance is present [148]. Another complication may arise when the control is premultiplied by control vector fields $g_i(x)$ as is done with more general systems of control affine form. In that case, control is not explicitly defined by Eq. (B.9). In these cases, inverse dynamics would require some method to invert the set of $g_i(x)$ over the right-hand-side. This may be possible if the $g_i(x)$ span all directions along the tangent space. If not, the control solution, u will not be guaranteed to meet the desired closed-loop dynamic response in Eq. (B.8). This is analogous to a B matrix in terms of constant g_i for linear system that does not provide control to enough directions for control. Therefore, inverse dynamics may not always be the best choice for control synthesis, even for control affine nonlinear systems.

Remark B.9 For spacecraft momentum control systems, we usually assume the absence of disturbances or at least disturbances with a nonzero mean value with respect to time. The reason we will not typically address such disturbances for spacecraft momentum control is because such disturbances cannot be offset with momentum control systems alone as they will saturate the actuators. Furthermore, in space, external disturbances are typically very small and take a long time to accumulate sufficient angular momentum to effect the control of the spacecraft. However, internal disturbances such as those cause by spacecraft structure excitation due to its flexible effects are problematic and usually are considered in the spacecraft design and provide constraints to the frequency at which attitude control can be applied.

Remark B.10 Spacecraft momentum control systems that utilize CMGs are not simple gyrostats and the very motion of the CMG assembly about the gimbal axes makes the inertia a function of the gimbal angles and therefore, the mass matrix a function of the gimbal angles as states if they are considered states of the system (see Appendix A). Unless the system is set up to have control at the level of current or voltage (depending on the motor type) to the gimbal motors, for practical control purposes, the gimbal angles, rates, or accelerations are considered as control inputs to the system. The effect of dynamic inertia from gimbal rotations can be lumped into the disturbances, but it ultimately depends directly on the gimbal angles.

B.5 Controllability and Observability

Controllability is the existence of or the ability of a specific control law to bring you from any point to any other point in the state space of the system in a finite time interval. Observability is the dual to the notion of controllability and is the ability to reconstruct the current state from measured outputs from past measurements over a finite time interval. Controllability is not stability, though controllability is required for certain stability results. Controllability only requires that there exists a control law to provide a path to get you between two points in the state-space.

B.5.1 Linear Time-Invariant System Controllability and Observability

Definition: (Linear Controllability and Observability for LTI Systems) For a system of dimension n, a controlled LTI system is controllable if the matrix

$$\mathcal{C} = \left[B \ BA \ BA^2 \ \cdots \ BA^{n-1}\right], \tag{B.10}$$

is full row rank and observable if

$$\mathcal{O} = \begin{bmatrix} C \\ CA \\ CA^2 \\ \vdots \\ CA^{n-1} \end{bmatrix}. \tag{B.11}$$

is full column rank.

For LTV systems, controllability and observability cannot be checked by the rank of controllability and observability matrices. In this case, Gramian's must be used to check for controllability and observability.

B.5.2 Linear Time-Varying System Controllability and Observability

Definition: (Linear Controllability and Observability for LTV Systems) For the state transition matrix $\phi(t, t_0)$ in Eq. (B.6), an LTV system of dimension n is controllable to a state, x_1 from a state x_0 within the interval $[t_0, t_1]$, if the Controllability Gramian

$$\mathcal{C}(t, t_0) = \int_{t_0}^{t_1} \phi(t, t_0) B(t) B(t)^T \phi(t, t_0)^T dt \tag{B.12}$$

is full column rank (i.e., dim(A)) and you are able to determine x_0 uniquely from succeeding states in the interval $[t_0, t_1]$ if

$$\mathcal{O}(t, t_0) = \int_{t_0}^{t_1} \phi(t, t_0) C(t) C(t)^T \phi(t, t_0)^T dt \tag{B.13}$$

is full rank.

Remark B.11 There are analogs to the LTI and LTV controllability and observability results for control affine systems. These analogs are local results and follow from Lie bracketing of the vector fields $g_i(x)$ with the drift vector field, $f(x)$ of the control affine system and rank conditions on these brackets. There are a large number of these local results for controllability and observability for control affine nonlinear systems. We will not discuss these results further, but more can be found on this topic in the reference (see reference [149]). For controllability of a CMG spacecraft momentum control system on the pair of attitude and angular velocity states can be found in reference [150].

B.6 Robustness

Robustness of a control system is the ability to maintain "good" behavior in the presence of system uncertainty and disturbances. It is basically the ability for the system to be insensitive to a degree to uncertainty, disturbances, time-delays, or other changes. Therefore, robustness is always measured with respect to something that is not ideal. Some examples of non ideal conditions are uncertain model parameters, model structure (i.e., different ODEs), noisy measurements, time-delays, actuator or sensor faults, and disturbances. Resilience which is the ability to recover from changes such as faults should not be confused with robustness. Furthermore, unlike robustness, resilience can not easily be measured or mathematically defined.

For LTI systems, there are a wealth of tools to measure robustness with respect to noise and disturbances. Some of these tools measure of robustness with respect to noise or disturbances by a gain and phase margin analysis of the closed-loop transfer function in frequency domain in the form of Bode plots [143]. For multi-input-multi-output LTI systems, the measure of robustness by gain and phase margin may be very conservative. For LTV systems, a notion of stability radius can be used for robustness analysis (See reference [151]). Furthermore, it is possible to group uncertainties and disturbances together as is done in the field of robust control and design controllers that are robust where robustness is quantified by H_∞ or some other Hardy space norm (e.g., H_1, H_2 norms) analysis of the closed-loop system response with respect to noise or disturbances (see reference [152]).

For nonlinear systems, small gain theorems can be utilized to measure robustness as discussed in references [144], [143], and [145]. Another methodology for nonlinear robustness analysis utilizing a distance between input-output trajectories known as the gap metric (see reference [153]). For continuous feedback control laws, there is always some level of robustness and analysis really needs to account for the level of uncertainty, faults, and disturbances that can appear and its effect on the control laws performance and stability.

B.7 Safety

Future space systems will most likely have a higher degree of autonomy. For that reason, specifications on safety will be a requirement and construction of safe controllers will be imperative. Safety can mathematically be cast as the distance from a "bad" set. Examples of bad sets for space systems are keep out cones of the payload from the sun, and constraints on actuator inputs and states. You could define these bad sets in the form of a formal specification on a formal logic symbolically, utilizing languages and their associated tools within the field of formal methods (see reference [154] to explore formal methods, model checking, and temporal logics). However, formal methods were primarily developed for software systems that are finite-state machines and are for the most part computationally intractable for verification and control synthesis for nonlinear controlled dynamical systems. This is because dynamical systems have an infinite number of states. Moreover, systems of interest for spacecraft attitude control have nonlinear dynamics and therefore, do not have an analytic or closed form expression for their solution (i.e., flows). Therefore, it is desired to construct these bad sets in a way that the choice of any control law can be verified to also be safe as well as stable and robust.

Recently, a generalization of a CLF known as a control barrier function (CBF) has been shown as an effective tool to check if a control law exists that is safe and synthesize such a control law. In fact, CBFs have also been shown to tightly approximate sets associated with safety specifications in linear temporal logic (see references [155] and [156]). With CBFs, Lyapunov type analysis tools can be utilized to verify safety of a control law and in some cases, synthesize a safe control law. In fact, in many cases, especially for CBFs of a nice form (e.g., polynomials) and dynamics in control affine form, computationally tractable programs can be developed to easily verify stability [157]. Control barrier functions utilize the idea of set invariance to guarantee safety. They are defined as follows

Definition: (Control Barrier Function [158]): (CBF): Assume you have a dynamical system of the form $\dot{x} = f(x)$ and the safe set $S = \{x|h(x) \geq 0\}$ with boundary $\partial S = \{x|h(x) = 0\}$ is defined as a super level set of a smooth function $h(x)$. A control barrier function (CBF) is this function $h(x) : \mathbb{R}^n \to \mathbb{R}$ such that

1. (Safe Invariant Set): $h(x) \geq 0$, $x \in \mathbb{R}^n$ and $\frac{\partial h}{\partial x} \neq 0$ for all $\partial S = \{x|h(x) = 0\}$.
2. (Invariance): There exists a u, such that S is invariant $\Leftrightarrow \dot{h}(x, u) \geq 0$ for all $x \in \partial S$.

Basically, the CBFs allow you to provide an analytic expression for bad sets and allow you to check for invariance of the controlled dynamical systems with respect to these bad sets. Control barrier functions have been transformative in enabling safety verification of complex dynamical systems with formal specifications. Before their popularity, the set of dynamical systems that could be verified safe was severely limited.

B.8 Summary

Attitude control of spacecraft is nonlinear because attitude trajectories evolve on a nonlinear space. For this reason, more complicated notions of spaces, representations of dynamical systems, and notions of stability, robustness, and safety needed to be covered. This appendix provide some abstract mathematical definitions and discussion required to make sense of some of the material present in the chapters.

Note that throughout this appendix, some concepts were only briefly discussed and references were made available for further reading. An in depth treatment of the material in this appendix would be too extensive and not completely applicable to the subject of the book. Therefore, the authors suggest that any reader that is interested to explore these details further, obtain the cited publications.

Appendix C
Numerical Linear Algebra

In this appendix, we will discuss some concepts of analytic and numerical linear algebra that will supplement Chaps. 5 and 9 on singularities and steering algorithms. It is important for the reader of this book to understand the theory of matrices, their subspaces, inverses, pseudoinverse, and decompositions that are practical for implementation on-board spacecraft. Practically, discussion will be based off of numerical stability, precision, floating point operations, and memory, all of which are constrained onboard spacecraft. The majority of the discussion in this appendix can be found for general numerical linear algebra references [159, 160].

C.1 Matrices

A map A takes one element from a space X known as the domain, $\text{dom}(A)$ to another space Y known as a codomain, $\text{cod}(A)$. For example, a map $f : x \to y$ where $x \in X$ and $y \in Y$. Matrices are specific maps that are linear maps that take matrices and vectors (as stacked elements in a column or row) to other matrices and vectors where dimensions are consistent with matrix multiplication, $(m \times n) * (n \times p) = m \times p$, where m, n, p are associated with the sizes of rows and columns of the three matrices.

Matrices are linear maps since for any matrix,

1. homogeneous, $cAx = Acx = cy$
2. they have zero as a fixed point, $A0 = 0$ and
3. they distribute over multiplication and addition, $A(x + y) = Ax + Ay$,

where c is a constant scalar, A is a matrix, and x, y are vectors.

Matrices show up throughout the study of spacecraft momentum control systems. They appear in steering algorithms (see Chap. 9) and inverse kinematics in general (i.e., Jacobian matrix), we see them throughout in attitude kinematics and rotational

© The Author(s), under exclusive license to Springer Nature Switzerland AG 2025

F. A. Leve et al., *Spacecraft Momentum Control Systems*, Space Technology Library 46, https://doi.org/10.1007/978-3-031-96165-6

dynamics (see Chap. 4) as well. Therefore, it behooves the reader to understand some general properties of matrices that will be needed throughout the book as well as the specific types of matrices utilized in the study of spacecraft momentum control systems.

Matrices associated with spacecraft momentum control systems are usually associated with the mapping from actuator states to angular momentum and internal torque. These matrices are rarely sparse and always contain 3 rows since torque and angular momentum are always vectors that lie in $\mathbb{R}^3$. Furthermore, the number of columns is almost always larger than 3 since spacecraft momentum control systems are usually required to have redundancy in actuators. In addition to enhancing spacecraft reliability, redundancy also allows us to bias RWA arrays or provide null-space to avoid kinematic singularities in CMG arrays. Therefore, these matrices are almost always $3 \times n$ matrices, where $n \geq 3$ is the number of actuators.

C.2 Range, Rank, Null-Space

Definition: Range of a matrix A, range(A) is the subset of points in the codomain that can be reached by the action of A on all points in the domain (i.e., range$(A) = \{y \in Y | y = Ax$, for some $x \in X\}$). This subset tells you exactly the effective reach your linear map has in the codomain. Maps where range$(A) =$ codomain(A) are known as surjective maps.

For RWA, the matrix W is the linear map for both the angular momentum and internal torque. It takes vectors in $\mathbb{R}^m$ (rotor rates or accelerations) to angular momentum and internal torque vectors in $\mathbb{R}^3$. However for CMGs, the matrix taking vectors of gimbal rates in $\mathbb{R}^m$ to vectors in angular momentum in $\mathbb{R}^3$ varies with gimbal angles, Δ. In theory, the map A is generally not surjective with codomain $\mathbb{R}^3$ due to kinematic singularities which change the matrix A range as a function of gimbal angle.

Definition: Rank of a matrix A, rank(A) is the minimum of the column or equivalently row rank of the matrix. This is the number of linear independent column or rows of the matrix.

Since it is the minimum of these two numbers, the rank$(A) \leq 3$ for spacecraft momentum control systems. Kinematic singularities are associated with rank$(A) < 3$ at specific gimbal angle configurations, Δ. Furthermore, it is a proven and well-known fact that no matter how many CMG actuators you have (i.e., the level or redundancy), you will always have kinematic singularities unless you restrict the domain to regular points (i.e., points in the domain such that A is full rank). This is a mathematical result, but it has practical implications and the spacecraft momentum control system design should take this in consideration.

Definition: The null-space of a matrix A, null(A) is the set of all vectors x, such that $Ax = 0$ (i.e., null$(A) = \{x \in X | Ax = 0\}$).

The null-space is fundamental to steering algorithm design. It provides an additional orthogonal set of solutions that allow the spacecraft momentum control system to possibly steer away from a singular A (i.e., path connected sets of Δ that are nonsingular for A) while fixing the result in the codomain (see Chap. 9).

C.3 Norms and Metrics

There are many norms and metrics associated with matrices and vectors. Distance metrics provide a notion of error or distance as a scalar number between points or vectors in a specific space. Calculus, analysis, and geometry are based off of metrics, since convergence, limits, and shape usually require this or a different notion of distance between points or vectors.

Distance metrics $d : M \times M \to \mathbb{R}$ have the following properties for any points/vectors $x, y, z \in M$ in a space M:

1. $d(x, x) = 0$,
2. (Symmetry) $d(x, y) = d(y, x)$,
3. (Triangle Inequality) $d(x, z) \leq d(x, y) + d(y, z)$.

Remark C.1 Points and vectors are not the same thing. In reality, a vector space requires a fixed distinguished point (e.g., $x = 0$) for which all vectors are based. Vectors lie in vector (sometimes called linear) spaces since they can be added, subtracted, and multiplied by scalars within a field. Points may exist in a space that is more general and not a vector space such as an affine space that does not have a distinguished point. As a consequence, such points do allow subtraction, but they do not allow addition. Furthermore, points do not have norms, but vectors do. Since we will be dealing mostly with the vector space $\mathbb{R}^n$, this is a nonissue.

Since metrics provide the mathematical construct for error, they are important for understanding error, accuracy robustness, convergence and stability, and safety of control solutions as well amount of control effort utilized. The majority of norms used in the development of attitude control and steering algorithms are the Euclidian space norm for vectors (i.e., the standard 2-norm).

Norms provide a nonnegative number to the length of vectors. Metrics can be induced by norms and defined as $d(x, y) = ||x - y||$ where 0 occurs only at $x = y$.

Remark C.2 Note, not all metrics are induced from norms.

Norms $||x||$, $x \in M$ and $\alpha \in \mathbb{R}$, have the following properties

1. (Nonnegativity) $||x|| \geq 0$,
2. (Positive Definiteness) $||x|| = 0$ if and only if $x = 0$,
3. (Absolute Homogeneity) for any $\alpha \in \mathbb{R}$, $||\alpha x|| = |\alpha|||x||$ where $|\cdot|$ is the absolute value,
4. (Triangle Inequality) $||x + y|| \leq ||x|| + ||y||$.

We will use the properties of norms and metrics discussed in the succeeding sections.

C.4 Least Squares Problem

The least squares problem searches for the solution, x that minimizes the 2-norm error $||b - Ax||$ given b and A. The problem of inverse kinematics is equivalent to solving this least squares problem when A is full rank. Clearly, there are cases when A is rank deficient and one wishes to minimize the least squares error. Since this problem is of the utmost importance for inverse kinematics and steering algorithms for spacecraft momentum control systems, we will now focus on this problem for the remainder of the discussion.

C.5 Inverse and Pseudo-Inverse

As stated previously, MCS Jacobian matrices are usually nonsquare, $3 \times n, n \geq 3$ size matrix. Such matrices, even when full rank, are not invertible and thus require a pseudoinverse in the place of an inverse to solve for the actuator commands from RWA or CMG arrays to meet a specific torque. The most natural pseudoinverse is the Moore-Penrose pseudoinverse, A^+. This inverse always exists and is defined uniquely by the four requirements

1. (Column Consistent) $AA^+A = A$,
2. (Weak Inverse) $A^+AA^+ = A^+$,
3. (Hermitian 1) $(AA^+)^* = AA^+$,
4. (Hermitian 2) $(A^+A)^* = A^+A$.

where A^* is the complex conjugate of A, but for our case of $A \in \mathbb{R}^{m\times n}$, $A^* = A^T$ is just the matrix transpose.

Remark C.3 A matrix satisfying only condition 1 is a generalized inverse and if it also satisfies condition 2, it is known as a generalized reflexive inverse. Note, that unlike the Moore-Penrose pseudoinverse, generalized inverses and even generalized reflexive inverses are not unique unless they satisfy all four properties. Generalized inverse have been used in steering algorithms as well (see Chap. 9). See reference [161] for more on generalized inverses.

The typical objective of trying to obtain A^+ is to solve equations of the form $y = Ax$ (i.e., $x = A^+y$) that come from steering algorithms and inverse kinematics. Even though the Moore-Penrose pseudoinverse exists, that does not mean $x = A^+y$ exactly, rather this pseudoinverse provides the minimum norm solution that minimizes the least squares problem of the 2-norm error $||y - Ax||$. The solution is exact (i.e., satisfies $||y - Ax|| = 0$) if and only if $AA^+y = y$. The general solutions to this least squares problem is

$$x = A^+y + (1 - A^+A)n = A^+y + Pn, \tag{C.1}$$

where $n \in \mathbb{R}^n$ is any arbitrary vector that is projected onto the null-space, null(A) by the orthogonal projection P. Since P is an orthogonal projection, we know by the properties of norms, that the general solution with $n = 0$ is the minimum norm solution.

C.6 Matrix Decomposition

Only in the case when A has full rank, can we represent $A^+ = A^T(AA^T)^{-1}$. More generally, $A^+ = A^T(AA^T)^{-1}$ when A is onto and $A^+ = (A^TA)^{-1}A^T$ is 1-1. However, it should be noted that for CMG arrays, $A = A(\Delta)$ can be rank deficient or near rank deficient (i.e., ill-conditioned) for specific Δ, and therefore, you would never actually use the matrix version of the Moore-Penrose pseudoinverse since it would not always be valid. Therefore, to solve for x from y, we need to resort to numerical methods, some of which come from matrix decompositions.

Some of the most well known matrix decomposition methods used for the solutions to the least squares problem are

1. Cholesky Decomposition
2. QR decomposition
3. Singular Value Decomposition (SVD)

The algorithms for these methods are taken directly from reference [159].

C.6.1 Cholesky Decomposition

The "normal equations"

$$A^*Ax = A^*y. \tag{C.2}$$

or

$$Ax = Py. \tag{C.3}$$

where P is the same orthogonal projection from Eq. (C.1), are a fundamental set of equations that can be solved for the least squares problem.

The standard algorithm for Cholesky Decomposition for the solution of the normal equations, Eq. (C.2) has the following steps

1. Compute Cholesky Decomposition of $A^*A = R^*R$
2. Solve $R^*w = A^*b$ for w, where R^* is lower triangular and $w = Rx$
3. Solve $w = Rx$ for x, where R is upper triangular.

C.6.2 QR Decomposition

The standard algorithm using QR Decomposition for the solution of the least squares problem, Eq. (C.2) has the following steps

1. Compute a reduced QR factorization for $A = \hat{Q}\hat{R}$
2. Form $\hat{Q}^*b = \hat{R}x$ and solve for x.

C.6.3 SVD Decomposition

The standard algorithm for SVD Decomposition for the solution of the least squares problem, Eq. (C.2) has the following steps

1. Compute a reduced SVD factorization $A = U_1\Sigma_1 V_1^*$
2. Compute U_1^*b
3. Solve the system $\Sigma_1 w = U_1^*b$ where Σ_1 is diagonal
4. Solve the system $x = V_1 w$.

C.7 Algorithms

According to reference [160], the first method is the fastest, but the least accurate and only valid for well conditioned matrices, the second method is slower, but more accurate but only valid for well conditioned matrices, and the third method is much slower than the first two, but it is actually the most useful for ill-conditioned matrices A.

A couple of things to mention about the Jacobian matrices associated with steering algorithms. First, for RWA arrays, the matrices W are essentially constant (i.e., assuming rigidly fixed RWA and spacecraft) so any method that is accurate only needs to be computed once and saved onboard. However, things are more complicated for CMG arrays since $A = A(\Delta)$ and any solution to the least squares or minimum error problem must be accurate, stable, and valid for rank deficient or close to rank-deficient A. Furthermore, since this is being done on a spacecraft with limited computation, the chosen method also must not be too computationally intensive and provide a solution for every Δ for each time deterministically. Luckily, the matrices for spacecraft momentum control systems are relatively small (i.e., 3 and rows and n columns). Therefore, Singular Value Decomposition (SVD) is the method of choice.

There are many different algorithms for the least squares and minimum error problems to either directly obtain the Moore-Penrose pseudoinverse or use the discussed matrix decompositions. For the spacecraft momentum control systems, the obvious choice to solve these problems is to use a variation of the SVD decomposition since it provides the most accurate solution for the rank-deficient matrices which surely show up in the case of a Jacobian of a CMG array in a singular configuration. However, when A is rank deficient or close to it, normal SVD is inaccurate. A version of SVD for rank deficient or close to rank deficient matrices needs to be used.

C.7.1 Exactly Rank Deficient Matrix

For a matrix A that is exactly rank deficient, we can solve the least squares directly by directly calculating the Moore-Penrose inverse uniquely from an SVD of the nonzero singular values. Furthermore, since SVD decomposition methods typically base their number of floating point operations on matrices where $m \geq n$ where m are the rows and n are the columns, for the case of spacecraft momentum control systems that are redundant, we will take the SVD of A^T and then form the Moore-Penrose pseudo-inverse. Assume that Σ_1 is the diagonal matrix of singular values that are nonzero. Then the new SVD of rank-deficient A^T is

$$A^T = \begin{bmatrix} U_1 & U_2 \end{bmatrix} \begin{bmatrix} \Sigma_1 & & \\ & \ddots & \\ & & 0 \end{bmatrix} \begin{bmatrix} V_1 & V_2 \end{bmatrix}^T = U_1 \Sigma_1 V_1^T \tag{C.4}$$

where Σ_1 is a diagonal $r \times r$ matrix of the singular values σ_i, all assumed to be nonnegative with $\sigma_{i+1} \leq \sigma_i$ down the diagonal, and U_1 and V_1 are matrices with r orthonormal columns but of n and m rows, respectively. Now the transpose of this matrix gives

$$A = V_1 \Sigma_1 U_1^T \tag{C.5}$$

The Moore-Penrose inverse in this case is

$$A^+ = U_1 \Sigma_1^{-1} V_1^T \tag{C.6}$$

It can be proven that this meets the criteria for a Moore-Penrose pseudoinverse and it is unique. It should be noted that the Moore-Penrose pseudoinverse least squares solution for $x = A^+ b$ is only the min-norm solution and the total solution is actually $x = U_1 \Sigma_1^{-1} V_1^T b + U_2 z$ for any arbitrary vector $z \in \mathbb{R}^{n-r}$ (i.e., no null solutions).

Note that in the CMG array case, the matrix A maps n to 3 and we assume $n \geq 3$. Therefore, the rank $r \leq 3$. For the case of the rooftop arrangements, it is very well possible that we can have ranks as low as $r = 1$ (see Chap. 6).

Remark C.4 For typical CMG arrays, the transpose of the Jacobian matrix, A^T has a number of rows m larger than its number of columns, n. In this case of Eq. (C.4), the SVD A was substituted with A^T. In this book as well as in other literature, there are cases where the SVD is defined on A rather than A^T. This is still valid as an SVD exists for any arbitrary matrix.

C.7.2 Near Rank Deficient Matrix

We run into a problem with Eq. (C.6) when A is near and not exactly rank deficient. When A is near rank deficient, we need to construct some sort of threshold for when to use $\frac{1}{\sigma_p} = 0$ appearing in A^+ for all $p > r$ that balances error of solution with distorted size of A^+. Luckily, for CMG arrays, under the assumption of uniform rotor speeds, the angular momentum of the array has an angular momentum scale factor h_0 that can be pulled out. The leftover matrix A now has orthonormal columns and thus, is bounded in norm at every Δ. This bound should be on order 1 and therefore, you can decide on the allowable change in the pseudoinverse solution A^+ by $\frac{1}{\sigma_p} = 0$ for all $\sigma_p \leq \epsilon$. Clearly, U and V are matrices with orthonormal columns and therefore, they do not affect (increase) the norm of A^+, only Σ. The 2-norm of the solution, $||x||_2 \geq |v_i^T b|/\sigma_{min}$ where σ_{min} is the smallest singular value, and $v_i \in V$ is the vector associated with σ_{min}. When we choose not to zero out the inverse of the singular values $\sigma_p \leq \epsilon$, the norm of the solution is inversely proportional to the smallest singular value and if $|v_i^T b| \neq 0$, when $\sigma_{min} \sim 0$ it is possible to have a distorted solution with an arbitrarily large norm.

When we choose to zero out the inverse singular values below a threshold, the effective error of our solution $b - Ax$ from setting $\frac{1}{\sigma_p} = 0$ for all $\sigma_p \leq \epsilon$ of A^+ is

$$\begin{aligned} e = b - Ax &= b - AA^+b \\ &= \left(1 - V(\Sigma^T\Sigma)(\Sigma^T\Sigma)^+V^T\right)b \\ &= V\left(1 - (\Sigma^T\Sigma)(\Sigma^T\Sigma)^+\right)V^Tb \\ &= V_2V_2^Tb \end{aligned} \tag{C.7}$$

since $VV^T = 1$ and $V = [V_1 \; V_2]$ where here we are associating V_2 with the span of left singular vectors of A associated with the $\sigma_p \leq \epsilon$. Therefore, by Eq. (C.7), the norm of the least squares error is $||e|| = ||V_2V_2^Tb||$ and is zero only when $b \in \text{Null}(V_2V_2^T)$. However, the case $e = 0$, is equivalent to $b \in \text{Span}(V_1)$ which is orthogonal to the singular direction of A and null-space of A^T since we took the SVD of the latter. Therefore, the only way to achieve a zero least squares error solution with the Moore-Penrose inverse at singularity is for b to have no components along the singular directions. This is absurd for the case of dynamic matrices that are rank deficient.

Remark C.5 If b has components along singular directions associated with $\text{Span}(V_2)$, we either get a large norm solution inversely proportional to the smallest singular values, or one on order the norm of b. We balance the norm of the solution $||x||$ and the error whether at or near singularity of A.

Ultimately, for arrays of CMGs, the need for accurate steering algorithms (See Chap. 9) combined with an arrangement that only exhibits manageable singularities (see Chaps. 5 and 6) is critical since it is impossible to have accurate SVD around singularities while still meeting the internal torque requirements sufficiently and accurately. The goal would be to utilize then null-motion to keep σ_{min} far enough away from zero so that an accurate SVD-based pseudoinverse could be used to map the gimbal rates accurately to the desired internal torque.

C.8 Performance

There are many attributes of algorithms used to find solutions to the least squares problems. Some of these desired attributes are accuracy, stability, and low computation or speed of computation. Typically, you cannot find an algorithm that is superior in achieving the best results of all attributes. Since the least squares problem is the objective of steering algorithms, we will focus on behavior of the previously discussed methods of solving least squares and their performance.

C.8.1 Accuracy

The accuracy of all of the proposed methods to solving least squares depends on the condition of the problem. The condition of the problem is usually based off the condition number of the matrix A. For nonsquare (i.e., noninvertible) matrices, the singular values define the condition number as $\kappa = \frac{\sigma_1}{\sigma_n}$ where σ_n is the smallest singular value of A. If A is rank deficient or close to rank deficient, $\sigma_n \sim 0$ and the problem is ill-conditioned. The best one can hope for in terms of accuracy for a digital computer is on the order of the smallest number

$$O(\epsilon_{machine}) \leq C\epsilon_{machine}$$

where $C > 0$ is some constant.

In addition to approximation errors from the algorithms themselves, there are floating point approximations rounding errors that effect not only the accuracy but also the stability. Floating point errors are a consequence of the digital computers carrying out the floating point calculations (i.e., flops). The standard floating point operations for $+, -, *, \div$ are $\oplus, \ominus, \circledast, \oslash$. Sometimes the square root operation, $\sqrt{\cdot}$ is also considered a flop.

If we assume for any $x, y \in \mathbb{R}$ that $\text{fl}(x) = x(1+\epsilon)$, $\text{fl}(y) = y(1+\epsilon)$ and $x \circledast y = \text{fl}(x * y) = x \circledast y(1+\epsilon)$ for some $|\epsilon| \leq \epsilon_{machine}$ then for a problem $f(x)$ and a backwards stable algorithm $\tilde{f}(x)$, the relative error

$$\frac{||f(x) - \tilde{f}(x)||}{||f(x)||} = O(\kappa(x)\epsilon_{machine}), \tag{C.8}$$

is on order of the condition number $\kappa(x)$ scaling the machine precision (see reference [159], Theorem 15.1). For SVD of a matrix A, $\kappa(x) = \frac{\sigma_1}{\sigma_{min}}$ which is infinity when A is rank deficient and very large when close to rank deficiency. Clearly, a very large condition number will destroy accuracy of any algorithm since it will make the problem itself ill-conditioned. For the case of the least squares problem that we need to solve for steering algorithms, we will hope to steer the Jacobian matrix of the actuator kinematics to be far enough away from a large condition number that the error of the algorithm will be guaranteed to be within our limits. As shown in Table 1.1 of Chap. 1, RWA and CMGs are accurate to around a thousandth of a degree. Furthermore, measurement accuracy of the gimbal angles and rates for CMGs needs also to be considered. Therefore, consideration of the floating precision of the processor hardware, and software, as well as the error that is smaller than a thousandth of a degree will provide the order of the condition number allowed for steering algorithms.

C.8.2 Stability

The idea of stability for numerical problems is given some errors in the algorithm data, you are still able to solve the problem close to the solution when error is absent. We consider the algorithm data of the least squares problem to be the matrices A, b and the solutions to be x, y where we wish to solve $y = Ax$ by solving the least squares solution minimizing $||b - Ax||$. The matrices b, y correspond to the desired output for the solution x to match verses the actual output $y = Ax$ produced by the solution.

There are two typical notions of stability for numerical algorithms, stability and backwards stability.

Definition (Stability): Given a problem $f(x)$ and an algorithm for that problem $\tilde{f}(x)$, we define an algorithm as being stable if for each $x \in X$,

$$\frac{||f(\tilde{x}) - \tilde{f}(x)||}{||f(\tilde{x})||} = O(\epsilon_{machine})$$

for some

$$\frac{||\tilde{x} - x||}{||\tilde{x}||} = O(\epsilon_{machine})$$

Therefore, this stability implies that you get almost the right answer to almost the right question.

Definition (Backwards Stability): Given a problem $f(x)$ and an algorithm for that problem $\tilde{f}(x)$, we define an algorithm as being backward stable if for each $x \in X$,

$$f(\tilde{x}) = \tilde{f}(x)$$

for some

$$\frac{||\tilde{x} - x||}{||\tilde{x}||} = O(\epsilon_{machine})$$

Therefore, this stability implies that you get exactly the right answer to almost the right question.

For our discussion purposes, we will focus more on backwards stability of algorithms since it is a common form of stability discussed for numerical linear algebra algorithms.

Remark C.6 Accuracy and stability are notions that are usually defined for well conditioned algorithms that correspond to matrices A that are full rank. For the case of steering algorithms, we wish to have methods that allow for a rank-deficient or more realistically a close to rank-deficient A. Furthermore, stability is usually defined by conditioning in terms of sensitivities of perturbed data to perturbed solutions. From a control theory perspective, these notions of sensitivities and conditioning appear like measures of robustness.

Sensitivities to the to perturbing the data, A, b on the solutions x, y are defined by Table 18.1 in reference [159]. This table is recreated in Table C.1
where the fit of the output is measured by

$$\theta = \cos^{-1}\left(\frac{||y||}{||b||}\right)$$

and $0 \leq \theta \leq \frac{\pi}{2}$, where the condition number of A,

$$\kappa = \frac{\sigma_1}{\sigma_{min}}$$

and $1 \leq \kappa \leq \infty$,

$$\eta = \frac{||A||||x||}{||y||} = \frac{||A||||x||}{||Ax||}$$

and $1 \leq \eta \leq \kappa$.

Table C.1 Sensitivities of solutions to perturbations in data for least squares problem

	y	x
b	$\frac{1}{\cos(\theta)}$	$\frac{\kappa}{\eta\cos(\theta)}$
A	$\frac{\kappa}{\cos(\theta)}$	$\kappa + \frac{\kappa^2\tan(\theta)}{\eta}$

Clearly, the sensitivities of solutions to perturbations of data in the least squares problem only make sense when the condition number is not infinity (i.e., when A is full rank). As in the case of steering algorithms for CMGs, there are times when you may encounter a rank deficient or more often an almost rank deficient A. In these cases, you may either need to constrain the set of matrices b that you can reach, or use null-motion solutions n to steer A toward being full rank. Perturbations to A could happen from an algorithm used to decompose it into useful pieces of information (e.g., SVD, QR), or from inaccurate gimbal angles that define this Jacobian matrix. It is inevitable the algorithm and the gimbal angle measurements and their estimates will incorporate some error from noise and other forms of uncertainty. Furthermore, the required precision of y directly depends on the steering algorithm's solution x and the perturbed A. Since Table C.1 provides us information on how sensitive y is to perturbations in A, we can combine this information with the performance of the steering algorithm's solution x to understand the allowable condition number κ as a distance from singularity that we can allow from the steering algorithms.

The majority of used algorithms for computing SVD and QR for the purpose of solving the least squares problem are backwards stable. However, this is not the case for the using the Cholesky decomposition to solve the normal equations. Solving the normal equations as a method of solving the least squares problem is not backwards stable.

C.8.3 Computation/Complexity

The final section of this chapter will just provide a comparison of the polynomial complexity of the previously discussed methods of solving the least squares problem. It should be noted that computation and complexity are not the same thing. For a least squares problem given an $m \times n$ matrix A, Table C.2 provides an approximate polynomial complexity for the standard methods of solving the least squares problem.

Clearly, the SVD has the highest complexity. However, since we assume that we will solve SVD for A^T and we are working with matrices that are not that large, there is no loss in using SVD as part of the steering algorithm implementation since it is backwards stable, accurate, and provides all of the information required for analysis.

Table C.2 Complexity in flops for standard methods of solving the least squares problem

Method	Complexity
Normal equations	$\sim mn^2 + \frac{1}{3}n^3$ flops
QR factorization	$\sim 2mn^2 - \frac{2}{3}n^3$ flops
SVD	$\sim 2mn^2 + 11n^3$ flops

Remark C.7 Complexity is typically measured in number of floating point operations, where computation (or computational complexity) may correspond to iterations that are not described by a polynomial of operation counts, function calls, and sorting that are used in the code used to implement the algorithm.

C.9 Appendix Summary

This appendix discusses some of the most important aspects of numerical algorithms that should be a precursor for any steering algorithm design and its implementation on a spacecraft momentum control system. The references [159] and [160] were relied upon heavily for this appendix, and focused discussion related to steering algorithms was added. Though also very important, no discussion of state-of-the-art spacecraft processors or their performance were discussed. It would behoove the reader to understand the current spacecraft processors available as well the sources of error such as noise from sensor measurements or inaccurate actuation that may add to losses in precision of control for spacecraft momentum control systems.

Appendix D
Nonlinear Attitude Control

Spacecraft attitude control is typically constrained to follow pre-generated attitude control trajectories provided by the satellite operator. For this reason, the control currently used for spacecraft attitude, angular rate, or angular momentum control can in most cases be approximated as linear (time-varying) and based off of small angle, angular rate, or angular momentum displacements with respect to a reference trajectory (or point) in terms of error coordinates (see Chap. 8). However, there are cases when the full use of nonlinear control may be beneficial or even required. This appendix will discuss some of the critical aspects of nonlinear control as it pertains to spacecraft attitude, angular rate, and angular momentum control by momentum control systems.

D.1 Error Kinematics

The error kinematics are critical elements of the control design. The error to be consistent, will have to have the same "form" of solutions to the kinematic differential equations in the attitude coordinates. What is meant by form is that the error state must be in the same space as the state of the kinematics (e.g., an error quaternion is still a quaternion). These differential equations will usually be defined as multiplicative in attitude and angular velocity coordinates. Since DCMs represent points on SO(3) that globally and uniquely describe attitude states, it behooves us to first derive the error kinematics for DCMs before we derive it for the other sets of coordinates. The attitude kinematics shown here were partially taken from the derivations in reference [162].

Euler angle error kinematics will not be discussed here since they are not a good choice of attitude coordinates for arbitrary large angle maneuvers.

© The Author(s), under exclusive license to Springer Nature Switzerland AG 2025
F. A. Leve et al., *Spacecraft Momentum Control Systems*, Space Technology Library 46, https://doi.org/10.1007/978-3-031-96165-6

D.1.1 Direction Cosine Matrices

The error kinematics for DCMs are formed from the error DCM. The error DCM for the attitude from the body to the inertial frame is simply $Q_e = Q_d^T Q$, where Q and Q_d are the current and desired attitudes. Considering the kinematics from Eq. (8.2) for Q and Q_d^T with the definition of Q_e, we would like to find the kinematics for the error DCM and error angular velocity. They are

$$\dot{Q}_e = \dot{Q}_d^T Q + Q_d^T \dot{Q} = -\omega_d^\times Q_d^T Q + Q_d^T Q \omega^\times = -\omega_d^\times Q_e + Q_e \omega^\times \tag{D.1}$$

and we need

$$Q_e^T \dot{Q}_e = \omega_e^\times = -Q_e^T \omega_d^\times Q_e + \omega^\times = -(Q_e^T \omega_d)^\times + \omega^\times = (\omega - Q_e^T \omega_d)^\times \tag{D.2}$$

using the identity $Q x^\times Q^T = (Qx)^\times$ for any vector x and DCM, Q. Therefore, the error angular velocity is, $\omega_e = \omega - Q_e^T \omega_d$. The error kinematics for Q_e is clearly,

$$\dot{Q}_e = Q_e \omega_e^\times \tag{D.3}$$

with the error angular velocity kinematics

$$\dot{\omega}_e = \dot{\omega} + \omega_e^\times Q_e^T \omega_d - Q_e^T \dot{\omega}_d \tag{D.4}$$

D.1.2 Euler Parameters (Unit Quaternions)

Error Euler Parameters (unit quaternions) are composed of eigen-axis components as $\hat{e} = [e_1\ e_2\ e_3]^T$ and a scalar element as e_4. Their error kinematics are

$$\dot{e} = \frac{1}{2} \begin{bmatrix} -\omega_e^\times & \omega_e \\ \omega_e^T & 0 \end{bmatrix} e = F(\omega_e) e \tag{D.5}$$

Note that the error quaternion, e can be computed similarly to DCMs, but with quaternion multiplication

$$e = q_d * q^{-1} = \begin{bmatrix} q_{d_4} & -q_{d_1} & -q_{d_2} & -q_{d_3} \\ q_{d_1} & q_{d_4} & -q_{d_3} & q_{d_2} \\ q_{d_2} & q_{d_3} & q_{d_4} & -q_{d_1} \\ q_{d_3} & -q_{d_2} & q_{d_1} & q_{d_4} \end{bmatrix} \begin{bmatrix} -q_1 \\ -q_2 \\ -q_3 \\ q_4 \end{bmatrix} \tag{D.6}$$

which is guaranteed to be a unit quaternion if both e and e_d are unit quaternions. The error DCM is derived from the error quaternions in Eq. (D.6)

$$Q_e(e) = \begin{bmatrix} (e_4^2 + e_1^2 - e_2^2 - e_3^2) & 2(e_1e_2 + e_3e_4) & 2(e_1e_3 - e_2e_4) \\ 2(e_1e_2 - e_3e_4) & (e_4^2 - e_1^2 + e_2^2 - e_3^2) & 2(e_2e_3 + e_1e_4) \\ 2(e_1e_3 + e_2e_4) & 2(e_2e_3 - e_1e_4) & (e_4^2 - e_1^2 - e_2^2 + e_3^2) \end{bmatrix} \quad \text{(D.7)}$$

D.1.3 Euler Rodrigues and Modified Rodrigues Parameters

The error ERP is

$$\rho_e = \frac{\hat{e}}{e_4} \quad \text{(D.8)}$$

and the error MRP is

$$\sigma_e = \frac{\hat{e}}{1 + e_4} \quad \text{(D.9)}$$

both as a function of the error quaternion parameters shown in Eq. (D.6). The error kinematics for the ERP is

$$\dot{\rho}_e = \frac{1}{2}\left(1 + \rho_e^\times + \rho_e\rho_e^T\right)\omega_e \quad \text{(D.10)}$$

The error kinematics for the MRP is

$$\dot{\sigma}_e = \frac{1}{4}\left((1 - |\sigma_e|^2)\mathbf{1} + 2\sigma_e^\times + 2\sigma_e\sigma_e^T\right)\omega_e \quad \text{(D.11)}$$

and the angular velocity dynamics has the same form as Eq. (D.4), substituting $Q(\rho_e)$ and $Q(\sigma_e)$ in place of Q_e where $Q(\rho_e)$ and $Q(\sigma_e)$ is the DCM formed by the error ERP and MRP. For DCMs in terms of the error ERP and MRP, see reference [163].

D.2 Closed-Loop Equations of Motion

The total set of matrix equations of motion that includes attitude, angular velocity, and angular momentum are of the form

$$\begin{bmatrix} \dot{Q} \\ \dot{\omega} \\ \dot{h} \end{bmatrix} = \begin{bmatrix} 0 & S(Q) & 0 \\ 0 & J^{-1}(H)^\times & 0 \\ 0 & 0 & 0 \end{bmatrix} \begin{bmatrix} Q \\ \omega \\ h \end{bmatrix} + \begin{bmatrix} 0 \\ (J^{-1}u + \varepsilon) \\ -u \end{bmatrix} \quad \text{(D.12)}$$

or

$$\begin{bmatrix} \dot{Q} \\ \dot{\omega} \\ \dot{h} \end{bmatrix} = \begin{bmatrix} 0 & S(Q) & 0 \\ 0 & -J^{-1}(\omega)^{\times} J & -J^{-1}(\omega)^{\times} \\ 0 & 0 & 0 \end{bmatrix} \begin{bmatrix} Q \\ \omega \\ h \end{bmatrix} + \begin{bmatrix} 0 \\ (J^{-1}u + \varepsilon) \\ -u \end{bmatrix} \tag{D.13}$$

where $H = J\omega + h$, $(\cdot)^{\times}$ refers to the linear skew-symmetric operation of turning a column matrix into a skew-symmetric matrix, and ε is the possible external torques present from the environment or from on-board spacecraft elements, $S(Q)$ is the Jacobian matrix that comes from the kinematic differential equations with Q being one of the chosen attitude coordinate representations from Table 8.1, h is the angular momentum contribution from the spacecraft momentum control system array where the control is produced. Note, that $\dot{Q} = S(Q)\omega = F(\omega)Q$ for coordinate representations that have bilinear kinematics (e.g., DCMs and Euler parameters). Also, h itself is linear in control inputs for the case of RWA, but is nonlinearly dependent on gimbal angles for the case of CMGs. Finally, Eq. (D.13) is set up in an almost state-space form that will help provide insight for the attitude control problems to be discussed later. Note, the gimbal angles or reaction wheel rates could also be considered as states in place of h. However, considering these as states rather than inputs for CMGs, would require a steering algorithm or pseudo-inverse to map from array angular momentum or internal torque to the actuator state (see Chap. 9).

In the literature, the closed-loop equations of motion have $u = \tau = -\dot{h} - \omega^{\times}h$ as the internal control torque. This is physically accurate, and is most likely used to provide an analogy with attitude control of a rigid body with external torques. However, describing the attitude control torque for the closed-loop dynamical system this way may confuse the reader into thinking that τ can be treated as an external torque. It also does not allow for simplification of control schemes by bringing $\omega^{\times}h$ into the dynamics when $H = J\omega + h = 0$. It should be noted that the any control $u = \dot{h}$, must not allow h to exceed the momentum capacity of the array or moment control actuators.

In the majority of practical situations, your model for your dynamics has uncertainty, you do not know the entire state of the system, and the states that you do have knowledge of, are measured by sensors corrupted by noise. Therefore, the closed-loop dynamical system (i.e., dynamical system along with the control system in a feedback loop) requires additional equations for representation. The total representation of such a system is typically a stochastic differential equation (SDE) of the form

$$\begin{aligned} dx(t) &= f(x(t), u(t), t)dt + g(x(t), u(t), t)d\epsilon(t) \\ y(t) &= \int_0^t h(x(s), u(s), s)ds + \nu(t) \end{aligned} \tag{D.14}$$

where the right hand side, $f(x(t), u(t), t)$ is closed-loop drift dynamics for combined angular velocity and attitude states, possibly an explicit function of time, u is the admissible controls, $g(x(t), u(t), t)$ is the closed-loop diffusion process premultiplying $\epsilon(t)$, a stochastic model uncertainty, assumed to be a small magnitude Brownian motion stochastic random process, and $y(t)$ are the measurement outputs, $h(x(t), u(t), t)$ is the output map corrupted by noise $\nu(t)$ also assumed to be a small magnitude Brownian motion stochastic random process (see Eq. 4.2 in reference [164]). Note $\epsilon(t)$ and $\nu(t)$ are assumed to be independent noise processes. Furthermore, it is typically assumed that model uncertainty and measurement noise are additive and small in comparison to the other terms on average. In practice, this is usually a valid assumption since sensors have their own models (dynamics), and are calibrated to an extent to exploit all structure in the signals from the sensors leaving only the minimal additive stochastic noise components.

Remark D.1 Stochastic differential equations rather than ordinary differential equations are considered when random processes such as Brownian motion are added to the state and output functions. This is because Brownian motion is nowhere differentiable. One can consider a non-stochastic output feedback model by conservatively bounding the uncertainty by a tube of error in the dynamics and the measurement noise.

To avoid discussing attitude estimation, as there are a very large set of adequate references on the subject (see reference [81]), we will only focus on the full-state feedback (i.e., complete state information) case of attitude control. For full-state feedback attitude control, we will assume that the attitude, angular velocity, and acceleration estimated perfectly and later on assume that the true measurement contains some small additive non-stochastic measurement error, model uncertainty, or perturbative disturbance torques.

The full-state feedback case is represented as (assuming zero model uncertainty)

$$\dot{x}(t) = f(x(t), u(t), t) \tag{D.15}$$

where the right hand side of the differential equation is the assumed to be the actual closed-loop dynamics for combined angular velocity and attitude states $x(t)$, possibly an explicit function of time, u is the admissible controls. The goal of attitude control is to find a specific function $u \in U$ where U is the admissible set of controls (e.g., intersection of sets associated with constraints in Table D.1), such that the states $x \rightarrow x_d(t)$ in a specific amount of time (could be infinity), where $x_d(t)$ is the desired attitude and angular velocity states. Therefore, the subject of attitude control is to find this associated function $u \in U$ while remaining close (*stability*), converge to the desired states (*asymptotic stability*), but also avoid bad states (*safety*), and being robust to uncertainty and disturbances and uncertainty (*robustness*).

The dynamics equations of motion for spacecraft rotation are the attitude kinematics equations discussed previously along with the Euler's rotational equations of motion that were developed in Chap. 4 and more generally in Appendix A. Therefore, we will label our entire state vector as x and it will be made up of components consisting the angular velocity in the body frame about the center of

Table D.1 Some typical sets of attitude parameters

Constraint	Form
Attitude keepin	$\mathbf{c} \cdot \mathbf{n} \leq \alpha$
Attitude keepout	$\mathbf{c} \cdot \mathbf{n} \geq \alpha$
Angular velocity	$\|\omega\| \leq \alpha$
Angular acceleration	$\|\dot{\omega}\| \leq \beta$
Angular jerk	$\|\ddot{\omega}\| \leq \gamma$
Rotor rate	$\|\Omega\| \leq \alpha$
Rotor acceleration	$\left\|\dot{\Omega}\right\| \leq \beta$
Gimbal rate	$\left\|\dot{\Delta}\right\| \leq \alpha$
Gimbal acceleration	$\left\|\ddot{\Delta}\right\| \leq \beta$

mass as well as the coordinates representing the attitude of the spacecraft body with respect to an inertial frame. Note, that the behavior and performance of attitude control systems is directly dependent on the choice of coordinates. We will not specify the attitude coordinates until we go into the specifics of some attitude control systems.

There are three typical assumptions for the design of control systems that will be discussed for all maneuver types.

1. There are external torques present and $H \neq 0$ (general).
2. There are no external torques present and $H \neq 0$ (momentum bias).
3. There are no external torques present and the magnitude of angular momentum for the entire spacecraft, $H = J\omega + h = 0$ (zero momentum bias).

Here, H is the total spacecraft bus+momentum control system angular momentum represented in the body coordinates.

D.3 Attitude Tracking and Point-to-Point Maneuvers

Many attitude tracking and point-to-point attitude and angular velocity maneuvers require the spacecraft to trace out large angles. For these objectives, linearizations of the dynamics about small angles is not practical for such cases and the majority of the information about the dynamics and the control is not contained in matrices A and B from Chap. 8. Furthermore, large angle maneuvers will more than likely require the CMGs to pass through kinematic singularities and therefore, steering algorithms will be required. However, attitude control will not be discussed at the actuator level here but only at the internal torque level. Choice of a good set of coordinates to parameterize attitude is imperative for this class of maneuvers. From here on, we will assume that Euler angles are an invalid set of coordinates for most large angle attitude maneuvers as they will more than likely require passage through a coordinate singularity.

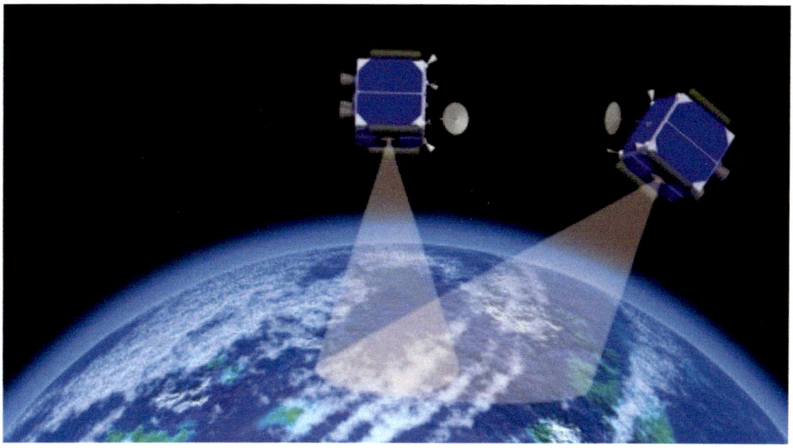

Fig. D.1 Satellite Earth imaging

Fig. D.2 Point-to-point maneuvers

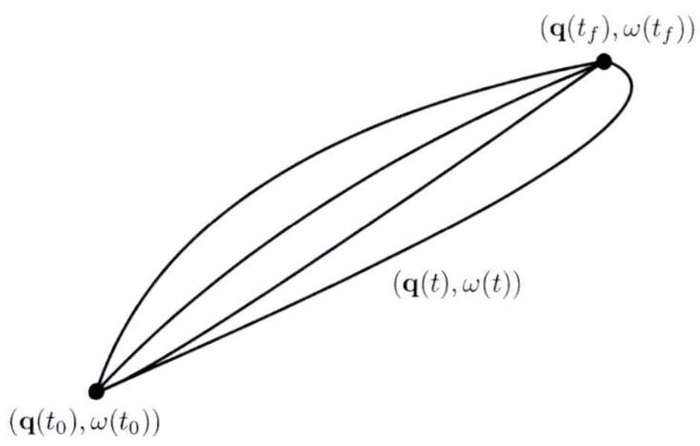

Attitude tracking maneuvers require you to track time dependent desired attitude and angular velocity states. Attitude tracking is usually a requirement for Earth imaging as shown in Fig. D.1.

Such maneuvers leave little wiggle room for error as any error in tracking a predefined attitude trajectory would negate the benefit of using spacecraft momentum control actuators, the most accurate attitude control actuators in the first place.

However, point-to-point maneuvers as shown in Fig. D.2 allow for some wiggle room in attitude control (i.e., the trajectory between the two boundary points is free), but they usually are not restricted to small angle attitude control, nor small gimbal angle excursions for a CMG array to meet the attitude control needs.

Therefore, for these maneuvers, avoidance or escape of CMG singularities will need to be considered either in the attitude control design, or in the steering algorithm (control allocation) design.

Controllability is the natural property to check for existence of a point-to-point maneuver since it provides you an answer if it is possible starting from an initial point in the state space if you can reach the terminal point in the state space in finite time, $T = (t_f - t_0)$ (see reference [165] for controllability results between thrusters and momentum control systems). However, for large angle slews, methods of checking nonlinear controllability must be used (see Appendix B). Moreover, for spacecraft momentum control systems, attitude and angular velocity is not controllable in general. That being said, in the absence of external torques, if we restrict controllability to the states that satisfy the condition of the conservation of angular momentum in Eq. (D.55), it is always possible to do point-to-point maneuvers, even with a single CMG [150]. Sadly, unlike a CMG, for an array of RWA with less than 2 actuators will not be controllable since you may not be able to reach desired $H(t_f)$ with components orthogonal to the spin axis of the single RWA since it only has a single axis of controllable angular momentum.

D.4 Nonlinear Control System Analysis

Inverse dynamics control design (see reference [166] and [167]) and Lyapunov methods (see reference [168]) are typical methodologies for construction and analysis of nonlinear control systems stability behavior (see Appendix B for details). Inverse dynamics approach to constructing control laws though applicable to many systems, is very prescriptive and may resort to very complex control laws as a consequence of requiring linear closed-loop error dynamics. However, linear systems tools can be used for the closed-loop error dynamics. Here we will consider Lyapunov methods. It should be noted, that the crux of using inverse dynamics or Lyapunov analysis is to define an error function as a function of the error "vector" we wish to converge to zero. This is trivial for linear dynamics with Euler angles but not so for nonlinear control where general attitude maneuvers are allowed.

Any typical Lyapunov function candidate for attitude control we will discuss here has the form

$$V = E(Q_e) + G(Q_e, \omega_e) + F(\delta\varepsilon) \tag{D.16}$$

where $E(Q_e) \geq 0$, is the error function in error attitude state, Q_e, $G(Q_e, \omega_e) \geq 0$ is a mixed function of both the error attitude and angular velocity states, and $F(\delta\varepsilon)$ is a term devoted to the error in an estimated unknown disturbance torque, and $\delta\varepsilon = \varepsilon - \hat{\varepsilon}$ is the error in estimate. The function V is assumed positive definite and zero only at our desired error state for attitude, angular velocity, and uncertainty/disturbance estimate. The reason we chose $G(Q_e, \omega_e)$ to represent a function of attitude and angular velocity error is angular velocity error is in explicitly dependent on attitude error from the kinematics.

For the remainder of this section, we will assume Lyapunov function candidates have this general form. We will also focus on the error coordinates, their kinematics, and $E(Q_e)$ for much of the discussion since defining this term is a precursor for any control design. Since there are many attitude control derivations for different choices of $E(Q_e)$ and $G(Q_e, \omega_e)$ and for different choices of attitude coordinates, we will not re-derive the results present in literature, but we will point to some inherent issues with attitude control as it pertains to the typically choices of these elements, and reference some of the latest results. Furthermore, we will not necessarily discuss how to chose $\dot{h}$, but we will discuss what controls are possible for spacecraft momentum control systems based off of our three assumptions, and point to some of the more recent attitude control systems for different sets of coordinates.

The unknown term $\delta\varepsilon$ poses a problem for stability and convergence. If the term is unknown, but constant, we may take the approach of adding a filter and using either direct or indirect adaptive control to construct a control solution that accounts for the uncertainty of ε and its effect on the stability and the convergence of derived controller (see references [169] for example of adaptive control solutions for attitude control in the presence of constant but unknown disturbances). On the other hand, we need to understand what is possible if we have a time varying and unknown ε.

With a good filter for ε, and for fixed disturbances, we may be able to prove global asymptotic stability regardless of estimate convergence to the true disturbance. For the time being, we will in some cases, assume that the disturbance torques ε can be modeled accurately in the large where only small variations are uncertain. This will allow us to exploit these torques for use in attitude control. We will discuss more on how we can mitigate the uncertainty in control systems later and for the time being, assume that ε is known and therefore, $\delta\varepsilon = 0$.

D.4.1 Direction Cosine Matrices

Since all attitude coordinate parameterizations have to account for an attitude trajectory on SO(3), we will first start our discussions with DCMs. Recall that attitude is naturally defined by DCMs as points on SO(3) as a manifold. Before discussion of control systems, it should be noted that, SO(3) is a compact Lie-group manifold, and therefore, there are well established results that show that it is impossible to construct a time-invariant continuous or smooth control system that is globally asymptotically stable [170]. This limitation is due to topological obstructions with the manifold SO(3) itself and cannot be removed by a change of coordinates and therefore, will be a problem for all choices of coordinates.

Typically, Lyapunov analysis (see Appendix B) is used to verify that a controller provides asymptotic stability. However, we have a problem here since Q_e is DCM error state equal to $\mathbf{1}$ at our desired attitude, and a Lyapunov function candidate requires that we have a scalar function of the error state (i.e., an error function) that is zero only when at the desired state.

Literature provides us multiple ways to choose $E(Q_e)$ and $G(Q_e, \omega_e)$ for a Lyapunov function candidate. Consider the Lyapunov function candidate

$$V = k\mathrm{tr}(B - BQ_e) + \frac{1}{2}\omega_e^T J \omega_e \tag{D.17}$$

where k is a positive gain, $E = k\mathrm{tr}(B - BQ_e)$, B is a diagonal matrix with distinct positive diagonal entries, and $G(Q_e, \omega_e) = \frac{1}{2}\omega_e^T J \omega_e$.

For $E(Q_e) = k\text{tr}(B - BQ_e)$, it has been proven away from the undesired equilibrium and their stable directions associated with the saddle points, B is required to have distinct positive diagonal elements to enforce convergence to the desired equilibrium of $\dot{E}(Q_e) = 0$, $Q_e = \mathbf{1}$ and not the other three are diagonal matrices $Q_e \in \{(-1, 1, 1), (1 - 1, 1), (-1, 1, -1)\}$. Therefore, the term $E(Q_e) = \text{tr}(B - BQ_e)$ is a valid choice to define the error between DCM representations of attitude that is consistent with zero error being the correct attitude state (i.e., $Q_e = \mathbf{1}$) [171] away from the undesired equilibrium.

Other ideas as in reference [169] do not form the error directly from Q_e, but use the fact that attitude is also defined by two orthogonal unit vectors attached to the rigid body of the spacecraft and represented on the 2-D unit sphere in 3 dimensions, $\mathbb{S}^2$ with the third vector found by a cross product. From this assumption, we can define

$$E(Q_e) = \sum_{i=1}^{2} \frac{1}{2} k_i \left\| B_i - Q^T r_{d_i} \right\|^2 \tag{D.18}$$

where k_i are positive gains, B_i are the actual vectors in the body frame related by, $r_i = QB_i$, and r_{d_i} are the desired vectors in the inertial frame. Both choices of $E(Q_e)$ suffer the same issue and are unable to provide a continuous time invariant control law that is globally stabilizing (see reference [170]). Note, there is no third component of the error for B_3 since it is already determined by the other two body-fixed vectors due to the orthogonality of B_1 and B_2. It has been shown that with a hybrid or time-varying controller using $E(Q_e) = \sum_{i=1}^{2} \frac{1}{2} k_i \left\| B_i - Qr_{d_i} \right\|^2$, that you can achieve even global exponential stability [169].

The Lyapunov function candidate component derivatives of Eq. (D.17) are

$$\dot{E}(Q_e) = \frac{d}{dt} k\text{tr}(B - BQ_e) = k\omega_e^T \left(\sum_{i=1}^{3} b_i (Q_e^T e_i)^\times e_i \right), \tag{D.19}$$

where b_i are the diagonal elements of B, and e_i are the standard vectors with zeros everywhere and 1 in the ith entry (see reference [172]), and

$$\dot{G}(Q_e, \omega_e) = \omega_e^T \left[-\omega^\times J\omega - \omega^\times h - \dot{h} + \varepsilon + J\left(Q_e^T \omega_e^\times \omega_d + Q_e^T \dot{\omega}_d \right) \right] \tag{D.20}$$

giving

$$\begin{aligned} \dot{V} = \omega_e^T \Bigg[& k \left(\sum_{i=1}^{3} b_i (Q_e^T e_i)^\times e_i \right) - \omega^\times J\omega - \omega^\times h - \dot{h} + \varepsilon \\ & + J\left(Q_e^T \omega_e^\times \omega_d + Q_e^T \dot{\omega}_d \right) \Bigg] \end{aligned} \tag{D.21}$$

Clearly, without even considering constraints at the moment, we would like to design an $\dot{h}$ such that $\dot{V} \leq 0$, preferably $\dot{V} < 0$ for asymptotic stability. Keep in mind that due to topological obstructions, and the fact that for SO(3) and therefore any chosen set of local coordinates, there is no continuous time-invariant control law that gives a global asymptotic result, and there are four critical points of the function, $E(Q_e) = \text{tr}(B - BQ_e)$, three of which where diagonal matrices $Q_e \in \{(-1, 1, 1), (1 - 1, 1)(-1, 1, -1)\}$. Results exist in the literature for DCMs that provide almost global and even local exponential stability of the errors and estimates for attitude tracking and point-to-point maneuvers as a special case (see references [169]). However, we could chose a control solution

$$\dot{h} = K\omega_e + k\left(\sum_{i=1}^{3} b_i (Q_e^T e_i)^\times e_i\right) - \omega^\times J\omega - \omega^\times h + \varepsilon + J\left(Q_e^T \omega_e^\times \omega_d + Q_e^T \dot{\omega}_d\right) \tag{D.22}$$

which will give us a negative semi-definite result,

$$\dot{V} = -\omega_e^T K\omega_e \tag{D.23}$$

Note, that this only proves that the angular velocity error decreases asymptotically to zero. We require that the attitude error also asymptotically decreases to zero away from the undesired equilibria. If we are tracking a prescribed trajectory that is a function of time, $V = V(x, t)$ is an explicit function of time (i.e., nonautonomous) we cannot use LaSalle's Invariance Principle in Appendix B. However, we can search for comparison functions for a V that is explicitly time-varying and/or use Barbalat's lemma to prove asymptotic stability of Q_e.

If we assumed point-to-point attitude maneuver and the absence of external torque disturbances, we can see that $\dot{h}$ reduces to

$$\dot{h} = K\omega_e + k\left(\sum_{i=1}^{3} b_i (Q_e^T e_i)^\times e_i\right) - \omega^\times J\omega - \omega^\times h \tag{D.24}$$

since $\dot{\omega}_d = 0$ which will give us the same negative semi-definite result. However, now we are only concerned with constants Q_d and ω_d and therefore, V is not explicitly time-varying. Therefore, here we can try to use LaSalle's Invariance Principle by looking at the invariant set $\dot{V} = 0$ and make sure it can only be zero at the desired attitude and angular velocity. In the absence of external torques, we must have $\dot{h} = 0$ and $\dot{V} = 0$ or we would not stay invariant. Here we run into a problem. Clearly, the first term of Eq. (D.24) is zero when $\dot{V} = 0$. However, it is nontrivial to infer the largest invariant set inside $\dot{V} = 0$ from the control, $\dot{h} = k\left(\sum_{i=1}^{3} b_i (Q_e^T e_i)^\times e_i\right) - \omega_d^\times (J\omega_d + h)$ at $\dot{V} = 0$. It makes more sense to consider a simpler case as an example.

If we are considering a special case of a rest-to-rest maneuver, $\omega_d = 0$ which gives

$$\dot{V} = -\omega^T K \omega \tag{D.25}$$

Then for Eq. (D.25), we can use LaSalle's Invariance Principle and set $\dot{V} = 0$ with the control in this invariant set, $\dot{h} = 0$ only allows for one solution away from the undesired equilibrium and that is the correct attitude and angular velocity.

It should be noted that there critical points of $E(Q_e) = \text{tr}(B - BQ_e)$ where $Q_e \neq \mathbf{1}$, are unstable equilibrium points (saddles) when B is chosen in the form discussed, are only weakly unstable (i.e., along the unstable directions associated with the saddle point). Therefore, if you end up near one of these points, even along unstable saddle directions, you will still be slow to converge to the desired equilibrium, $Q_e = \mathbf{1}$. To circumvent this slow convergence, and guarantee global asymptotic stability for attitude control, discontinuous hybrid control schemes have been used at the cost of adding discontinuity to the control scheme.

There has been successful design of almost (modulo the three spurious critical points of E) global asymptotic stable controllers in for SO(3) and hybrid (switched) global exponential stability (see references [169, 173, 174]) for different choices of $\dot{h}$ and $G(Q_e, \omega_e)$. Many of these results are robust to uncertainty and disturbances.

D.4.2 Euler Parameters (Unit Quaternions)

The closed-loop equations of motion for our system in Euler parameters are equivalent to those in Eq. (D.13), with $S(Q)$ now linear in elements of Q.

In the literature on large angle attitude control with Euler parameters, one of the most considered Lyapunov function candidates is

$$V = E(e) + G(e, \omega_e) = \hat{e}^T \hat{e} + (1 - e_4)^2 + \frac{1}{2}\omega_e^T J \omega_e \tag{D.26}$$

where $E(e) = (\hat{e}^T \hat{e} + (1 - e_4)^2)$, and the term $(1 - e_4)^2$ is added to ensure that only the solution of the Euler parameters with $e_4 = +1$ provides a $V = 0$. The second component of the Lyapunov function candidate is

$$G(e, \omega_e) = \frac{1}{2}\omega_e^T J \omega_e = \frac{1}{2}(\omega - Q_e(e)^T \omega_d)^T J \left(\omega - Q_e(e)^T \omega_d\right) \tag{D.27}$$

The time derivative in Eq. (D.26) along system trajectories gives

$$
\begin{aligned}
\dot{V} &= 2e^T\left(\frac{1}{2}e^\times\omega_e + \frac{1}{2}e_4\omega_e\right) - 2(1-e_4)\left(-\frac{1}{2}e^T\omega_e\right)\\
&\quad + \omega_e^T J J^{-1}\left(-\omega^\times J\omega - \omega^\times h - \dot{h} + \varepsilon\right)\\
&\quad - \omega_e^T J\left(Q_e(e)^T\omega_e^\times\omega_d + Q_e(e)^T\dot{\omega}_d\right)\\
&= \omega_e^T\left[\hat{e} - \omega^\times J\omega - \omega^\times h - \dot{h} + \varepsilon - J\left(Q_e(e)^T\omega_e^\times\omega_d + Q_e(e)^T\dot{\omega}_d\right)\right]
\end{aligned}
\tag{D.28}
$$

If we assumed $\dot{\omega}_d = 0$ here and if we use the control law

$$
\dot{h} = K\omega_e + \hat{e} - \omega^\times J\omega - \omega^\times h + \varepsilon - J\left(Q_e(e)^T\omega_e^\times\omega_d + Q_e(e)^T\dot{\omega}_d\right) \tag{D.29}
$$

we arrive at

$$
\dot{V} = -\omega_e^T K\omega_e \tag{D.30}
$$

which is a negative semi-definite result. Clearly, if we assume a rest-to-rest maneuver with $\varepsilon = 0$, we can use Lasalle's Invariance Principle again to prove that the control law in $\dot{h}$ in Eq. (D.29) is almost globally asymptotically stabilizing. However, this result is not robust to noise and suffers from unwinding and in fact all continuous time-invariant control laws for quaternion based attitude control have this problem (see Example D.1).

D.4.3 Modified Rodrigues Parameters

Euler Rodrigues and Modified Rodrigues parameters could provide a Lyapunov function candidate of the form

$$
V = E(\sigma_e,\omega_e) + G(\sigma_e,\omega_e) = 2k\ln(1+\sigma_e^T\sigma_e) + \frac{1}{2}\omega_e^T J\omega_e \tag{D.31}
$$

The Lyapunov function candidate component derivatives are

$$
\dot{E}(\sigma_e) = k\omega_e^T\sigma_e \tag{D.32}
$$

and

$$
\dot{G}(\sigma_e,\omega_e) = \omega_e^T\left[-\omega^\times J\omega - \omega^\times h - \dot{h} + \varepsilon - J\left(Q_e(\sigma_e)^T\omega_e^\times\omega_d + Q_e(\sigma_e)^T\dot{\omega}_d\right)\right] \tag{D.33}
$$

Clearly, with a attitude control choice

$$\dot{h} = k\sigma_e + K\omega_e - \omega^\times J\omega - \omega^\times h + \varepsilon - J\left(Q_e(\sigma_e)^T\omega_e^\times\omega_d + Q_e(\sigma_e)^T\dot{\omega}_d\right) \tag{D.34}$$

leads to

$$V = -\omega_e^T K\omega_e \tag{D.35}$$

which is a negative semi-definite result in which the control law in Eq. (D.34) can also assume to be almost globally asymptotically stabilizing for rest-to-rest maneuvers in the absence of external torques.

Note, the results you would get from the described Lyapunov analysis would be almost global for most attitude maneuvers such as attitude tracking. This is because the point $\phi = 2\pi$ for the MRPs is a singularity. Therefore, even though MRP attitude control can be continuous and handle large maneuvers between $\phi \in [0, 2\pi)$, it can not handle maneuvers that have angles larger than a single revolution. Furthermore, the use of MRPs for large angle slews is highly nonlinear and not a transparent choice for verification or analysis.

There exists results where an MRP based attitude controller was derived using inverse dynamics and checked using Lyapunov analysis to satisfy a more prescriptive error dynamics behavior (see references [167]).

Under assumption 2 mentioned earlier, we can take $H \neq 0$ and $\varepsilon = 0$ for all time. Note, as in assumption 1, it may not even be possible to achieve $\omega = 0$ for this case if it does not satisfy $\omega = J^{-1}({}^{\mathrm{N}}Q^{\mathrm{B}}H_0 - H) = 0$ which would require that the CMG or RWA array can achieve $H = {}^{\mathrm{N}}Q^{\mathrm{B}}H_0$ for at least the terminal attitude, ${}^{\mathrm{N}}Q^{\mathrm{B}}(t_f)$. Assuming that the chosen terminal points attitude and angular velocity are feasible, we can again consider the control from the three different choices for attitude coordinate parameterizations. The equations for $\dot{\omega_e}$, $\dot{E}$, $\dot{G}$, and $\dot{V}$ follow by simplification of those equations in assumption 1.

D.4.4 Quaternion Unwinding

This subsection in the absence of external torques, will address an issue with Euler parameter (unit quaternion)-based attitude control known as unwinding.

The double covering property of Euler parameters (unit quaternions) for SO(3), means that the space the represents the points of Euler parameters, $\mathbb{S}^3$ has two solutions corresponding to each point of SO(3) (i.e., each DCM). We can think of this correspondence as a one-to-two map. The additional solutions present here inherently cause problems with the robustness of global asymptotic stability and jumping back and forth between these solutions for any attitude control law that is required to be continuous, smooth, or time-invariant. This phenomena is known as quaternion unwinding. Basically, when you get to the zero error quaternion, you have two solutions $\hat{e} = 0$, $e_4 = \pm 1$. These solutions may seem far apart, but if we

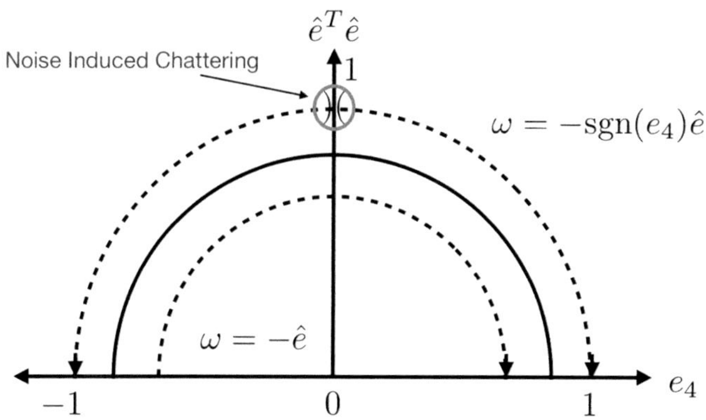

Fig. D.3 Quaternion unwinding

add a small amount of noise around $e_4 = 0$ (see Fig. D.3), it is possible to have an arbitrarily small perturbation to e_4 that change its sign of the controller's direction.

When this happens, the control law discussed (and all control laws that are either continuous, smooth, and/or time invariant), $\dot{h}$ will have a sign change in torque applied. This sign change causes the controller to request a torque that takes the long arc back to $\hat{e} = 0$, $e_4 = +1$. This can go back and forth around $e_4 \sim 0$ for eternity and requires controllers that are either discontinuous or more generally hybrid (see reference [174]) with both continuous and discontinuous components, or time-varying (see references [80]).

Example D.1 The following analysis is reintroduced from reference [174] to focus on the assumptions for attitude control restricted to momentum control actuators. For the moment, let's assume that ε and $H = 0$. One can look at a simpler Lyapunov function candidate for this problem when rest-to-rest maneuvers are considered, namely

$$V = 2(1 - e_4) \tag{D.36}$$

$e_4 = 1$ corresponds to the desired zero error solution and is the only zero of V. The time derivative of V is

$$\dot{V} = \hat{e}^T\omega \tag{D.37}$$

where $\omega_e = \omega$ since we wish to perform a rest-to-rest maneuver. We wish to control at the angular velocity level here since torque does not show up, so we would like to chose ω such that we get the desired result of $\hat{e} = 0$ and $e_4 = 1$.

(continued)

If we chose $\omega = -\hat{e}$ we get the result

$$\dot{V} = -\hat{e}^T \hat{e} \tag{D.38}$$

In reality, we chose $\dot{h}$ such that H gives us exactly this ω. Under our assumption that $H = 0$. this is simply $h = -J\omega = J\hat{e}$ which gives

$$\dot{h} = J\dot{\hat{e}} = \frac{1}{2}J(\hat{e}^{\times}\omega + e_4\omega) = -\frac{1}{2}Je_4\hat{e} \tag{D.39}$$

where we substituted $\omega = -\hat{e}$. Clearly, $\omega = -\hat{e}$ provides a $\dot{V}$ with two zero points, $e_4 = \pm 1$ and this poses a problem when any noise or disturbance is added to the system that may change the direction of $\hat{e}$ that will then change the direction of the control taking the long arc. Furthermore, the internal torque control, $\dot{h} = -\frac{1}{2}Je_4\hat{e} = 0$ where small noise at $e_4 \sim 0$ can flip the sign of e_4 and thus, change the direction of the torque. Furthermore, at $\omega = 0$ and $e_4 = 0$ or π rad away from the correct attitude, the system remains stuck in a local minimum and even around this point, convergence is very slow. Therefore, the unwinding that shows up with control at the level of angular velocity, and at the internal control torque level and causes a nil control solution π rad away from the proper attitude and near this point, will lead to slower convergence to the desired attitude. To solve the issue of unwinding, we can chose a discontinuous control law of the form $\omega = \text{sgn}(e_4)\hat{e}$ with the locally Lipschitz (i.e., not longer differential) Lyapunov function candidate of the form

$$V = 2(1 - |e_4|) \tag{D.40}$$

Note, $\text{sgn}(e_4) = -1$ for $e_4 < 0$ else $\text{sgn}(e_4) = 1$. The result for $e_4 = 1$ here is clearly globally asymptotically stable since at $e_4 = -1$, the controller $\omega = \text{sgn}(e_4)\hat{e}$ will redirect you on the shortest arc back to $e_4 = 1$ at the cost of possible chatter. Therefore, with the purely a discontinuous control solutions, the switching surface at the top of Fig. D.3 is infinitesimal and the controller is not robust since it will chatter back and forth between solutions even as this chattering gets smaller in magnitude. However, a hybrid solution such as the one in reference [174] allows for a robust solution to quaternion unwinding. This is made possible by expanding the switching surface in Fig. D.3 to a region of hysteresis larger than the magnitude of noise to avoid chattering back and forth between solutions. Note, that this is a more practical result since there will always be noise and no control designer should design a controller that controls to a precision finer than the knowledge of the state.

D.4.5 Underactuated Nonholonomic Attitude Control

Underactuated spacecraft momentum control is a fascinating subject. The fact that the constraint of the conservation of total angular momentum of the system, the inertia of the spacecraft, and orientation of the actuators allow for control where there appears to be none. The idea of exploiting either time-scale separation or the nonlinearities associated with the system and the systems geometric phase may allow control along axes without obvious control authority at the expense of oscillations. For example, in reference [175], the authors started with an underactuated spacecraft with only two RWA and constructed discontinuous oscillating switching control scheme for the controllable axes where RWA were present providing control along the axes orthogonal to these where no apparent control torque exists. This scheme was able to perform rest-to-rest maneuvers, usually only in a local neighborhood whenever the terminal state was viable as it related to H. In fact, as long as the component of angular momentum along the uncontrolled axis was zero, this was always possible, and when it is nonzero, there is still the ability to control to a neighborhood of the correct attitude. Of course, for CMGs, it is already known that it is possible to control to any terminal attitude and angular velocity that satisfies H (see reference [150]) even with a single CMG. Even so, the knowledge that there exists a control solution, does not permit the straightforward construction of one and the use of geometric phase can be proven to work at the expense of oscillations for an underactuated system of spacecraft momentum control actuators.

Unlike assumption 1 and 2, zero and constant total angular momentum not only simplifies $\dot{\omega_e}$, $\dot{E}$, $\dot{G}$, and $\dot{V}$, but also allows for the maximum capacity of angular momentum for control. Furthermore, the point of zero angular momentum is a point that is fixed in all frames and therefore, the total angular momentum is no longer attitude dependent. With this simplification, the control may now be designed at the level of the angular momentum directly, since $\omega = J^{-1}H$ as long as you know ω that converges in attitude (see Example D.1). However, this may only be appropriate for RWA. For the case of CMGs, there still need to be a mapping from h to Δ and since the gimbal angles are embedded in nonlinear expressions, it is better to control at the level of $\dot{h}$ and utilize a steering algorithm to map to $\dot{\Delta}$.

D.5 Angular Rate Tracking Maneuvers

We can derive the attitude control law for angular rate tracking, directly from the angular velocity error. For example, to track angular velocity perfectly, we need $\omega = Q_e^T \omega_d$. Clearly, the control law in Eq. (D.29) for $\dot{h}$ provides us the desired result since we are only concerned with asymptotically decreasing ω_e to zero and this is exactly what is achieved. Since we only care about matching and angular velocity here, the result in Eq. (D.23) proves exponential stability of angular rate tracking. Therefore, one can think of angular rate tracking as a special case of attitude tracking where you are satisfied with only converging to the correct angular velocity and are unconcerned with precisely tracking and attitude trajectory. Clearly, if you are tracking an attitude trajectory, you are also tracking an angular velocity, but not vice versa.

D.6 Momentum Tracking Maneuvers

We can treat the tracking of the total spacecraft angular momentum as the combination of two subproblems.

1. Track a specific set of RWA rotor angular velocities, Ω_d or CMG gimbal angles, Δ_d associated with a desired array angular momentum h_d (array angular momentum tracking).
2. Track a specific fixed angular velocity of the spacecraft, ω_d associated with a desired spacecraft bus angular momentum $J\omega_d$ (spacecraft bus angular momentum tracking).

We will first assume that we have the spacecraft momentum control system and an external torque to change to help modify the total angular momentum.

D.6.1 RWA Momentum Management with External Control Torques

Consider the first Lyapunov function candidate

$$V_1 = \frac{1}{2} \left\| \Omega_e \right\|^2 \tag{D.41}$$

with time derivative

$$\dot{V}_1 = \Omega_e^T \dot{\Omega} \tag{D.42}$$

If we match $\dot{\Omega} = -\Omega_e$, we get

$$\dot{V}_1 = -\left\| \Omega_e \right\|^2 = -2V_1 \tag{D.43}$$

which is an exponentially stable (i.e., $V_1(t) = \exp(-2t)V_1(0)$) result for matching the desired RWA rotor speeds. Now we need to design the control to match the angular velocities. Consider

$$V_2 = \frac{1}{2}\omega_e^T J\omega_e \tag{D.44}$$

with time derivative

$$\dot{V}_2 = \frac{1}{2}\omega_e^T J \omega_e =$$

$$\omega_e^T J \left(J^{-1}(-\omega^\times J\omega - \omega^\times h + \varepsilon - I_w W \Omega_e + \tau_e) + Q_e^T \omega_e^\times \omega_d + Q_e^T \dot{\omega}_d \right) \tag{D.45}$$

where τ_e are the external torques we have for control.

If we pick,

$$\tau_e = K\omega_e + \omega^\times J\omega + \omega^\times h - \varepsilon + I_w W \Omega_e - J(Q_e^T \omega_e^\times \omega_d + Q_e^T \dot{\omega}_d) \tag{D.46}$$

we get

$$\dot{V}_2 = -\omega_e^T K \omega_e \tag{D.47}$$

which provides us an exponential stability result for the angular velocity tracking. Therefore, with these choices of $\dot{\Omega}$ and τ_e, we have an exponentially stable result for momentum tracking. See reference [176] for good results on attitude tracking using RWA and even momentum dumping in the presence of unmodeled disturbances and reference [177] for momentum desaturation of RWA with constraint satisfaction. Note, this is not a good momentum management law. In practical implementation, you wish to conserve propellant and if thrusters are used purely to dump angular momentum, you would want to use as little external torques as possible. Furthermore, you may wish to control directly to $H = 0$ to maximize the control capacity of the RWA. Therefore, you would assume that you are absorbing only the external disturbances and excess angular momentum from the system and that the RWA will compensate for the rest of the states while effectively helping you minimize propellant. Also, we would most likely have control bandwidth differences between the propulsions system used to dump angular momentum and the RWA array.

D.6.2 CMG Momentum Management with External Control Torques

Consider the first Lyapunov function candidate

$$V_1 = \frac{1}{2}\left\|\dot{\Delta}_e\right\|^2 \tag{D.48}$$

with time derivative

$$\dot{V}_1 = \dot{\Delta}_e^T \dot{\Delta}. \tag{D.49}$$

If we match $\dot{\Delta} = -\dot{\Delta}_e$, we get

$$\dot{V}_1 = -\left\|\dot{\Delta}_e\right\|^2 \tag{D.50}$$

which is an exponentially stable result for matching the CMG gimbal angles. Now we need to design the control to match the angular velocities, but this can be done utilizing the knowledge we gained from the last section with $\dot{h} = -h_0 A \dot{\Delta}_e$. Consider

$$V_2 = \frac{1}{2}\omega_e^T J \omega_e \tag{D.51}$$

with time derivative

$$\begin{aligned}\dot{V}_2 = \frac{1}{2}\omega_e^T J \omega_e = \\ \omega_e^T J \left(J^{-1}(-\omega^\times J\omega - \omega^\times h + \varepsilon - h_0 A\dot{\Delta}_e + \tau_e) + Q_e^T \omega_e^\times \omega_d + Q_e^T \dot{\omega}_d\right)\end{aligned} \tag{D.52}$$

where τ_e are the external torques we have for control. If we pick,

$$\tau_e = K\omega_e + \omega^\times J\omega + \omega^\times h - \varepsilon + h_0 A\dot{\Delta}_e - J(Q_e^T \omega_e^\times \omega_d + Q_e^T \dot{\omega}_d) \tag{D.53}$$

we get

$$\dot{V}_2 = -\omega_e^T K \omega_e \tag{D.54}$$

which provides us an exponential stability result for the angular velocity tracking. Therefore, with these choices of $\dot{\Delta}$ and τ_e, we have an exponentially stable result for momentum tracking. We wish to use a minimal amount of external torques to dump excess angular momentum and offset disturbance torques. Also, we would most likely have control bandwidth differences between the propulsions system used to dump angular momentum and the CMG array.

Just as it was shown for control purposes in reference [86], it is possible to use external disturbance torques, ε to dump angular momentum from spacecraft (see reference [84] for desaturation of CMGs). It could be imagined that the solar panels of a spacecraft could utilize specific attitudes with respect to the sun, while spinning down the rotors of RWA to desired speeds or zeroing the gimbal rates to land a set of desired gimbal angles. However, this would typically require an accurate model of the disturbance torque as a function of attitude.

D.7 Constraints

The attitude control discussion in the previous sections has not addressed constraints. However, practically, spacecraft are constrained in their rotational and translational motion. Angular jerk, acceleration, velocity and attitude are constrained for many reasons. The angular rates, accelerations, jerk etc. may be constrained based off the resonant excitation frequencies or other structural attributes associated with the spacecraft, saturation of the IMUs, or saturation of the momentum control system array. Attitude may be constrained by keep out cones from the sun or Earth that may be required to avoid destroying the sensors of the payload. Keep in mind that momentum control systems are also constrained by the conservation of angular momentum in the absence of external torques. Therefore, if angular velocity is constrained, then by definition, so is the angular momentum of the entire system (see Chap. 4). With this constraint, in the absence of torque, the entire total angular momentum vector in a inertial reference frame

$$H^{\mathrm{I}} = {}^{\mathrm{N}}Q^{\mathrm{B}}(J\omega + h) = \text{constant} \tag{D.55}$$

is constant, where ${}^{\mathrm{N}}Q^{\mathrm{B}}$ is the rotation matrix from the body to inertial coordinate frames, $J\omega$ is the contribution of angular momentum about the center of mass of the spacecraft modulo that generated by the spacecraft momentum control system in body coordinates, H. Recall, that in body coordinates, the components of angular momentum vary with the attitude of the spacecraft with respect to the inertial reference frame and only the magnitude is invariant with attitude (i.e., rotation matrices leave magnitude of vectors invariant). However, the point of zero angular momentum allows for invariance of the total angular momentum vector in any frame. The constraint of conservation of angular momentum is already handled in the equations of motion and is not needed as a separate constraint if the proper equations of motion are considered for planning.

A typical set of the total set of constraints to be considered for the attitude control problems with momentum control devices is shown in Table D.1. Note, that the admissible set of attitude control inputs to the spacecraft, must lie inside the intersection of the sets posed by the constraints in Table D.1 where α, β, γ are constants, $\mathbf{n}$ is the axis of the keepin/keepout cone, and $\mathbf{c}$ is an axis associated with a payload or sensor boresight that needs to be within a cone or away from a cone. This makes it clear that is it quite complicated to design an implementable attitude control system for momentum control devices for the application of attitude control on real spacecraft.

One can think of the total angular momentum H as defining a point for the case where $H = 0$ and a sphere when $H \neq 0$ but still constant in magnitude since rotation is allowed of H in the body frame with respect to an inertial frame as in Eq. (D.55). In the absence of an external torque, this stringently constraints the attitude, angular velocity, and angular momentum trajectories that spacecraft are allowed to traverse. When external torques are added, the radius of this sphere changes with time. This is the clear difference between attitude control with momentum control devices versus control with external torques. Furthermore, by choosing a specific momentum control system arrangement (see Chap. 6) and steering algorithm (see Chap. 9), along with constraints on rotor and gimbal rates and accelerations, there are additional constraints that need to be considered. Moreover, attitude control algorithms must ask for momentum vectors that avoid impassable singularities while also satisfying the actuator constraints on the rotors and gimbals. Therefore, the choice of arrangement and momentum control system sizing, steering algorithm, and attitude control algorithm must be chosen simultaneously considering the desired maneuver types required to meet mission objectives.

D.8 Supervisory Control

Any control system that is considered for use, must satisfy constraints. Control that monitors the control system satisfying these constraints is known as supervisory control. Constraints provide one of the most substantial computational roadblocks to control systems. Rotational control of spacecraft is fairly slow when compared to rotational control or robots or aircraft. However, the processors used for computation on spacecraft are required to be rad-hardened which makes them slower than typical off-the-shelf commercial processors. Therefore, given that spacecraft are typically ill-equipped for large computations at high speed, control systems that are computationally tractable for solutions in near real-time for spacecraft attitude control maneuvers need to be considered. There are many ways to provide controls systems given constraints. Some of these methods will be described here for the general problem of spacecraft attitude control.

D.8.1 Optimal Control

The most explicit way of providing a control solution is by solving an optimal control program. There has been many works and even some inflight experimental results from optimal control utilizing spacecraft momentum control systems, specifically RWA (see references [53]). However, the problem of spacecraft attitude control in the general setting is nonlinear and therefore, generally nonconvex. The nonconvexity of the problem shows up in the nonlinear dynamics of the various forms of the equations of motion and in the possibly nonconvex constraints in Table D.1. Nonconvexity poses a problem of making the problem NP-hard and therefore, the time-to-solution is unknown. This adds an inherent risk to the use of optimal control for use in spacecraft attitude control where avoiding risk is common.

There are ways of sidestepping nonconvexity when relaxations of the problem are possible. For example, you can solve the full nonlinear program using a direct method by discretizing the cost function, constraints, and dynamics associated with the optimal control problem and solving the problem in finite dimensions [54]. However, such methods do not usually provide a guaranteed convergence time and error bound and thus, are not very amenable to risk-adverse spacecraft attitude control. Also, it is very hard to guarantee the solutions admissibility and optimality between the discretization points when interpolating. Other relaxations rely on lossless (i.e., not loss in optimality) convexification that resorts in finding ways of transforming the constraints, dynamics, and objective function of the problem into convex functions and sets that solve for an optimal one-step ahead control solution [178]. Disappointedly, lossless methods usually are not always possible for specific problems. Other methods allow for minimal loss of optimality while maintaining feasibility (i.e., a solution that satisfies all constraints exists). Examples, of these methods that have even been proven on real platforms are those that utilize successive convexification that solves a hierarchy of convex programs in the place of a nonconvex program [179, 180]. These methods are very promising, but may still be too computationally intensive or sophisticated for implementation on-board spacecraft.

D.8.2 Model Predictive Control

A promising approach to spacecraft attitude control with momentum actuators is known as model predictive control. Model predictive control (MPC) solves a finite time-horizon optimal control problem every time-step and utilizes the computed first time-step as the current control. Model predictive control has shown a lot of success in the chemical process industry and for control of automobiles [181]. The mathematical theory of nonlinear model predictive control is sophisticated and usually based off of the Bellman's principal of dynamic programming. Still it may not guarantee a computationally tractable control solution since such methods when relying on dynamic programming [182]. However, linear model predictive control has shown a lot of promise in application for many systems. Linear model predictive control either starts with linear (really affine) closed-loop dynamics and reformulates constraints to solve a convex program in discrete time valid only locally in state and over small time horizons. However, if the chosen horizon length, and the frequency of solutions is fast enough when compared to the speed of the dynamics of the system and required control, linear MPC can be a valuable solution to spacecraft attitude control with momentum actuators. Moreover, MPC (respectively linear MPC) variants allow for existence of solutions and provability of controllability, stability, and robustness direct by construction (see reference [183] for spacecraft momentum attitude control).

D.8.3 Reference Governors

Reference governors modify the reference to track for the controller such that constraints are met by the control solution. Such method have shown to be very computationally efficient and useful in the automobile industry [177, 184]. These controllers are very amenable to traditional control engineering as it allows any heritage spacecraft attitude controller to be used and only the reference need be modified. Therefore, for control engineering, this is a natural and acceptable choice.

D.8.4 Safety Filters and Control Barrier Functions

It may be the case that constraints are not of the form in Table D.1. You may be required to satisfy constraints described by commands such as "visit all points of a set of the state space within a specific amount of time" or "wait until a specific amount of time before you go here". Such constraints are best represented as formal logic specifications in formal methods (e.g., linear temporal logic, signal temporal logic). Though such constraints may seem impossible to use in a control theoretic framework, there has been a large amount of recent progress in tightly approximating some of these constraints into functions that are associated to "nice" sets in the state-space and time through. One such method utilizes control barrier functions [157]. These functions provide constraints of the form in Table D.1. Other methods utilize solvers for formal method to check all constraints including the ones represented by formal logic and have a type of supervisory constraint engine for to modify the control solution to satisfy these constraints [185]. Constraint engines though very general, may be more relevant to use for higher levels of autonomy than spacecraft attitude control.

D.9 Dealing with Uncertainty

Recall, that the original form of a Lyapunov function candidate in Eq. (D.16) that we considered contained a term $F(\delta\varepsilon)$ associated with the estimate error between a true and estimated external disturbance torque. In fact, any uncertain parts of the dynamics (e.g., other torques, flexible effects, moments of inertia and actuator alignments) can be grouped into $F(\delta\varepsilon)$ and contribute to part of the analysis and control construction. If disturbance torques are assumed fixed (e.g., constant), then we can treat it as an uncertain parameter and utilize adaptive control, either direct or indirect (see references [148, 186, 187]). Examples of fixed uncertain parameters are scale factors of unknown disturbance torques, alignments of actuators and sensors, mass properties, material properties, etc. If the dynamics are linear in these parameters, then the adaptive control can be easily designed. Indeed, consider for example the rest-to-rest attitude control problem and the Lyapunov candidate function using quaternions, but with uncertainty linear in the parameters ε with $\dot{\varepsilon} = 0$,

$$V = E(Q_e)+G(Q_e,\omega)+F(\delta\varepsilon) = \hat{e}^T\hat{e}+(1-e_4)^2+\frac{1}{2}\omega^T J\omega+\frac{1}{2}\delta\varepsilon^T\delta\varepsilon \tag{D.56}$$

with time derivative

$$\begin{aligned}\dot{V} &= 2e^T\left(\frac{1}{2}\hat{e}^\times\omega + \frac{1}{2}e_4\omega\right) - 2(1-e_4)\left(-\frac{1}{2}\hat{e}^T\omega + \delta\varepsilon^T\dot{\hat{\varepsilon}},\right) \\ &+ \omega^T J J^{-1}(-\omega^\times h + \hat{e} - \dot{h} + \varepsilon) + \delta\varepsilon^T\dot{\hat{\varepsilon}} \\ &= \omega^T\left(-\omega^\times h + \hat{e} - \dot{h} + \varepsilon\right) + \delta\varepsilon^T\dot{\hat{\varepsilon}} \\ &= -\omega^T\left(\hat{e} - \dot{h} + \varepsilon\right) + \delta\varepsilon^T\dot{\hat{\varepsilon}}\end{aligned} \tag{D.57}$$

We can simultaneously design the attitude control $\dot{H}$ and the adaptive update law $\dot{\hat{\varepsilon}}$ as

$$\dot{h} = K\omega + \hat{e} + \hat{\varepsilon} \tag{D.58}$$

and

$$\dot{\hat{\varepsilon}} = \omega. \tag{D.59}$$

Notice, that by substitution of Eqs. (D.58) and (D.59) into Eq. (D.57), we get the result

$$\dot{V} = -\omega^T K\omega \tag{D.60}$$

Therefore, the adaptive update law provides the cross terms needed to remove the uncertain parameter ε from $\dot{V}$. The result from Eq. (D.60) is only negative semi-definite, but can be proven to be almost globally asymptotically stable in attitude and angular velocity. However, we cannot say anything about the convergence of the parameter error which can only be guaranteed if a persistence of excitation condition is satisfied. This is a typical result of direct adaptive control. Do note a phenomena known as bursting can happen when persistence of excitation is lost (see reference [188]).

Even though assuming a constant disturbance torque made the analysis simpler, such a torque is detrimental to a spacecraft momentum control system since it would saturate the momentum control array indefinitely. If the uncertain contribution is not constant, we could use nonlinear robust control to minimize its effect of this uncertainty on the performance of the attitude control solution (see reference [189]). Robust control usually provides a stability result to a ball of error unless the disturbance has some structure that can be exploited.

D.10 Learning-Based Attitude Control

It is inevitable that there will be uncertainty present in a space system and its environment. However, sometimes it may be possible to learn specific parameters or contributions from the dynamics while you are controlling the spacecraft. There have been many advances in learning for control over the last decade, and there have even been a set of conferences devoted each year to the subject known as Learning For Dynamics and Control (L4DC). Many of the results associated with learning for control either set up an adaptive update law as in conventional adaptive control, rely on training of neural network elements directly present in the attitude control law, or as part of the model learning (see references [190], [191]). Some methods even avoid the design of the control law and resort to using reinforcement learning directly from data. Learning is helpful for many conventional control methods that rely on having a model. For example, learning a model is required for model predictive control. Recently, theory has been developed that has shown that it is possible to learn-on-the-fly safely while controlling a system (see reference [192]). With the recent advances, it is now true that learning has become an important element of control system design and has become computational tractable and data efficient to incorporate in spacecraft attitude control. A controller with learning elements can typically be broken up into three parts

$$u = u_0 + u_a + u_n, \tag{D.61}$$

where u_0 is the nominal controller that is usually of a proportional derivative form possibly with nonlinear terms, u_a is an adaptive control law that is accounting for uncertainty that can be parameterized in a model, and u_n is the control made up of neural network or elements that are for unmodeled contributions. Typically, in terms of function norm, the contributions of the components for this controller are mostly from u_0, less from u_a, and the least from u_n. The reason why the contributions are done in this way is to guarantee the behavior of the controller in the large and avoid relying completely on a blackbox result with unknown behavior like with u_n.

D.11 Multiple Rigid-Body Translation-Attitude Control

A very important aspect of momentum control systems is that in theory, their internal control torque is only dependent on their orientation with respect to the spacecraft body frame, and invariant to location within the spacecraft body. This is under the assumption that the spacecraft is a rigid body. However, nothing is in fact rigid body, and for large space structures, this is even more so not be the case. Placement of the actuators anywhere in the satellite may in fact deform the satellite, especially for high torque actuators like CMGs. If the spacecraft is deformed, its center of mass will move and the orientation of each rotor axis of each RWA, $\hat{\mathbf{s}}_i$ or

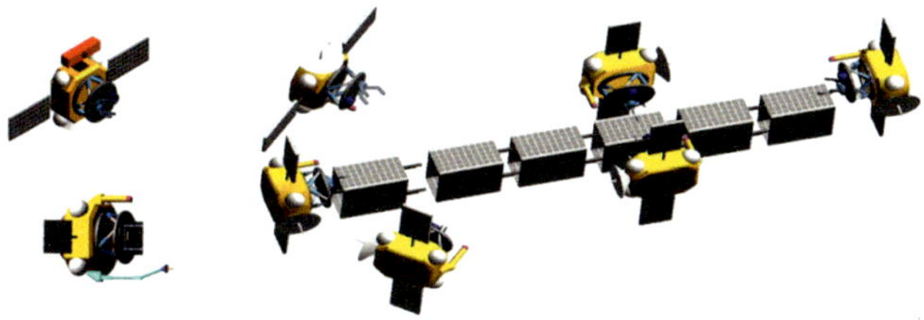

Fig. D.4 Multiple spacecraft orienting a large space structure

gimbal axis of each CMG $\hat{\mathbf{g}}_i$, will also be misaligned from its nominal position. A centralized and larger spacecraft momentum control system made up of contained off-the-shelf systems is definitely a possibility if it is rigidly held together. However, if it is required to be distributed around a spacecraft or as separate component momentum control systems on many spacecraft orienting a large structure as in Fig. D.4, isolation and flexible effects need to be considered.

The centralized m-RWA array on a perfectly rigid body would have the equation for the effective internal torque

$$\dot{h} = \begin{bmatrix} I_{w_1} W_1 & I_{w_2} W_2 & \cdots & I_{w_m} W_m \end{bmatrix} \begin{bmatrix} \dot{\Omega}_1 \\ \dot{\Omega}_2 \\ \vdots \\ \dot{\Omega}_m \end{bmatrix} \tag{D.62}$$

where I_{w_i}, W_i, and $\dot{\Omega}_i$ are the ith RWA array rotor inertia (assumed uniform for each component array), matrix of rotor axes in body coordinates, and rotor accelerations for each component array. Similarly for a centralized m-CMG array,

$$\dot{h} = \begin{bmatrix} h_{0_1} A_1 & h_{0_2} A_2 & \cdots & h_{0_m} A_m \end{bmatrix} \begin{bmatrix} \dot{\Delta}_1 \\ \dot{\Delta}_2 \\ \vdots \\ \dot{\Delta}_m \end{bmatrix} \tag{D.63}$$

where h_{0_i}, A_i, and $\dot{\Delta}_i$ are the ith CMG array rotor angular momentum (assumed uniform for each component array), matrix of gimbal axes in body coordinates, and gimbal rates for each component array. If a distributed set of component momentum control systems is considered on a nonrigid spacecraft or the case of separate satellites with their own momentum control systems orienting a large space structure, misalignments from compliance issues and the inaccurate attitude knowledge for each spacecraft attitude with respect to the space structure frame adds to the uncertainty of the attitude control problem. For example, consider an effective m-RWA array made up of either m spacecraft with RWA arrays onboard separate spacecraft orienting large space structure, or rigidly attached to it would have the internal torque

$$\dot{h} = \begin{bmatrix} Q_1 & Q_2 & \cdots & Q_m \end{bmatrix} \begin{bmatrix} I_{w_1} W_1 & & \\ & \ddots & \\ & & I_{w_m} W_m \end{bmatrix} \begin{bmatrix} \dot{\Omega}_1 \\ \dot{\Omega}_2 \\ \vdots \\ \dot{\Omega}_m \end{bmatrix} \tag{D.64}$$

and similarly for a distributed m-CMG array

$$\dot{h} = \begin{bmatrix} Q_1 & Q_2 & \cdots & Q_m \end{bmatrix} \begin{bmatrix} h_{0_1} A_1 & & \\ & \ddots & \\ & & h_{0_m} A_m \end{bmatrix} \begin{bmatrix} \dot{\Delta}_1 \\ \dot{\Delta}_2 \\ \vdots \\ \dot{\Delta}_m \end{bmatrix} \tag{D.65}$$

where Q_i are the error DCMs of the expected orientation of the m-RWA and m-CMG arrays with respect to the space structure. Assuming the rotor axes in the RWA component arrays and gimbal axes in the CMG component arrays are rigidly fixed to that array, the uncertainty in attitude of the momentum control systems come purely from the DCMs, Q_i where $Q_i = \mathbf{1}, \forall i$ is ideal. If the space structure is flexible and/or there is not in perfect compliance among the spacecraft and the space structure, Q_i will not be a constant DCM and depend on the torque from the component array, compliance, and even location on the hypothetical space structure. Such problems are two complex to deal with in this current edition of the book. For a good reference on modeling and simulating flexible space structures, see reference [193].

D.12 Appendix Summary

Nonlinear spacecraft attitude control derived for spacecraft momentum control systems was discussed. It was shown that much of general attitude control is similar to spacecraft momentum control systems, but external disturbance torques could be detrimental to spacecraft momentum control systems. Furthermore, much of the attitude control results that are found in literature are of the proportional derivative type with some additional nonlinear terms. This is no coincidence, but falls directly out of the analysis an actually makes them more practical for verification. Choice of coordinates, error vectors/matrices, and error functions for Lyapunov analysis is a key part of the control design. In fact, no matter what coordinates are chosen, the attitude control problem is never globally asymptotically stabilizing for a continuous time-invariant control solution. This is an important general fact that follows from the mathematics and was reiterated during the entire appendix. However, specific coordinate choices, error vector/matrix and error function choices provide for stronger results. Though detailed analysis was not performed here, references were provided for recent results. It is expected that the reader of this book will read the recent results in spacecraft attitude control and those specific to spacecraft momentum control systems and decide for themselves which results are the most practical.

References

1. B. Aerospace, *Ball Aerospace-built Worldview-1 satellite launched from Vandenberg*. Press Release (2007)
2. W. Bialke, Low cost attitude control system development final report, NASA-CR-191332. Tech. rep., NASA (1991)
3. S. Loewenthal, Reactionless, momentum compensated payload positioner (1998). US Patent 5,751,078
4. S. Zammit, Control and dynamics simulation facility at Hughes space and communications, in *AIAA Modeling and Simulation Technologies Conference, New Orleans, LA* (1997), pp. 3807
5. J. Smay, Universal spacecraft attitude steering control system (1997). US Patent 5,692,707
6. M. Yuan, A. Wittmann, Tilting momentum wheel for spacecraft (1992). US Patent 5,112,012
7. R. Stribling, Hughes 702 concentrator solar array, in *Photovoltaic Specialists Conference, 2000. Conference Record of the Twenty-Eighth IEEE* (IEEE, 2000), pp. 25–29
8. B. Bialke, E. Stromswold, Reaction wheel actuator with two Newton-meter torque capability for increased spacecraft agility, in *Proceedings of the 6th International ESA Conference on Guidance, Navigation and Control Systems* (2006)
9. W. Lim, J. Salvatore, Method and system for maintaining communication with inclined orbit geostationary satellites (2010). US Patent App. 12/657,188
10. D. Poli, E. Angiuli, F. Remondino, Radiomeric and geometric analysis of worldview-2 stereo scenes. Int. Arch. Photogramm. Remote Sens. Spatial Inf. Sci. **38**(1), 1 (2010)
11. N. Bedrossian, S. Bhatt, M. Lammers, L. Nguyen, Y. Zhang, First ever flight demonstration of zero propellant manuevertm attitude control concept, in *Guidance, Navigation, and Control Conference, vol. AIAA-2007-6734,(Hilton Head, South Carolina)* (2007), pp. 1–12
12. L3 Space and Navigation, Double-gimbal cmg datasheet (2015). http://www2.l-3com.com/spacenav/pdf/datasheets/
13. J. Foust, Emerging opportunities for low-cost small satellites in civil and commercial space, in *24th AIAA USU Small Satellite Conference* (2010)
14. J. Gersh, M. Peck, Violet: A high-agility nanosatellite for demonstrating small control-moment gyroscope prototypes and steering laws, in *AIAA Guidance, Navigation and Control Conference* (2009), pp. 10–13
15. Air Force Materiel Command, AFRL/RVKV - Kirtland AFB. Plug and Play (PnP) spacecraft technologies (TacSat-5) Broad Agency Announcement (BAA-RV-09-01) (2009)
16. M. McMickell, P. Davis, F. Leve, C. Schudrowitz, A momentum control system for agile responsive space satellites, in *8th Responsive Space Conference* (AIAA, 2010)

© The Author(s), under exclusive license to Springer Nature Switzerland AG 2025

F. A. Leve et al., *Spacecraft Momentum Control Systems*, Space Technology Library 46, https://doi.org/10.1007/978-3-031-96165-6

17. B. Hamilton, Turnkey cmg-based momentum control for agile spacecraft, in *AAS Advances in the Astronautical Sciences*, vol. 149 (2013)
18. B. Hamilton, B. Underhill, Modern momentum systems for spacecraft attitude control. Adv. Astronaut. Sci. **125**, 57 (2006)
19. P. Davis, Momentum system concepts and trades for the new class of smaller lower cost satellites. Adv. Astronaut. Sci. **125**, 13 (2006)
20. M. Long, A. Richards, D. Hastings, On-orbit servicing: a new value proposition for satellite design and operation. J. Spacecraft Rockets **44**(4), 964 (2007)
21. N. Strange, T. Landau, G. Lantoine, T. Lam, M. McGuire, M. Burke, L. Martini, J. Dankanich, Overview of mission design for NASA asteroid redirect robotic mission concept, in *33rd International Electric Propulsion Conference, The George Washington University, Washington, DC* (2013)
22. M. Carpenter, M. Peck, Reducing base reactions with gyroscopic actuation of space-robotic systems. IEEE Trans. Robot. **25**(6), 1262 (2009)
23. D. Brown, Energetics of control moment gyroscopes in robotic joint actuation. Ph.D. thesis, Cornell University (2009)
24. L. Brennan, Means for imparting stability to unstable bodies. US Patent 796,893 (1905)
25. P.C. Hughes, *Spacecraft Attitude Dynamics* (Courier Corporation, 2012)
26. J. Stuelpnagel, On the parametrization of the three-dimensional rotation group. SIAM Rev. **6**(4), 422 (1964)
27. M. Shuster, A survey of attitude representations. Navigation **8**(9) (1993)
28. F. Markley, J. Crassidis, *Spacecraft Attitude Determination and Control* (Springer, 2014)
29. P. Hughes, *Spacecraft Attitude Dynamics* (J. Wiley, 1986)
30. H. Kurokawa, A geometric study of single gimbal control moment gyros (singularity problems and steering law). Tech. Rep. 175, Agency of Industrial Technology and Science, Japan (1998)
31. N. Bedrossian, Classification of singular configurations for redundant manipulators, in *Robotics and Automation, 1990. Proceedings., 1990 IEEE International Conference on* (IEEE, 1990), pp. 818–823
32. M. Meffe, G. Stocking, Momentum envelope topology of single-gimbal cmg arrays for space vehicle control, in *Guidance and Control 1987*, vol. 1 (1987), pp. 19–32
33. F.A. Leve, Homological invariants for classification of kinematic singularities. Automatica **111**, 108611 (2020)
34. G. Margulies, J. Aubrun, Geometric theory of single-gimbal control moment gyro systems. AAS J. Astronaut. Sci. **26**(2), 159 (1978)
35. E. Tokar, V. Platonov, Singular surfaces in unsupported gyrodyne systems. Cosmic Res. **16**, 547 (1979)
36. N.S. Bedrossian, J. Paradiso, E.V. Bergmann, D. Rowell, Redundant single gimbal control moment gyroscope singularity analysis. AIAA J. Guidance Control Dyn. **13**(6), 1096 (1990)
37. B. Wie, Singularity analysis and visualization for single-gimbal control moment gyro systems. AIAA J. Guidance Control Dyn. **27**(2), 271 (2004)
38. T. Yoshikawa, Dynamic manipulability of robot manipulators, in *Robotics and Automation. Proceedings. 1985 IEEE International Conference on*, vol. 2 (IEEE, 1985), pp. 1033–1038
39. N. Bedrossian, Steering law design for redundant single gimbal control moment gyroscpoes. Master's thesis, MIT (1987)
40. K. Yamada, I. Jikuya, Directional passability and quadratic steering logic for pyramid-type single gimbal control moment gyros. Elsevier Acta Astronaut. **102**, 103 (2014)
41. H. Yoon, P. Tsiotras, Singularity analysis of variable-speed control moment gyros. AIAA J. Guidance Control Dyn. **27**(3), 374 (2004)
42. G. Leve, F. Boyarko, N. Fitz-Coy, Optimization in choosing gimbal axis orientations of optimization in choosing gimbal axis orientations of a cmg attitude control system, in *AIAA Infotech@Aerospace Conference April 6–10, Seattle WA* (2009)

43. T. Sands, J. Kim, B. Agrawal, 2h singularity-free momentum generation with non-redundant single gimbaled control moment gyroscopes, in *45th IEEE Conference on Decision and Control* (2006), pp. 1551–1556
44. J. Dominguez, B. Wie, Computation and visualization of control moment gyroscope singularities, in *AIAA Guidance, Navigation, and Control Conference and Exhibit* (2002)
45. H. Schaub, J. Junkins, Singularity avoidance using null motion and variable-speed control moment gyros. J. Guidance Control Dyn. **23**(1), 11 (2000)
46. D. Kim, F. Leve, N. Fitz-Coy, W. Dixon, New startup method using internal momentum management of variable-speed control moment gyroscopes. AIAA J. Guidance Control Dyn. **35**(5), 1472 (2012)
47. F. Markley, R. Reynolds, F. Liu, K. Lebsock, Maximum torque and momentum envelopes for reaction wheel arrays. AIAA J. Guidance Control Dyn. **33**(5), 1606 (2010)
48. W. Chubb, H. Kennel, C. Rupp, S. Seltzer, Flight performance of skylab attitude and pointing control system. AIAA J. Spacecraft Rockets **12**(4), 220 (1975)
49. D. Cunningham, G. Driskill, A torque balance control moment gyroscope assembly for astronaut maneuvering, in *NASA. Ames Research Center 6th Aerospace Mechanics Symposium (SEE N72-26377 17–15)* (1972), pp. 121–126
50. litmotors. http://litmotors.com/c1/ (2015)
51. J. Bodden, Attitude control system lectures (1980). Lecture Notes
52. H. Kurokawa, Survey of theory and steering laws of single-gimbal control moment gyros. AIAA J. Guidance Control Dyn. **30**(5), 1331 (2007)
53. M. Karpenko, S. Bhatt, N. Bedrossian, I.M. Ross, Flight implementation of shortest-time maneuvers for imaging satellites. J. Guidance Control Dyn. **37**(4), 1069 (2014)
54. I.M. Ross, M. Karpenko, A review of pseudospectral optimal control: From theory to flight. Annu. Rev. Control **36**(2), 182 (2012)
55. H. Kojima, Singularity analysis and steering control laws for adaptive-skew pyramid-type control moment gyros. Elseview Acta Astronaut. **85**, 120 (2013)
56. J.J. Bonn, M.A. Peck, Dynamic cmg array and method. US Patent 7,561,947 (2009)
57. D. Cornick, Singularity avoidance control laws for single gimbal control moment gyros, in *AIAA Guidance, Navigation and Control Conference* (1979)
58. H. Schaub, J. Junkins, CMG singularity avoidance using VSCMG null motion (Variable Speed Control Moment Gyroscope), in *AIAA/AAS Astrodynamics Specialist Conference and Exhibit, Boston, MA* (1998), pp. 213–220
59. J. Roes, Electro-mechanical energy storage system for space application. Prog. Astronaut. Aeronaut. (United States) **3** (1961)
60. V. Babuska, S. Beatty, B. deBlonk, J. Fausz, A review of technology developments in flywheel attitude control and energy transmission systems, in *Proceedings of the IEEE Aerospace Conference*, vol. 4 (2004)
61. J. Notti, A. Cormack, W. Klein, Integrated power/attitude control system (IPACS). AIAA J. Spacecraft Rockets **12**(5), 485 (1975)
62. J. Fausz, D. Richie, Flywheel simultaneous attitude control and energy storage using a VSCMG configuration, in *Proceedings of the 2000 IEEE International Conference on Control Applications, 2000* (2000), pp. 991–995
63. P. Tsiotras, H. Shen, C. Hall, Satellite attitude control and power tracking with energy/momentum wheels. AIAA J. Guidance Control Dyn. **24**(1), 23 (2001)
64. H. Yoon, P. Tsiotras, Singularity analysis and avoidance of variable-speed control moment gyros–part II: Power constraint case, in *AIAA Guidance, Navigation, and Control Conference, Providence, RI* (2004)
65. K.L. McLallin, R.H. Jansen, J. Fausz, R.D. Bauer, Aerospace flywheel technology development for IPACS applications. Tech. Rep. TM-2001-211093, NASA (2001)
66. M.D. Shuster, D.S. Pitone, G.J. Bierman, Batch estimation of spacecraft sensor alignments, I. relative alignment estimation. J. Astron. Sci. **39**(4), 519 (1991)
67. M.D. Shuster, D.S. Pitone, Batch estimation of spacecraft sensor alignments. II-absolute alignment estimation. J. Astronaut. Sci. **39**, 547 (1991)

68. M.E. Pittelkau, Kalman filtering for spacecraft system alignment calibration. J. Guidance Control Dyn. **24**(6), 1187 (2001)
69. J.L. Crassidis, F.L. Markley, Y. Cheng, Survey of nonlinear attitude estimation methods. J. Guidance Control Dyn. **30**(1), 12 (2007)
70. J.L. Crassidis, M.S. Whorton, Multi-user system for earth sensing spacecraft attitude calibration using international space station attitude information, in *AIAA Scitech 2020 Forum* (2020), p. 1607
71. M.A. Peck, *Space-system Parameter Estimation Applied to Attitude Determination* (University of California, Los Angeles, 2001)
72. M.C. Norman, M.A. Peck, D.J. O'Shaughnessy, In-orbit estimation of inertia and momentum-actuator alignment parameters. J. Guidance Control Dyn. **34**(6), 1798 (2011)
73. R.R. Bate, D.D. Mueller, J.E. White, W.W. Saylor, *Fundamentals of Astrodynamics* (Courier Dover Publications, 2020)
74. J.A. Beck, C.D. Hall, Relative equilibria of a rigid satellite in a circular Keplerian orbit. J. Astron. Sci. **46**, 215 (1998)
75. D. Nychka, P. Ma, D. Bates, A conversation with Grace Wahba (2020)
76. S. Tanygin, T. Williams, Mass property estimation using coasting maneuvers. J. Guidance Control Dyn. **20**(4), 625 (1997)
77. A.Y. Lee, J.A. Wertz, In-flight estimation of the Cassini spacecraft's inertia tensor. J. Spacecraft Rockets **39**(1), 153 (2002)
78. S.C. Solomon, B.J. Anderson, The messenger mission: science and implementation overview. Mercury the View after MESSENGER (2018)
79. M.E. Pittelkau, D.J. O'Shaughnessy, Gyro misalignment decomposition applied to messenger calibration. Adv. Astronaut. Sci. **134**, 1527 (2009)
80. B. Wie, Space vehicle dynamics and control (AIAA) (2008)
81. F.L. Markley, J.L. Crassidis, *Fundamentals of Spacecraft Attitude Determination and Control*, vol. 1286 (Springer, 2014)
82. H. Schaub, J.L. Junkins, *Analytical Mechanics of Space Systems* (AIAA, 2003)
83. J.R. Wertz, *Spacecraft Attitude Determination and Control*, vol. 73 (Springer Science & Business Media, 2012)
84. N.S. Bedrossian, S. Bhatt, W. Kang, I.M. Ross, Zero-propellant maneuver guidance. IEEE Control Syst. Mag. **29**(5), 53 (2009)
85. K.A. Larson, K. McCalmont-Everton, C.A. Peterson, S. Ross, J. Troeltzsch, D. Wiemer, in *Space Operations: Inspiring Humankind's Future* (Springer, 2019), pp. 615–633
86. C.D. Petersen, F. Leve, M. Flynn, I. Kolmanovsky, Recovering linear controllability of an underactuated spacecraft by exploiting solar radiation pressure. J. Guidance Control Dyn. **39**(4), 826 (2016)
87. S.B. Howell, *The NASA Kepler Mission* (IOP Publishing, 2020)
88. O. Tekinalp, E. Yavuzoglu, A new steering law for redundant control moment gyroscope clusters. Aerospace Sci. Technol. **9**(7), 626 (2005)
89. N. Bedrossian, J. Paradiso, E. Bergmann, D. Rowell, Steering law design for redundant single-gimbal control moment gyroscopes. AIAA J. Guidance Control Dyn. **13**(6), 1083 (1990)
90. B. Wie, D. Bailey, C. Heiberg, Singularity robust steering logic for redundant single-gimbal control moment gyros. AIAA J. Guidance Control Dyn. **24**(5), 865 (2001)
91. F. Leve, Evaluation of steering algorithm optimality for single-gimbal control moment gyroscopes. IEEE Trans. Control Syst. Technol. **22**(3), 1130 (2014)
92. J. Leve, F. Munoz, N. Fitz-Coy, Gimbal-lock escape via orthogonal torque compensation, in *Proc. AIAA Guidance, Navigation, Control Conf* (2011), pp. 1–10
93. A. Azzalini, *Matrix Inversion Lemma* (Wiley Online Library, 2006)
94. J. Havill, J. Ratcliff, A twin-gyro attitude control system for space vehicles. Tech. rep., NASA (1964)
95. L. Jones, R. Zeledon, M. Peck, Generalized framework for linearly constrained control moment gyro steering. AIAA J. Guidance Control Dyn. **35**(4), 1094 (2012)

96. D.S. Elliott, *Momentum Control Systems and Their Application in Robotic Systems* (Cornell University, 2019)
97. D.S. Elliott, M.A. Peck, I. Nesnas, Control momentum gyroscope steering law for box-90 array with performance guarantees, in *AIAA Scitech 2020 Forum* (2020), p. 1101
98. S. Yahya, M. Moghavvemi, H.A. Mohamed, Geometrical approach of planar hyper-redundant manipulators: Inverse kinematics, path planning and workspace. Simul. Model. Pract. Theory **19**(1), 406 (2011)
99. S. Yahya, M. Moghavvemi, S. Yang, H.A. Mohamed, Motion planning of hyper redundant manipulators based on a new geometrical method, in *2009 IEEE International Conference on Industrial Technology* (IEEE, 2009), pp. 1–5
100. F. Leve, N. Fitz-Coy, Hybrid steering logic for single-gimbal control moment gyroscopes. AIAA J. Guidance Control Dyn. **33**(4), 1202 (2010)
101. J. Munoz, F. Leve, Artificial potential steering for angular momentum exchange devices, in *AAS/AIAA Spaceflight Mechanics Meeting* (2012)
102. S. Krishnan, S. Vadali, An inverse-free technique for attitude control of spacecraft using CMGs. Elsevier Acta Astronaut. **39**(6), 431 (1996)
103. A. Pechev, Feedback-based steering law for control moment gyros. AIAA J. Guidance Control Dyn. **30**(3), 848 (2007)
104. F. Dueri, D. Leve, B.i.p. Acikmese, Minimum error dissipative power reduction control allocation via lexicographic convex optimization for momentum control systems. IEEE Trans. Control Syst. Technol. **24**, 678 (2015)
105. S. Boyd, L. Vandenberghe, *Convex Optimization* (Cambridge University Press, 2004)
106. J. Fisher, R. Bhattacharya, S. Vadali, Spacecraft momentum management and attitude control using a receding horizon approach, in *Proc. AIAA Conf. on Guidance, Navigation, and Control* (2007)
107. D. Garg, M. Patterson, W. Hager, A. Rao, D. Benson, G. Huntington, A unified framework for the numerical solution of optimal control problems using pseudospectral methods. Elsevier Automatica **46**(11), 1843 (2010)
108. F. Fahroo, I. Ross, Pseudospectral methods for infinite-horizon nonlinear optimal control problems. AIAA J. Guidance Control Dyn. **31**(4), 927 (2008)
109. W. Shapiro, F. Murray, R. Howarth, R. Fusaro, Space mechanisms lessons learned study. NASA technical memorandum 107046. Tech. rep. (1995)
110. S.A. Howard, C. DellaCorte, M.J. Dube, Magnetic levitation for long-life space mechanisms: Technology assessment and remaining challenges, NASA technical memorandum 2019-220052. Tech. rep. (2019)
111. B.J. Hamrock, W.J. Anderson, Rolling-element bearings, NASA reference publication 1105. Tech. rep. (1983)
112. A. Leveille, J. Murphy, Determination of the influence of static loads on the output torque of instrument ball bearings, in *Proceedings of the Charles Stark Draper Lab. International Ball Bearings Symposium* (1973)
113. J.W. Miles, On structural fatigue under random loading. J. Aeronaut. Sci. **21**(11), 753 (1954)
114. W. Bialke, Low cost attitude control system reaction wheel development, nasa-cr-191332. Tech. rep. (1991)
115. J. Archard, Contact and rubbing of flat surfaces. J. Appl. Phys. **24**, 981 (1953)
116. B.J. Hamrock, D. Dowson, Isothermal elastohydrodynamic lubrication of point contacts: Part 1 - theoretical formulation. ASME J. Lubr. Technol. **98**, 223 (1976)
117. M.J. Dube, J. Fisher, S. Loewenthal, P. Ward, Recovery and operational best practices for reaction wheel bearings, in *Proceedings of the 45th Aerospace Mechanisms Symposium, Houston, TX, USA* (2020), p. 285
118. E. Stromswold, B. Bialke, Behavior of reaction wheels near zero speed. Adv. Astronaut. Sci. **118**, 1 (2004)
119. D. Busse, J. Erdman, R.J. Kerkman, D. Schlegel, G. Skibinski, System electrical parameters and their effects on bearing currents, in *Proceedings of Applied Power Electronics Conference. APEC'96*, vol. 2 (IEEE, 1996), pp. 570–578

120. W. Bialke, E. Hansell, A newly discovered branch of the fault tree explaining systemic reaction wheel failures and anomalies, in *Proceedings of the European Space Mechanisms and Tribology Symposium* (2017), pp. 20–22
121. O. Bruegman, K. Thakore, S. Loewenthal, J. Cymerman, Suomi-NPP mission on-orbit experience with toroid ball bearing retainers under unidirectional and reversing motion, in *43rd Aerospace Mechanisms Symposium, NASA/CP-2016-219090* (2016), p. 79
122. T.B. Corporation, ATM CMG bearing failure analysis final report, nasa-cr-120607. Tech. rep. (1975)
123. P. McMahon, R. Laven, Results from 10 years of reaction/momentum wheel life testing, sbn 92-9092- 902-2, in *11th European Space Mechanisms and Tribology Symposium, ESMATS 2005*, vol. 591 (2005)1, pp. 299–305
124. P. McMahon, R. Seiler, A. Accomazzo, S. Lodiot, P. Van Put, R. Porta, In-orbit maintenance of the Rosetta reaction wheels, in *11th European Space Mechanisms and Tribology Symposium, ESMATS* (2017)
125. M. Pantaleoni, P. Chapman, R. Harris, M.G. Kirsch, R. Kresken, J. Martin, P. McMahon, A. Mcdonald, F. Schmidt, T. Strandberg, et al., Curing XMM-Newton's reaction wheel cage instability: the in-flight re-lubrication experience, in *SpaceOps 2014 Conference* (2014), p. 1875
126. W.E. Bialke, J.I. Minow, R.M. Meloy, Correlations of space-charging environments with aerospace mechanism friction anomalies, in *Proceedings of the 15th Spacecraft Charging Technology Conference* (IEEE, 2018)
127. A.E. Tungseth, DSP wheel recent experience (satellite spin axis momentum cancellation), in *Proceedings of the 17th Annual Rocky Mountain Guidance and Control Conference*, vol. 86 (1994), pp. 593–606
128. N.Y. Times, Skylab's gyroscope has worst seizure; breakdown feared (1974)
129. L. Losik, Using analog telemetry to measure spacecraft equipment usable life upgrading factory ATP, in *2012 IEEE Aerospace Conference* (IEEE, 2012), pp. 1–21
130. Inside GNSS, Farewell to a great generation: GPS IIA, in *Inside GNSS - global navigation satellite systems, engineering, policy, and design* (2020, 15 April)
131. J.M. Hacker, J. Ying, P.C. Lai, Reaction wheel friction telemetry data processing methodology and on-orbit experience. J. Astronaut. Sci. **62**(3), 254 (2015)
132. C. Gurrisi, R. Seidel, S. Dickerson, S. Didziulis, P. Frantz, K. Ferguson, Space station control moment gyroscope lessons learned, in *Proceedings of the 40th Aerospace Mechanisms Symposium* (2010)
133. NASA, Nasa-hdbk-4002b, mitigating in-space charging effects - a guideline. Tech. rep. (2022)
134. W. Bialke, A discussion of friction anomaly signatures in response to electrical discharge in ball bearings, in *Aerospace Mechanics Symposium*, vol. 44 (2018), pp. 55–68
135. M. Bampton, R. Craig, Jr., Coupling of substructures for dynamic analyses. AIAA J. **6**(7), 1313 (1968)
136. P. Dahl, A solid friction model. Tech. rep., Aerospace Corporation Report TOR-158(3107-18) (1968)
137. J. Schwartz, M. Peck, C. Hall, Historical review of air-bearing spacecraft simulators. AIAA J. Guidance Control Dyn. **26**(4), 513 (2003)
138. B. Hamilton, Honeywell's momentum control system testbed, in *AAS Advances in the Astronautical Sciences*, vol. 151 (2014)
139. R.M. James, Topology. Prentice Hall of India Private Limited, New Delhi (2000)
140. R. Abraham, J.E. Marsden, T. Ratiu, *Manifolds, Tensor Analysis, and Applications*, vol. 75 (Springer Science & Business Media, 2012)
141. B. O'neill, *Elementary Differential Geometry* (Elsevier, 2006)
142. J. Guckenheimer, P. Holmes, *Nonlinear Oscillations, Dynamical Systems, and Bifurcations of Vector Fields*, vol. 42 (Springer Science & Business Media, 2013)
143. T. Kailath, *Linear Systems*, vol. 156 (Prentice-Hall, Englewood Cliffs, 1980)
144. H.K. Khalil, J.W. Grizzle, *Nonlinear systems*, vol. 3. (Prenticehall, Upper Saddle River, 2002)

145. S. Sastry, *Nonlinear Systems: Analysis, Stability, and Control*, vol. 10 (Springer Science & Business Media, 2013)
146. E.D. Sontag, in *Open Problems in Mathematical Systems and Control Theory* (Springer, 1999), pp. 211–216
147. Y.S. Ledyaev, E.D. Sontag, A Lyapunov characterization of robust stabilization. Nonlinear Anal. Ser. A Theory Methods Ser. B Real World Appl. **37**(7), 813 (1999)
148. W.E. Dixon, A. Behal, D.M. Dawson, S.P. Nagarkatti, *Nonlinear Control of Engineering Systems: A Lyapunov-based Approach* (Springer Science & Business Media, 2003)
149. A.M. Bloch, A.M. Bloch, *Nonholonomic Mechanics*. Nonholonomic Mechanics and Control (Springer, New York, 2015), pp. 235–313
150. S.P. Bhat, K.T. Pawan, Controllability of spacecraft attitude using control moment gyroscopes. Trans. Autom. Control **54**(3), 585 (2009)
151. D. Hinrichsen, A. Ilchmann, A.J. Pritchard, Robustness of stability of time-varying linear systems. J. Differential Equations **82**(2), 219 (1989)
152. K. Zhou, J.C. Doyle, *Essentials of Robust Control*, vol. 104 (Prentice Hall, Upper Saddle River, 1998)
153. T.T. Georgiou, M.C. Smith, Robustness analysis of nonlinear feedback systems: An input-output approach. IEEE Trans. Autom. Control **42**(9), 1200 (1997)
154. C. Baier, J.P. Katoen, *Principles of Model Checking* (MIT Press, 2008)
155. L. Niu, A. Clark, Control barrier functions for abstraction-free control synthesis under temporal logic constraints, in *2020 59th IEEE Conference on Decision and Control (CDC)* (IEEE, 2020), pp. 816–823
156. L. Lindemann, D.V. Dimarogonas, Control barrier functions for signal temporal logic tasks. IEEE Control Syst. Lett. **3**(1), 96 (2018)
157. A.D. Ames, X. Xu, J.W. Grizzle, P. Tabuada, Control barrier function based quadratic programs for safety critical systems. IEEE Trans. Autom. Control **62**(8), 3861 (2016)
158. A.D. Ames, S. Coogan, M. Egerstedt, G. Notomista, K. Sreenath, P. Tabuada, Control barrier functions: Theory and applications, in *2019 18th European Control Conference (ECC)* (IEEE, 2019), pp. 3420–3431
159. L.N. Trefethen, D. Bau, *Numerical Linear Algebra*, vol. 181 (SIAM, Philadelphia, 2022)
160. J.W. Demmel, *Applied Numerical Linear Algebra* (SIAM, Philadelphia, 1997)
161. A. Ben-Israel, T.N. Greville, *Generalized Inverses: Theory and Applications*, vol. 15 (Springer Science & Business Media, 2003)
162. A. Bani Younes, D. Mortari, Derivation of all attitude error governing equations for attitude filtering and control. Sensors **19**(21), 4682 (2019)
163. H. Schaub, J.L. Junkins, Matlab toolbox for rigid body kinematics, in *Spaceflight Mechanics 1999* (1999), pp. 549–560
164. V.S. Borkar, S.K. Mitter, A note on stochastic dissipativeness, in *Directions in Mathematical Systems Theory and Optimization* (Springer, 2003), pp. 41–49
165. P. Crouch, Spacecraft attitude control and stabilization: Applications of geometric control theory to rigid body models. IEEE Trans. Autom. Control **29**(4), 321 (1984)
166. B. Widrow, E. Walach, *Adaptive Inverse Control: A Signal Processing Approach* (John Wiley & Sons, 2008)
167. H. Schaub, M.R. Akella, J.L. Junkins, Adaptive control of nonlinear attitude motions realizing linear closed loop dynamics. J. Guidance Control Dyn. **24**(1), 95 (2001)
168. J. Baillieul, T. Samad, *Encyclopedia of Systems and Control* (Springer, 2021)
169. T. Lee, Global exponential attitude tracking controls on so(3). IEEE Trans. Autom. Control **60**(10), 2837 (2015)
170. S.P. Bhat, D.S. Bernstein, A topological obstruction to continuous global stabilization of rotational motion and the unwinding phenomenon. Syst. Control Lett. **39**(1), 63 (2000)
171. D.E. Koditschek, The application of total energy as a Lyapunov function for mechanical control systems. Contemp. Math. **97**, 131 (1989)

172. A. Sanyal, A. Fosbury, N. Chaturvedi, D.S. Bernstein, Inertia-free spacecraft attitude tracking with disturbance rejection and almost global stabilization. J. Guidance Control Dyn. **32**(4), 1167 (2009)
173. T. Lee, Exponential stability of an attitude tracking control system on SO (3) for large-angle rotational maneuvers. Syst. Control Lett. **61**(1), 231 (2012)
174. C.G. Mayhew, R.G. Sanfelice, A.R. Teel, On quaternion-based attitude control and the unwinding phenomenon, in *Proceedings of the 2011 American Control Conference* (IEEE, 2011), pp. 299–304
175. C.D. Petersen, F. Leve, I. Kolmanovsky, Underactuated spacecraft switching law for two reaction wheels and constant angular momentum. J. Guidance Control Dyn. **39**(9), 2086 (2016)
176. E.A. Hogan, H. Schaub, Three-axis attitude control using redundant reaction wheels with continuous momentum dumping. J. Guidance Control Dyn. **38**(10), 1865 (2015)
177. M. Castroviejo-Fernandez, I. Kolmanovsky, Constrained reaction wheel desaturation and attitude control of spacecraft with four reaction wheels. Preprint. arXiv:2305.05817 (2023)
178. Y. Mao, B. Acikmese, P.L. Garoche, A. Chapoutot, Successive convexification for optimal control with signal temporal logic specifications, in *Proceedings of the 25th ACM International Conference on Hybrid Systems: Computation and Control* (2022), pp. 1–7
179. D. Malyuta, T.P. Reynolds, M. Szmuk, T. Lew, R. Bonalli, M. Pavone, B. Açıkmeşe, Convex optimization for trajectory generation: A tutorial on generating dynamically feasible trajectories reliably and efficiently. IEEE Control Syst. Mag. **42**(5), 40 (2022)
180. B. Açıkmeşe, J.M. Carson, L. Blackmore, Lossless convexification of nonconvex control bound and pointing constraints of the soft landing optimal control problem. IEEE Trans. Control Syst. Technol. **21**(6), 2104 (2013)
181. L. Del Re, F. Allgöwer, L. Glielmo, C. Guardiola, I. Kolmanovsky, *Automotive Model Predictive Control: Models, Methods and Applications*, vol. 402 (Springer, 2010)
182. L. Grüne, J. Pannek, L. Grüne, J. Pannek, *Nonlinear Model Predictive Control* (Springer, 2017)
183. C.D. Petersen, F. Leve, I. Kolmanovsky, Model predictive control of an underactuated spacecraft with two reaction wheels. J. Guidance Control Dyn. **40**(2), 320 (2017)
184. E. Garone, S. Di Cairano, I. Kolmanovsky, Reference and command governors for systems with constraints: A survey on theory and applications. Automatica **75**, 306 (2017)
185. Y. Zhao, Y. Li, L. Sentis, U. Topcu, J. Liu, Reactive task and motion planning for robust whole-body dynamic locomotion in constrained environments. Int. J. Robot. Res. **41**(8), 812 (2022)
186. K.J. Åström, B. Wittenmark, *Adaptive Control* (Courier Corporation, 2008)
187. A. Astolfi, D. Karagiannis, R. Ortega, *Nonlinear and Adaptive Control with Applications*, vol. 187 (Springer, 2008)
188. B.D. Anderson, Adaptive systems, lack of persistency of excitation and bursting phenomena. Automatica **21**(3), 247 (1985)
189. R. Freeman, P.V. Kokotovic, *Robust Nonlinear Control Design: State-space and Lyapunov Techniques* (Springer Science & Business Media, 2008)
190. W. MacKunis, F. Leve, P. Patre, N. Fitz-Coy, W. Dixon, Adaptive neural network-based satellite attitude control in the presence of CMG uncertainty. Aerospace Sci. Technol. **54**, 218 (2016)
191. W. MacKunis, K. Dupree, N. Fitz-Coy, W. Dixon, Adaptive satellite attitude control in the presence of inertia and CMG gimbal friction uncertainties. J. Astronaut. Sci. **56**, 121 (2008)
192. U. Topcu, F. Leve, Learning for dynamical systems when data are scarce (2022). https://sinews.siam.org/Details-Page/learning-for-dynamical-systems-when-data-are-scarce
193. P.W. Likins, Analytical dynamics and nonrigid spacecraft simulation. Tech. rep. (1974)

Index

© The Author(s), under exclusive license to Springer Nature Switzerland AG 2025

F. A. Leve et al., *Spacecraft Momentum Control Systems*, Space Technology Library 46, https://doi.org/10.1007/978-3-031-96165-6

MIX
Papier aus verantwortungsvollen Quellen
Paper from responsible sources
FSC® C105338

If you have any concerns about our products,
you can contact us on
ProductSafety@springernature.com

In case Publisher is established outside the EU,
the EU authorized representative is:
Springer Nature Customer Service Center GmbH
Europaplatz 3, 69115 Heidelberg, Germany

Printed by Libri Plureos GmbH
in Hamburg, Germany